Advances in Intelligent Systems and Computing

Volume 1113

The series "Advances in Intelligent Systems and Computing" contains publications on theory, applications, and design methods of Intelligent Systems and Intelligent Computing. Virtually all disciplines such as engineering, natural sciences, computer and information science, ICT, economics, business, e-commerce, environment, healthcare, life science are covered. The list of topics spans all the areas of modern intelligent systems and computing such as: computational intelligence, soft computing including neural networks, fuzzy systems, evolutionary computing and the fusion of these paradigms, social intelligence, ambient intelligence, computational neuroscience, artificial life, virtual worlds and society, cognitive science and systems, Perception and Vision, DNA and immune based systems, self-organizing and adaptive systems, e-Learning and teaching, human-centered and human-centric computing, recommender systems, intelligent control, robotics and mechatronics including human-machine teaming, knowledge-based paradigms, learning paradigms, machine ethics, intelligent data analysis, knowledge management, intelligent agents, intelligent decision making and support, intelligent network security, trust management, interactive entertainment, Web intelligence and multimedia.

The publications within "Advances in Intelligent Systems and Computing" are primarily proceedings of important conferences, symposia and congresses. They cover significant recent developments in the field, both of a foundational and applicable character. An important characteristic feature of the series is the short publication time and world-wide distribution. This permits a rapid and broad dissemination of research results.

**** Indexing: The books of this series are submitted to ISI Proceedings, EI-Compendex, DBLP, SCOPUS, Google Scholar and Springerlink ****

More information about this series at http://www.springer.com/series/11156

Mykola Nechyporuk · Vladimir Pavlikov ·
Dmitriy Kritskiy
Editors

Integrated Computer Technologies in Mechanical Engineering

Synergetic Engineering

Springer

Editors
Mykola Nechyporuk
National Aerospace University
"Kharkiv Aviation Institute"
Kharkov, Ukraine

Vladimir Pavlikov
National Aerospace University
"Kharkiv Aviation Institute"
Kharkov, Ukraine

Dmitriy Kritskiy
National Aerospace University
"Kharkiv Aviation Institute"
Kharkov, Ukraine

ISSN 2194-5357 ISSN 2194-5365 (electronic)
Advances in Intelligent Systems and Computing
ISBN 978-3-030-37617-8 ISBN 978-3-030-37618-5 (eBook)
https://doi.org/10.1007/978-3-030-37618-5

This Springer imprint is published by the registered company Springer Nature Switzerland AG
The registered company address is: Gewerbestrasse 11, 6330 Cham, Switzerland

Contents

Robotics and UAV

Smart Energy and Grids

Software Engineering and IT-infrastructure

Aerospace Engineering

A Data-Driven Approach to the Prediction of Plasticity in Composites

Gennadiy Lvov(✉) and Olga Kostromytska

NTU "Kharkiv Polytechnic Institute", 2, Kyrpychova str., Kharkiv 61002, Ukraine
lvovdpm@ukr.net

Abstract. The work is devoted to the application of the data-driven approach to determining the yield criterion of unidirectionally reinforced composite materials. This condition has the form of a function of the average coordinate stresses. In contrast to the traditional phenomenological approach, where the form of this function is postulated with the inclusion of a limited number of material parameters, the new approach involves the construction of a grid form of the yield criterion. A numerical determination of the yield strengths of the composite based on the known properties of the matrix and fibers leads to a typical problem of analysis of big data. Based on the finite element analysis of the representative cell, a technique has been developed for determining the yield strengths of the composite under various load paths. Illustrative results are given for special cases of combinations of coordinate stresses.

Keywords: Data-driven approach · Yield criterion · Composite materials

1 Introduction

Data-driven approach is a new area of computational analysis, which differs from the classical approach of mathematical modeling of the phenomena of the world around us. A common feature of the traditional methodology in the mechanics of a deformable body is the formulation of certain models. Such models include a number of arbitrary parameters or functions to be determined as a result of physical experiments. Even for homogeneous materials, the traditional approach encounters complications caused by a significant scatter of experimental data. And for heterogeneous materials, problems are exacerbated by the random nature of the heterogeneity of the internal structure and a large number of basic experiments. In particular, the production technology of modern composite materials does not allow obtaining components of their structure with stable physical characteristics. The use of statistical processing methods in such cases leads to the incorrectness of the identification problems of the material parameters of the models.

Using the new computational paradigm allows you to directly use large amounts of experimental data for the numerical solution of boundary value problems of the continuum mechanics. This eliminates subjectivity in the formulation of material models, errors and incorrect modeling of material properties. In the present work, this approach is used to find the yield criterion of unidirectionally reinforced composites.

M. Nechyporuk et al. (Eds.): *Integrated Computer Technologies in Mechanical Engineering*, AISC 1113, pp. 3–10, 2020.
https://doi.org/10.1007/978-3-030-37618-5_1

2 Literature Review

The traditional approach to determining the yield criterion of composite materials is to replace the heterogeneous structure of the composite with an equivalent homogeneous material. An analytical dependence is formulated a priori for such a material, which determines the surface of the onset of plasticity in a multidimensional space of medium stresses. The first plasticity criterion for the general case of homogeneous material was proposed by Mises [1] using a second-order function:

$$A_{ijkl}\sigma_{ij}\sigma_{kl} = 1 \quad (i, j, k, l = 1, 2, 3), \tag{1}$$

where A_{ijkl} are the components of the fourth-rank tensor, whose values are determined by the yield strengths of the material in different directions. For materials with different tensile and compression yield strengths, the tensor D_{ij} of the second rank is also included in the yield criterion:

$$A_{ijkl}\sigma_{ij}\sigma_{kl} + D_{ij}\sigma_{ij} = 1. \tag{2}$$

In order to reduce the number of material parameters determined from experiments, a large number of modifications of such criteria have been proposed [2–8]. The choice of the form (2) as the criteria of the relations implies certain forms of the yield surfaces for composite materials. Such a phenomenological approach requires a large number of experiments to determine the yield strengths for various types of complex stress state. The number of such independent experiments should correspond to the number of material parameters included in the equation of the yield surface. Carrying out such experiments is a difficult and not always feasible task, especially for biaxial and triaxial stress states. A review of the methods of physical testing of samples from layered composites is given in [9]. For unidirectionally reinforced fiber composites, tests are carried out in particular cases of biaxial stresses. In [10–12], the results for carbon fiber composite under transverse compression and shear in the plane of the sample are presented. A characteristic feature of the presented results is an acute shortage of the amount of experimental data. In the best case, it is possible to obtain the minimum number of points necessary to uniquely determine the equation of the yield surface. Even the question of matching, the theoretically predicted behavior of the material under other loading paths, with experimental results is open. These circumstances determine the relevance of mathematical modeling of elastoplastic deformation of composites based on experimental data on the properties of fibers and matrix. Theoretical modeling makes it possible to generate large amounts of information about the yield surfaces of composite materials and use the data driven computing approach for computer analysis of structural elements from composite materials. In [13, 14] reviews of methods for the direct use of experimental arrays for solving static and dynamic problems of computer mechanics are given.

3 Creation of Discrete Yield Functions of Composites

A geometric interpretation of yield functions is to define a hypersurface in a six-dimensional space of average stresses. The classical forms of these conditions (1), (2) suggest that the equations of such hypersurfaces are of the second order. The validity of this assumption can be established experimentally. Obtaining the results of a physical experiment to determine the yield strengths of composite materials with all kinds of loading paths is not feasible. An alternative is the theoretical prediction of such limits by numerical simulation based on the known properties of the components of the composites. In the present work, the experimental data are replaced by numerical investigation provided by the finite element simulations.

For regular structures, it is possible to select a periodically repeating representative cell. The boundaries conditions at the cell should provide such a stress state within its area that corresponds to a certain type of average uniform stress state of the composite. As an example, a unidirectionally reinforced fiber composite with an orthogonal packaging scheme is considered. Equivalent homogeneous material is orthotropic. The mean stresses are related to the mean strains of the Hooke law, which in matrix form has the form:

$$\begin{bmatrix} \langle\sigma_x\rangle \\ \langle\sigma_y\rangle \\ \langle\sigma_z\rangle \\ \langle\tau_{xy}\rangle \\ \langle\tau_{yz}\rangle \\ \langle\tau_{xz}\rangle \end{bmatrix} = \begin{bmatrix} b_{11} & b_{12} & b_{13} & 0 & 0 & 0 \\ b_{21} & b_{22} & b_{23} & 0 & 0 & 0 \\ b_{31} & b_{32} & b_{33} & 0 & 0 & 0 \\ 0 & 0 & 0 & b_{44} & 0 & 0 \\ 0 & 0 & 0 & 0 & b_{55} & 0 \\ 0 & 0 & 0 & 0 & 0 & b_{66} \end{bmatrix} \cdot \begin{bmatrix} \langle\varepsilon_x\rangle \\ \langle\varepsilon_y\rangle \\ \langle\varepsilon_z\rangle \\ \langle\gamma_{xy}\rangle \\ \langle\gamma_{yz}\rangle \\ \langle\gamma_{yz}\rangle \end{bmatrix}. \tag{3}$$

The homogenization technique for this class of composites, which allows analysis based on the solution of two-dimensional problems, is described in [15]. The minimum representative cell is shown in Fig. 1. A numerical study of the stress state was performed as part of a linear static analysis in the ANSYS software package.

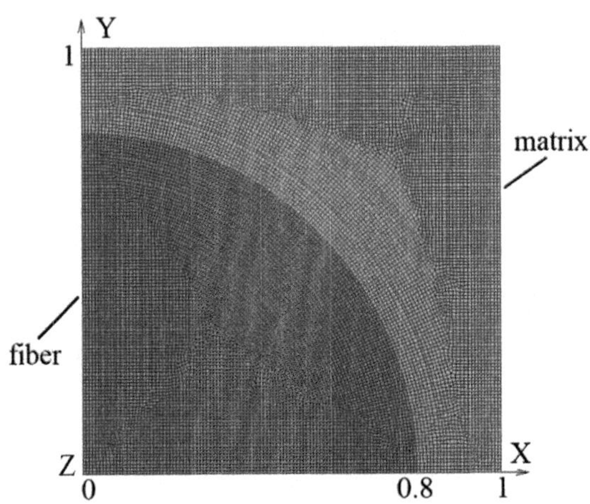

Fig. 1. Finite element model of representative cell.

The calculations were performed for composite, the properties of the components of which are given in Table 1.

Table 1. Mechanical properties of composites components.

Characteristic	Fiber	Matrix
Axial elastic modulus, GPa	$E_z = 238$	$E = 3.3$ (isotropic)
Transverse elastic modulus, GPa	$E_x = E_y = 13$	
Longitudinal shear modulus, GPa	$G_{xz} = G_{yz} = 13$	
Poisson ratios	$\nu_{xy} = 0.22, \nu_{yz} = \nu_{xz} = 0.2$	$\nu = 0.35$
Yield strength, MPa		$\sigma_{ys} = 2.7 \cdot 10^7$

The matrix of elastic constants (3) contains 9 nonzero components. In the case of a square fiber packing scheme, the number of independent components is reduced to six. To determine them by method [15], the following basic experiments were numerically simulated: longitudinal and transverse tensions, longitudinal and transverse shears. Values of effective elastic constants of composite material are given in Table 2.

Table 2. Effective elastic constants of composite.

$b_{11} = b_{22}$, MPa	b_{33}, MPa	b_{12}, MPa	$b_{13} = b_{23}$, MPa	b_{44}, MPa	$b_{55} = b_{66}$, MPa
8.14	122.93	3.10	3.03	2.08	3.01

The procedure for constructing of a discrete yield hypersurface consists in numerically modeling the stress state of a representative cell for numerous loading paths. Each path is determined by a certain ratio of the average coordinate stresses. For the chosen path, cell boundary conditions are set that allow us to accurately simulate the stress state within the representative cell, which occurs for a given combination of average stresses in an unlimited array of the composite. The condition for reaching the yield strength of the composite for a given loading path is the appearance of the first plastic deformations in the matrix. The Mises condition (1) is accepted as a criterion for the onset of plasticity for an isotropic matrix.

For any loading path determined by a given combination of average coordinate stresses, the point of reaching the yield hypersurface can be found. Array of these points determine discrete yield function of composite.

In the three-dimensional subspace of normal stresses, the construction of a regular grid defining the yield is carried out according to the following procedure. Static analysis was performed within the framework of generalized plane deformation. To specify biaxial transverse tension – compression of the composite at the boundaries of the representative cell, the following conditions were specified. On the lines x = 0 and y = 0, the symmetry conditions were used, while on the others, displacements were specified as

$$\begin{aligned} x = 1 \quad & u_x = C \cdot \cos\theta_i, \quad \tau_{xy} = 0; \\ y = 1 \quad & u_y = C \cdot \sin\theta_i, \quad \tau_{xy} = 0. \end{aligned} \tag{4}$$

To illustrate Fig. 2 shows the distribution of the equivalent (according to Mises) stresses for the whole representative cell and the matrix area at the equal stresses σ_x and σ_y without longitudinal stress at $C = 1 \cdot 10^{-3}$ and $\theta_i = 45^\circ$ (i = 10).

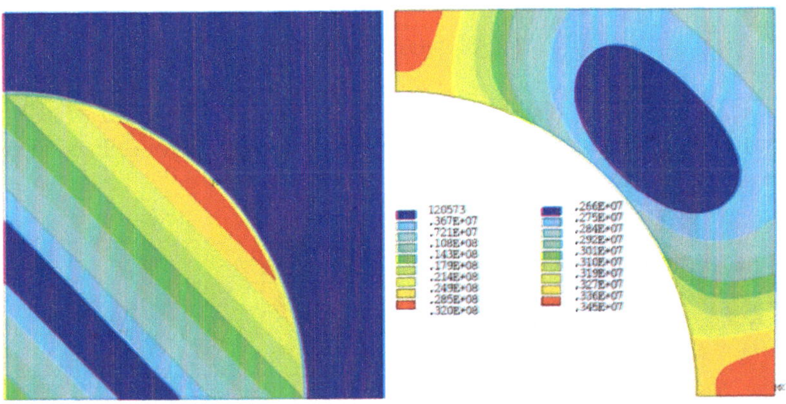

Fig. 2. Equivalent von Mises stresses in whole representative cell and the matrix at simultaneous tension along axis x and y, Pa.

A series of calculations $0 < \theta_i < 2\pi$ allows one to obtain a single section of the discrete surface of plasticity with the plane $\sigma_x\, \sigma_y$. Changing the longitudinal stress with a certain step makes it possible to construct a series of other such sections. Figure 3 shows the results of calculations at $C = 10^{-3}$, and variants of the relations between σ_x and σ_y were set with a uniform step for θ_i $(i = 1, 2, \ldots 36)$.

By a similar algorithm, calculations were performed under simultaneous loading of the composite by tangents τ_{xy} and longitudinal stresses σ_z. To construct a series of loading paths, the following boundary conditions were specified:

$$\begin{aligned} x = 0\ u_y = 0, \sigma_x = 0; \\ y = 0\ u_x = 0, \sigma_y = 0; \\ x = 1\ u_y = 0, \sigma_x = 0; \\ y = 1\ u_x = C \cdot \cos\theta_i, \sigma_y = 0. \end{aligned} \tag{5}$$

The average stress in the generalized plane strain analysis was set according to the following dependence $\sigma_z = N_z \cdot \sin\theta_i$. For one loading path at the equal average

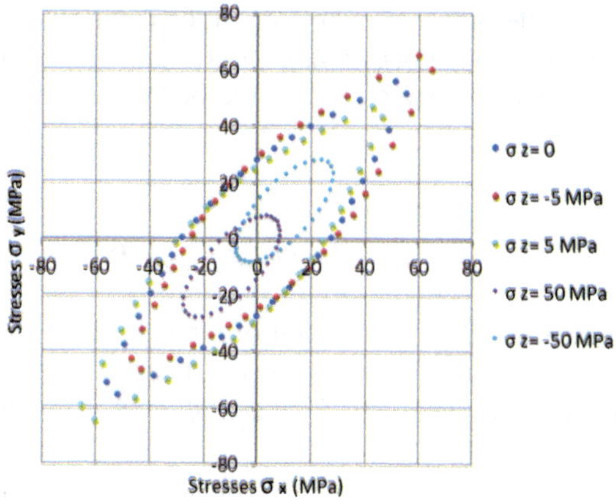

Fig. 3. Stresses σ_x and σ_y.

stresses τ_{xy} and σ_z, the distributions of the equivalent stresses at $\mathrm{C} = 3 \cdot 10^{-4}$, $N_z = 1 \cdot 10^7$ Pa and $\theta_i = 45^\circ$ (i = 10) for the whole representative cell and the matrix area are shown on Fig. 4.

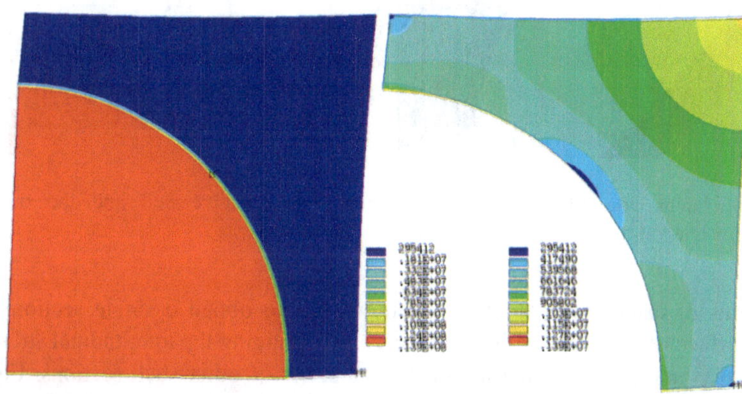

Fig. 4. Equivalent von Mises stresses in whole representative cell and the matrix at simultaneous shear in plane xy and tension along axis z, Pa.

A discrete change in the relationship between the longitudinal stress σ_z and the tangential stress τ_{xy} was ensured by setting the loading parameters for the next cycle $0 < \theta_i < 2\pi$, were (i = $1, 2, \ldots 72$). The cross section of the discrete surface of plasticity by the τ_{xy} σ_z plane is shown in Fig. 5.

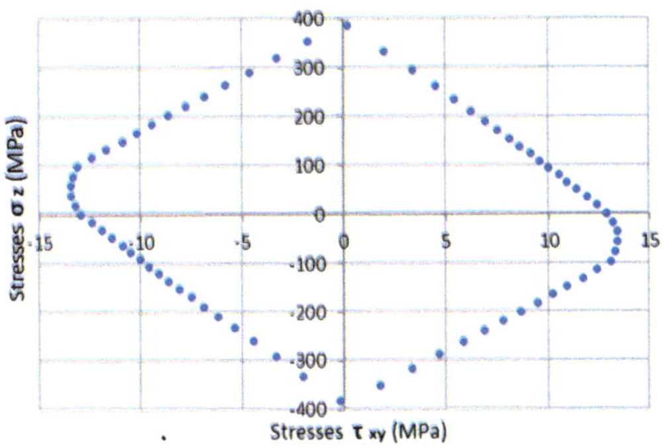

Fig. 5. Stresses τ_{xy} and σ_z.

4 Conclusion

Proposed method allows constructing a discrete plasticity function with any degree of detail and directly using it in the analysis of structures made of composites without involving phenomenological yield criterion. An analysis of the calculation results shows that the quadratic criteria widely used for homogeneous materials are not applicable for composites with significant anisotropy of mechanical properties. A particularly significant mismatch is observed for section of the yield surface by the plane τ_{xy} σ_z, (Fig. 5) which does not correspond to the ellipse. For engineering applications it is enough 100 points for one section, for hypersurface in a six-dimensional space of stresses about 100 GB of data will be required.

References

1. Von Mises, R.: Mechanik der plastischen formänderung von kristallen. ZAMM-Z Angew. Math. Mechanik **8**, 161–185 (1928)
2. Hill, R.: A theory of the yielding and plastic flow of anisotropic metals. Proc. R. Soc. London Ser. A **193**, 281–297 (1948)
3. Tsai, S.W., Wu, E.M.: A general theory of strength for anisotropic materials. J. Compos. Mater. **5**, 58–80 (1971)
4. Hashin, Z.: Failure criteria for unidirectional fiber composites. J. Appl. Mech. **47**(2), 329–334 (1980)
5. Hansen, A.C., Blackketter, D.M., Walrat, D.E.: An invariant-based flow rule for anisotropic plasticity applied to composite materials. J. Appl. Mech. **58**, 881–888 (1991)
6. Vogler, M., Rolfes, R., Camanho, P.P.: Modeling the inelastic deformation and fracture of polymer – part I: plasticity model. Mech. Mater. **59**, 50–64 (2013)
7. Melro, A.R., Camanho, P.P., Andrade Pires, F.M., Pinho, S.T.: Micromechanical analysis of polymer composites reinforced by unidirectional fibres: part I - constitutive modelling. Int. J. Solids Struct. **50**, 1897–1905 (2013)
8. Van der Meer, F.P.: Micromechanical validation of a mesomodel for plasticity in composites. Eur. J. Mech. A-Solid **60**, 58–69 (2016)

9. Olsson, R.: A survey of test methods for multiaxial and out-of-plane strength of composite laminates. Compos. Sci. Technol. **71**(6), 773–783 (2011)
10. Koerber, H., Xavier, J., Camanho, P.: High strain rate characterisation of unidirectional carbon-epoxy IM7-8552 in transverse compression and in-plane shear using digital image correlation. Mech. Mater. **42**(11), 1004–1019 (2010)
11. Gonzalez, C., LLorca, J.: Mechanical behavior of unidirectional fiber reinforced polymers under transverse compression: microscopic mechanisms and modeling. Compos. Sci. Technol. **67**(13), 2795–2806 (2007)
12. Pae, K., Rhee, K.: Effects of hydrostatic pressure on the compressive behavior of thick laminated 45 and 90 unidirectional graphite fiber/epoxy-matrix composites. Compos. Sci. Technol. **53**(3), 281–287 (1995)
13. Kirchdoerfer, T., Ortiz, M.: Data-driven computing in dynamics. Int. J. Numer. Meth. Eng. **113**(11), 1697–1710 (2018)
14. Kirchdoerfer, T., Ortiz, M.: Data-driven computational mechanics. Comput. Methods Appl. Mech. Eng. **304**, 81–101 (2016)
15. Zadeh, S.D., L'vov, G.I.: Numerical procedure of determining the effective mechanical characteristics of an aligned fiber composite. Strength Mater. **47**(4), 536–543 (2015)

Fiber-Optic Pressure Instrument Transducers

Control Systems and Engineering

N. D. Koshevoy[1], O. V. Zabolotnyi[1], I. I. Koshevaya[1], V. V. Muratov[1(✉)], and T. G. Rozhnova[2]

[1] National Aerospace University "Kharkiv Aviation Institute", Kharkiv, Ukraine
vmuratov77@gmail.com
[2] Kharkiv National University of Radio Electronics, Kharkiv, Ukraine

Abstract. Fiber-optic pressure instrument transducers with elastic sensors were developed. Functional circuits and the way of proposed fiber-optic pressure instrument transducers work, protected with utility model patents of Ukraine, are described. Developed fiber-optic pressure instrument transducers can be applied in automatic control and regulating systems for different objects.

Keywords: Fiber-optic instrument transducers · Pressure · Measurement range · Sensor · Functional circuit · Patent

1 Introduction

Pressure instrument transducers with elastic sensors are widely used in modern automatic control and regulating systems [1–4]. Errors, connected with hysteresis, elastic lag, vibration influence, hits and temperature are relevant to them.

Pressure instrument transducers [3] with fiber-optic converters are well-known. Fiber-optic cable usage provides high resolving power, but errors, relevant to elastic elements, are not compensated in instrument transducers.

Known fiber-optic instrument transducer [Pat. №2010106417/28, RF] used to measure weight of mobile objects, machines, train carriages in the way of mobile object dynamic loadings on the instrument transducer measurements. Disadvantages of such an instrument transducer are its low manufacturability, universality and insufficient pressure measuring range, because to enlarge it is necessary to project and manufacture grooves with different corrugated surface in the base.

Fiber-optic pressure instrument transducer with dynamically tuned measurement range [Pat. №24565663(13), RF] has such disadvantages: limited measurement range, insufficient sensitivity, reliability and repairability.

That's why there is a task to increase pressure measurement accuracy, enlarge the range of measurement, simplify the design, increase manufacturability, reliability and repairability of instrument transducers.

M. Nechyporuk et al. (Eds.): *Integrated Computer Technologies in Mechanical Engineering*, AISC 1113, pp. 11–23, 2020.
https://doi.org/10.1007/978-3-030-37618-5_2

2 Basic Research Materials

2.1 Fiber-Optic Instrument Transducers with Elastic Sensors

Fiber-optic pressure instrument transducer, based on the method of measurement [Pat. №91917, Ukraine] when membrane's surface is scanned with light signals and reflected signals are detected with photodetector. At Fig. 1 we can see the design of fiber-optic pressure instrument transducer [Pat. №92188, Ukraine], and Fig. 2 contains electrical functional circuit.

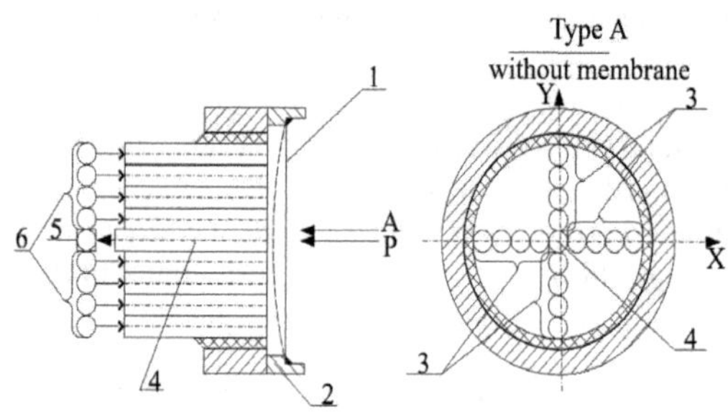

Fig. 1. Fiber-optic pressure instrument transducer design

Fiber-optic pressure instrument transducer, that has n = 16 transmitting LEDs, operates in such a way. Microchip 8 performs consequent switching on of LEDs 6 by sending digital signals to LED driver 11. Driver converts received signal into LED number, that should be connected, and provides corresponding LED with the current of fixed value. LED in its turn converts that current into light flux.

Transmitting optical fibers 3 consequently light the surface of membrane 1, that is fixed in the case 2 and detects pressure P, in correspondence with a program. Lightning, for example, can be performed at first at horizontal, and then at vertical membrane symmetry axes (from outermost optical fibers 3 to central receiving optical fiber 4).

Membrane, under the influence of pressure P, temperature and vibration changes its shape. Light signals, reflected from the membrane surface, fall into the receiving optical fiber 4 zone and through it on the photodetector 5, which converts light intensity into voltage.

Received voltage values are transduced into digital code in ADC converter 7.

After microcontroller had received 16 voltage values it performs calculation of the current pressure value using the algorithm of digital signal processing and, besides, its correction using the data form digital temperature sensor 9.

Under these circumstances pressure value can be defined $P_{cor} = P_{av} + \Delta P$, where P_{av} - average pressure value; $\Delta P = f_T(T)$ – correction data for the temperature influence T, that can be taken from the microcontroller memory; P_{hor} – pressure value which can be defined after the measured results of reflected optic signals, coming from two transmitting LEDs which butts are placed on membrane horizontal symmetry axes (to

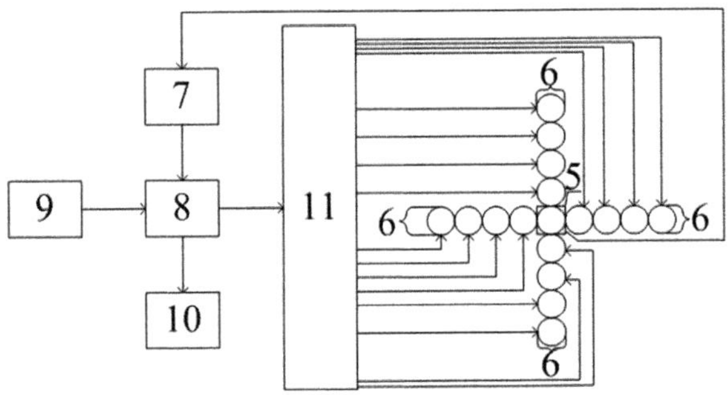

Fig. 2. Electrical functional circuit of the instrument transducer

the left and to the right) on the similar distances form coordinates origin; P_{vert} – pressure value which can be defined after the measured results of reflected optic signals, coming from two transmitting LEDs which butts are placed on membrane vertical symmetry axes (up and down) on the similar distances form coordinates origin. Besides $P_{hor} = f_{hor}(U_{av.hor})$ and $P_{vert} = f_{vert}(U_{av.vert})$ can be chosen from microcontroller's memory by average voltage values, received after reflected light signals conversion, obtained from corresponding transmitting optical fibers.

Received pressure value P_{cor} is transmitted to the digital numerical indicator 10.

Due to membrane surface scanning with light signals and microcontroller processing of the converted reflected signals, errors from temperature influence, vibrations and other factors are decreased. During the membrane testing standard pressure values are supplied to its input, and in microcontroller membrane deflections, defined using converted reflected optical signals, are compared with standard signals, stored in microcontroller's memory. According to the results of comparison data about membrane's operability is given.

Application of suggested fiber-optic pressure instrument transducer allows to increase accuracy of pressure measurement, enlarge its functional abilities by providing the function of membrane testing to estimate it's operability.

For further increase of pressure measurement accuracy and membranes operability estimation by using mathematical models, received during the process of the device functioning, fiber-optic instrument transducer [Pat. №98866, Ukraine, Pat. №98865, Ukraine] was developed.

Figure 3 illustrates fiber-optic pressure instrument transducer construction and Fig. 4 – it's electrical functional circuit.

Fiber-optic pressure instrument transducer consists of membrane sensor 1, fixed in the case 2. Nine transmitting optical fibers 3 and receiving optical fiber 4 with photodetector 5 on it's output are placed in front of membrane. Nine LEDs 6 are placed on the transmitting optical fibers inputs.

Butts of LEDs are optically connected with membrane sensor and placed in XY system of coordinates in correspondence with the points of orthogonal central compositional planning and the butt of LED, connected with photodetector, is placed at the beginning of coordinate system.

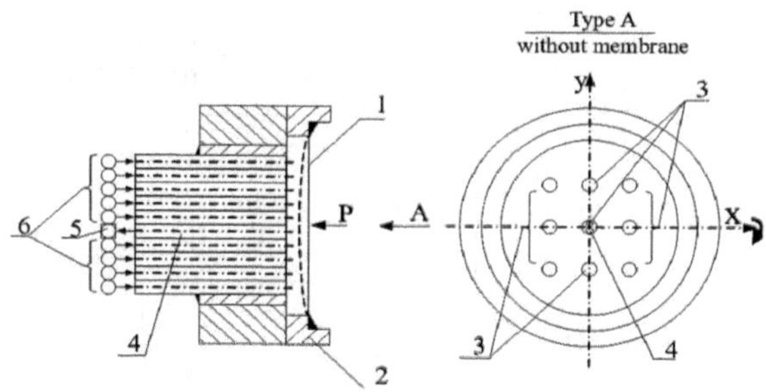

Fig. 3. Fiber-optic pressure instrument transducer construction

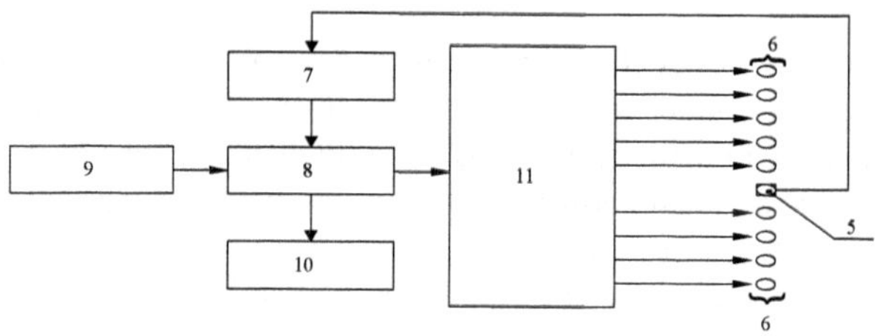

Fig. 4. Electrical functional circuit of the instrument transducer

Photodetector is connected to microcontroller 8 through ADC converter 7. Input of microcontroller is connected with temperature sensor 9, its first output is connected to indicator 10, second output – to LED driver 11. LED driver is connected with nine LEDs, which are connected with nine optical fibers.

Fiber-optic pressure instrument transducer operates in such a way. Microcontroller 8 consequently turns on LEDs 6 by transmitting digital signals to the LED driver 11. LED driver converts received signal into the LED number that should be connected, and provides it with the current of fixed value. LED, in its turn, converts this current into light flux. Transmitting optical fibers 3 consequently lights up the surface of membrane 1 in correspondence with a program, and membrane is fixed in the case 2 and receives pressure P. Output butts lighting up, for example, can be done from left to right in each line, starting from the upper line.

Membrane 1 under the influence of pressure P, temperature and vibrations changes its shape. Light signals, reflected from the membrane surface, get to the zone of receiving optical fiber 4 and trough it – to the photodetector 5, that converts light flux intensity into voltage.

Received voltage values are converted into digital code in ADC converter 7. After microcontroller receives nine voltage values, it calculates current pressure values P_i for each of nine points in the plan of orthogonal central compositional planning (OCCP) with the digital signal processing algorithm application and their correction by the temperature sensor 9 data on the correction value ΔP.

Taking into consideration current pressure values $P_i + \Delta P$ received with the help of experimental data processing algorithm using OCCP plan, mathematical model can be received $P = b_0 + b_1x + b_2y + b_{12}xy + b_{11}x^2 + b_{22}y^2$, where b_0, b_1, b_2, b_{12}, b_{11}, b_{22} – mathematical model coefficients; x, y – coordinates of the sensor point, where the pressure is being calculated.

Using this mathematical model it is possible to find pressure values in different membrane points, after what average pressure is calculated and transmitted to the digital indicator 10.

Due to building the mathematical model P = F(x, y) and pressure values averaging, received with the help of this model, errors, connected with temperature, vibration and other influences can be reduced.

During the process of membrane testing standard pressure value is being transited to its input, and membrane fluxes f_i, received in microcontroller after converted reflected optical signals, are processed using the OCCP plan algorithm.

After the mathematical model in a form $f = b_0 + b_1x + b_2y + b_{12}xy + b_{11}x^2 + b_{22}y^2$ fluxes in different points of coordinate system can be found and compared with standard values, stored in microcontroller memory. After the results of comparison information about membrane workability is given.

Usage of the proposed fiber-optic instrument transducer provides the possibility to define pressure value with better accuracy and to estimate membrane workability by using mathematical models, received during the process of instrument transducer work.

Disadvantages of such a device are the necessity to receive and store in microcontroller's memory correction values ΔP taking into account temperature influence and to calculate corrected pressure values P_i.

Taking that into account fiber-optic instrument transducer which provides automatic temperature error compensation during the process of pressure measurement was developed [Pat. №101771, Ukraine].

Usage of the packing 12 in the form of ring piezoelectric element, linked through ADC converter 13 with the third input of microcontroller 8, and setting membrane sensor 1 on this packing allows automatic temperature error compensation during the process of pressure measurement (Figs. 5 and 6).

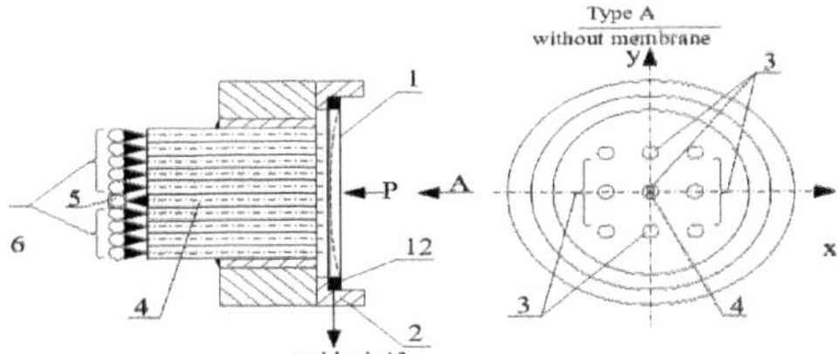

Fig. 5. Fiber-optic pressure instrument transducer construction

Fiber-optic pressure instrument transducer operating is described below. Microcontroller 8 through ADC converter 13 supplies ring piezoelectric element 12 with voltage to compensate temperature error ΔP and fulfills consequent LEDs 6 turning on by sending digital signals to LED driver 11.

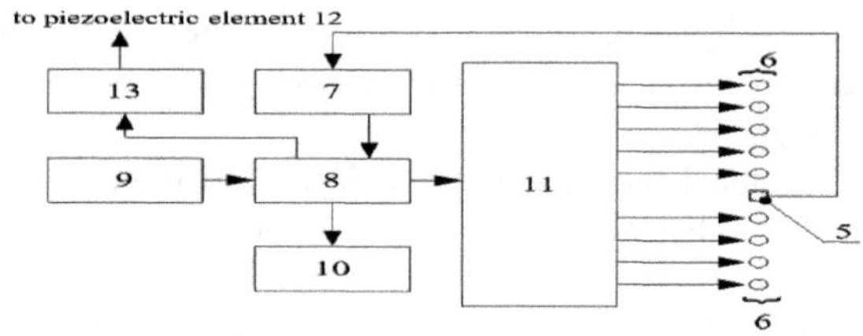

Fig. 6. Electrical functional circuit of the instrument transducer

LED driver 11 converts received signal into LED's number, which is necessary to connect, and supplies it with the electric current of fixed value. LED in its turn converts this current into light flux.

Transmitting optical fibers 3 consequently, in correspondence with a program, consequently lights up the surface of membrane 1, which is fixed in the case 2 and receives pressure P. Output butts lighting up, for example, can be done from left to right in each line, starting from the upper line.

Membrane 1 under the influence of pressure P and vibrations changes its shape. Light signals, reflected from the membrane 1 surface, get to the zone of light receiving fiber 4 and trough it – to the photodetector 5, which converts light flux intensity into voltage.

Received voltage values are converted into digital code in ADC converter 7. After microcontroller 8 receives nine voltage values, it calculates current pressure values P_i for each of nine points in the OCCP plan with the digital signal processing algorithm application.

Using mathematical model that was mentioned above it is possible to find pressure values in different membrane points, after what average pressure P_{av} is calculated and transmitted to the digital indicator 10.

Due to automatic temperature error ΔP compensation, building the mathematical model $P = F(x, y)$ and pressure values averaging, received with the help of this model, errors, connected with temperature, vibration and other influences can be reduced.

Usage of the proposed fiber-optic instrument transducer provides the possibility to simplify the process of pressure values P_i correction by automatic temperature error ΔP implementation.

To provide sufficient reliability and possibility to fulfill the device control and diagnostics during its work fiber-optic pressure instrument transducer [Pat. №102828, Ukraine] was developed, where quartz piezoelectric element 1 is used as a sensor and connected to the second input of ADC converter 7 trough amplifier 12 (Figs. 7 and 8).

Fiber-optic pressure instrument transducer operates in a way described below. Microcontroller 8 consequently turns on LEDs 6 by sending digital signals to LED driver 11. Driver 11 converts received signals into the LED number which is necessary to connect, and supplies it with the electric current of fixed value. LED, in its turn, converts this current into light flux.

Transmitting optical fibers 3 consequently, in correspondence with a program, consequently lights up the surface of sensor 1, which is fixed in the case 2 and receives

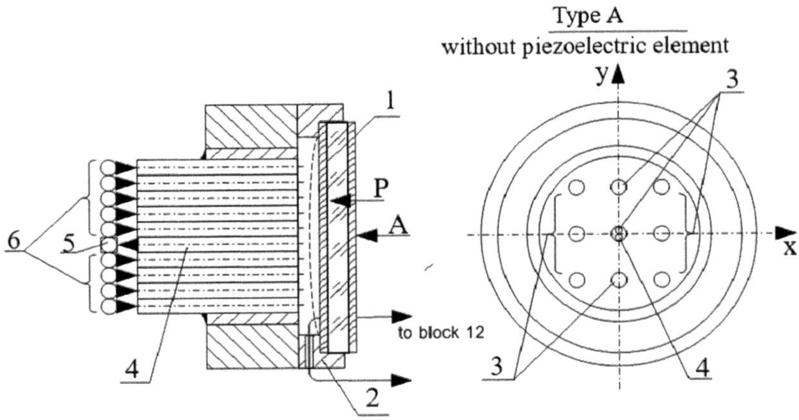

Fig. 7. Fiber-optic pressure instrument transducer construction

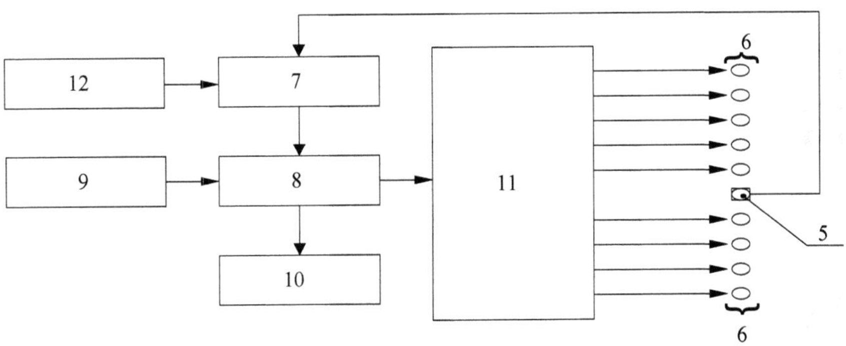

Fig. 8. Electrical functional circuit of the instrument transducer

pressure P. Output butts lighting up, for example, can be done from left to right in each line, starting from the upper line.

Sensor 1 under the influence of pressure P, temperature and vibrations changes its shape. Light signals, reflected from the membrane surface, get to the zone of light receiving fiber 4 and trough it – to the photodetector 5, which converts light flux intensity into voltage.

Received voltage values are converted into digital code in ADC converter 7. After microcontroller 8 receives nine voltage values, circuitry of the device operates like described for the transducer on Fig. 3.

Due to the mathematical model P = F(x, y) building and pressure values averaging, received with the help of this model, errors, connected with temperature, vibration and other influences can be reduced.

Voltage from piezoelectric element 1, proportional to the pressure, is amplified in amplifier 12 and converted into digital code in ADC converter 7.

Pressure value P is calculated in microcontroller 8 after this code taking into account correction value ΔP, which is transmitted to the digital indicator 10. Microcontroller 8 also compares P and P_{av} pressure values. If the difference overcomes permissible value P_{add} indicator 10 gives information about one of the measuring channels malfunction.

To find the measuring channel with malfunction standard pressure value is being sent to the sensor 1 input, and obtaining P and P_{av} pressure values on the indicator 10, channel with malfunction can be defined.

Usage of the proposed fiber-optic pressure instrument transducer allows to increase reliability of the device and to provide its control and diagnostics during the process of operation.

2.2 Fiber-Optic Instrument Transducers with Dynamically Tuned Measurement Range

Sensitivity increasing and measurement range widening can be provided due to introduction into the prototype [Pat. №24565663(13), RF] (n − 1) isolated pads, created from piezoelectric material with electric contacts and switch block, which input is connected with data processing block and outputs are connected to electric contacts of isolated pads, where n – number of pressure measurement subranges [Pat. №119363, Ukraine].

Fiber-optic instrument transducer with dynamically tuned measurement range (Fig. 9) operates like described below.

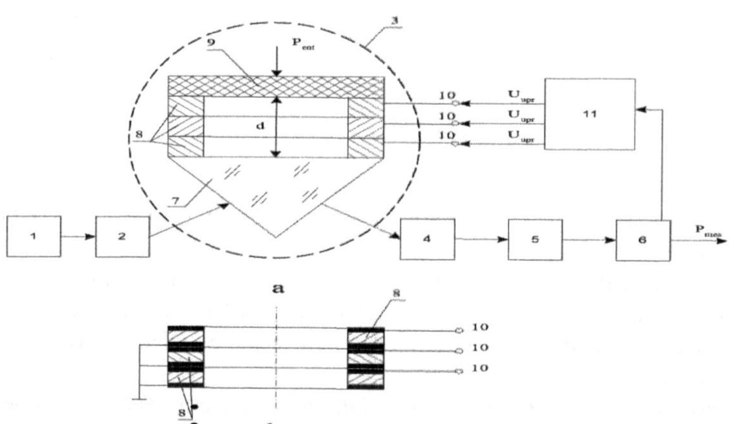

Fig. 9. Fiber-optic pressure instrument transducer: a – functional circuit; b – electric contacts of isolated pads connection circuit; 1 – emission source, 2, 4 – optical fibers, 3 – sensor, 5 – receiver, 6 – data processing block, 7 – prism of complete internal reflection, 8 – isolated pads, 9 – reflecting membrane, 10 – switch block

Optical emission source 1 generates optical emission of prescribed power and sends it to optical fiber 2, which transmits optical emission to the prism of complete internal reflection 7. Gap d between the membrane 9 and the base of a prism 7 can be changed under the pressure P_{ent} influence, what leads to the change of optical emission reflected part. Gap d decreasing leads to bigger optical emission penetration into membrane 9 and optical energy absorption. Reflected emission trough the prism die 7 is transmitted by the optical fiber 4 to the optical emission receiver 5, which converts it into electric signal. Data processing block 6 converts electric signal into pressure value P_{mes}.

When maximal permissible pressure value P_{ent} is exceeded, gap d between the reflecting membrane 9 and prism 7 is completely reduced, and sensor loses its operability what can be defined after the value of optical emission receiver 5 output signal.

To wide instrument transducer measuring range control signal U_{up} from the data processing block 6 is given to the contacts 10 of a lower pad 8. Under these circumstances lower pad 8 becomes deformed and the gap between prism 7 and membrane 9 becomes bigger.

Then, when the measurement range of a pressure P_{ent} changes, control signal U_{up} is given to the contacts 10 of another isolated from the lower pad 8. In this circumstances number of measurement subranges is defined with the number n of isolated pads 8.

So, gap control between the prism and membrane allows widening pressure measuring range and increasing instrument transducer sensitivity. To increase fiber-optic instrument transducer reliability and repairability by using embedded control the block of comparison 13 is additionally added, which inputs 12 are connected to electrical contacts of isolated pads 8 and output of the data processing block 6 [Pat. №124929, Ukraine] (Fig. 10).

To estimate pressure instrument transducer workability comparative block 13 receives signals from data processing block 6 and electrical contacts 12 of pads 8. Equality of these signals ($\Delta P_{mes} = 0$) points that pressure measuring channels of the instrument transducer are workable. The case when $\Delta P_{mes} \neq 0$ points that one of the measuring channels are unworkable.

So, fiber-optic pressure instrument transducer allows reliability and repairability increasing due to embedded control implementation.

2.3 Fiber-Optic Pressure Instrument Transducers for Mobile Objects Weight Measurement

Pressure measurement range widening and instrument transducer's [Pat. №2010106 417/28, RF] manufacturability and universality are provided by placing balls in the grooves of fluted surface, which diameter corresponds chosen pressure range, and optical fiber in the slot is pressed to that balls [Pat. №118842, Ukraine].

Figure 11 illustrates the fiber-optic pressure instrument transducer's sensor design.

In the lower part 1 of the case shaft there is a bandwidth channel 2, where optical fiber 3 is placed. Channel 2 contains a least al last one region to place optical fiber 3 in parallel with the base, which is designed like a slot 4 with a fluted surface 5.

Balls 6, created form rubber or plastic, are placed in the grooves of a fluted surface. Cover 7 has a hollow 8 under the slot 4 to place the plate 9, created from the heat-resistant rubber, which dimensions corresponds the dimensions of hollow 8.

During the process of sensor manufacturing optical fiber 3 is run through the bandwidth channel 2 and pressed to the balls 6 by the plate 9. Input and output regions

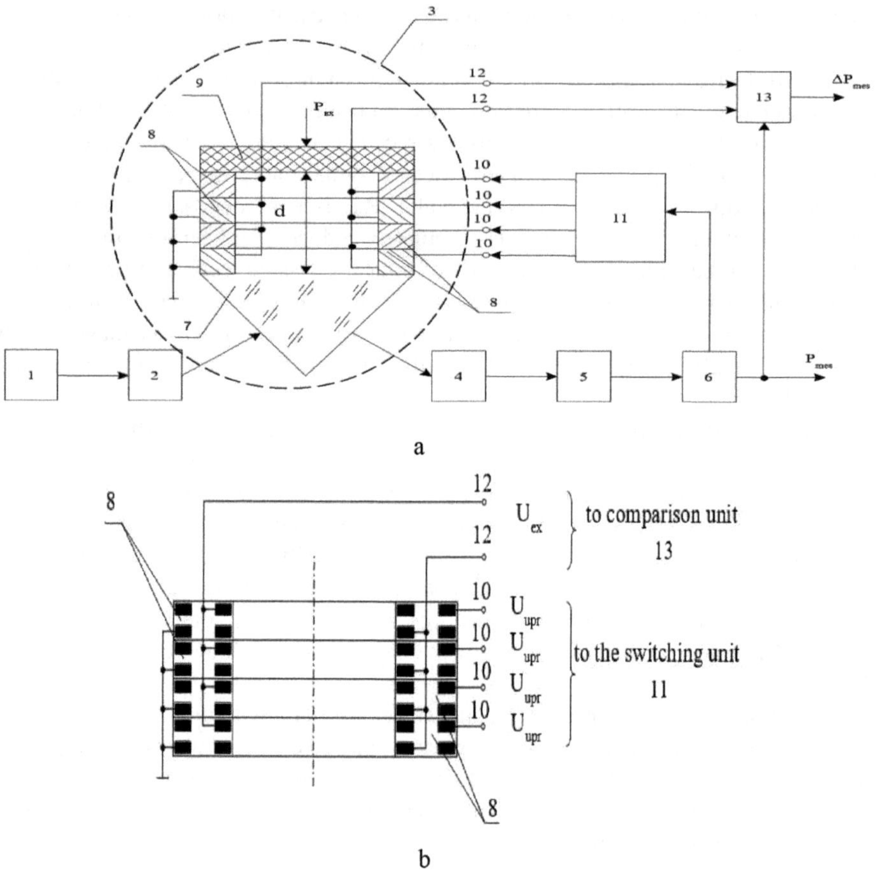

Fig. 10. Fiber-optic instrument transducer with embedded control: a – functional circuit, b – electrical contacts of isolated pads connection circuit

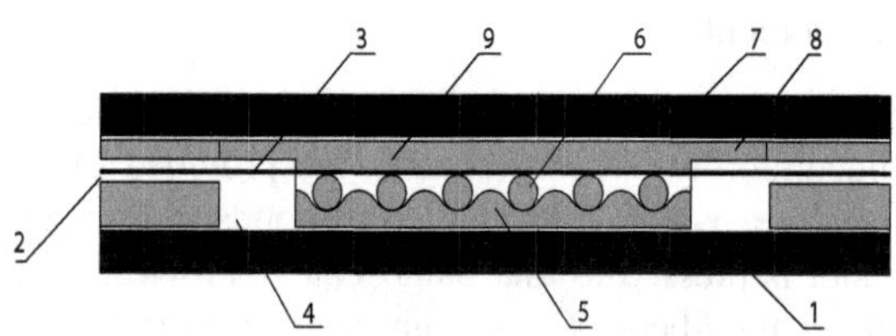

Fig. 11. Fiber-optic pressure instrument transducer's sensor design

of optical fiber are placed in a metallic sleeve (absent on the drawing). After that lower part of the case 1 is hermetically covered with cover 7.

To make the range of measured pressure values wider the balls 6 diameter is defined to take into account controlled loadings. So, by changing the balls 6 diameter, which placed in the grooves of fluted surface 5, necessary pressure measuring range can be set.

Fiber-optic pressure instrument transducer operates like described below.

Input region of the optical fiber 3 is connected to the source of emission, and output – to the measuring system (is absent on the drawing). During the process of loading plate 9 from heat-resistant rubber makes pressure on the optical fiber 3 and presses it to the balls 6, placed in the grooves of a slot's 4 fluted surface 5. Under that circumstances optical fiber 3 bends down and leads to the change of a signal, which passes the fiber. Optical fiber 3 bursting value is proportional to the measured pressure.

So, developed fiber-optic pressure instrument transducer provides increasing of its manufacturability and universality due to usage of balls, placed in the fluted surface grooves, and to make wider the loadings measurement range by choosing the balls diameter.

To simplify the instrument transducer's design and to increase its manufacturability [Pat. №124927, Ukraine] slot 4 is designed with a flat surface 5 in its base, and balls 6 are placed in the slot in a raw (Fig. 12).

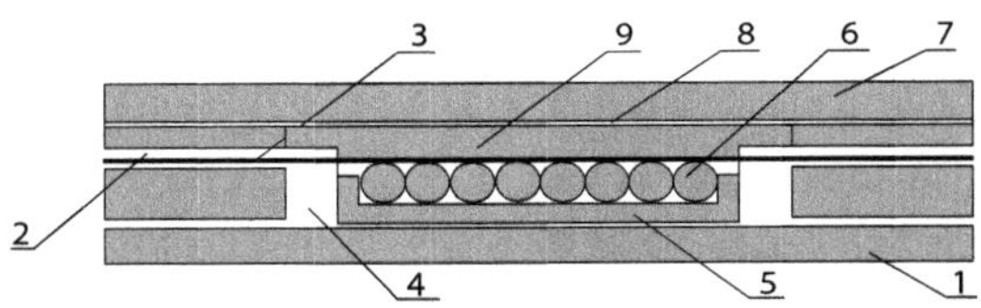

Fig. 12. Simplified sensor's design of a fiber-optic pressure instrument transducer

By changing the ball's 6 diameter, placed in a raw on the flat surface 5, we can set necessary pressure measurement range.

So, presented fiber-optic pressure instrument transducer allows simplification of the design and increasing manufacturability by using balls, placed in a raw on the flat surface, and to make the loadings measurement range wider by choosing the diameter of the balls.

Using of suggested fiber-optic pressure instrument transducers allows design simplification, manufacturability and universality increasing and provides the possibility to make the range of measurements wider.

3 Results and Discussion

Fiber-optic pressure instrument transducers with elastic sensors, dynamically tuned range were developed to measure pressure of mobile objects.

Fiber-optic instrument transducer, where 16 transmitting optical fibers are placed in front of membrane along X and Y axes, allows pressure measurement accuracy increasing, enlarge its functional possibilities providing the function of membrane testing to estimate their workability.

Pressure measurement accuracy is increased by reducing the errors form temperature influence, vibrations and other factors due to membrane surface scanning with light signals and converted reflected signals microcontroller processing.

To increase accuracy of pressure values definition and to estimate membranes workability by using mathematical models, obtained during the working process of the device, fiber-optic instrument transducer was developed, where 9 transmitting optical fibers are placed in front of membrane in coordinate system XY and in correspondence with the points of orthogonal central compositional planning (OCCP).

In this device it is necessary to receive and store in microcontroller's memory corrections ΔP with taking into account temperature action and calculate corrected pressure values P_i what is a disadvantage.

That's why fiber-optic instrument transducer was developed, which provides automatic temperature error compensation in the process of pressure measurement due to usage of the packing in the form of ring piezoelectric element and setting the membrane sensor on it. Besides, packing is linked through ADC converter with one of microcontroller inputs.

To increase device's reliability and to provide its control and diagnostics during its work fiber-optic pressure instrument transducer was developed, where quartz piezoelectric element, connected to the second input of ADC converter trough amplifier, is used as a sensor.

Sensitivity increase and pressure measurement range widening is provided in a fiber-optic instrument transducer by introduction of piezoelectric isolated pads and switch block, which outputs are connected to electric contacts of isolated pads and input – to data processing block.

For the fiber-optic instrument transducers to measure weight of mobile objects measuring range widening and accuracy increase is provided by placing balls in the grooves of fluted surface, which diameter corresponds with chosen pressure range.

For further instrument transducer manufacturability increase and sensor's design simplification balls are placed in a raw in a slot with flat surface.

4 Conclusions

Fiber-optic pressure instrument transducers with elastic sensors, dynamically tuned measurement range, to measure weight of mobile objects were proposed. Functional circuits of developed fiber-optic instrument transducers and the description of their work are given. Fiber-optic instrument transducers are protected with utility model patents of Ukraine.

Developed fiber-optic pressure instrument transducers can be widely applied in information-measuring systems and in control systems for different objects.

References

1. Petrishin, I.S., Safronov, B.M.: Measurements of pressure: navch. pos. Ivano-Frankivsk, Fakel, 269 p. (2004)
2. Sharapov, V.M., Polishchuk, E.S., Koshevoy, N.D., et al.: Instrument transducers. In: Sharapov, V.M., Polishchuk, E.S. (eds.) Technosphere, 624 p. (2012)

3. Koshevoy, N.D., Kostenko, E.M., Koshevaya, I.I., Rozhnova, T.G.: Pressure instrument transducers with a digital output: monographiya, 108 p. National Aerospace University of N.E. Zhukovsky "Kharkiv Aviation Institute", Kharkiv (2017)
4. Koshevoy, N.D., Gordienko, V.A., Koshevoy, O.N., Rozhnova, T.G.: Comparative analysis of pressure instrument transducers with digital output. Telecommun. Radio Eng. **60**(3, 4), 154–156 (2003). https://doi.org/10.1615/TelecomRadEng.V60.i34.180

Methodology for the Experimental Calculation the Coefficients of the Functional Dependencies Electrical Circuits Plasma Substitution

Yuliia Bilokonska(✉), Mariia Breslavets, Serhii Firsov, and Andrii Boyarkin

National Aerospace University "KhAI", Chkalova Street, 17, Kharkiv 61070, Ukraine
{y.bilokonska,a.boyarkin}@khai.edu, breslavec.marija@gmail.com, sn.firsof@gmail.com

Abstract. The article discusses the electrical equivalent circuit for installing a plasma generator in order to highlight the functional dependencies of the steady state of the control system. Experimentally established effect coating mechanism using tungsten carbides and rare earth impurities to the process of increasing resistance parts, which depends specifically introduced into the plasma generation system of controlled variables of internal resistances. This allows the use of a functionally stable control system for a non-stationary plasma medium. Experimental studies have confirmed that plasma formation occurs with the necessary parameters due to voltage stabilization at the target cathode and anode. Also, in order to stabilize the process, it is necessary to correct the current of the ionized flow, and equalize the energy potential of charged particles of the ion-plasma medium. To maintain the specified parameters within specified limits, it is necessary to introduce the corresponding resistance variables into the plasma generation system. Resistance variables are combined into a block of ballast resistors, which ensures the stability of the plasma formation and affects the value of current or voltage. These derived variable coefficients help maintain the functional stability of the specified parameters within the specified limits. This suggests that it is possible to control the flow of ionized particles by controlling their internal resistance. Due to this, the deposition of the tungsten carbide mixture on the workpiece is controlled. Coating using tungsten carbides and rare earth impurities is one of the most common in the production of tools that require high hardness, corrosion resistance and wear resistance.

Keywords: Electrical circuit · Coefficients of functional stability · Functional stability · Current stabilization · Internal resistance

1 Introduction

Many factors influence the working conditions and aircraft carrying capacity. A special factor is not only the usefulness of control systems for such devices, but also the satisfactory condition of the components that make up the main structure.

Material, shape, internal structure effect the efficiency of the system as a whole. Therefore, it is necessary that the physical and chemical characteristics of the elements

M. Nechyporuk et al. (Eds.): *Integrated Computer Technologies in Mechanical Engineering*, AISC 1113, pp. 24–34, 2020.
https://doi.org/10.1007/978-3-030-37618-5_3

of the control system ensure reliability and correspond to the quality of transition processes. This study is relevant due to the given growing of demand for increased wear resistance of parts.

The method which able to obtain uniform coatings on relatively complex surfaces without moving small products is the vacuum-condensation coating method. It involves the use of a particulate flux of a substance at the level of atoms, molecules, ions and the interaction of this flux with the surface of a solid. The consequence of this interaction is condensation - the deposition of a substance on a coating surface or saturation of a surface layer with a substance. All processes take place under vacuum. This choice caused by plasma methods advantages as environmentally friendly impact, safe and less energy-intensive action [1].

The aim of the work is to design a modern control system to form a stable directed plasma flow of the parts of argon, an increase in the corrosion-resistant part is processed.

To achieve this goal, it is necessary to choose the optimal option for applying an ionized coating, it is necessary to analyze the parameters and characteristics of the sputtering of particles, consider the modes of obtaining a coating by ion-plasma technology, conduct an experiment with the operation of the internal resistance control system, and ensure current stability in the gas environment of the plasma generator chamber.

1.1 Discharge Mechanism, Parameters and Characteristics of Magnetron Sputtering Systems

Anomalous scattering in magnetron sputtering systems occur at the intersection of electric and magnetic fields. Electrons moving toward the target by ion bombardment are captured by a magnetic field and perform complex cycloid motion along closed paths near the target surface. As a result of numerous collisions of electrons with the working atoms of the gas, the degree of plasma ionization increases sharply. At the same time, the ion current density increases by about 100 times in comparison with diode sputtering systems without a magnetic field. This leads to an increase in the spraying rate of the target material (50...100 times).

Since magnetic induction increases the path of electrons, the number of collisions with atoms of the working gas increases simultaneously. That is why the application of a magnetic field is equivalent to an increase in gas pressure.

The presence of inhomogeneous crossed electric and magnetic fields in magnetron sputtering systems does not allow us to fully describe all discharge parameters using existing theories. As a result of studying magnetron sputtering systems in a characteristic region for magnetic inductions of 0.03...0.1 T and pressures of 0.1...10 Pa, a model of magnetron sputtering has been developed. The model based on analytical expressions for current-voltage characteristics describes, to a first approximation, the processes occurring in magnetron sputtering systems. A method has also been developed for calculating magnetron sputtering systems, which relates the sputtering rate to the process parameters: magnetic induction, working gas pressure, voltage on the target, ion current density, an experimental method for calculating magnetron sputtering systems has been developed. This model is suitable for the approximate calculation of magnetron sputtering systems in the initial period of a pronounced erosion zone [2].

The main advantages of the magnetron sputtering system are:

(1) universality of the process, allows to obtain a layer of metals, alloys, semiconductors and dielectrics;
(2) high level of deposition (up to several microns/min.) And the possibility of its regulation over a wide range;
(3) preservation of stoichiometry when spraying complex substances;
(4) high purity of layers;
(5) high adhesion of the substrate layers;
(6) the possibility of changing the structure and properties of the layers due to the displacement potential on the substrate, the pressure and composition of the gaseous medium, the synchronous sputtering of several targets and other methods;
(7) low porosity of the layers even with a small thickness;
(8) radiation and thermal effects on the treated structure are lower than in conventional diode atomization systems;
(9) the possibility of carrying out the process in a reactive medium, which allows to obtain layers of carbides, nitrides, oxides and other compounds;
(10) the ability of a number of materials at high current density to the target to spray independently;
(11) high energy efficiency of the process compared to diode and triode spray systems;
(12) the ability to automate the process;
(13) process inversion, which allows it to be used for deposition and etching of a wide class of materials.

2 The Operating Mode of the Vacuum System of the Plasma Generator

In general, the vacuum system has the form shown in Fig. 1 and consists of a hollow tank 1 from which air is pumped out, pressure gauges 2, valve 3, pipeline 4 and pump 5.

Before the pump starts, the pressure in the entire vacuum system is the same and the gas is stationary. With the beginning of the pump, the gas begins to move from the hollow tank 1 by pipeline 4 to the pump 5. At the same time, the amount of gas in the vacuum system is continuously decreasing, and since the volume and temperature of the gas remain almost unchanged, a decrease in pressure in the vacuum system occurs. The pH pressure at the inlet to the pump becomes lower than the pressure p at the outlet of the evacuated hollow tank 1. Thus, a p-pH pressure difference is created, which is due to the presence in the system of a pipeline, a tap and other elements that resist gas flow. The difference in p-pH is called the pressure difference, carries out the movement.

In connection with the difference in the rate of pressure reduction in the hollow tank 1 and at the pump inlet (the need to consider when synthesizing control and selecting elements of the vacuum system), the concepts of gas pumping speed from the hollow tank and the speed of the pump are distinguished [3].

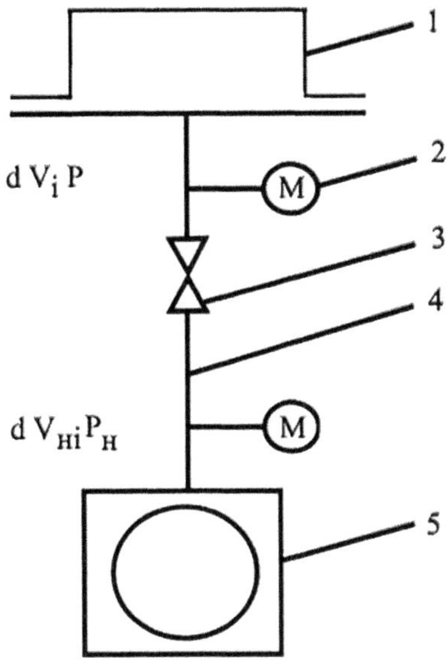

Fig. 1. General view of the vacuum system

3 Method of Ensuring Functional Stability

The empirical physical law (1) of the relationship of voltage, current and resistance (Ohm's law) makes it possible to identify the functional dependencies of the control object

$$I = \frac{\varepsilon}{R + r} \tag{1}$$

where ε – the electromotive force of the voltage source;
I – the value of the current of the electrical circuit;
R – the resistance of all external elements of the electrical circuit;
r – internal resistance of the power source.

Since a voltage source is used to power the technological installation of the plasma generator, we have:

$$R >> r \tag{2}$$

According to (2) write formula (1) so:

$$I = \frac{\varepsilon}{R} \tag{3}$$

The voltage across the entire power supply remains unchanged when internal electrical processes are unstable:

$$U = U_{con} + U_{discon} + U_{\sim} = const \tag{4}$$

where U – power source voltage;
U_{con} – value of voltage drop on analog elements;
U_{discon} – value of voltage drop on discrete elements;
$U_{\sim}$ – AC voltage of the non-stationary area of the system.

The stability of the technological process of applying a chrome coating in the framework of the use of magnetron sputtering systems using section-targeted cathodes is possible with the retention of electrical processes in a given region of the current-voltage characteristics of the gas discharge.

Ensuring the stable state of the sputtering process is also directly related to the stabilization of the current, which acts as a driving and regulating force for the directional motion of electric charges in the plasma environment.

This means that in the case of alternating uncontrolled voltage on the one hand, and due to the task of stabilizing the gas discharge gas on the other, the only control parameter of the system is resistance:

$$U = f(U_{\sim}) = f(R_{add)} \tag{5}$$

This means that if there is uncontrolled pressure from one side, then in order to create functional stability and increase the corrosion resistance of the workpiece, it is necessary to create a functional stable control system with controlled internal resistance [4], which can stabilize the voltage on the target cathode and anode, stabilize the current of the ionized stream and normalize the energy potential of charged particles Fig. 2.

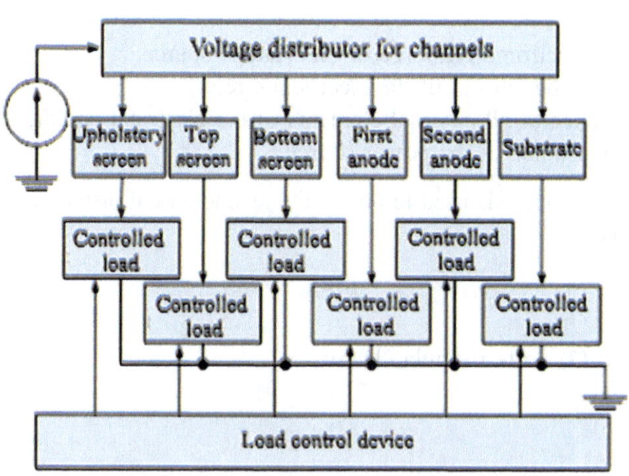

Fig. 2. Generalized structure for controlling spraying modes

The size of the controlled load is selected taking into account the voltage and currents that are realized during the operation of the plasma generator, taking into account the current-voltage characteristics of the main discharge [5]. This indicates that the flow of ionized particles can be controlled by their internal resistance.

3.1 Manual Resistance Control

In the first phase of the plasma generator control system study, alternating resistance relays were used to manually adjust the settings of work zone [6]. This means that to change the mode of application of the coating, it was necessary to stop the generator each time. It took a lot of time to change the characteristics of the system, not to mention the energy costs of the plasma-forming gas.

The implementation of a functionally stable system for controlling a plasma generator will make it possible to level out time and energy expenditures by getting the main coefficients of the system and their functional dependences [7] with the physical parameters of the plasma, namely, current, resistance, and voltage.

To solve this problem, initial experimental studies of the installation of a plasma generator were carried out and the current-voltage characteristics of the anode and cathode were obtained.

The current-voltage characteristics of the cathode and the anode under these operating conditions are shown in Figs. 3 and 4 respectively.

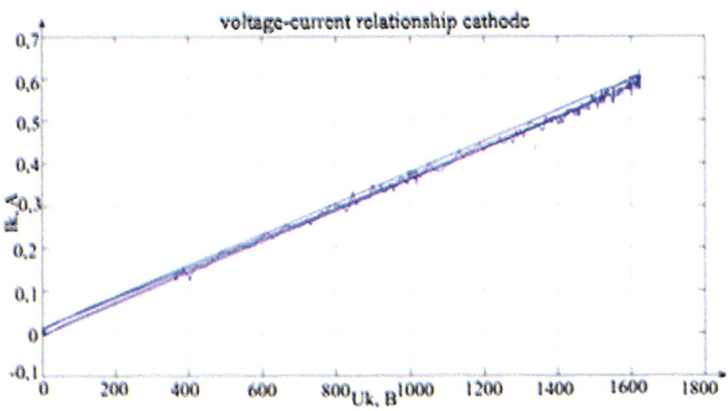

Fig. 3. Current-voltage characteristics of the cathode

These volt-current characteristics of the channels of the control system of the plasma generator make it possible to detect the jump transitions from one value to another, which does not allow to smoothly control the application of the coating to the workpiece. Such an unmanaged process gives an uneven thickness of the protective layer, which leads to a shift of the center of pressure of the workpiece, the loss of a high-quality index of surface wear resistance and, as a consequence, a rapid failure.

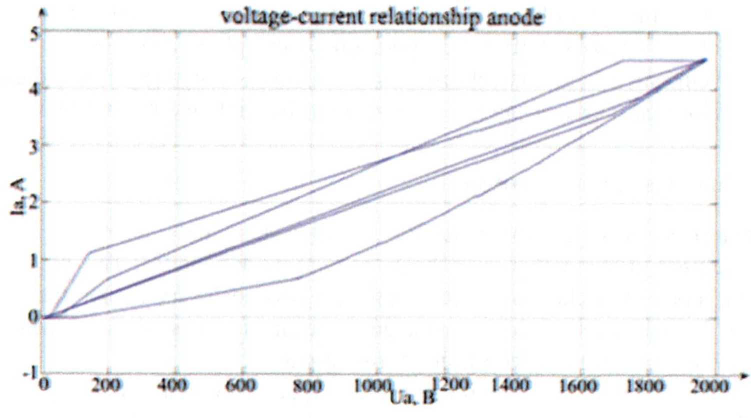

Fig. 4. Voltage-current relationship anode

3.2 Calculation of Functional Resistance Coefficient

It is quite difficult to describe plasma as a typical unit of a control system, since plasma does not have geometric parameters and, in this case, certain, regular laws of behavior.

In this regard, it is proposed to describe the sources of plasma, namely the cathode and anode. We represent the cathode and anode in the form of variable nonlinear resistors. Functionally, this can be represented as follows Fig. 5.

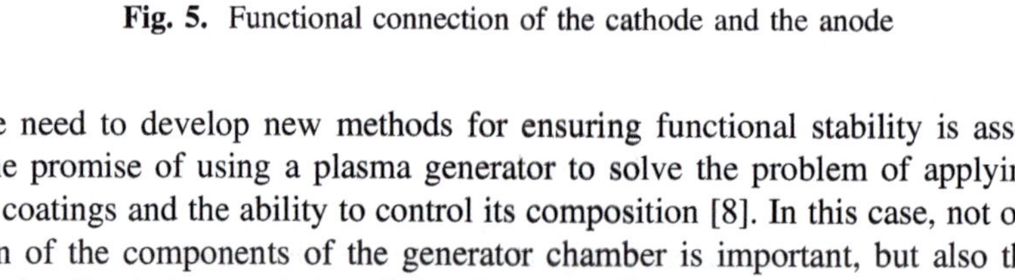

Fig. 5. Functional connection of the cathode and the anode

The need to develop new methods for ensuring functional stability is associated with the promise of using a plasma generator to solve the problem of applying protective coatings and the ability to control its composition [8]. In this case, not only the location of the components of the generator chamber is important, but also the provision of optimal characteristics of the plasma medium.

To achieve this research goal, a simplified electrical circuit of the installation under study was obtained (Fig. 6).

In the figure, respectively, R1 and R2 are included in the target cathode circles, R3 and R4 are included in the circle of the upper and lower end screens, and R5 is included in the side circle, the resistance of which changes from 1 Step to 4095 Ohms, depending on the task and voltage control on switch gear inlet.

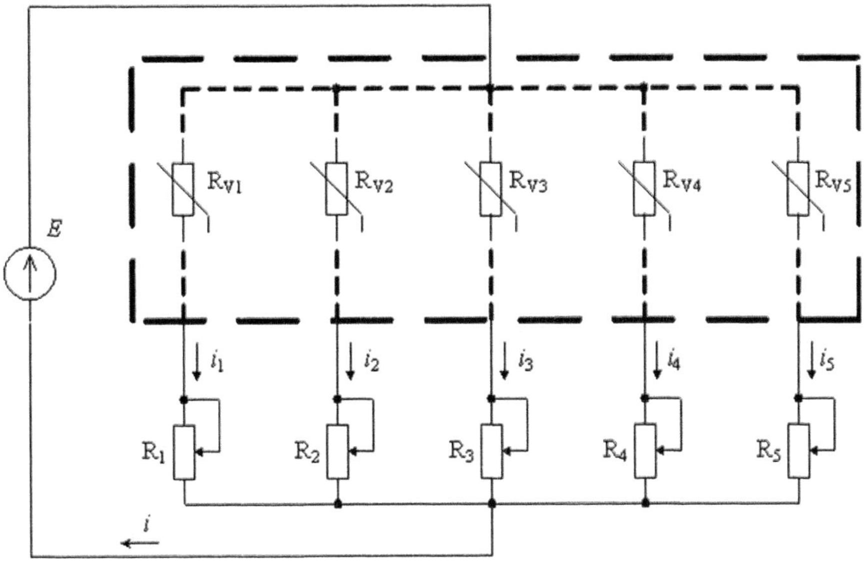

Fig. 6. Schematic diagram of the replacement of the plasma generator model

In this case, a set of resistances is formed in such a way that the nominal resistance is selected in accordance with the polynomial:

$$R_{\Sigma} = a_0 2^0 (1\Omega) + a_1 2^1 (1\Omega) + \ldots + a_k 2^k (1\Omega) = \left(a_0 + 2a_1 + \ldots + a_k 2^k\right)\Omega, \quad (6)$$

where $a_i = [0; 1]$, i.e. a_i takes 0 or 1.

Based on the equivalent scheme using the second Kirchhoff law, a system of equations was obtained:

$$\begin{cases} I_1 R_{V1} + I_1 R_1 = E; \\ I_2 R_{V2} + I_2 R_2 = E; \\ I_3 R_{V3} + I_3 R_3 = E; \\ I_4 R_{V4} + I_4 R_4 = E; \\ I_5 R_{V5} + I_5 R_5 = E. \end{cases} \quad (7)$$

In accordance with the conditions of providing a functionally stable mode $E = const$, $I_1 = const$, $I_2 = const$, $I_3 = const$, $I_4 = const$, $I_5 = const$, where $I_1 = \frac{E}{R_{V1} + R_1}, \ldots, I_5 = \frac{E}{R_{V5} + R_5}$.

The plasma formation process is unstable due to this RV1...RV5 has a variable resistance. To ensure direct current, it is necessary that R1...R5 varied so as to satisfy the conditions:

$$R_{V1} + R_1 = const.$$

If the resistances R1...R5 are independent variables, and the resistances RV1...RV5 are dependent, then we obtain the following equation:

$$\begin{aligned} R_{V1} &= f(R_1, R_2, R_3, R_4, R_5); \\ R_{V2} &= f(R_1, R_2, R_3, R_4, R_5); \\ R_{V3} &= f(R_1, R_2, R_3, R_4, R_5); \\ R_{V4} &= f(R_1, R_2, R_3, R_4, R_5); \\ R_{V5} &= f(R_1, R_2, R_3, R_4, R_5); \end{aligned}$$

After the differentiation get:

$$\begin{cases} dR_{V1} = \dfrac{\partial R_{V1}}{\partial R_1} dR_1 + \dfrac{\partial R_{V1}}{\partial R_2} dR_2 + \dfrac{\partial R_{V1}}{\partial R_3} dR_3 + \dfrac{\partial R_{V1}}{\partial R_4} dR_4 + \dfrac{\partial R_{V1}}{\partial R_5} dR_5; \\ dR_{V2} = \dfrac{\partial R_{V2}}{\partial R_1} dR_1 + \dfrac{\partial R_{V2}}{\partial R_2} dR_2 + \dfrac{\partial R_{V2}}{\partial R_3} dR_3 + \dfrac{\partial R_{V2}}{\partial R_4} dR_4 + \dfrac{\partial R_{V2}}{\partial R_5} dR_5; \\ dR_{V3} = \dfrac{\partial R_{V3}}{\partial R_1} dR_1 + \dfrac{\partial R_{V3}}{\partial R_2} dR_2 + \dfrac{\partial R_{V3}}{\partial R_3} dR_3 + \dfrac{\partial R_{V3}}{\partial R_4} dR_4 + \dfrac{\partial R_{V3}}{\partial R_5} dR_5; \\ dR_{V4} = \dfrac{\partial R_{V4}}{\partial R_1} dR_1 + \dfrac{\partial R_{V4}}{\partial R_2} dR_2 + \dfrac{\partial R_{V4}}{\partial R_3} dR_3 + \dfrac{\partial R_{V4}}{\partial R_4} dR_4 + \dfrac{\partial R_{V4}}{\partial R_5} dR_5; \\ dR_{V5} = \dfrac{\partial R_{V5}}{\partial R_1} dR_1 + \dfrac{\partial R_{V5}}{\partial R_2} dR_2 + \dfrac{\partial R_{V5}}{\partial R_3} dR_3 + \dfrac{\partial R_{V5}}{\partial R_4} dR_4 + \dfrac{\partial R_{V5}}{\partial R_5} dR_5; \end{cases} \tag{8}$$

If replace the differentials with small increments $A = (\partial R_{V1})/(\partial R_1)$ obtain the variables:

$$\begin{cases} \Delta R_{V1} = A_{11}\Delta R_1 + A_{12}\Delta R_2 + A_{13}\Delta R_3 + A_{14}\Delta R_4 + A_{15}\Delta R_5; \\ \Delta R_{V2} = A_{21}\Delta R_1 + A_{22}\Delta R_2 + A_{23}\Delta R_3 + A_{24}\Delta R_4 + A_{25}\Delta R_5; \\ \Delta R_{V3} = A_{31}\Delta R_1 + A_{32}\Delta R_2 + A_{33}\Delta R_3 + A_{34}\Delta R_4 + A_{35}\Delta R_5; \\ \Delta R_{V4} = A_{41}\Delta R_1 + A_{42}\Delta R_2 + A_{43}\Delta R_3 + A_{44}\Delta R_4 + A_{45}\Delta R_5; \\ \Delta R_{V5} = A_{51}\Delta R_1 + A_{52}\Delta R_2 + A_{53}\Delta R_3 + A_{54}\Delta R_4 + A_{55}\Delta R_5. \end{cases} \tag{9}$$

where

$$A_{11} = \frac{\Delta R_{V1}}{\Delta R_1};$$

when $R_2 = const$, $R_3 = const$, $R_4 = const$, $R_5 = const$.

$$A_{21} = \frac{\Delta R_{V2}}{\Delta R_1};$$

when $R_2 = const$, $R_3 = const$, $R_4 = const$, $R_5 = const$.

$$A_{55} = \frac{\Delta R_{V5}}{\Delta R_5};$$

when $R_1 = const$, $R_2 = const$, $R_3 = const$, $R_4 = const$.

The experimental data allow to obtain the values of voltage and current in the circles of cathodes and screens with variable supports R1...R5, we can obtain the parameters A11, A21, ..., A55.

4 Conclusion

This article reveals the idea of obtaining mathematical dependencies by replacing the physical processes of a plasma generator with an equivalent electrical circuit. The main processes: getting current-voltage characteristic, obtaining of functional dependences for the electrical equivalent circuit of plasma processes and getting a simplified mathematical model of the investigated system. The next step is a formation of a control law which ensures the functional stability of plasma process and maintain stable effect of the current.

The mathematical system of functional dependencies allows reveal the dependence and influence of the coefficients on the deposition process and give a possibility to develop a control law formation system based on these dependencies, which should ensure the functional stability of plasma formation and maintain a stable current effect.

References

1. Tahar, H., Yoshimura, N., Koshiro, Y.: Spraying using electromagnetically accelerated plasma. In: Designing of Interfacial Structures in Advanced Materials and their Joints, vol. 127, pp. 319–324 (2007). https://doi.org/10.4028/www.scientific.net/SSP.127.319
2. Baranov, O., Gorbenko, S.: Plasma-ion methods for changing the performance properties of surface layers of the workpiece. KHADI, № 82, pp. 62–67 (2018). (in Russian)
3. Fisrov, S., Breslavets, M., Bilokonska, Y.: The control system of an internal variable resistance conductive gas (argon) plasma generator chamber. In: Innovate Approaches to the Development of Science [Text]: Materials of International Scientific and Practical Conference, Dublin, Ireland, 1 June 2018. For the production Hold-enblat M.A. NGO "European Scientific Platform", part 2, pp. 145–150 (2018)
4. Firsov, S., Kochuk, S., Breslavets, M., Bilokonska, Y., Slusar, D.: Functional stability of the control system of a plasma generator with sectioned cathode units in the mode of ion-plasma deposition of multicomponent nanostructured materials. Open Access Peer-reviewed J. Sci. Rev. **2**(9) (2018). (in Russian), vol. 1, Scientific Edition, pp. 38–42. ISSN 2544-9346, 2544-9443
5. Firsov, S., Boyarkin, A., Breslavets, M., Bilokonska, Y.: Functionally stable current control of the working chamber of the plasma generator. Instrum. Technol. Sci.-Tech. J. **1**, 17–20 (2019). (in Russian)

6. Bilokonska, Y., Breslavets, M., Fisrov, S.: Automated plasma generator system. Synergetics, mechatronics, telematics of road machines and systems in educational process and science, № 1, pp. 36–39 (2018). (in Russian)
7. Firsov, S.N.: Formation of fault-tolerant flywheel engine units in satellite stabilization and attitude control systems [text]. J. Comput. Syst. Sci. Int. **53**(4), 601–609 (2014). Article title. Journal 2(5), 99–110 (2016)
8. Slusar, D., Kolesnik, V., Litovchenko, L., Stepanushkin, N., Garin, V.: Control composition functionally gradient coatings in the transition zone. Aerosp. Eng. Technol. **6**(123), 58–63 (2015). (in Russian)

Nonlinear Postbuckling Behavior of a Simply Supported, Uniformly Compressed Rectangular Plate

Sergii G. Kravchenko[1] and Vitalii Myntiuk[2(✉)]

[1] Purdue University, West Lafayette, USA
[2] National Aerospace University "Kharkiv Aviation Institute", Kharkiv, Ukraine
vitalii.myntiuk@khai.edu

Abstract. The geometrically nonlinear deformation theory is used for the far post-buckling analysis of a uniformly compressed rectangular plate. The plate geometrically nonlinear deformation theory is developed using the right stretch tensor and the Biot stress tensor. The plate is assumed thin such that the Kirchhoff hypothesis are applied. The plate material is assumed linear elastic. The simply supported boundary conditions are assumed for the plate deflection function. The in-plane displacements of a plate are not constrained except for the prescribed uniform shortening along the opposite edges. The weak solution is constructed using the Ritz method. The basis functions in the displacement function approximations are assumed as Legendre polynomials and their linear combinations. The nonlinear simultaneous equations are solved by the Newton method. Three equilibrium paths are determined, which originate from the first three bifurcation points; two additional equilibrium paths are found to appear during the post-buckling deformation, which correspond to the formation of the buckle-waves (wrinkles) along the non-loaded edges. Stable and unstable branches of the equilibrium paths are determined, as well the bifurcation and limit points. The variation of the potential energy is demonstrated and the possible points of jump-like transitions between the adjacent buckled configurations are identified, including those with where wrinkles are formed. The high accuracy and convergence rate of the numerical solution are demonstrated.

Keywords: Simply supported rectangular plate · Postbuckling · Equilibrium paths

1 Introduction

The approaches to performance analysis of aerospace structures require the consideration of specific deformation traits inherent to the thin-walled load-carrying components. One of such traits is the ability of thin plates and shells to keep their load-bearing capacity after initial buckling [1, 2]. The challenge in the analysis of post-buckling behavior is caused by the fact that the post-buckling equilibrium state can change abruptly. This phenomenon has been previously found experimentally [3–6], while its analysis and interpretation have been undertaken by many researchers [7–13]. The susceptibility of an elastic system to the secondary structural instabilities largely

M. Nechyporuk et al. (Eds.): *Integrated Computer Technologies in Mechanical Engineering*, AISC 1113, pp. 35–44, 2020.
https://doi.org/10.1007/978-3-030-37618-5_4

depends on the membrane and bending boundary conditions, geometrical dimensions, material properties, but the origin of secondary structural instabilities is probably in the existence of several equilibrium paths with various corresponding potential energy. The main focus of the present treatise is on (i) the investigation of the far post-buckling behavior and (ii) the new stable equilibrium states of a simply supported, compressed rectangular plate, which had not been previously studied.

The post-buckling analysis of plates must be performed using the geometrically non-linear deformation theories. The typically utilized (in academia) Föppl–von Kármán plate theory is not well-suited for that purpose [14, 15], this is because the accuracy of that theory is deemed acceptable for the deflections comparable with plate thickness. Using the Föppl–von Kármán plate theory in the far post-buckling analysis may lead not only to the quantitative errors but also to the qualitatively incorrect results. In some cases, the Föppl–von Kármán plate theory doesn't allow to find the limit points on the equilibrium paths, which correspond to the critical loads. Under certain conditions, these incorrectly determined critical loads might be underestimated with respect to the Euler critical loads, which is unacceptable for safety reasons [16].

The authors herein give preference to the plate nonlinear deformation theory derived by means of the Biot stress tensor and the right stretch tensor [17–24]. These two tensors possess an advantage of having an explicit physical meaning. Another most often used plate theory is built upon the application of the Cauchy–Green deformation tensor and the second Piola–Kirchhoff stress tensor, which do not have an explicit physical meaning but have simpler displacement-strain relations. Note that both theories provide the close matching results when strains are small [15].

2 Geometrically Nonlinear Theory of a Plate

The plate is assumed linear elastic and isotropic and having a uniform thickness. The strain energy density function (specific strain energy) of a deformed plate has a well-known form [25] given by Eq. (1):

$$\begin{aligned} dU = \frac{Eh}{2(1-\nu^2)} \Big\{ & (\varepsilon_{11}+\varepsilon_{22})^2 + 2(1-\nu)\left(\varepsilon_{12}^2 - \varepsilon_{11}\varepsilon_{22}\right) \\ & + \frac{h^2}{12}\left[(\kappa_{11}+\kappa_{22})^2 + 2(1-\nu)\left(\kappa_{12}^2 - \kappa_{11}\kappa_{22}\right)\right] \Big\} \end{aligned} \tag{1}$$

where E, ν are the elastic modulus and Poisson's ratio, respectively; h is the plate thickness; $\varepsilon_{\alpha\beta}$, $\kappa_{\alpha\beta}$ are the mid-plane membrane (in-plane) strains and curvatures of a plate, $\alpha = 1,2$ $\beta = 1,2$.

The plate forces and moments are obtained as derivatives of the specific strain energy, $N_{\alpha\beta} = \frac{\partial dU}{\partial \varepsilon_{\alpha\beta}}$, $M_{\alpha\beta} = \frac{\partial dU}{\partial \kappa_{\alpha\beta}}$.

Let the mid-plane of an undeformed plate lie in the x_1x_2-plane of the Cartesian coordinate system and $u_i(x_1, x_2)$ $(i = 1, 2, 3)$ are the displacement functions of the plate mid-plane. Let us use the notation $(*)_{,\alpha} \equiv \frac{\partial *}{\partial x_\alpha}$ to define a derivative, then according to [17], the in-plane strains $\varepsilon_{\alpha\beta}$ and the curvatures $\kappa_{\alpha\beta}$ are determined by the following chains of computations:

- tangent vectors ($\mathbf{r}_1$, $\mathbf{r}_2$) and normal vector ($\mathbf{n}$) to the deformed mid-plane of a plate: $\mathbf{r}_1 = \{1 + u_{1,1}, u_{2,1}, u_{3,1}\}^T$; $\mathbf{r}_2 = \{u_{1,2}, 1 + u_{2,2}, u_{3,2}\}^T$; $J = |\mathbf{r}_1 \times \mathbf{r}_2|$; $\mathbf{n} = \frac{\mathbf{r}_1 \times \mathbf{r}_2}{J}$;
- components of the metric tensor of the deformed middle plane: $g_{\alpha\beta} = \mathbf{r}_\alpha \cdot \mathbf{r}_\beta$;
- eigenvalues of the metric tensor: $e_{1/2} = \frac{1}{\sqrt{2}}\sqrt{g_{11} + g_{22} \pm \sqrt{(g_{11} - g_{22})^2 + 4g_{12}^2}}$;
- right stretch tensor: $\lambda_{\alpha\beta} = \frac{g_{\alpha\beta} + J\delta_{\alpha\beta}}{e_1 + e_2}$;
- the Biot strain tensor:

$$\varepsilon_{\alpha\beta} = \lambda_{\alpha\beta} - \delta_{\alpha\beta} \tag{2}$$

  where $\delta_{\alpha\beta}$ is the Kronecker delta;
- orthonormal vectors of the rotated original basis: $\mathbf{q}_1 = \frac{\lambda_{22}\mathbf{r}_1 - \lambda_{12}\mathbf{r}_2}{J}$, $\mathbf{q}_2 = \frac{\lambda_{11}\mathbf{r}_2 - \lambda_{12}\mathbf{r}_1}{J}$;
- symmetric components of the curvature tensor:

$$\begin{aligned}
\kappa_{11} &= \mathbf{q}_1 \cdot \mathbf{n}_{,1} + \overline{\lambda}_{12}(\mathbf{q}_2 \cdot \mathbf{n}_{,1} - \mathbf{q}_1 \cdot \mathbf{n}_{,2});\\
\kappa_{22} &= \mathbf{q}_2 \cdot \mathbf{n}_{,2} + \overline{\lambda}_{12}(\mathbf{q}_1 \cdot \mathbf{n}_{,2} - \mathbf{q}_2 \cdot \mathbf{n}_{,1});\\
\kappa_{12} &= \kappa_{21} = \overline{\lambda}_{11}\mathbf{q}_1 \cdot \mathbf{n}_{,2} + \overline{\lambda}_{22}\mathbf{q}_2 \cdot \mathbf{n}_{,1},
\end{aligned} \tag{3}$$

  where $\overline{\lambda}_{\alpha\beta} \equiv \frac{\lambda_{\alpha\beta}}{\lambda_{11} + \lambda_{22}}$.

Remark 1. The symmetry of the curvature tensor (3) obtained in [17] has a clear advantage over the expressions $\widetilde{\kappa}_{\alpha\beta} = \mathbf{q}_\alpha \cdot \mathbf{n}_{,\beta}$ previously reported in [19–24]. The curvatures in (3) are obtained as the linear part of the right stretch tensor series expansion by the degraded coordinate, $\frac{\partial \mathbf{\Lambda}}{\partial z}\Big|_{z=0} \equiv \kappa_{\alpha\beta}\mathbf{e}_\alpha \otimes \mathbf{e}_\beta$. The curvatures obtained as $\mathbf{Q}^T \cdot \frac{\partial \mathbf{F}}{\partial z}\Big|_{z=0} \equiv \widetilde{\kappa}_{\alpha\beta}\mathbf{e}_\alpha \otimes \mathbf{e}_\beta$ must be further subjected to the symmetrization $\overline{\kappa}_{\alpha\beta} = \overline{\kappa}_{\alpha\beta} = \frac{1}{2}(\widetilde{\kappa}_{\alpha\beta} + \widetilde{\kappa}_{\alpha\beta})$. In the above, $\mathbf{F}$ is the deformation gradient tensor of the layer of a plate at a small distance z from the plate mid-plane, $\mathbf{\Lambda}$ and $\mathbf{Q}$ are the right stretch tensor and a proper orthogonal tensor, respectively; using the polar decomposition theorem, $\mathbf{F} = \mathbf{Q} \cdot \mathbf{\Lambda}$.

Remark 2. The strains of the plate theory based on the Cauchy-Green tensor and the strains of Föppl–von Kármán plate theory can be obtained as particular cases (3) and (4). To get the Cauchy-Green strains, $\varepsilon_{\alpha\beta}^{CG} = \frac{1}{2}(g_{\alpha\beta} - \delta_{\alpha\beta})$, $\kappa_{\alpha\beta}^{CG} = \mathbf{r}_\alpha \cdot \mathbf{n}_{,\beta}$, the expressions from (3) and (4) need to be expanded into Taylor series keeping the quadratic terms of the displacement derivatives. To get the Föppl–von Kármán strains, $\varepsilon_{\alpha\beta}^{FK} = \frac{1}{2}(u_{\alpha,\beta} + u_{\beta,\alpha} + u_{3,\alpha}u_{3,\beta})$, $\kappa_{\alpha\beta}^{FK} = -u_{3,\alpha\beta}$, only the $u_{3,\alpha}u_{3,\beta}$ quadratic terms must be withheld.

3 Numerical Solution

The numerical solution for the rectangular plate is obtained by the Ritz method. The unknown displacement functions are approximated by the series expansions (the following notations will be used $x \equiv x_1$, $y \equiv x_2$, $u \equiv u_1$, $v \equiv u_2$, $w \equiv u_3$)

$$\begin{aligned} u &= u_0(\mathrm{x}, \mathrm{y}) + \sum_{m=0,1,\ldots} \sum_{n=0,1,\ldots} U_{mn} \Xi^u_m(x) \Psi^u_n(y) \\ v &= \sum_{m=0,1,\ldots} \sum_{n=0,1,\ldots} V_{mn} \Xi^v_m(x) \Psi^v_n(y) \\ w &= \sum_{m=0,1,\ldots} \sum_{n=0,1,\ldots} W_{mn} \Xi^w_m(x) \Psi^w_n(y) \end{aligned} \quad (4)$$

The unknown coefficients U_{mn}, V_{mn} и W_{mn} have to be determined. The known basis functions $\Xi(x)$ и $\Psi(y)$ must at least satisfy the essential boundary conditions. The essential boundary conditions for a plate under consideration (Fig. 1) are given in the following: (i) the deflections $w|_{\bar{x}=\pm 1} = w|_{\bar{y}=\pm 1} = 0$; (ii) the displacement $u|_{\bar{x}=\pm 1} = \mp \Delta$. Herein, the dimensionless coordinates are used $\bar{x} = \frac{2x}{a}$ и $\bar{y} = \frac{2y}{b}$ (Fig. 1); Δ is the prescribed plate end shortening.

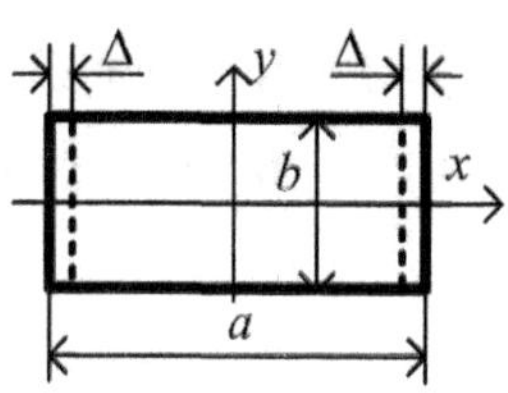

Fig. 1. Uniformly compressed simply supported plate.

The essential boundary conditions (i) will be satisfied is the basis functions are $\Xi^w_m = P_m(\bar{x}) - P_{m+2}(\bar{x})$ and $\Psi^w_n = P_n(\bar{y}) - P_{n+2}(\bar{y})$, where P_i are the Legendre polynomials. To satisfy the essential conditions (ii), let us assume $u_0 = -\Delta P_1(\bar{x})$ and $\Xi^u_m = P_{2m+1}(\bar{x}) - P_{2m+3}(\bar{x})$. Since the remaining in-plane displacements $u|_{\bar{y}=\pm 1}$, $v|_{\bar{x}=\pm 1}$ and $v|_{\bar{y}=\pm 1}$ on the edges are not constrained, let the basis functions be $\Psi^u_n = P_{2n}(\bar{y})$, $\Xi^v_m = P_{2m}(\bar{x})$ and $\Psi^v_n = P_{2n+1}(\bar{y})$. Even and odd Legendre polynomials are selected considering the symmetry of the problem.

Note that the basis functions are linearly independent, form a complete system, and prevent the rigid body motions. The deflection function in Eq. (4) contains all the terms in the expansion series, but the number of terms can be reduced if the obviously zero terms W_{mn} are eliminated. In case of a symmetric deflection function, only even terms are to be kept; in case of asymmetric deflection function, only odd terms need to be preserved.

The unknown coefficients in Eq. (4), U_{mn}, V_{mn} and W_{mn}, can be written as a vector

$$\mathbf{c} = \{C_i\} = \{U_{00}, U_{01}, U_{10}, \ldots, V_{00}, V_{01}, V_{10}, \ldots, W_{00}, W_{01}, W_{10}, \ldots\}^T \quad (5)$$

and determined from a system of simultaneous nonlinear equations obtained from the principle of stationary total potential energy $\frac{\partial \Pi}{\partial C_i} = 0$, $\Pi = \int\limits_{-b/2}^{b/2} \int\limits_{-a/2}^{a/2} dUdxdy.$

The system of equations is solved via the Newton method. Note that the transition from an initial, undeformed state of an ideally flat plate to the deformed configuration is typically achieved by prescribing either an initial deflection or an additional perturbation load. In such case, the bifurcation point becomes a limit point on the equilibrium path diagram, i.e. the equilibrium path shows a sharp increase in the deflection instead of branching (bifurcating) the equilibrium curves. In the current approach, the deflected shape corresponding to liner buckling is assumed for the first iteration of the Newton method. Next, the procedure of incremental loading is used allowing for the control of both the loading increment and the magnitude of the Euclidian vector norm $\|\mathbf{C}\| = \sqrt{\sum C_i^2}$. Hessian matrix $\left[\frac{\partial^2 \Pi}{\partial C_i \partial C_j}\right]$ condition number (maximum to minimum eigenvalue fraction) is evaluated at every loading increment. The sign of the condition number allows to conclude either the current equilibrium state is stable or unstable. The solution algorithm is numerically implemented in C++.

4 Numerical Example

Let us consider an isotropic, simply supported, uniformly compressed rectangular plate having the in-plane dimensions of $a \times b = 100 \times 100$ and thickness of $h = 1$, the Young's modulus of $E = 7200$ and Poisson's ratio of $\nu = 0.3$. Euler critical buckling load of the plate is $p_{kr} = \frac{4\pi^2 D}{b^2} = 2.60297$, where $D = \frac{Eh^3}{12(1-\nu^2)} = 659.341$ is the plate flexural stiffness. Correspondingly, $\Delta_{kr} = \frac{p_{kr} a}{2Eh} = 0.0180762$.

Figure 2 shows the equilibrium paths (top) and the corresponding values of the potential energy (bottom). The abscissa axis is used for the normalized plate end shortening $\overline{\Delta} = \frac{\Delta}{\Delta_{kr}}$. The ordinate axis represents the Euclidian vector norm $\|\mathbf{C}\|$ (5), which characterizes the total displacement of the plate mid-plane points. The bottom part of Fig. 2 has the total potential energy along the ordinate axis. The unstable equilibrium paths are shown as dash lines. Figure 2 also reports the bifurcation points, limit points, and the deformed configurations of a buckled plate.

The initial planar equilibrium state becomes unstable at point A_1 and the first buckling mode shape is seen to have a single buckle-wave. The next two bifurcation points B_1 ($\overline{\Delta} = \frac{25}{16} \approx 1.56$) and C_1 ($\overline{\Delta} = \frac{25}{9} \approx 2.78$) correspond to the more complex buckling mode shapes featuring two and three buckle-waves, respectively. Note, that the third branch of the equilibrium path curve becomes stable only at the load exceeding the magnitude of 6.54 (limit point C_2). The beginning of the second branch of the equilibrium path curve is only slightly different from the bifurcation point B_1.

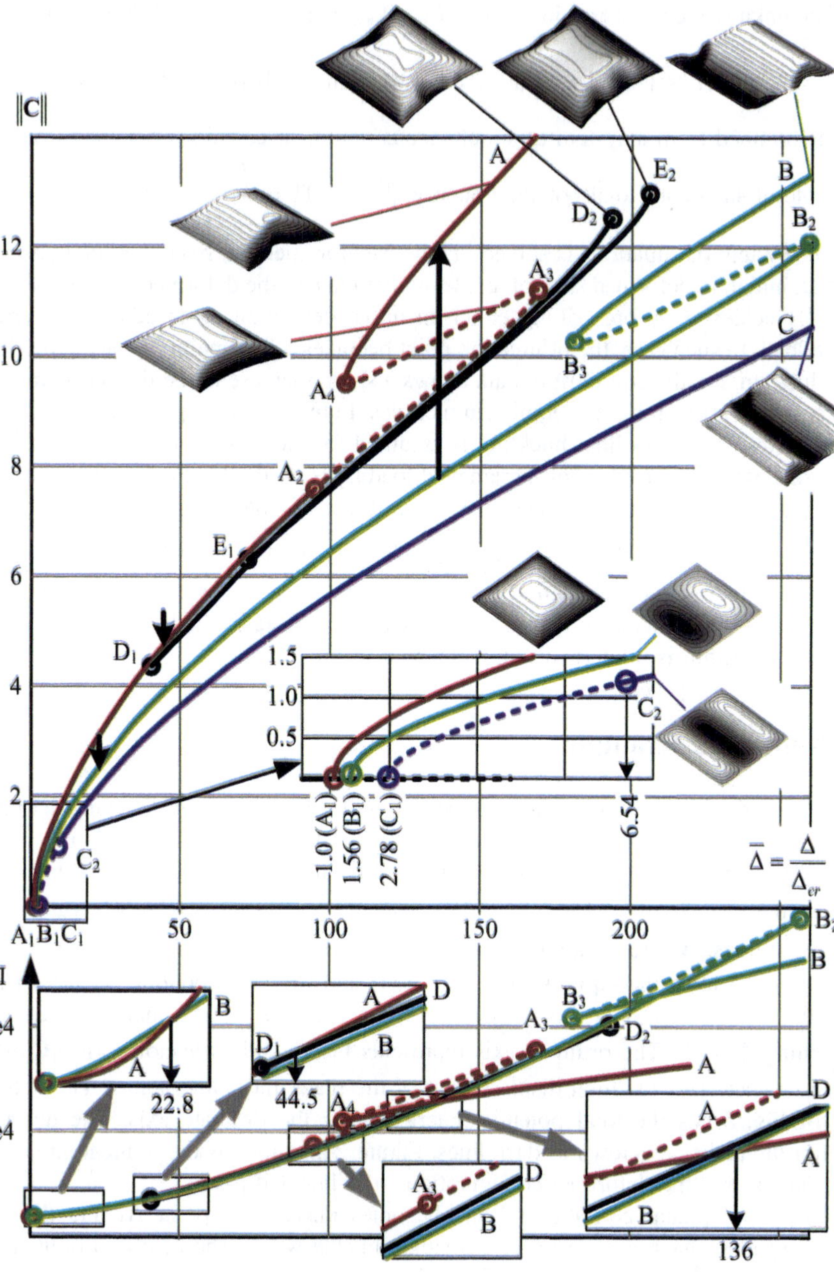

Fig. 2. Equilibrium paths and potential energy of a simply supported plate.

The first branch of the equilibrium path curve is stable between the bifurcation point A_1 to the limit point A_2 ($\overline{\Delta} = 92.4$). Next, this branch becomes unstable until the point A_4. Then, a stable part of this branch follows again, which corresponds to the formation of wrinkles at the edges that are not loaded.

The second branch of the equilibrium path curve is stable up until the point B_2, then an unstable region of the branch follows (B_2–B_3), wherein the wrinkles form. The stable segment of the branch pass beyond the point B_3 corresponds to the increase of the wrinkle.

Besides the discussed branches of the equilibrium path curve starting at the bifurcation points of the initial equilibrium state, more branches, D_1–D_2 and E_1–E_2, have also been found. These branches correspond to the formation of additional buckle-waves at the non-loaded edges of a plate. The branch D_1–D_2 is stable in the interval of $40.7 < \overline{\Delta} < 194.5$ and corresponds to the formation of one extra buckle-wave along the edges. The equilibrium branch E_1–E_2 corresponds to the formation of the three buckle-waves on each side and is stable in the interval of $72.3 < \overline{\Delta} < 205.5$. These two branches, D_1–D_2 and E_1–E_2, have not been previously reported in the literature, but similar buckling modes were derived [2, 11, 26].

It is difficult to judge which of the discussed equilibrium paths would be actually realized in a physical plate, since the buckled equilibrium configurations are heavily influenced by a variety of factors related to the inherent imperfections of the physical structural system. Nevertheless, the proposed methodology allows to follow the transitions between the stable and unstable equilibrium configurations, as long as it is believed that the principle of stationary total potential energy holds true for the plate [27].

The minimum potential energy is realized within the first branch A_1–A_2 in the interval of $1 < \overline{\Delta} < 22.8$. Next, in the interval of $22.8 < \overline{\Delta} < 136$ and right up to the possible wrinkle formation (point A_3), the plate with two buckle-waves has the minimum potential energy (segment B_1–B_2). The sharp decrease of potential energy occurs on the segment A_3–A_4 of the equilibrium path curve, followed by a slow increase of the potential energy along the segment A_4–A. For $\overline{\Delta} > 136$, these stable equilibrium configurations have significantly lower levels of potential energy.

The energy of the equilibrium states along the branches C and E is always greater than along the other branches A, B, and D.

Therefore, in the interval of $22.8 < \overline{\Delta} < 92.4$, the jump transition between the braches A and B is possible if the potential barrier can be overcome. If the transition didn't happen, the jump transition between from the branch A to the branch D would become possible in the interval of $44.5 < \overline{\Delta} < 92.4$. In any case, the jump transition from the branch A has to take place in the interval of $22.8 < \overline{\Delta} < 92.4$.

5 Discussion on the Solution Accuracy

The accuracy of the solution (4) depends on the number of series expansion terms. If all series expansions in Eq. (4) have the same number of terms N, then the total number of

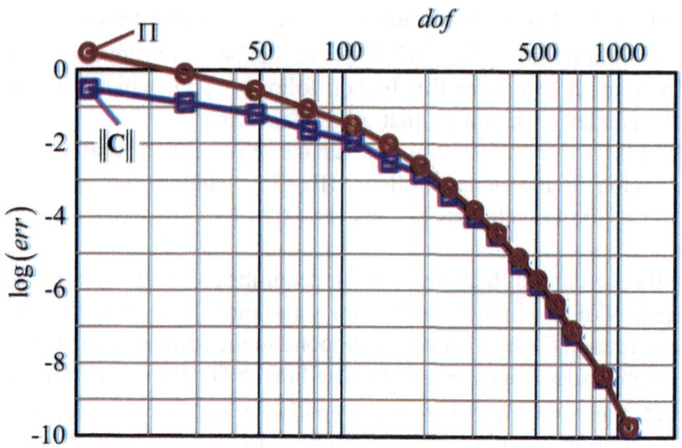

Fig. 3. Convergence of the displacement and potential energy magnitudes.

unknowns (i.e. the size of vector **C**) is $dof = 3(N+1)^2$. Let us use $N = 20$ ($dof = 1323$) to obtain a reference solution. The relative change of the displacement and potential energy magnitudes with respect to the increase in the number of terms in Eq. (4) is evaluated for the branch A at $\overline{\Delta} = 80$. The reference values are $\|\mathbf{C}\|_{1323}=$ 6.8543914039 (maximum deflection 9.9067) and $\Pi_{1323} = 5767.5616202$. The convergence of these quantities is shown in Fig. 3.

The convergence rate is such that the accuracy is increased by an order of magnitude when two terms are added to the series expansions in Eq. (4). It is enough to consider 10 terms for each expansion (total 300 unknowns) to achieve the displacement accuracy of 0.01%.

6 Conclusions

The theoretical-computational framework for the far post-buckling analysis of a simply supported, uniformly compressed rectangular plate is developed using the geometrically nonlinear deformation theory. It is determined that equilibrium states that don't originate from the linear bifurcation points can occur at large post-buckling loads. These equilibrium states may have a smaller potential energy and, therefore, can be realized.

High accuracy must be ensured in the computational analysis, otherwise the qualitatively wrong results can be obtained. For example, if only six terms (or less) are considered in the series expansions in Eq. (4), the limit point A2 will not be discovered, and the current branch of the equilibrium path diagram would falsely appear stable. Figure 2 does not show all the equilibrium paths, some of the discovered unstable branches are not shown to improve the readability of the data. The stability analysis of the equilibrium configurations under a given limiting potential energy level remains a challenging problem.

References

1. Anderson, M.S.: Design of panels having postbuckling strength. In: AIAA/ASME/ASCE/AHS/ASC Structures, Structural Dynamics and Materials Conference, pp. 2407–2413. AIAA, New York (1997)
2. Boni, L., Fanteria, D., Lanciotti, A.: Post-buckling behaviour of flat stiffened composite panels: experiments vs. analysis. Compos. Struct. **94**, 3421–3433 (2012)
3. Stein, M.: Loads and deformations of buckled rectangular plates. NASA Technical report R-40, National Aeronautics and Space Administration (1959)
4. Stein, M.: The phenomenon of change of buckling patterns in elastic structures. NASA Technical report R-39, National Aeronautics and Space Administration (1959)
5. Arnold, R.R.: A correlative analysis of northrop metal compression panels MCl-MC4. Report No. 84.033, lC1, Anamet Laboratories (1984)
6. Namdar, E., Darendeliler, H.: Buckling, postbuckling and progressive failure analyses of composite laminated plates under compressive loading. Compos. Part B **120**, 143–151 (2017)
7. Nakamura, T., Uetani, K.: The secondary buckling and post-buckling behaviours of rectangular plates. Int. J. Mech. Sci. **21**, 265–286 (1979)
8. Suchy, H., Troger, H., Weiss, R.: A numerical study of mode jumping of rectangular plates. ZAMM Z. Angew. Math. Mech. **65**(Z), 71–78 (1985)
9. Golubitsky, M., Schaeffer, D.: Singularities and Groups in Bifurcation Theory. Applied Mathematical Sciences, vol. I. Springer, New York (1985)
10. Riks, E., Rankin, Ch.C., Brogan, F.A.: On the solution of mode jumping phenomena in thin-walled shell structures. Comput. Methods Appl. Mech. Eng. **136**, 59–92 (1996)
11. Hofmeyer, H., Courage, J.: Analytical and finite element modelling of long plate mode jumping behavior. Thin-Walled Struct. **73**, 101–111 (2013)
12. Chen, H., Virgin, L.N.: Finite element analysis of post-buckling dynamics in plates. Part II: a non-stationary analysis. Int. J. Solids Struct. **43**(13), 4008–4027 (2006)
13. Vescovini, R., Bisagni, C.: Two-step procedure for fast post-buckling analysis of composite stiffened panels. Comput. Struct. **128**, 38–47 (2013)
14. Ciarlet, P.G.: Plates and Junctions in Elastic Multi-Structures. Springer, New York (1990)
15. Myntiuk, V.B.: Post-buckling of uniformly compressed simply supported plate with free inplane translating edges. J. Appl. Ind. Math. **14**(1) (2020, Printed)
16. Khalilov, S.A., Myntiuk, V.B.: Postbuckling analysis of flexible elastic frames. J. Appl. Ind. Math. **12**(1), 28–39 (2018)
17. Myntiuk, V.B.: Biot stress and strain in thin-plate theory for large deformations. J. Appl. Ind. Math. **12**(3), 501–509 (2018)
18. Novozhilov, V.V., Chernykh, K.F.: On 'true' stress and strain measures in nonlinear solid mechanics. Izv. Akad. Nauk SSSR, Mekh. Tverd. Tela **5**, 73–80 (1987)
19. Bufler, H.: The Biot stresses in nonlinear elasticity and the associated generalized variational principles. Ing. Arch. **55**, 450–462 (1985)
20. Nayfeh, A.H., Pai, P.F.: Linear and Nonlinear Structural Mechanics. Wiley, New York (2004)
21. Alumäe, N.A.: On the presentation of fundamental correlations of the nonlinear theory of shells. Prikl. Mat. Mekh. **20**(1), 136–139 (1956)
22. Atluri, S.N.: Alternate stress and conjugate strain measures and mixed variational formulations involving rigid rotations for computational analyses of finitely deformed solids with application to plates and shells. Comput. Struct. **18**(1), 93–116 (1983)

23. Bufler, H.: The Biot stresses in nonlinear elasticity and the associated generalized variational principles. Ing. Arch. **55**, 450–462 (1985)
24. Pietraszkiewicz, W.: Geometrically nonlinear theories of thin elastic shells. Adv. Mech. **12** (1), 51–130 (1989)
25. Reddy, J.N.: Theory and Analysis of Elastic Plates and Shells. CRC Press, Taylor and Francis, Boca Raton (2007)
26. Chai, H.: Contact buckling and postbuckling of thin rectangular plates. J. Mech. Phys. Solids **49**, 209–230 (2001)
27. Shin, D.K., Griffin, O.H., Gürdal, Z.: Postbuckling response of laminated plates under uniaxial compression. Int. J. Non-Linear Mech. **28**(1), 95–115 (1993)

New Possibilities of Creating the Efficient Dimensionally Stable Composite Honeycomb Structures for Space Applications

A. Kondratiev, V. Gaidachuk, T. Nabokina, and A. Tsaritsynskyi(✉)

National Aerospace University Kharkiv Aviation Institute, Chkalova Str., 17, Kharkiv 61070, Ukraine
a.kondratiev@khai.edu, tsaritsynskyy.a.a@gmail.com

Abstract. Today there is a great need in the optimal designing and manufacturing of dimensionally stable precision structures for satellite communication systems and sensing systems, in order to implement the international space programs successfully. It is well known that sandwich panels with load-bearing skins made of polymeric composites based on carbon, organic or glass fibers and honeycomb filler of aluminum foil or other materials providing combined action of skins, feature high dimensional stability. Optimization of design parameters was performed on a specimen of operating section of the composite solar panel with honeycomb filler. Results of the analysis of various reinforcement patterns for the load-bearing skins and rational distribution of material for several loading cases of the solar panel are presented. Technological warpage of these panels was analyzed and assessed. Causes of defects generated during the manufacturing process in the form of continuous and discrete strips of thin load-bearing skins of CFRP (carbon fiber reinforced plastic) were also investigated. With the use of the analytical methods supported by finite element method, the integrated study of the adhesive joint's bearing capacity of solar panel composite skins with the honeycomb filler at transverse avulsion for the main technological methods of applying the adhesive (as a continuous layer and targeted dosing to the ends of the honeycomb) was carried out. The results obtained allow predicting the fracture behavior of the load-bearing skins' bonding with the honeycomb filler depending on parameters and properties of the honeycomb cell material and the adhesive layer.

Keywords: Honeycomb structures · Optimal designing · Space applications

1 Introduction

At present time, the international space projects more actively employ the programs on launching new satellite communication systems and sensing systems into space. Creation of spacecrafts (SC) of new generation which put forward higher demands to quality, reliability and competitiveness in the global market of space services requires the development of new structural layouts (SL) or modernization of existing ones. It is also worth noting that speed of data transmission and accuracy of location of SC-serviced ground objects and near-to-Earth objects is increasing constantly.

M. Nechyporuk et al. (Eds.): *Integrated Computer Technologies in Mechanical Engineering*, AISC 1113, pp. 45–59, 2020.
https://doi.org/10.1007/978-3-030-37618-5_5

Specific feature of structures operating in the conditions of open space and meant for precision coordination of SC relationship with ground objects is the necessity of meeting rather strict requirements including provision of thermal dimensional stability thereby [1].

It is known that high and stable stiffness is featured by sandwich structures with fillers of various types among which the honeycomb fillers are the most widely used in rocket and space technology (RST) [2]. Unique set of strength, technological and operational characteristics of such structures has shown their priority among other classes of sandwich structures and predetermined their extensive use in rocket and space technology [3].

A promising direction in the creation of precision honeycomb structures (HS) is using of polymeric composite materials (PCM) based on carbon fibers and polymeric matrices in such structures [4]. Necessity of continuous improvement of their efficiency promoted growth of theoretical and experimental investigations aimed at development of the methods of optimal designing of products of that class and creation of pilot-design units to be operated in the conditions of open space [5]. To a certain extent, it was promoted by the works of the team of contributors aimed at enhancing of the scientific support of creation of PCM honeycomb structures for RST results of which are summarized in monographs [6] and a number of our reports such as [7, 8].

2 Analysis of Various Bearing Skins' Reinforcement Patterns and Rational Distribution of Material of Solar Panel at Its Multiple-Factor Loading

Usually, solar panels are experiencing high overloads in the process of launch vehicle start, as well as vibration impacts and thermo-cycling. At optimal solar panel designing, with a view to meeting strict requirements asserted therefor it is necessary to obtain the most accurate values of parameters of stress-strain distribution (SSD) in its structural elements for the whole spectrum of external impacts specified.

Ample opportunities of the modern software systems implementing the finite element method (FEM) allow ensuring to the full extent and within reasonable time the targeted optimal designing of that class of structures [9].

The authors previously initiated the studies in this area, and by now the contours of the present-day concept of optimization of basic parameters of RST structures made of PCM are outlined [6]. Within the framework of this concept, implementing of comprehensive approach to creation of precision HS for space applications continues, and the essence of the above approach consists in:

- theoretical prediction of maximum possible reduction of surface mass of products at the current and future levels of the material and technical base;
- subsequent analysis of technological capabilities of manufacturing solar panels with minimal surface mass taking into account the progress of materials used;
- revealing typical defects associated with the prospective level of solar panel production, determining their potential risk for the normal functioning of products in service, and finding the ways to avoid it.

Below, by the example of the operating section of composite solar panel with the honeycomb filler (HF) the results of theoretical prediction of maximum possible reduction of surface mass of solar panel at multiple-factor loading thereof are given [6].

Geometrical parameters of SC solar panel are shown in Fig. 1.

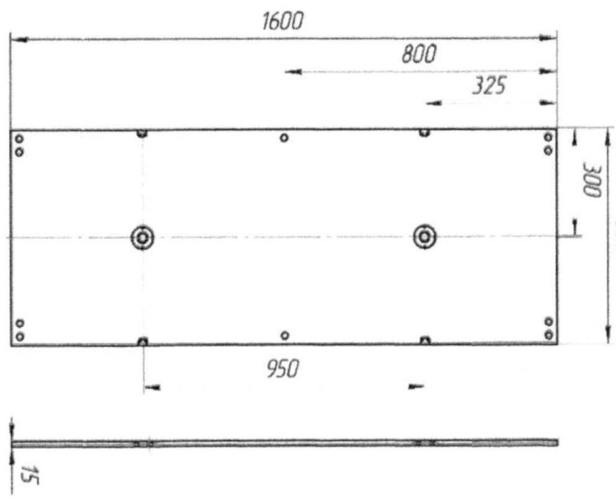

Fig. 1. Geometrical parameters of SC solar panel

Investigations were carried out as based on the requirements that the solar panel is to maintain its characteristics in the absence of mechanical damage after its exposure to:

- loads during acceptance tests by bending;
- quasi-static loads of the launch phase: longitudinal load of 15 g (along the longer side of the panel); lateral load of 10 g (along the shorter side and normally to the panel surface);
- acoustic loads;
- temperature field from minus 160 °C to plus 100 °C.

Calculation of SSD of solar panel was made with the use of software package of finite-element analysis. At its discretization to the finite element grid, four-unit multilayer shell-type finite element was used, with the bending and membrane properties. Base layers (BL) were simulated in the form of packages of the finite number of monolayers with the relevant physical and mechanical characteristics (PhMC). HF was represented as a conventional, uniform by volume layer of the finite element being used. Finite element model of the solar panel is represented by 10400 elements.

For evaluation of stress state of the base layer, Von-Mises–Hill energy failure criterion for layered PCM was used.

The diagrams of fixation and loading of the solar panel during acceptance tests, rated values of deflection, boundary conditions corresponding to them in the finite element analysis software system and examples of the patterns of strained state are represented in Fig. 2.

Below stated mass loads are adopted as quasi-static loads in the launch phase:

- longitudinal load of 15 g;
- lateral load of 10 g.

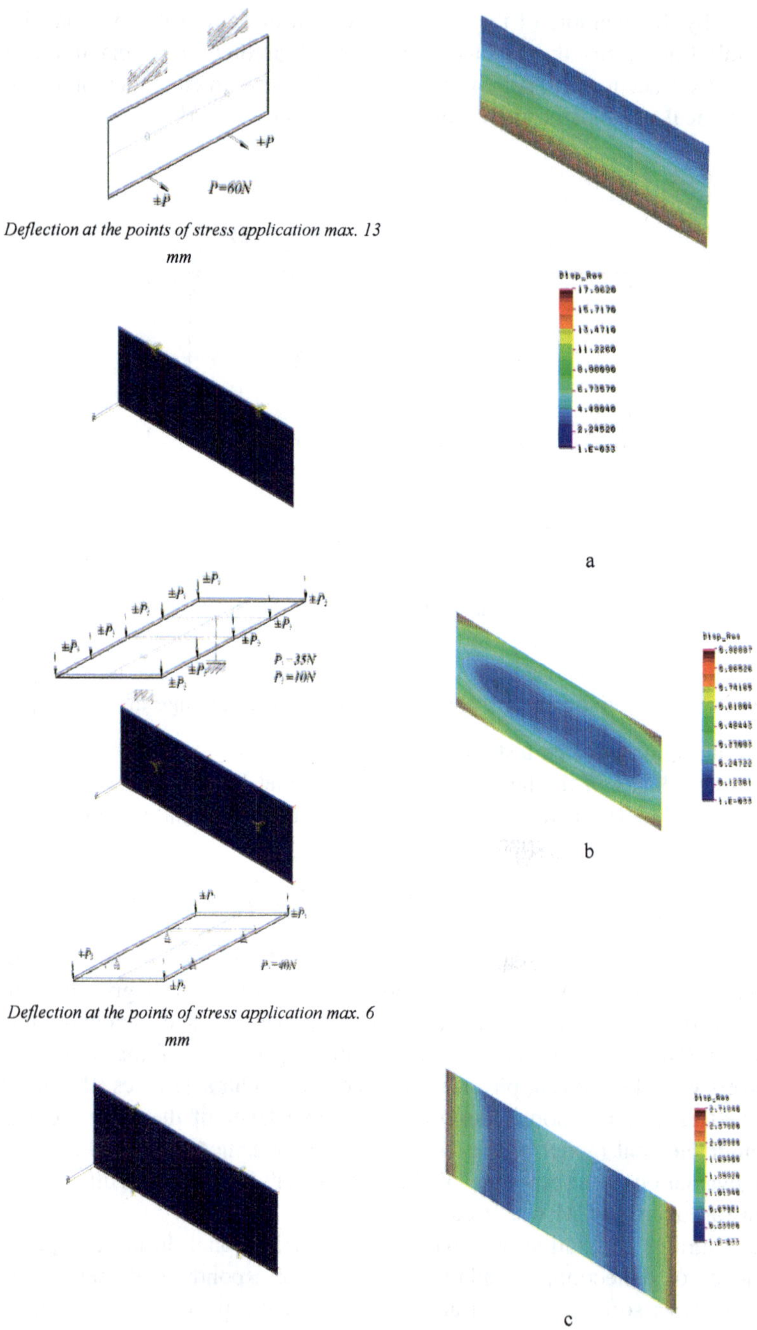

Fig. 2. Diagrams of static bending tests of the solar panel during acceptance tests, rated values of deflection, boundary conditions corresponding to them in the finite element analysis software system, patterns of strained state (mm)

Example of the result of calculation of the solar panel model under action of quasi-static loads is given in Fig. 3.

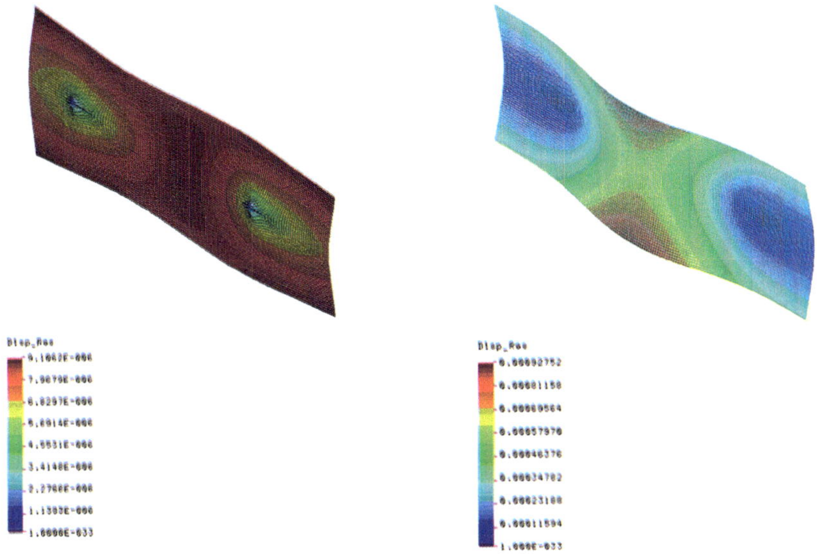

Longitudinal load 15g   Lateral load 10g

Fig. 3. Pattern of strained state of the solar panel under action of quasi-static loads (m)

Evaluation of solar panel acoustic strength was performed for the first three modes of its self-oscillations, results of calculations for which are given in Table 1.

Table 1. First three calculated self-oscillation modes for solar panel

Oscillation mode	Value
1	90,03 Hz
2	106,40 Hz
3	145,09 Hz

Taking into account that acoustic loads act on the solar panel in "closed" state when it is launched into orbit, we considered the supporting conditions similar to the second acceptance tests' diagram (Fig. 2, b).

Under action of the variable acoustic loads (acoustic pressure), there occur forced oscillations with the continuous set of frequencies. Table 2 gives frequencies of forced oscillations close to obtained self-oscillation values of the solar panel, with the values of acoustic and amplitude pressures relevant thereto.

Table 2. Frequencies of forced oscillations of the solar panel close to the values of self-oscillations

Basic frequency of 1/3 octave band, Hz	Level of acoustic pressure (regarding $2 \cdot 10^{-5}$ Pa), dB	$\bar{p}$, Pa	p_{max}, Pa
80	132	79	112
100	133	89	126
125	134	100	141
160	135	112	159

To enable further evaluation of the effect of resonant phenomena, we calculated the solar panel model under action of maximum possible static pressure of the frequency range considered $p_{max} = 159\,\text{Pa}$.

The given value corresponds to the acoustic pressure of 135 dB with the basic frequency of 1/3 octave band, 160 Hz. Figure 4 represents the results of calculations of the solar panel under action of static pressure of $p_{max} = 159\,\text{Pa}$ equivalent to the acoustic pressure of 135 dB with the basic frequency of 1/3 octave band, 160 Hz. For calculation of the solar panel model under action of the forced oscillations, we've carried out the dynamic harmonic analysis [6].

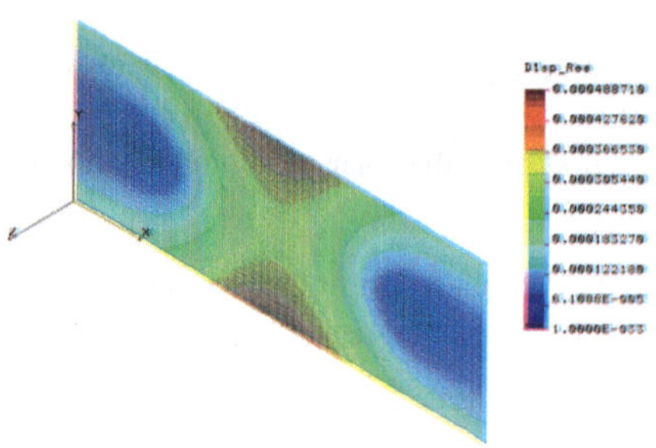

Fig. 4. Pattern of strained state of the panel under action of the static pressure thereon, equivalent to acoustic pressure of 135 dB

We have investigated the range of forced acoustic frequencies of 80...160 Hz which completely covered the range of the relevant self-frequencies of the solar panel model. Figure 5 shows the resulting graphic chart of the amplitude-frequency characteristics of the solar panel.

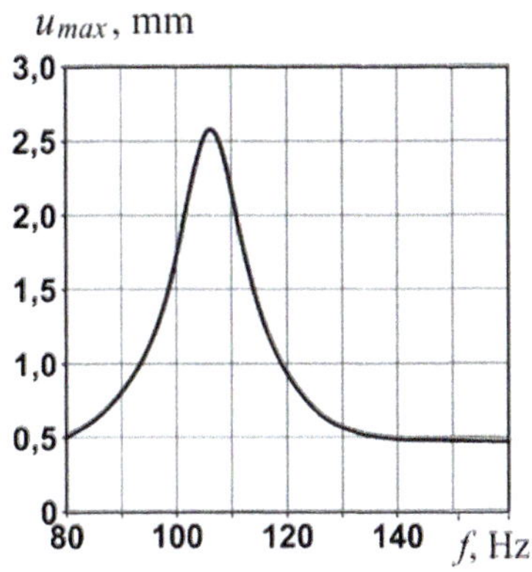

Fig. 5. Graphic chart of the amplitude-frequency characteristic for maximum resulting movement of the solar panel

Besides, the solar panel is exposed to the temperature impact. The range of temperatures is from minus 160 °C to plus 100 °C. It is evident that the given loading case is typical for the normal operation of the solar panel in the open space in its expanded state. This being the case, the temperature regime of the solar panel is stipulated by the mode of its operation: upper surface of the panel is heated since photo-converters are absorbing the solar energy, while the lower one is always in the shade.

With a view to evaluating the load-bearing capacity of the solar panel under action of maximal possible temperature drop thereon the relevant temperature gradient was found. Results of the calculation of solar panel model under action of the temperature gradient are shown in Fig. 6.

When searching for the rational distribution of the base layer material, two potential materials have been investigated: conventional carbon fiber of TAIRFUL type and high-modulus one Kulon P-VK36RT.

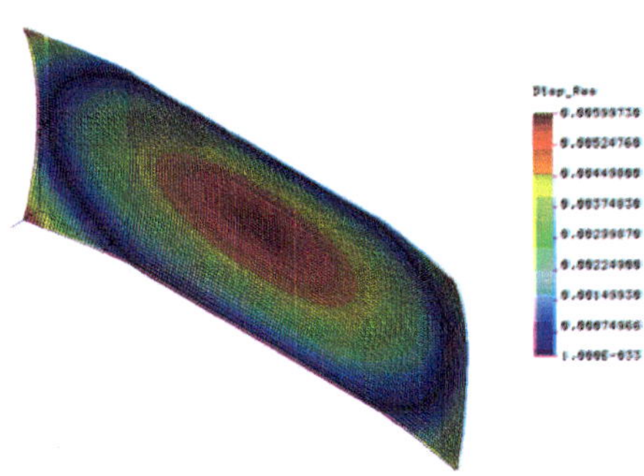

Fig. 6. Pattern of strained state of the solar panel under action of the temperature gradient thereon (temperature drop is from minus 160 to plus 100 °C)

As based on the technological capabilities of producing the carbon fibers of the above types, we considered a number of possible thicknesses of prepreg (mono-layer) in the range of: $t = 0,04\ldots0,08$ mm. During the search of the rational distribution of the base layer material, a number of reinforcement patterns thereof were considered.

The rational parameters of reinforcement patterns of base layers of the solar panels were obtained by superimposition of zones of their use which are feasible by strength and stiffness. Obtained rational variants of the solar panel base layer reinforcement pattern are summarized in Table 3.

Table 3. Obtained rational parameters of the solar panel

Two layers for the base layer	
Kulon P-VK36RT	TAIRFUL
t = 0,05 mm; mass 0,5336 kg	t = 0,07 mm; mass 0,6488 kg
Reinforcement pattern	Reinforcement pattern
±60°	±60°, −60°
Three layers for the base layer	
Kulon P-VK36RT	TAIRFUL
t = 0,04 mm; mass 0,5915 kg	t = 0,05 mm; mass 0,6773 kg
Reinforcement pattern	Reinforcement pattern
0°, ±75°	0°, 90°, 90°

We have also considered possible methods of the rational distribution of material in the solar panel model:

- usage of advanced technological operations on decreasing the base layer thickness (flattening-out) or its discretization;
- choice of the structural layout providing for reinforcement between the embedded elements for mounting with the use of locks, hinge assemblies and stops.

The above results of the comparative analysis of the various options of reinforcement patterns and distribution of base layer material gave the reason to expect a possibility in principle to reduce the specific weight of the solar panel for the rated loading cases to the level of 0,6 kg/m^2.

However, provision of the minimal weight of solar panels at specified operating conditions is connected with a number of technological peculiarities and production problems not occurring with the creation of similar structures: possibility of formation of skins which thickness is comparable with thicknesses of elementary reinforcing fibers of PCM; provision of minimal application of adhesive during the panels' forming; development of the operating procedures of solar panels' production which ensure the specified quality and service life of products.

Analysis of continual and discrete contractions of thin base layers of solar panels

In the process of experimental investigations of the capabilities of making solar panels with thin base layers on the pilot models of solar panel fragments some technological problems were revealed [6]. Typical defects for solar panels with thin base layers are as follows: uniformly distributed on the surface (continual) contractions of base layers ("rippled surface") (Fig. 7) and considerable random discrete contractions of skins above individual cells of honeycombs ("dips").

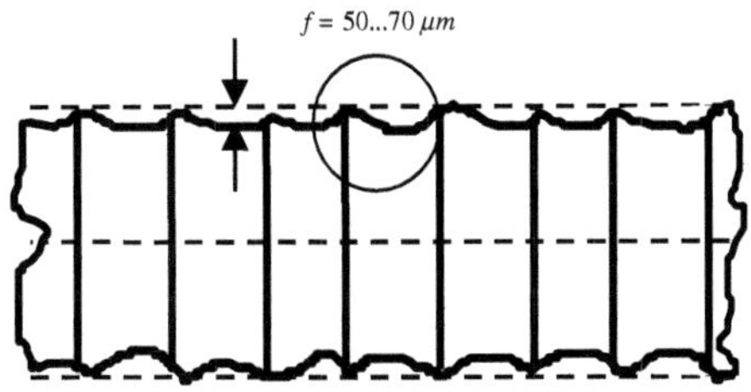

Fig. 7. Diagram of continual contractions of thin base layers of solar panels

Apart from these new defects, the known effect of technological warpage of the solar panel connected with the thermal instability of skins was revealed. Continual contractions of base layers of the solar panels were investigated on the basis of the mathematical model taking into account the kinetics of the process of skins-to-honeycomb bonding. Mathematical model was offered which took into account the thermal expansion of the base layer and honeycomb filler at the temperature of panel bonding, with the subsequent cooling-down and fixation of the adhesive in the heated state, and further cooling of the bonded panel to the normal temperature [6].

Solving of this task allowed obtaining the value of continual contraction of f depending on parameters varying in the process of bonding:

– shrinkage $\varepsilon^y(\tau)$:

$$\varepsilon_y(\tau) = \begin{cases} \varepsilon^y_{max} \cdot tg^n\left(\frac{\tau-\tau_0}{\tau_n}\right) & at\ \varepsilon_y \leq \varepsilon^y_{max} \\ \varepsilon^y_{max} & at\ \varepsilon_y \leq \varepsilon^y_{max} \end{cases} \quad (1)$$

– coefficient of linear thermal expansion (CLTE) of adhesive $\alpha(\tau)$:

$$\alpha(\tau) = \begin{cases} \alpha_c\left[0,2 + \left(\frac{\tau}{\tau^*}\right)^{\frac{\ln 0,8}{\ln\left(\frac{\tau_3}{\tau^*}\right)}}\right] & at\ \alpha \leq \alpha_c \\ \alpha_c & at\ \alpha > \alpha_c \end{cases} \quad (2)$$

– adhesive elasticity modulus $E(\tau)$:

$$E(\tau) = \begin{cases} E_c \cdot \sqrt{\frac{\tau-\tau_2}{\tau_4}} & at\ E \leq E_c \\ E_c & at\ E > E_c \end{cases} \quad (3)$$

where τ - current time of polymerization process; τ_0 - incubation period from which the shrinkage starts ($\tau_0 = \tau_1$); τ_n - time of completion of intensive polymerization

($\tau_n = \tau_3$); $\tau^* > \tau_I + \tau_{II} + \tau_{III}$; τ_2 и τ_4 - time of completion of the second and fourth stages of bonding process, accordingly; ε^y_{max}, α_c, E_c - relevant parameters of cured adhesive; n - empiric constant depending on the rate of temperature rise.

Our calculations showed that taking the kinetics of adhesive curing into account leads to the considerable reduction of residual technological stresses in the solar panel.

It was established that the reason of discrete contractions of single cells of the solar panel is the absence of perforations at faces which results in occurrence of pressure difference (rarefaction) inside the defective cell and perforated cells surrounding it. On the basis of our investigations, the recommendations were given on reduction of deflections by means of the rational combination of technical capabilities to form the panel of thermally instable skins with the operating conditions thereof [6].

Analysis of possible reasons of occurrence of the technological warpage of solar panel in rather full scope forms the complex of engineering support of the given class of products' manufacturing.

3 Analysis of Load-Bearing Capacity of the Adhesive Joint of HS Base Layer

One of the deficiencies of honeycomb structures is the potential tear-off of the bearing skins from the honeycomb filler because of excessive pressure inside. There are various methods of application of the adhesive layer on skins: from the solution or melt onto ends of HF and by means of adhesive films [6]. Each of these methods provides for its own thickness of the adhesive layer which results in HS with the various surface mass. Optimization of the adhesive layer thickness is one of the ways of weight improvement of the RST honeycomb structures. However, less adhesive applied causes reduction of the load-bearing capacity of the product as well. Therefore, there is an urgent need to solve the task of determining the load-bearing capacity of the adhesive joint of HS skins at transversal tear-off for the basic technological methods of adhesive application. In the process of making the HF with BL using liquid or film-forming adhesive of η_{ad} thick, depending on the pressure of panel forming there is varying relative depth of penetration of ends of HF faces into the adhesive film $\bar{\eta} = \eta^*/\eta_{ad}$ (Fig. 8).

When considering the typical element of the honeycomb block with the cell of irregular hexagonal shape (Fig. 9) it was found that failure of the joint on foil would occur with fulfillment of in equation:

$$\frac{p_t a_c K(K\,cos\,\beta + 1)\,sin\,\beta}{(K+1)\delta_c} \leq \sigma_{v\,max}, \tag{4}$$

where p_t - pressure at transversal tear-off.

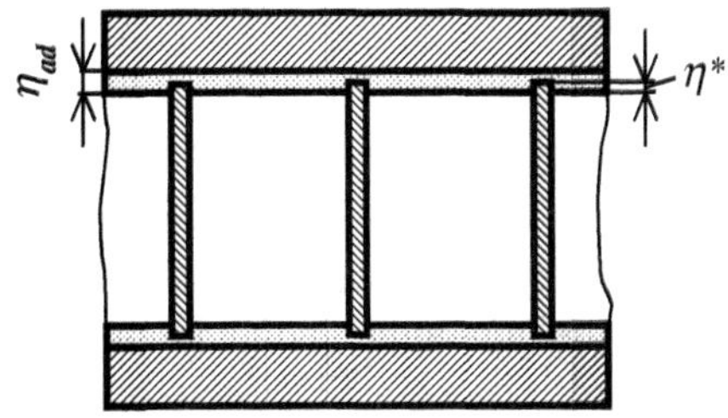

Fig. 8. Version of HF with BL bonding with the use of adhesive film

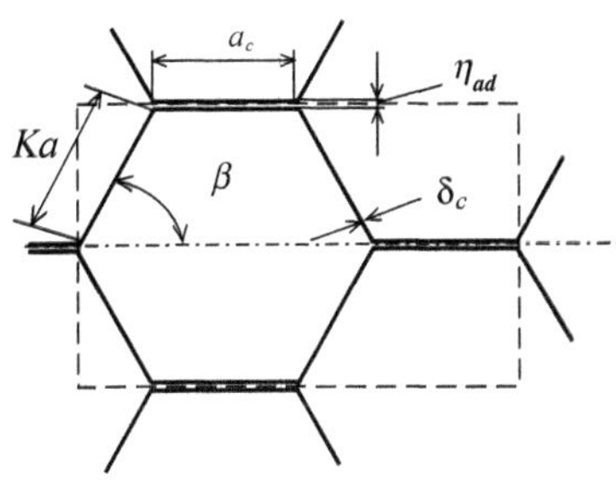

Fig. 9. Typical element of honeycomb block

For determining of load-bearing capacity of the joint on the adhesive finite-element model was used. Investigations were carried out for the plain strain at stresses in the aluminium foil of $\sigma_в$=1 MPa of $\delta_c = 30$ μm thick with the elasticity modulus of E_f = 70 GPa at Poisson ratios of the foil and adhesive $\mu_f = \mu_\eta$ = 0,3 for adhesives with elasticity moduli of $E_{\eta 1}$ = 3,5 GPa, $E_{\eta 2}$ = 7 GPa and $E_{\eta 3}$ = 15,6 GPa and thicknesses of the adhesive layer of η_1 = 0,05 mm, η_2 = 0,12 mm and η_3 = 0,2 mm.

An example of the pattern of strained state of the finite-element model of the fragment of honeycomb block typical element at depth of penetration of HF faces' ends $\bar{\eta}^*$ = 0,7 is shown in. Figure 10.

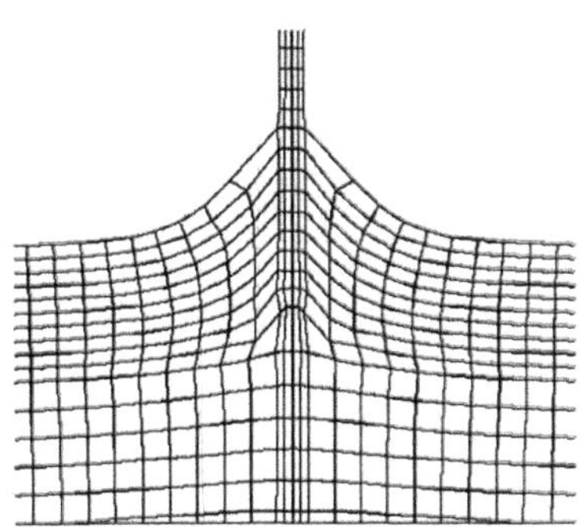

Fig. 10. Example of strained state of the model of honeycomb block typical element's fragment

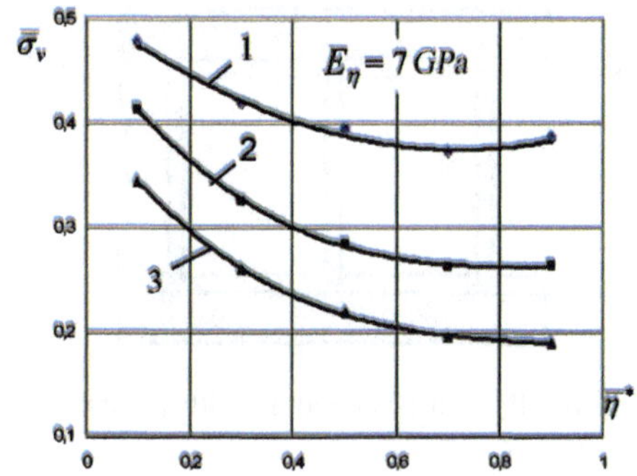

1 – η = 0,05 mm; 2 – η = 0,12 mm; 3 – η = 0,2 mm

Fig. 11. Graph of variance of the relative maximum equivalent stresses on the relative depth of the gap

Figure 11 shows the graphs of variance of the relative maximum equivalent stresses on the relative depth of the gap for the adhesive elasticity modulus E_η = 7 GPa.

Using similar method, investigations for the honeycombs of Nomex polymeric paper were carried out [6]. We have found the distinguishing features of behavior of adhesive joints with aluminium foil and Nomex polymeric paper under the load which are to be taken into account in designing and manufacturing HS.

One of the ways of the further reduction of HS surface mass is the implementation of technology of targeted dozed application of adhesive onto ends of honeycombs, which excludes its passive weight filling the inter-cell surface, not involved in ensuring the load-bearing capacity of the adhesive joint at transversal tear-off of the base layers [6]. We have carried out the analysis of analytical and finite-element methods of evaluation of the HS bearing capacity at transversal tear-off for the case of targeted dozed application of adhesive onto ends of HF faces (Fig. 12).

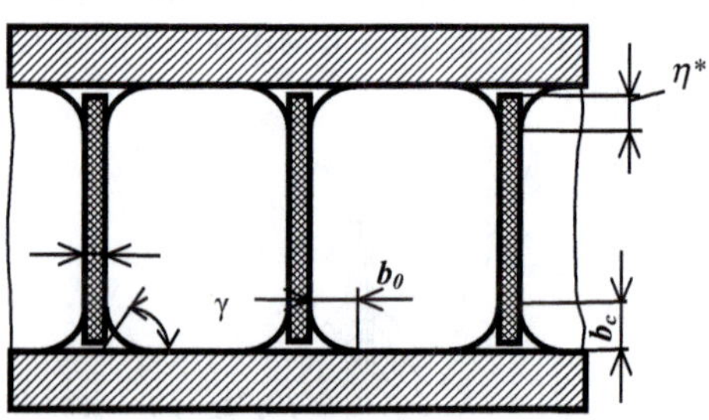

Fig. 12. Bonding of HF with skins of sandwich type with the use of adhesive applied onto ends

In the analytical solution we obtained the formula for medium equivalent stresses in the adhesive leg

$$\sigma_{v\,max} = \frac{a_s K(K\cos\beta+1)\sin(\beta)\,p_{вн}\left(\xi^2_{\sigma,\mu}+tg^2\gamma\right)\sqrt{\cos^2\gamma+3\sin^2\gamma}}{2(K+1)\,b_0\xi_{\sigma,\mu}\sqrt{1+tg^2\gamma}\left(\xi_{\sigma,\mu}+tg\gamma-\sqrt{2\xi_{\sigma,\mu}tg\gamma}\right)},$$

where $\xi_{\sigma,\mu} = b_c/b_0$ depends on parameters of surface tension at the boundary of adhesive with BL and material of honeycombs, viscosity of the adhesive and parameters of wetting of the interface.

Figure 13 shows the pattern of distribution of equivalent stresses in the adhesive fillet obtained on the basis of the finite-element model, and Fig. 14 gives the comparison of results of calculation of relative maximum equivalent stresses on both of these two models.

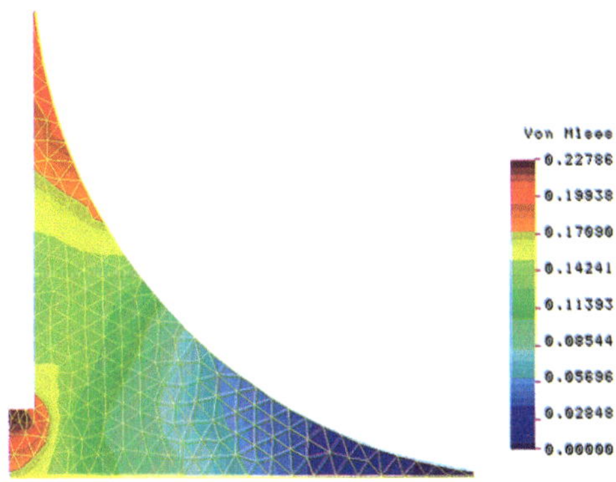

Fig. 13. Pattern of distribution of equivalent stresses in the adhesive fillet obtained on the basis of the finite-element model

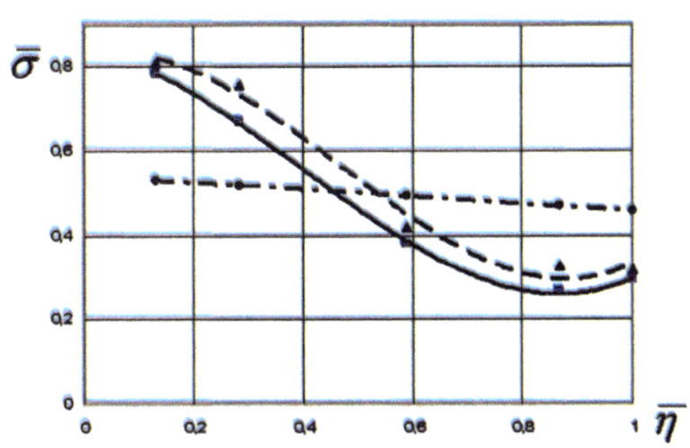

Fig. 14. Dependence $\bar{\sigma}_v = f(\bar{\eta})$ for various adhesives at: ——— E_η = 15,6 GPa; — — E_η = 7 GPa; — · — · analytical dependence

As it is seen from graphs, analytical model gives the result agreeing with the result of the finite-element model only with the relative depth of penetration of honeycomb ends into adhesive equal to 0,5. The finite-element model is also sensitive to the modulus of adhesive elasticity which influence, evidently, is not significant.

On the basis of results of the analysis it is concluded that bonding of skins with honeycombs is recommended to be performed at temperature of the forming pressure which ensures the relative depth of penetration of honeycomb ends into adhesive exceeding 0,5.

4 Conclusions

1. Theoretical prediction of the maximum possible reduction of the surface mass of composite solar panel with the honeycomb filler gave the reason to rely on the possibility in principle to reach its surface mass within the limits of 0,55… 0,6 kg/m^2.
2. We have established the rational diagrams of BL reinforcement of two and three mono-layers of solar panels of PCM based on carbon fibers TAIRFUL and Kulon P-VK36 RT with honeycomb fillers featuring various cell size which provide for minimal weight thereof.
3. We have developed the mathematical models of forming continual and discrete contractions of thin base layers of the solar panels in the process of their bonding with the honeycomb filler, on the basis of which analytical dependencies for determination of maximal depth of contractions are obtained.
4. The method of analysis of honeycomb structures' load-bearing capacity at transversal load is offered, allowing predicting with the reasonable accuracy the nature of their failure depending on parameters of HF cell, adhesive layer, PhMC and strength of the adhesive for specified temperature and pressure of bonding which determine one or another relative depth of penetration of HF faces' ends into the adhesive layer. Therefore, the above results in a complex form a scientific basis for the creation of modern and promising solar panels and can be used in manufacturing of the other precision products for space and conversion applications.

References

1. Webb, G., Da Silva Curiel, A.: Is access to space really a hurdle? Proceedings of the International Astronautical Congress, IAC 59, Glasgow, United Kingdom, 28 September–3 October 2008. Our World Needs Space, vol. 7, pp. 4064–4077 (2008)
2. Nunes, J.P., Silva, J.F.: Sandwiched composites in aerospace engineering. In: Advanced Composite Materials for Aerospace Engineering, pp. 129–174 (2016)
3. Herrmann, A.S.: Design and Manufacture of Monolithic Sandwich Structures with Cellular Cares, 274 p. Technomic Publishing Company, Stockholm (1999)
4. Slyvynskyi, V.I., Alyamovskyi, A.I., Kondratjev, A.V., Kharchenko, M.E.: Carbon honeycomb plastic as light-weight and durable structural material. In: 63rd International Astronautical Congress, IAC 2012, vol. 8, pp. 6519 – 6529. Curran, Red Hook (2012)
5. Ganguli, R.: Optimal design of composite structures: a historical review. J. Indian Inst. Sci. **93** (4), 557–570 (2013)

6. Gaydachuk, A.V., Karpikova, O.A., Kondratiev, A.V., Slivinskiy, M.V.: Sotovyie zapolniteli i panelnyie konstruktsii kosmicheskogo naznacheniya. Tehnologicheskie nesovershenstva sotovyih zapolniteley i konstruktsiy, p. 279. National Aerospace University Kharkiv Aviation Institute Publ., Kharkiv (2012). ISBN 978-966-662-273-3
7. Kondratiev, A., Gaidachuk, V.: Weight-based optimization of sandwich shelled composite structures with a honeycomb filler. East.-Eur. J. Enterp. Technol. **1/1**(97), 24–33 (2019). https://doi.org/10.15587/1729-4061.2019.154928
8. Slyvynskyi, V.I., Sanin, A.F., Kharchenko, M.E., Kondratyev, A.V.: Thermally and dimensionally stable structures of carbon-carbon laminated composites for space applications. In: Proceedings of the International Astronautical Congress, Our World Needs Space, vol. 8, pp. 5739–5751 (2014)
9. Mackerle, J.: Finite element analyses of sandwich structures: a bibliography (1980–2001). Eng. Comput. **19**(2), 206–245 (2002). https://doi.org/10.2514/2.991

Light Civil Turboprop Airplane Take-Off Weight Preliminary Design Estimation Method

A. G. Grebenikov(✉), A. M. Gumenniy, L. Y. Buival, A. S. Chumak, and A. A. Sobolev

National Aerospace University "Kharkiv Aviation Institute", Kharkiv, Ukraine
l.buival@khai.edu

Abstract. A method for preliminary design estimation of civil light turboprop airplane take-off weight, taking into account the requirements of aviation rules Part 23 "Airworthiness standards for civil light airplane" AP-23 (CS-23, FAR-23), has been developed. The analysis of statistical parameters and characteristics of civil light turboprop airplane has been held. New parameters changing ranges are established. Statistical parameters and characteristics of existing civil light airplanes are specified. A method for light civil turboprop airplane take-off weight preliminary design estimation in three approximations according to the Technical Requirements Specification and recommendations for its implementation is presented. Minimum take-off weight is the accepted as effectiveness criterion. An algorithm for light civil airplane take-off weight calculation in first approximation has been developed. To implement the take-off weight calculation method, software has been developed. It allows studying the influence of the wing geometric parameters (such as aspect ratio (λ), taper ratio (η), airfoil relative thickness ($\bar{c}$) and sweep angle (χ)) on aerodynamic performance, power-to-weight ratio and airplane mass characteristics and parameters. The software was tested in the calculation of modern light airplane, namely: A-Viator, Rysachok, King Air C90 GTx, Cessna 441, as well as An-14 and was used in the development of the preliminary design of the KhAI-90 civil light turboprop airplane.

Keywords: Civil light airplane · Take-off weight · Preliminary design

1 Introduction

Light airplanes are used for transportation of passengers, goods and mail, patrol the terrain and communications, and providing medical services to the population; initial pilot training, performing training flights, as well as for aerial acrobatics, providing various types of leisure activities, etc. The main characteristic feature of such airplanes is their operation from both paved and unpaved runways, especially in small and remote places, without airports and equipped runways. Because of rapid development of science and technology which is strongly dictated by modern requirements of the industry, affecting the development of small airplane international market statistical analysis of the parameters and performances of civil light airplanes which is carried out.

M. Nechyporuk et al. (Eds.): *Integrated Computer Technologies in Mechanical Engineering*, AISC 1113, pp. 60–74, 2020.
https://doi.org/10.1007/978-3-030-37618-5_6

Airplane application in the wide range of human activity areas leads to the search for new ways to improve the efficiency of aviation technology (AT). At the same time, the AT is constantly being upgraded on the basis of accumulated experience, engineering and design investigations, modern science computer integrated technologies that allow providing high quality design, production preparation, engineering analysis, testing, certification, and information support for the life cycle engineering. The degree of their implementation in the design process, the development of new methods and the improvement of existing ones lead to the integration of technical, humanitarian, natural sciences and modern technology, theory and practice, which is a determining factor in safety, ergonomics and economy at civil light airplane operating.

The aim of this article is to develop take-off weight estimation method for light civil turboprop airplane with a take-off weight from 2,200 to 5,700 kg and a payload from 600 to 2,000 kg, respectively.

The evolution of the airplanes basic parameters and characteristics was accompanied by constant complication and detailed elaboration of their structure, aerodynamic, internal and load-charring layout of units, systems and equipment.

Significant changes in socio-economic activity, information and statistical support of research as well as new analytical capabilities for processing statistics have contributed to demand for existing statistical airplane data clarification from organizations that are related to the creation, development and operation of aviation complex.

Summary and systematization of light civil airplane parameters and characteristics changes [5–7, 11] and conducted researches [3, 7] contributed to identification of distinctive and essential light civil airplane design features.

Based on modern requirements of the aviation industry and its fields of application, the light civil turboprop airplane with a take-off weight from 2,200 to 5,700 kg and a payload from 600 to 2,000 kg respectively was chosen as the type of airplane for the investigations [3, 4].

2 Light Civil Turboprop Airplane Take-Off Weight Preliminary Design Estimation Method Development

At the preliminary design stage information about the designed airplane is mainly limited by performance requirements and existing similar purpose airplane designing experience. Semi-empirical methods for calculating the take-off weight of light civil airplanes are used, which are based on simplified theoretical analytical expressions and approximations that display the most important parametric dependencies. To improve results accuracy correction coefficients obtained from the processing of statistical and experimental data are introduced into these formulas [5–7, 9].

Estimating the take-off weight of an airplane and its components is one of the central tasks of the design process. Results of this stage allow to clarify the mass, energy, and geometric parameters of the airplane and, in the iterative process, coordinate them according the airplane existence equation, as well as the specified requirements and restrictions [15].

The developed method for light civil turboprop airplane take-off weight estimation on preliminary design stage is shown in the Fig. 1.

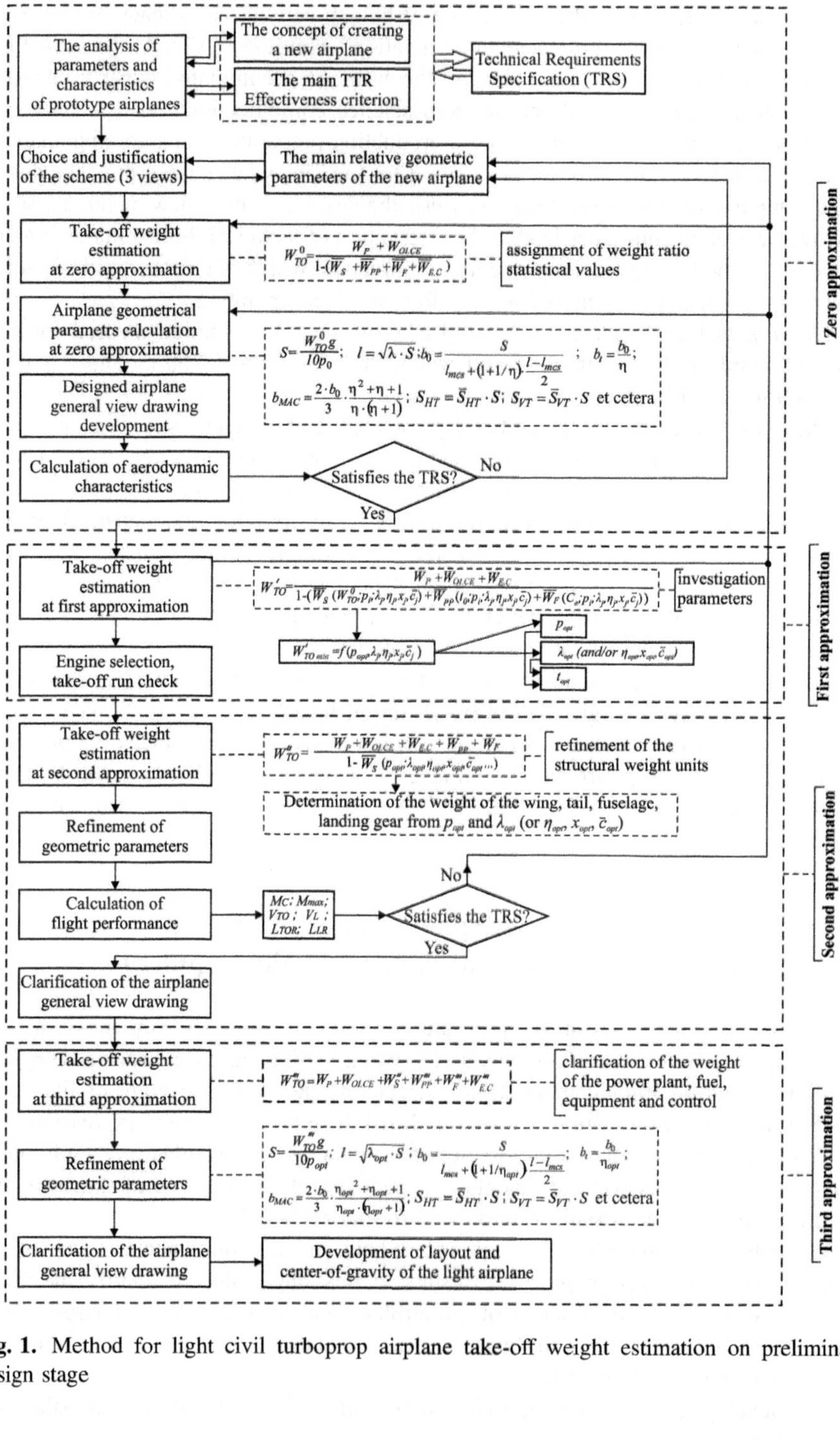

Fig. 1. Method for light civil turboprop airplane take-off weight estimation on preliminary design stage

2.1 Zero Approximation Stage

The initial data for the new light civil airplane design are:

- Technical Requirements Specification (TRS);
- the results of the prototype airplanes statistical data analysis;
- the concept of new airplane creating.

Technical Requirements Specification. In the TRS the following data are set:

- airplane designation;
- general requirements for engines, equipment, components, operational documentation, training aids and training programs for flight and engineering personnel, processing and analysis of flight information;
- expected operating (meteorological) conditions;
- requirements for airplane performance and economic characteristics: cruising flight speed V_C; cruising flight altitude H_C; ceiling H_{max}; range L; payload W_P; take-off run L_{TOR}; life cycle T;
- special requirements for airplane structure;
- requirements for functional systems and equipment;
- environmental requirements;
- physiological and hygienic requirements;
- requirements for operational manufacturability, testability, technical repair;
- ground support equipment and;
- operational documentation requirements;
- requirements for means of evacuation from damaged airplane;
- promotion of safety;
- requirements for metrological support for the airplane development, manufacture, testing and operation;
- requirements for operation of the airplane at international airports.

Statistics Analysis. Prototype airplanes statistical data analysis allows:

- new competitive airplane creating concept developing;
- complementing technical task principal tactical and technical requirements (TTR) to satisfy the requirements for a modern civil light airplane;
- establish expected operational factors.

At this stage, a statistical research of more than 30 civil light airplanes' parameters and characteristics has been conducted. Some of considered airplanes are: Cessna 441 Conquest II, Cessna 425 Corsair, Commander Jetprop 840, EMB-121 Xingu, Merlin IIIB, Mu-2B-60 Marquise, PA-42 Cheyenne III, Beech Model 200 Super King Air, King Air F90, Cessna Caravan, TBM-850, Rysachok, Piaggio P-166, M 101T Gzhel, SM-92 T Turbo-Finist, Piper Cheyenne I (PA-31T-500I), Piper Meridian 500, A-Viator, etc. It consists of three stages: "Statistical observation"; "Grouping of statistical data"; "Analysis of statistics" [3, 9–11]. Bar graphs, pie charts, scatter charts and the methods of mathematical statistics were used to analyze the parameters and characteristics dependencies.

The features of light civil airplanes aerodynamic layouts and their engines with the prevailing percentage of use in a given research range have been considered accordingly to the ongoing aviation science and technology development stage. They are low wing – 60%; two tractor propeller engines, normal aerodynamic layout, horizontal tail attached to the rear part of the fuselage, retractable landing gear – 80%; high lift devices consist of flaps and ailerons – 60%; control surfaces equipped with horn balance – 60%.

The statistical ranges of the wing parameters (Table 1), fuselage, horizontal and vertical tail (Tables 2 and 3) were determined. The wing loading range p_0 is from 98.7 to 313.85 daN/m^2, initial power-to-weight ratio is t_{TO} = 0.2585 (± 0.1055), payload ratio $\overline{W}_C$ equals to 0.271 (± 0.192), fuel weight ratio $\overline{W}_F$ equals to 0.255 (± 0.171) [3].

Table 1. Statistical ranges the wing geometric parameters [3, 14]

λ	η	χ_{LE}, degree	$\bar{c}$	S, m^2	$\overline{S}_{flap}$,	Dihedral wing V_{wing}	
						For low-winged airplane $V_{low\text{-}wing}$, degree	For high-winged airplane $V_{high\text{-}wing}$, degree
7.2...12.3	1.0...3.24	0...5	0.12...0.19	16...33	0.0177...0.362	+4...+8	−1...+4

The statistical range of the fuselage height is $h = (1.38...2.0)$ m , and its width is $b = (1.22...2.0)$ m.

Table 2. Statistical ranges of the horizontal tail relative geometric parameters [3, 14]

λ_{HT}	η_{HT}	$\chi_{LE\,HT}$, degree	$\bar{c}_{HT}$	$\overline{S}_{HT}$	$\overline{L}_{HT}$	A_{HT}
3.68...6.8	1.0...6.88	0...42	0.09...0.12	0.177...0.345	2.816...4.719	0.562...1.155

The value of dihedral horizontal tail V_{HT} depending on the place of its location has different values: for an airplane with horizontal tail located on the rear part of the fuselage, (deck tail), – 0... + 11°; for an airplane with horizontal tail located on the fin (at different distance from the fuselage axis), – 0, + 13°; for a T-tail airplane – 0, –5°.

Table 3. Statistical ranges of the vertical tail relative geometric parameters [3, 14]

λ_{VT}	η_{VT}	$\chi_{LE\,VT}$, degree	$\overline{S}_{VT}$	$\overline{L}_{VT}$	A_{VT}
0.765...1.78	1.34...3.68	0...49	0.09...0.319	0.241...0.548	0.0268...0.116

The statistical ranges of the control surfaces relative area are for elevator – $\overline{S}_E = 0.34\ldots0.472$; for rudder – $\overline{S}_R = 0.211\ldots0.491$ [3, 14].

In case of airplanes comparison at constant values of range, cruise speed, payload, etc. economic criteria are transformed into the generally accepted and more accessible criterion – the airplane take-off weight. Determination of the criteria making possible objective evaluation of the design results at each hierarchical level, finding parameters of each unit and system of the light airplane ensuring high efficiency of the airplane as a whole, is an important stage in the formalization of the design task [5, 8, 12, 13, 15].

Take-Off Weight Approximation. Light civil airplane Take-off weight zero approximation is calculated by the formula [3, 16]

$$W_0^0 = \frac{W_P + W_{OI.CE}}{1 - \left(\overline{W}_S + \overline{W}_{PP} + \overline{W}_F + \overline{W}_{E.C}\right)}, \tag{1}$$

where

W_P – payload weight, kg;
$W_{OI.CE}$ – operational items, crew and equipment weight, kg;
$\overline{W}_{E.C}$ – zero approximation equipment and control weight ratio;
$\overline{W}_S$ – zero approximation structure weight ratio;
$\overline{W}_{PP}$ – zero approximation power plant weight ratio;
$\overline{W}_F$ – zero approximation fuel weight ratio.

Zero approximations of structure weight ratio $\overline{W}_S$, power plant weight ratio $\overline{W}_{PP}$, fuel weight ratio $\overline{W}_F$, equipment and control weight ratio $W_{OI.CE}$ are assumed accordingly to the statistical range [3, 6].

The initial data are taken from requirements specified in aviation regulations, technical task (TT), tactical and technical requirements (TRS) and researches recommendations.

The numerical values of the payload weight W_P are taken from TRS or calculated by the formula

$$W_P = \left(W_{pas} + \Delta W_{lug}\right) \cdot n_{pas}, \tag{2}$$

where

W_{pas} – accepted passenger weight according to AP-23 (FAR-23, CS-23) [1, 2];
ΔW_{lug} = 14 kg – accepted passenger luggage weight.

Operational items, flight crew and equipment weight calculated by the formula [1, 2]

$$W_{OI.CE} = W_{crew} \cdot n_{crew} + \Delta W_{OI}, \tag{3}$$

where

W_{crew} – accepted crew weight according to AP-23 (FAR-23, CS-23) [1, 2], kg;
$n_{эк}$ – number of crew, pers.;
ΔW_{OI} – operational items and equipment weight, kg.

The operational items weight ΔW_{OI} is set by the designer according to the requirements of the Customer and the airworthiness standards of light civil airplane AP-23 (FAR-23, CS-23) [1, 2]. ΔW_{OI} includes personal items of flight personnel; unusable fuel reserve, technical fluids; seat covers, first-aid kit, literature for passengers, inboard repair tools, covers for units; rescue equipment, flares, containers for luggage, cargo, mail, etc. [16].

On this basis, the development of the designed airplane general view and its aerodynamic characteristics calculation can be considered as results of the zero approximation. Thus, it will make possible the most important decision to continue working on the project if technical requirements are met or to change basic relative geometric parameters if they are not satisfied.

2.2 Light Civil Turboprop Airplane Take-Off Weight First Approximation

The take-off weight of civil light airplane first approximation is made by studying the influence of its geometric parameters on the aerodynamic, energy and weight characteristics and parameters in order to determine the minimum airplane take-off weight and its optimal parameters (wing loading; aspect ratio, taper ratio, sweep, airfoil thickness ratio; power-to-weight ratio, etc.) at first approximation (see Fig. 2).

The airplane take-off weight at this stage is calculated by the formula

$$W'_{TO} = \frac{W_P + W_{OI.CE} + W'_{E.C}}{1 - \left(\overline{W}'_S\left(W^0_{TO}, p_i, \lambda_j, \eta_j, \chi_j, \bar{c}_j \ldots\right) + \overline{W}'_{PP}\left(t_{TO}, p_i, \lambda_j, \eta_j, \chi_j, \bar{c}_j \ldots\right) + \overline{W}'_F\left(C_e, p_i, \lambda_j, \eta_j, \chi_j, \bar{c}_j \ldots\right)\right)}, \quad (4)$$

where

$\overline{W}'_S$ – structural weight ratio at first approximation;
$\overline{W}'_{PP}$ – power plant weight ratio at first approximation;
$\overline{W}'_F$ – fuel weight ratio at first approximation;
$W'_{E.C}$ – equipment and control weight at first approximation, kg.

Accordingly to the developed method the calculation is performed in preliminary defined ranges of the geometric parameters, while the change in the output values is checked by the statistical ranges, actual values and behavior of the output curves of the corresponding dependencies [3, 4].

Achieving the criterion of minimum mass is possible by developed take-off weight approximation method. The introduction of the statistical correction coefficients, allows obtaining results close to the actual values of the modern civil light airplane parameters [4, 5].

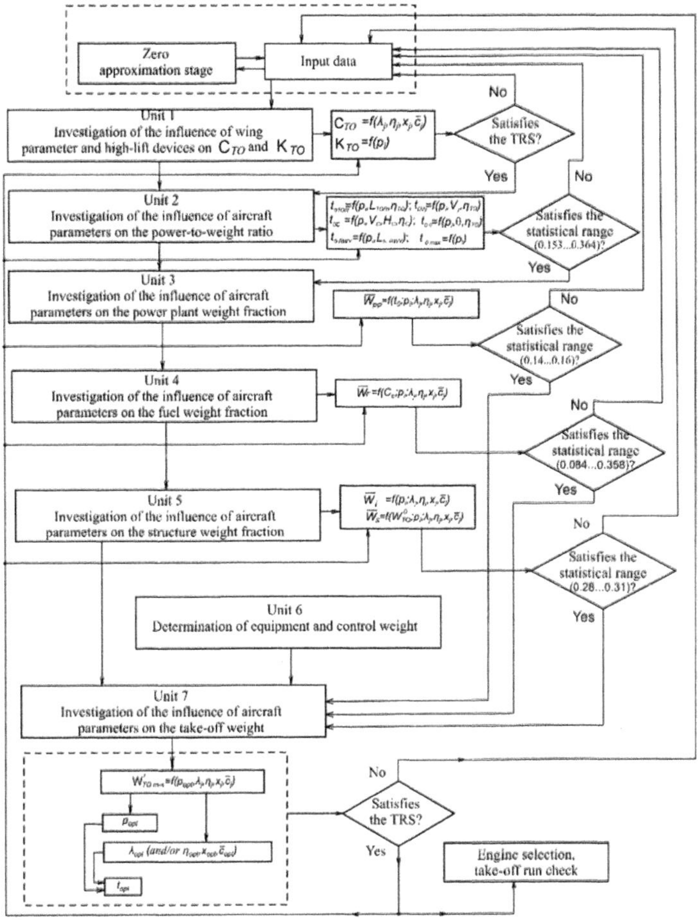

Fig. 2. The algorithm for calculation of light civil turboprop airplane take-off weight at first approximation

2.3 Estimation of the Take-Off Weight of Civil Light Airplane at Second Approximation [14]

The second approximation airplane take-off weight is determined by the formula

$$W''_{TO} = \frac{W_P + W_{OI.CE} + W'_{E.C} + W'_{PP} + W'_F}{1 - \overline{W}''_S\left(p_{opt}, \lambda_{opt}, \eta_{opt}, \chi_{opt}, \bar{c}_{opt} \ldots\right)}, \tag{5}$$

where

$W'_{E.C}$ – equipment and control weight at first approximation, kg;
$\overline{W}''_S$ – second approximation structural weight ratio;
W'_{PP} – power plant weight at first approximation, kg;
W'_F – fuel weight at first approximation, kg.

The estimation of the take-off weight of civil light airplane at second approximation is to clarify the relative masses of the units of the airplane structure. According to the graphical dependencies constructed at the first approximation $\overline{W}'_w = f(p)$, $\overline{W}'_t = f(p)$, $\overline{W}'_f = f(p, \lambda_f, d_f)$, obtain the mass parameters of the wing, tail, fuselage and landing gear at the optimal values of the wing loading p_{opt} and the investigated wing optimal geometric parameters λ_{opt} (and/or η_{opt}, $\chi_{LE\ opt}$, $\bar{c}_{opt}$). Further it is necessary to clarify the second approximation structural weight $\overline{W}''_S = \overline{W}''_w + \overline{W}''_f + \overline{W}_t + \overline{W}''_{l.g}$ and take-off weight.

2.4 Estimation of the Take-Off Weight of Civil Light Airplane at Third Approximation

The method of estimation the take-off weight of civil light airplane at third approximation refines weight of equipment and control, power plant weight and fuel.

Airplane take-off weight W'''_{TO} is determined by the formula [14, 17]

$$W'''_{TO} = W''_S + W'''_{PP} + W'''_F + W'''_{E.C} + W_P + W_{OI.CE}. \tag{6}$$

Weight of Equipment and Control [17]

$$W'''_{E.C} = k_C \cdot k_E^{e.l} \left[500 + 0.12 \cdot W''_{TO} + 0.04 \cdot W''_{TO} \cdot \sqrt{\frac{n_{pas} + n_{crew}}{100}} - 0.0011 \cdot \left(W''_{TO}\right)^{1.333} \right], \tag{7}$$

where

$k_C = 1 - 0.015 \cdot \Delta t$ – weighting factor improvement equipment,
Δt – future review period;
$k_E^{e.l}$ – factor depending on the engine layout.

Weight of Power Plant. Weight of power plant W'''_{PP} is determined by the formula

$$W'''_{PP} = R \cdot \mathrm{n}_e \cdot (m_e + m_b \cdot \mathrm{n}_b + m_h), \tag{8}$$

where R – coefficient taking into account the increase in power plant weight compared with the weight of engines and propellers combined. The value of coefficient R is determined by the formula [4, 14, 17]

$$R = 1.3 + \frac{1.5 \cdot \left(0.1 + \frac{0.9}{\sqrt[3]{N_0}}\right)}{\gamma_e}, \tag{9}$$

where

N_0 – one engine initial power, kW;

γ_e – engine specific weight, daN/kW [17].

$$\gamma_e = \frac{W_e + W_b \cdot \mathrm{n}_b + W_h}{N_0}, \tag{10}$$

where

W_e – one engine weight, kg;
W_b – blade weight, kg;
n_b – number of blades on one propeller, pieces;
W_h – propeller hub weight, kg.

Weight of Fuel. It is determined by the formula

$$W_F''' = \overline{W}_F''' \cdot W_{TO}'', \tag{11}$$

where

W_{TO}'' – second approximation take-off weight, kg;
$\overline{W}_F'''$ – fuel weight ratio at third approximation, which is determined by the formula [17]:

$$\overline{W}_F''' = \overline{W}_{F\,TOC}''' + \overline{W}_{F\,DL} + \overline{W}_{F\,ER} + \overline{W}_{F\,GTT} + \overline{W}_{FC}''', \tag{12}$$

where $\overline{W}_{F\,TOC}'''$ – fuel weight ratio, consumed during take-off and climb [14, 17]

$$\overline{W}_{F\,TOC}''' = 0.00477 \cdot \frac{C_e \cdot V_C \cdot \sqrt{A \cdot C_{d0\,C}} \cdot \tau_{ER}}{\eta_{b\cdot C}}, \tag{13}$$

where

C_e – specific fuel consumption at cruise flight mode, kg/kW h;
V_C – cruise speed, km;
A – drag-due-to-lift factor;
$C_{d0\,C}$ – airplane drag coefficient at zero lift for M_C corresponding cruising speed;
τ_{ER} – estimated time for the usable en-route fuel reserve, h;
$\eta_{b.\,C}$ – propeller efficiency at cruise flight mode.

Fuel weight ratio for the engine ground run-up, taxiing and trapped fuel $\overline{W}_{F\,GTT} = 0.006$.

$\overline{W}_{F\,DL}$ – fuel weight ratio for the descent and landing;
$\overline{W}_{F\,ER}$ – fuel weight ratio for the en-route fuel reserve [17]:

$$\overline{W}_{F\,ER} + \overline{W}_{F\,DL} = 0.00833 + 0.00144 \cdot H_C + 0.000222 \cdot H_C^2, \tag{14}$$

where

H_C – cruise altitude, km;
$\overline{W}'''_{FC}$ – fuel weight ratio consumed during cruise flight mode [14, 17]

$$\overline{W}'''_{FC} = k_F \cdot \frac{C_e \cdot L_{cal} \cdot \sqrt{A \cdot C_{d0\,C}}}{\eta_{b.\,C}}, \tag{15}$$

here k_F – statistical coefficient taking into account the influence of the calculated range to the value of fuel weight ratio consumed in cruise flight mode is determined from the dependence [14, 17]

$$k_F = 6.0379 \cdot L_{cal}^{-0.852}, \tag{16}$$

$$L_{cal} = L - L_{CD} = L - \left(2.6 \cdot H_C^2 + 24.8 \cdot H_C - 10\right), \tag{17}$$

where

L – range, km;
L_{cal} – calculated range, km.

3 Testing the Method for Estimation of Take-Off Weight of Civil Light Turboprop Airplane on the Preliminary Design

To implement the method for estimation of take-off weight of civil light turboprop airplane, the "CLA-TOW" (Civil Light Airplane – Take-off Weight) software of a cyclic steps has been developed [4, 14]. It allows to set the source data, edit it during the process, perform calculations, display the results in *xml* format and move them to tabular processors, build graphical dependencies for direct evaluation of the source data during calculation process.

The software package was tested by the authors with calculating modern light airplanes: A-Viator, Rysachok, King Air C90 GTx, Cessna 441.

A comparison (Table 4) of the actual and calculated minimum take-off weights, optimal aspect ratios and optimal wing payload, according to the developed method of the KhAI-90 new civil light turboprop airplane and prototypes airplanes has been made.

The method for estimation of take-off weight of civil light turboprop airplane taking into account the most important features of integrated design, was tested in the development of an advance project of the KhAI-90 new civil light turboprop airplane.

Table 4. Actual and calculated values of take-off weights, optimal aspect ratios and optimal wing payload of prototypes airplanes and KhAI-90

Airplane name	Name of quantity	Initial data	Estimated data according to the developed method	Δ, %
A-Viator	λ_{opt}	7.74	9	16.28
	$W'''_{TO\,min}$, kg	3000	3600	20.00
	p_{opt}, daN/m^2	158.2	160	1.10
Rysachok	λ_{opt}	9	10.6	15.09
	$W'''_{TO\,min}$, kg	6820	5800	17.00
	p_{opt}, daN/m^2	170	186.5	8.90
King Air C90 GTx	λ_{opt}	8.11	9	10.97
	$W'''_{TO\,min}$, kg	4756	4600	3.00
	p_{opt}, daN/m^2	141	160	19.20
Cessna 441	λ_{opt}	9.6	11	14.58
	$W'''_{TO\,min}$, kg	4468	4650	4.00
	p_{opt}, daN/m^2	160.1	200	24.90
An-14	λ_{opt}	12.15	13	7.00
	$W'''_{TO\,min}$, kg	3500	4000	14.00
	p_{opt}, daN/m^2	82.33	90	9.30
KhAI-90	λ_{opt}	9.6	8.8	9.00
	$W'''_{TO\,min}$, kg	2914	3123	7.00
	p_{opt}, daN/m^2	143.6	150	4.45

The main TTR, which are the most consistent with the Airworthiness Standards of civil light airplane AP-23 (CS-23, JAR-23) [1, 2], are given in Table 5.

Table 5. The main TTR of the KhAI-90 civil light turboprop airplane

M_{max}	L_{max}, km	n_{pas}, people	n_{crew}, people	V_C, km/h	t_{climb}, min	H_C, m	H_{max}, m	L_{TOR}, m	T, h
0.35	1 500	5–6	1–2	350	5	3 500	7 000	300	20 000

According to the results of the method for estimation of take-off weight of KhAI-90 civil light turboprop airplane, the general drawing was developed (Fig. 3).

The KhAI-90 civil light turboprop airplane is a freestanding monoplane with a high wing, single vertical tail, horizontal tail located in the rear of the fuselage, and a retractable tricycle landing gear with a nose strut.

The power plant of the designed airplane consists of two turboprop tractor engines mounted under the wing. By the decision of the Customer, one of the considered two types of power plants with modern engines can be installed. They are the AI-450C gas turbine engine of the Ukrainian production of “Motor Sich” enterprise with take-off

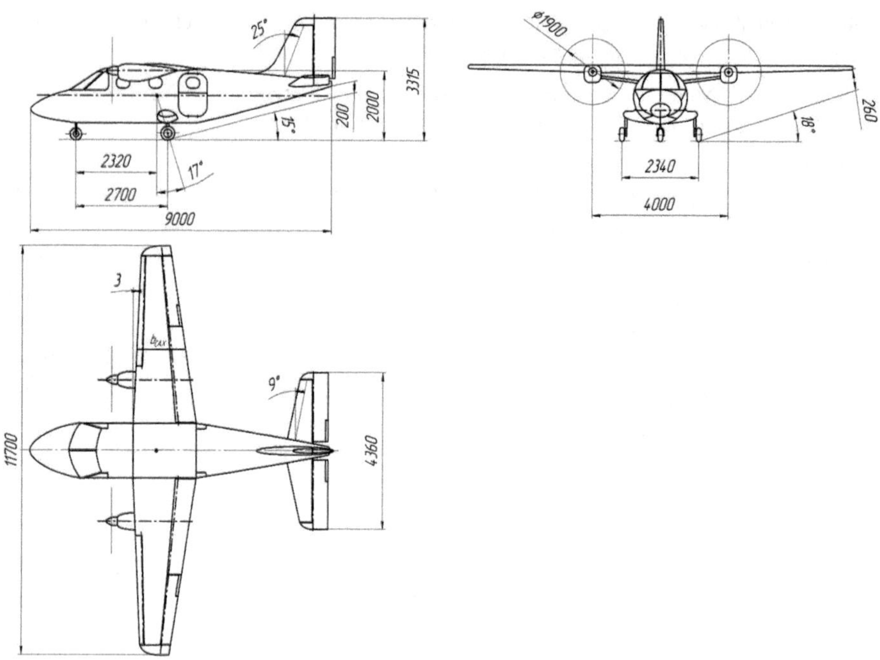

Fig. 3. Fragment of a general view drawing of the KhAI-90 airplane

power N_0 = 450 hp each and the RR 250-B17F turboprop engine of British company "Rolls-Royce" with take-off power N_0 = 420 hp each.

The application of the classical scheme of the KhAI-90 airplane with high lift devices (ailerons, flaps, slats) and controls (elevators and rudders) provides high stability and controllability in all flight modes.

The implementation of the high-wing airplane with engines located in the wing, allows operating both concrete, and grass or dirt runway surfaces. In this case, the elements of the power plant are protected from contaminants and small particles from the ground surface, which increases the life cycle and reliability of their operating. The use of modern engines designed for light airplane, reduces the specific fuel consumption and noise level on the ground.

4 Conclusions

The method for estimation the take-off weight of civil light turboprop airplane with a take-off weight from 2,200 to 5,700 kg and a payload from 600 to 2,000 kg, respectively has been developed.

The calculation at the zero approximation is based on existing methods developed by Arepiev (Moscow, 2006) [7], Badyagin - Mukhamedov (Moscow, 1978) [6], Chumak - Kryvokrysenko (Moscow, 1991) [18] and taking into account compliance with the Airworthiness Standards of civil light airplane AP-23 (FAR-23, CS-23) [1, 2, 12, 13].

The first approximation stage allows us to calculate the optimal wing payload p_{opt}, the power-to-weight ratio $t_{0\,opt}$ and the optimal geometric wing parameters λ_{opt} (or η_{opt}, $\chi_{LE\,opt}$, $\bar{c}_{opt}$), corresponding to the minimum take-off weight $W'_{\mathrm{TO}\,min}$, taking into account restrictions on landing speed $p^{V_L}_{limit}$ and normal g-factor during flying in turbulent atmosphere $p^{n_y}_{limit}$.

Based on the results of the investigation, graphical dependencies as $t_{0\,max} = f(p)$, $\overline{W}_{PP} = f(p)$, $\overline{W}_S = f(p)$, $\overline{W}_w = f(p)$, $\overline{W}_t = f(p)$, $\overline{W}_f = f(p, \lambda_f, d_f)$, $\overline{W}_F = f(p)$, $C_{TO} = f(p)$, $K_{TO} = f(p)$ are built, according to which the energy, weight and aerodynamic parameters of a civil light airplane are determined, namely, the structural units weight ratio are specified at the second approximation stage.

At third approximation weight of equipment and control, power plant weight and fuel are refined. According to these the W'''_{TO} take-off weight of civil light airplane at third approximation are determined.

The main design features of a civil light turboprop airplanes at the preliminary design stage are identified as:

- The analysis of changes parameters and characteristics of civil light airplanes made it possible to establish and clarify their limits of change and statistical ranges;
- The most important decisions in designing a light airplane on the continuation of the project makes it possible to get the results of the preliminary stage of the approach developed by the method of estimation the take-off weight;
- $W_{TO} \rightarrow$ min minimum take-off weight is the accepted effectiveness criterion. Its value is achieved by investigation the influence of its geometric parameters on the aerodynamic, energy and weight characteristics and parameters;
- The introduction of correction statistical coefficients into the investigation using the method of successive approximations [3, 4], expressed by regression equations, allows us to obtain the values of the parameters of a new civil light airplane that satisfy the TRS;
- The geometrical, aerodynamic, energy and weight characteristics and parameters of a civil light airplane at three approximations are determined using the developed cyclic "CLA-TOW" (Civil Light Airplane – Take-off Weight) software;
- The testing of method for estimation the take-off weight was carried out during the design and development of preliminary design of KhAI-90 new civil light turboprop airplane. The value of its take-off weight is 1.3 times less than that of the prototypes, which corresponds to nearly 4%. The average error of calculation by the developed method relative to the actual values of prototype airplane is nearly 12%.

References

1. Aviation Rules. Airworthiness standards for civil light aircraft, Part 23, IAC (2014)
2. EASA Certification Specifications for Normal, Utility, Aerobatic, and Commuter Category Aeroplanes, CS–23 (2012)

3. Buival, L.Y., Gumenniy, A.M.: Statistical research of parameters and characteristics of civil light turboprop aircraft. Open Inf. Comput. Integr. Technol. **71**, 30–45 (2016)
4. Buival, L.Y., Gumenniy, A.M., Grebenikov, A.G.: Algorithm and program for improvement of take-off weight calculation in the first approximation of civil light turboprop aircraft. Open Inf. Comput. Integr. Technol. **73**, 166–179 (2016)
5. Sheynin, V.M., Kozlovskiy, V.I.: Vesovoye proyektirovaniye i effektivnost' passazhirskikh samoletov. Tom 1. Mashinostroyeniye, Moscow (1977)
6. Badyagin, A.A., Mukhammedov, F.A.: Proyektirovaniye legkikh samoletov. Mashinostroyeniye, Moscow (1978)
7. Arep'yev, A.I.: Voprosy proyektirovaniya legkikh samoletov. Vybor skhemy i osnovnykh parametrov. MAI, Moscow (2001)
8. Torenbeek, E.: Advanced Aircraft Design: Conceptual Design, Analysis and Optimization of Subsonic Civil Airplanes. Delft University of Technology, Netherlands (2013)
9. Borovikov, V.: Statistica. Iskusstvo analiza dannykh na komp'yutere, 2nd edn. Piter, SPb (2003)
10. Stepanova, N.I.: Statistika. (statistika grazhdanskoy aviatsii), Part 2. MGTU GA, Moscow (2002)
11. Torenbeek, E.: Synthesis of Subsonic Airplane Design. Kluwer Academic Publishers, London (1982)
12. Roskam, J., Anemaat, W.A.: General Aviation Aircraft Design Methodology in a PC Environment. SAE International (1996)
13. Raymer, D.P.: Aircraft Design: A Conceptual Approach. Scopus (1989)
14. Buival, L.Y., Gumenniy, A.M., Grebenikov, A.G.: Method for determination of take-off weight of civil light turboprop aircraft. Open Inf. Comput. Integr. Technol. **78**, 18–35 (2018)
15. Grebenikov, A.G.: Metodologiya integrirovannogo proyektirovaniya i modelirovaniya sbornykh samoletnykh konstruktsiy. National aerospace university Kharkov Aviation Institute, Kharkov (2006)
16. Yeger, S.M., Mishin, V.F., Liseytsev, N.K.: Proyektirovaniye samoletov, 3rd edn. Mashinostroyeniye, Moscow (1983)
17. Grebenikov, A.G., Zheldochenko, V.N., Kobylyanskiy, A.A.: Osnovy obshchego proyektirovaniya samoletov s gazoturbinnymi dvigatelyami, Part 2. National aerospace university Kharkov Aviation Institute, Kharkov (2003)
18. Chumak, P.I., Kryvokrysenko, V.F.: Raschet, proyektirovaniye i postroyka sverkhlegkikh samoletov. Mashinostroyeniye, Moscow (1991)

The Nonuniform in Width Stressed State of the Lap Adhesive Joint

S. S. Kurennov, A. G. Poliakov(✉), K. P. Barakhov, and D. V. Dvoretska

National Aerospace University "Kharkiv Aviation Institute", Kharkiv, Ukraine
kurennov.ss@gmail.com, kpbarakhov@gmail.com,
o.poliakov@khai.edu

Abstract. The bearing capacity of construction is usually determined by the strength of the joints, where the stress state is irregular. Most existing mathematical joint models are one-dimensional, and imply a uniform stress distribution in width. However, there are constructions for which classical models are not applicable. To calculate the stress state of such joints, it is necessary to take into account the nonuniformity of stresses not only in the length, but also in the width of the joint. To solve such problems, a simplified two dimensional model of the lap adhesive joint of rectangular plates has been proposed. The simplification is that we consider the movement of the layers only along the applied load. Stresses are assumed to be distributed uniformely over the thickness of the layers, and the adhesive layer works only on shear. These simplifications made it possible to obtain an analytical solution to the problem. The stressed state problem for the adhesive joint of two rectangular plates, one of which is rigidly fixed along one of the sides, and the second plate is loaded with an nonuniform normal load at one of the butts, is solved. The problem is reduced to a system of second-order partial differential equations relatively to the longitudinal displacements of the joined layers. The solution is built using the variables separation method, and is a functional series consisting of eigenfunctions. The boundary conditions on the side are satisfied exactly. Satisfying the boundary conditions at the butts leads to a system of linear equations for the unknown coefficients of the functional series. The convergence of the obtained solution is proved. The model problem is solved and the numerical results are compared with the results of calculations performed using the finite element method. It is shown that the proposed approach is accurate enough for design problems.

Keywords: Adhesive joint · Two dimensional model · Analytical solution · Variables separation · Rectangular plate

1 Introduction

Most of the lap adhesive joint mathematical models that allow us to describe the deflected mode of a joint in an analytical form are one-dimensional [1–4]. And they assume a uniform distribution of stresses in the width of the joint, and an a priori given stress distribution in the thickness of the adhesive layer and the layers to be joined (usually uniform or linear). There are several directions for improvement and

M. Nechyporuk et al. (Eds.): *Integrated Computer Technologies in Mechanical Engineering*, AISC 1113, pp. 75–85, 2020.
https://doi.org/10.1007/978-3-030-37618-5_7

development of joint mathematical models. One of these approaches is the study of the stress distribution in the thickness of the joint, i.e. generation of two-dimensional lap joint models [5, 6]. In this case, the stress distribution over the joint width is assumed to be uniform. However, when calculating the stress state of some constructions, such as the joints of bearing elements or repairing patches with plating, it is necessary to take into account the nonuniformity of the deflected mode of the glued plates not only in the length, but also in the width of construction. To study the two dimensional stress state of joints, numerical methods, such as the finite element method or the finite difference method, are most often used [7–11]. An analytical solution to the problem of the joint stress state in a general two dimensional (in the adhesive plane) formulation is not yet known. Therefore, to solve problems, two simplified models are proposed that allow you to obtain a solution to the problem in an analytical form:

(1) to study the effect of transverse deformations due to the Poisson ratios of the joined plates on the stresses in the joint, the tangential stresses in the plates are assumed to be zero [12–15]. The load on the lateral sides of the plates in this case is assumed to be uniform, and the connected plates are completely shear compliant;
(2) to solve the stress state problem of the joint under a nonuniform load, the movements of the base layers in the joint plane in the transverse direction are assumed to be zero. I.e. the connected plates are assumed to be absolutely rigid in the direction transverse to the applied longitudinal load [16, 17].

The hypothesis that the transverse displacements are equal to zero was previously used to create an improved theory of beams [18]. However, it is noted that the proposed hypothesis system does not implement the law of paired tangential stresses in joined plates. This leads to the fact that the stresses in the neighborhood of the plate corners are calculated with some error, and stresses in the glue in the transverse direction are not taken into account. However, the calculations show that the proposed approach is accurate enough for engineering calculations, and can be used to solve a number of new problems.

The aim of this work is to obtain an analytical solution to the stress state problem of the lap adhesive joint of two rectangular plates, to one of which a load is applied in the longitudinal direction, and other plate has a rigidly fixed lateral side. I.e. the load is transmitted from the butt of one plate to the lateral side of the other. The obtained analytical solution was verified by comparing the results of calculations with calculations performed using the finite element method (FEM).

2 Formulation of the Problem

Let us consider a symmetrical adhesive joint of two rectangular plates of size $a \times b$, it is shown in Fig. 1. To reduce the influence of bending in the joint plane on the stresses in the bearing layers, let us assume that $a > b$. The longitudinal load is applied to the side $x = a$ of the first plate (of the first bearing layer). The second plate is rigidly fixed along the lateral side $y = 0$.

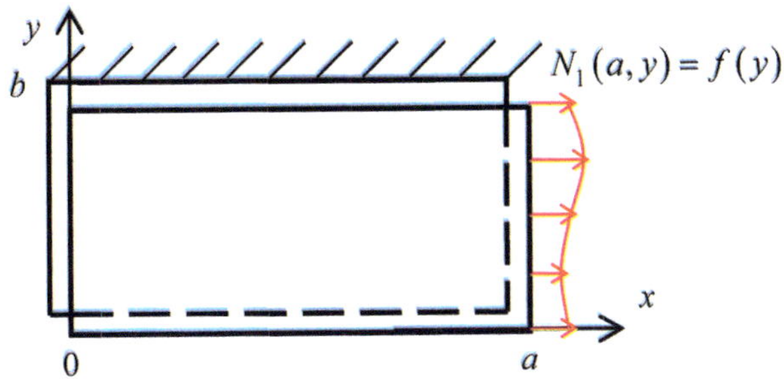

Fig. 1. Adhesive joint design

The proceeding of the problem solution based on the following hypothesizes:

- the stresses are uniformly distributed over the thickness of the layers;
- the adhesive layer works only in shear;
- there is no bend;
- the base layers are assumed to be absolutely rigid in the direction of the coordinate axis y, i.e. all construction points move only parallel to the fixed side.

3 Constructing the Solution

The equilibrium equations for the base layers elements are [8, 12, 13]

$$\tau + \frac{\partial N_1}{\partial x} + \frac{\partial q_1}{\partial y} = 0, \quad -\tau + \frac{\partial N_2}{\partial x} + \frac{\partial q_2}{\partial y} = 0, \tag{1}$$

where N_m, q_m – are normal (in the longitudinal direction) and tangent forces in the base layer m, $m = 1, 2$;

τ – are tangent stresses in the adhesive layer in the longitudinal direction.

The tangential stresses in the adhesive layer are assumed to be proportional to the difference of the layers displacement

$$\tau = P_0(U_2 - U_1), \tag{2}$$

where $U_m(x, y)$ – are longitudinal shift of the m-th layer;
P_0 – shift rigidity of the adhesive layer; as a rule, in engineering computations the rigidity is computed as $P_0 = \frac{G_0}{\delta_0}$, G_0 – is a shear modulus of the adhesive;
δ_0 – an adhesive layer thickness.

However, there are more accurate joint shear rigidity models [19].

The forces in the plates on condition that the transversal displacements are equal to zero are calculated as follows:

$$N_m = \delta_m E_m \frac{\partial U_m}{\partial x},\ q_m = \delta_m G_m \frac{\partial U_m}{\partial y}, m = 1, 2, \tag{3}$$

where δ_m, E_m and G_m – are the thickness, modulus of elasticity and m-th layer's shear modulus correspondingly (in the plane xOy).

So, the equilibrium equations (1) can be reduced to the system of equations relatively to the longitudinal displacements of the layers:

$$\begin{cases} \alpha_1\left(\frac{\partial^2 U_1}{\partial x^2} + \mu_1 \frac{\partial^2 U_1}{\partial y^2}\right) - U_1 + U_2 = 0; \\ \alpha_2\left(\frac{\partial^2 U_2}{\partial x^2} + \mu_2 \frac{\partial^2 U_2}{\partial y^2}\right) + U_1 - U_2 = 0, \end{cases} \tag{4}$$

where $\alpha_m = E_m \delta_m \frac{\delta_0}{G_0}$, $\mu_m = \frac{G_m}{E_m}$.

Boundary conditions:

$$\left.\frac{\partial U_1}{\partial y}\right|_{y=0} = \left.\frac{\partial U_2}{\partial y}\right|_{y=0} = \left.\frac{\partial U_1}{\partial y}\right|_{y=b} = 0, U_2(x, b) = 0, \tag{5}$$

$$N_1(a, y) = E_1 \delta_1 \left.\frac{\partial U_1}{\partial x}\right|_{x=a} = f(y), \tag{6}$$

$$\left.\frac{\partial U_1}{\partial x}\right|_{x=0} = \left.\frac{\partial U_2}{\partial x}\right|_{x=0} = \left.\frac{\partial U_1}{\partial x}\right|_{x=a} = 0. \tag{7}$$

The conditions (5) are conditions on the plate lateral sides – rigid fastening or equality to zero of the tangential stresses. The conditions (6) and (7) – are conditions on the butts of a plates, exactly the given forces on the butt of the first layer (6) and the sides free of load (7).

From the first equation of the system (4) it follows:

$$U_2 = U_1 - \alpha_1 \left(\frac{\partial^2 U_1}{\partial x^2} + \mu_1 \frac{\partial^2 U_1}{\partial y^2}\right), \tag{8}$$

Substituting (8) into the second equation of the system (4), we obtain:

$$\beta_1 \frac{\partial^4 U_1}{\partial x^4} + \beta_2 \frac{\partial^4 U_1}{\partial x^2 \partial y^2} + \beta_3 \frac{\partial^4 U_1}{\partial y^4} - \beta_4 \frac{\partial^2 U_1}{\partial x^2} - \beta_5 \frac{\partial^2 U_1}{\partial y^2} = 0, \tag{9}$$

where $\beta_2 = \alpha_1 \alpha_2$, $\beta_2 = (\mu_1 + \mu_2)\alpha_1 \alpha_2$, $\beta_2 = \mu_1 \mu_2 \alpha_1 \alpha_2$, $\beta_3 = \alpha_1 + \alpha_2$, $\beta_4 = \alpha_1 \mu_1 + \alpha_2 \mu_2$.

In [16] it was obtained the solution of the Eq. (9) by the variables separation method on conditions that $\mu_1 = \mu_2$. At the same time instead of zero condition for lateral side displacements $U_2(x, b) = 0$ in [16] it was used the condition of equality to zero of the tangential stresses on the all lateral sides of the plates. The obtained solution has the form $U_m = W_m(x) + V_m(x, y)$. Where $W_m(x)$ is a classical one dimensional Wolkersen solution [1], and $V_m(x, y)$ is a functional series, built of partial solutions (9), which are can be represented as a linear combinations of the functions of the form $e^{\pm\lambda x}\sin ky$ or $e^{\pm\lambda x}\cos ky$.

From the condition that the longitudinal displacements along the one lateral side of the one of the plates (5) are zero, it follows that $W_2(x) = 0$. From the system (4) it follows that $W_1(x) = 0$. I.e. the one dimensional solution (4) is identical zero. The partial solutions of the Eq. (9) we are finding also as $e^{\pm\lambda x}\sin ky$ (or $e^{\pm\lambda x}\cos ky$). Substituting given equation in (9), we will get an algebraic equation, which connects λ and k:

$$\beta_3 k^4 + (\beta_5 - \beta_2\lambda^2)k^2 + \beta_1\lambda^4 - \beta_4\lambda^2 = 0, \tag{10}$$

from here it follows, that for each value $\pm\lambda$ it corresponds four values k, which can be represented in the form $\pm k_1(\lambda)$ and $\pm k_2(\lambda)$. Therefore, the partial solution (9), which corresponds, for example, to the positive λ has form

$$U_1^* = e^{\lambda x}(S_1 \sin k_1 y + C_1 \cos k_1 y + S_2 \sin k_2 y + C_2 \cos k_2 y),$$

where C_m, S_m – are arbitrary constants.

From (8) we find

$$U_2^* = e^{\lambda x}(S_1\gamma_1 \sin k_1 y + C_1\gamma_1 \cos k_1 y + S_2\gamma_2 \sin k_2 y + C_2\gamma_2 \cos k_2 y),$$

where $\gamma_m = 1 - \alpha_1\left(\lambda^2 - \mu_1 k_m^2(\lambda)\right)$, $m = 1, 2$.

The partial solutions must satisfy the homogeneous boundary conditions (5). From the equality to zero of the tangential stresses in the both plates on the sides $y = 0$ it follows that $S_1 = S_2 = 0$. The partial solutions are correct for the negative values λ also. Therefore we can right

$$U_m^* = X^{(m)}(x)Y^{(m)}(y),$$

where $X^{(m)}(x) = \cosh\lambda x + C\ \sinh\lambda x$,

$$Y^{(1)}(y) = C_1 \cos k_1 y + C_2 \cos k_2 y,\ Y^{(2)}(y) = C_1\gamma_1 \cos k_1 y + C_2\gamma_2 \cos k_2 y. \tag{11}$$

Here, in-turn C, C_1, C_2 – are arbitrary constants.

Let us assume, that the boundary conditions on the sides $x = 0$ are accurate. From here it follows that $C = 0$. In addition, we assume that the boundary conditions on the sides $y = b$ are also accurate.

Hence,

$$\left.\frac{dY^{(2)}}{dy}\right|_{y=b} = 0,\ Y^{(2)}(b) = 0.$$

This two conditions bring us to the homogeneous system of linear equations

$$\mathbf{A} \cdot \mathbf{C} = 0, \tag{12}$$

where $\mathbf{A} = \begin{pmatrix} -k_1 \sin k_1 b & k_2 \sin k_2 b \\ \gamma_1 \cos k_1 b & \gamma_2 \cos k_2 b \end{pmatrix}$, $\mathbf{C} = \begin{pmatrix} C_1 \\ C_2 \end{pmatrix}$.

The system (12) has nontrivial solution if

$$\det(\mathbf{A}) = 0. \tag{13}$$

The Eq. (13) has infinite roots countable set λ_n. Let us denote $k_{m,n} = k_m(\lambda_n)$, $\gamma_{m,n} = \gamma_m(\lambda_n)$. From the system (13) we find the coefficients $C_{1,n}$ and $C_{2,n}$, corresponding to the numbers λ_n. This coefficients, and hence the functions $Y_n^{(m)}$ can be found accurate to an arbitrary factor. To simplify the convergence analysis of the obtained solution we introduce the normalization condition

$$\int_0^b \left[Y_n^{(1)} \right]^2 dx + \int_0^b \left[Y_n^{(2)} \right]^2 dx = 1. \tag{14}$$

So, we obtain

$$U_m = \sum_{n=1}^{\infty} B_n \frac{\cosh \lambda_n x}{\lambda_n \sinh \lambda_n a} \cdot Y_n^{(m)}, \tag{15}$$

where B_n – are coefficients, obtained from the boundary conditions on the sides $x = a$ of the both plates; $\lambda_n \sinh \lambda_n a$ – normalizing factors; $Y_n^{(m)}$ – normalizing functions, where coefficients $C_{1,n}$ and $C_{2,n}$ satisfy the Eqs. (12) and (14).

From the formulas (3) we obtain:

$$N_m = E_m \delta_m \sum_{n=1}^{\infty} B_n \frac{\sinh \lambda_n x}{\sinh \lambda_n a} \cdot Y_n^{(m)}, \tag{16}$$

$$q_m = G_m \delta_m \sum_{n=1}^{\infty} B_n \frac{\cosh \lambda_n x}{\lambda_n \cosh \lambda_n a} \cdot Z_n^{(m)},$$

where $Z_n^{(m)} = \frac{dY_n^{(m)}}{dy}$.

Boundary conditions at the joint's butts $x = a$ are

$$\sum_{n=1}^{\infty} B_n Y_n^{(1)} = \frac{f(y)}{E_1 \delta_1}, \sum_{n=1}^{\infty} B_n Y_n^{(2)} = 0.$$

The system of functions $Y_n^{(1)}$ and $Y_n^{(2)}$ on the interval $y \in [0;\ b]$ are not orthogonal. To find the coefficients we use the least-squares method – we minimize the root-mean-square deviation of the forces (16) on the sides $x = a$ from the given boundary conditions (6) and (7). On all other sides, the boundary conditions are satisfied automatically. The summation in the functional series (14) is limited to a certain number of terms N.

$$J = \int_0^b \left[\left(\sum_{n=1}^{N} B_n Y_n^{(1)}(y) - \frac{f(y)}{E_1 \delta_1} \right)^2 + \left(\sum_{n=1}^{N} B_n Y_n^{(2)} \right)^2 \right] dy \to \min.$$

The extremum conditions $\frac{\partial J}{\partial B_n}$ leads us to the linear equation system:

$$\sum_{n=1}^{N} B_n \int_0^b \left(Y_n^{(1)} Y_j^{(1)} + Y_n^{(2)} Y_j^{(2)} \right) dy = \int_0^b \frac{f(y)}{E_1 \delta_1} Y_n^{(1)} dy, j = 1, .., N. \tag{17}$$

Owing to the introduced normalization (14), the diagonal coefficients of the system matrix (17) are equal to one, and the off-diagonal coefficients are less than one and decrease as the distance from the main diagonal increases. It is proved that these system coefficients tend to zero with an unlimited increasing of any index.

4 Calculation Example

Let us consider the adhesive joint of two plates with the following parameters: $a = 60$ mm, $b = 30$ mm, $\delta_1 = 2$ mm, $\delta_2 = 2$ mm, $E_1 = E_2 = 70$ GPa, $G_1 = G_2 = 27,3$ GPa. The glue layer parameters are: $\delta_0 = 0,1$ mm, $G_0 = 0,3$ GPa. The applied load will be considered uniform in the width $f(y) = P = \text{const}$.

In Figs. 2 and 3 the graphs of the first four normalized eigenfunctions $Y_n^{(1)}(y)$ and $Y_n^{(2)}(y)$ are shown (11).

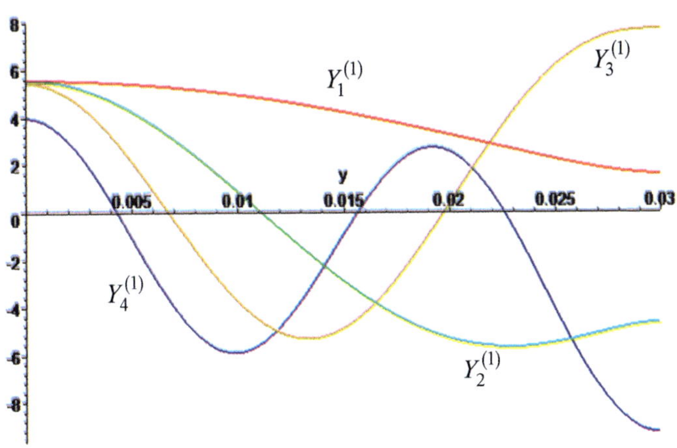

Fig. 2. Graphs of functions $Y_n^{(1)}(y)$.

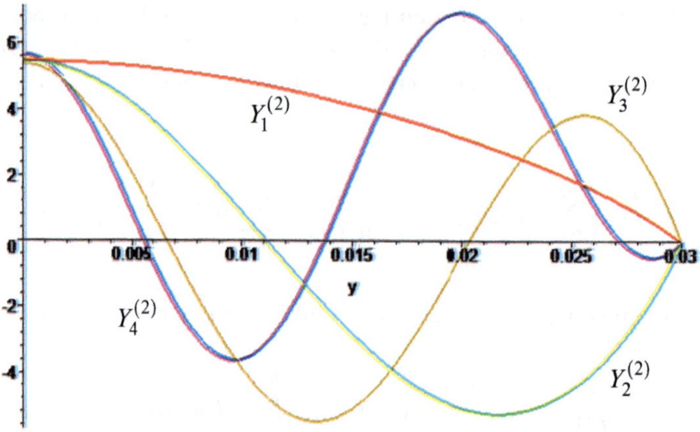

Fig. 3. Graphs of functions $Y_n^{(2)}(y)$.

In Fig. 4 it is shown the distribution of the tangential stresses (2) in the adhesive layer. Stresses are presented in a dimensionless form, as the ratio of actual stresses to hypothetical stresses that would occur if the applied forces were uniformly distributed in the joint plane. In fact, the graph in Fig. 4 shows the stress concentration in the joint.

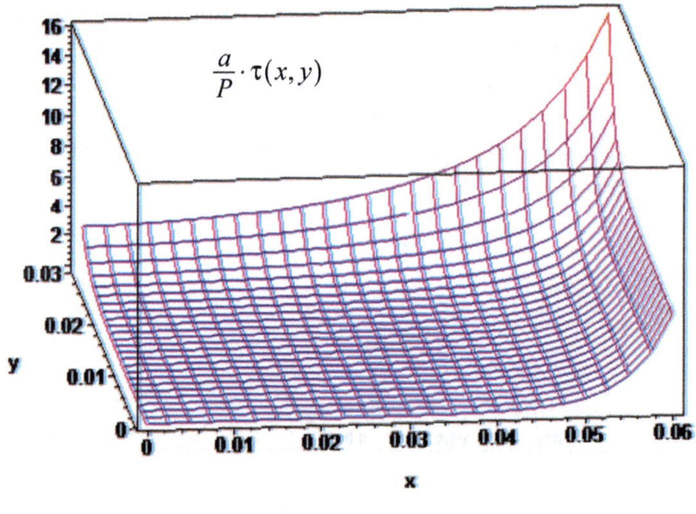

Fig. 4. Tangential stresses in the adhesive

From the given graph it is obvious that the tangential stresses reach a maximum at the corner point, at the intersection of the rigidly fixed side and the side to which the load is applied.

To verify this solution, it was produced the comparison of the computed results which was obtained by the model given above, with the computations performed by the finite element method (FEM).

A three-dimensional finite element model is used. The applied load is distributed uniformly along the corresponding side. The boundary conditions on the outer sides of the joined plates and on the lateral side of the first plate, which is near to the rigidly fixed lateral side of the second plate – a roller. I.e. transverse displacements of both plates on the sides $y = b$ are absent.

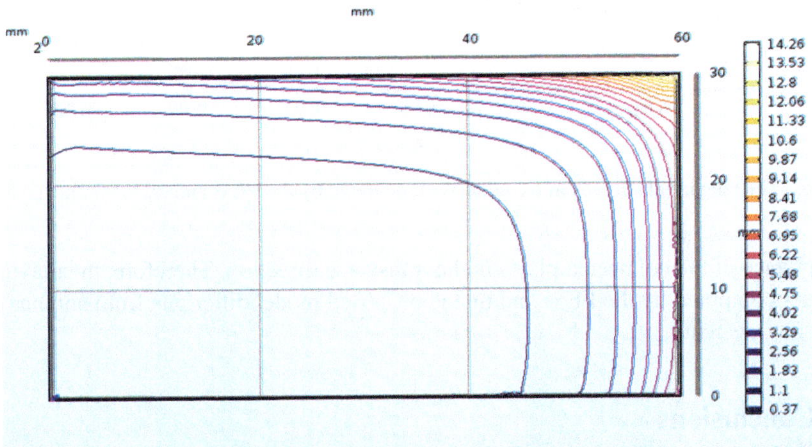

Fig. 5. Tangential stresses in the adhesive

In Fig. 5 it is shown the distribution of the tangential stresses $\left(\frac{a}{P}\sqrt{\tau_{xz}^2 + \tau_{yz}^2}\right)$ in the middle plane of the adhesive line, computed by FEM.

In Fig. 6 it is shown the graphs of the tangential stresses distribution in the adhesive, calculated by the proposed model (solid line) and by FEM: in the middle of the adhesive joint (dashed line), along the fixed side $y = b$ (curves 1) and along the middle of the layers to be glued $y = 0.5b$ (curves 2).

The graphs given above show that the stress values computed by the proposed in this paper simplified method and by the FEM are very close. The computations also show that the tangential stresses in the adhesive layer middle plane in the direction of the axis y, which the proposed model does not take into account, at these boundary conditions (a roller on the loaded layer's lateral side) do not exceed 10% of the stresses in the longitudinal direction. Therefore, the maximum resultant stresses are only slightly superior than the maximum stresses in the longitudinal direction. However, the proposed model gives us some exceeded values of the tangential stresses in the neighborhood of the joint edge, because it does not take into account the shear

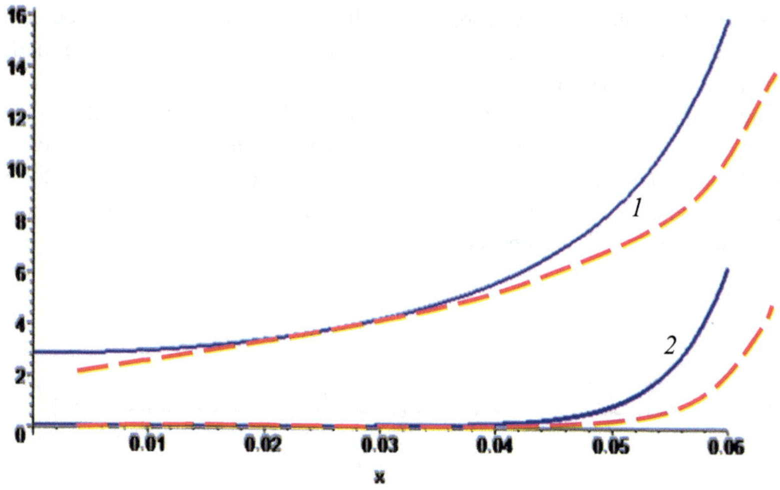

Fig. 6. The tangential stresses in the adhesive calculated by the FEM and by the proposed model

compliance of the connected plates in the transverse direction. Therefore, the maximum stresses computed by the FEM and by the proposed model differ one from another only by a few percent.

5 Conclusions

In this paper it was obtained an analytical solution for the two dimensional stressed state problem of the adhesive joint of the two rectangular plates, one of them is loaded by longitudinal forces, and the other is fixed on the lateral side. To construct the solution, we have used the high rigidity hypothesis of the base layers in the transverse direction. The solution which was obtained, has a form of a functional series expansion by eigenfunctions. To satisfy the boundary conditions, the least squares method is used. The model problem was solved. Comparison of the computations performed by the proposed model and by FEM shows us that the proposed model has high accuracy and can be used in the solution of the joint design and optimization problems. The proposed method can be used in the construction of analytical solutions for the patched plates stressed state problems, stringer joints with skin, etc.

References

1. da Silva, L.F.M., das Neves, P.J.C., Adams, R.D., Spelt, J.K.: Analytical models of adhesively bonded joints. Part I: literature survey. Int. J. Adhes. Adhes. **29**, 319–330 (2009)
2. Wang, J., Zhang, C.: Three-parameter elastic foundation model for analysis of adhesively bonded joints. Int. J. Adhes. Adhes. **29**, 495–502 (2009). https://doi.org/10.1016/j.ijadhadh.2008.10.002

3. Amidi, S., Wang, J.: Three-parameter viscoelastic foundation model of adhesively bonded single-lap joints with functionally graded adherends. Eng. Struct. **170**, 118–134 (2018). https://doi.org/10.1016/j.engstruct.2018.05.076
4. Budhe, S., Banea, M.D., de Barros, S., da Silva, L.F.M.: An updated review of adhesively bonded joints in composite materials. Int. J. Adhes. Adhes. **72**, 30–42 (2017). https://doi.org/10.1016/j.ijadhadh.2016.10.010
5. Liu, J., Sawa, T.: Stress analysis and strength evaluation of single-lap band adhesive joints subjected to external bending moments. J. Adhes. Sci. **13**(6), 729–749 (1999). https://doi.org/10.1163/156856199x00965
6. Kurennov, S.S., Koshevoi, A.G., Polyakov, A.G.: Through-thickness stress distribution in the adhesive joint for the multilayer composite material. Russian Aeronautics (Iz. VUZ) **58** (2), 145–151 (2015). https://doi.org/10.3103/s1068799815020026
7. Andruet, R.H., Dillard, D.A., Holzer, S.M.: Two- and three-dimensional geometrical nonlinear finite elements for analysis of adhesive joints. Int. J. Adhes. Adhes. **21**, 17–34 (2001). https://doi.org/10.1016/s0143-7496(00)00024-5
8. Rapp, P.: Mechanics of adhesive joints as a plane problem of the theory of elasticity. Part II: displacement formulation for orthotropic adherends. Arch. Civ. Mech. Eng. **15**(2), 603–619 (2015). https://doi.org/10.1016/j.acme.2014.06.004
9. Barut, A., Hanauska, J., Madenci, E., Ambur, D.R.: Analysis method for bonded patch repair of a skin with a cutout. Compos. Struct. **55**(88), 277–294 (2002). https://doi.org/10.1016/s0263-8223(01)00158-1
10. Grishin, V.I., Dzyuba, A.S., Dudarkov, Yu.I.: Strength and stability of elements and compounds of aircraft structures made of composites, 272 p. Publishing House of Physical and Mathematical Literature, Moscow (2013). (in Russian)
11. Okafor, C., Singh, N., Enemuoh, U.E., Rao, S.V.: Design, analysis and performance of adhesively bonded composite patch repair of cracked aluminum aircraft panels. Compos. Struct. **71**, 258–270 (2005). https://doi.org/10.1016/j.compstruct.2005.02.023
12. Adams, R.D., Peppiat, N.A.: Effect of Poisson's ratio strains in adherends on stresses of an idealized lap joint. J. Strain Anal. Eng. Des. **8**(2), 134–139 (1973)
13. Mathias, J.D., Grédiac, M., Balandraud, X.: On the bidirectional stress distribution in rectangular bonded composite patches. Int. J. Solids Struct. **43**, 6921–6947 (2006). https://doi.org/10.1016/j.ijsolstr.2006.02.016
14. Kessentini, R., Klinkova, O., Tawfiq, I., Haddar, M.: Transient hygro-thermo-mechanical stresses analysis in multi-layers bonded structure with coupled bidirectional model. Int. J. Mech. Sci. **150**, 188–201 (2019). https://doi.org/10.1016/j.ijmecsci.2018.10.004
15. Kurennov, S.S.: An approximate two-dimensional model of adhesive joints. Analytical solution. Mech. Compos. Mater. **50**(1), 105–114 (2014). https://doi.org/10.1007/s11029-014-9397-z
16. Kurennov, S.S.: A simplified two-dimensional model of adhesive joints. Nonuniform load. Mech. Compos. Mater. **51**(4), 479–488 (2015). https://doi.org/10.1007/s11029-015-9519-2
17. Kurennov, S.S.: Determining stresses in an adhesive joint with a longitudinal unadhered region using a simplified two-dimensional theory. J. Appl. Mech. Tech. Phys. **60**(4), 740–747 (2019). https://doi.org/10.1134/s0021894419040199
18. Vasil'ev, V.V.: Stress tensor symmetry and singular solutions in the theory of elasticity. Mech. Solids **45**(2), 205–213 (2010)
19. Karpov, Ya.S.: Jointing of high-loaded composite structural components. Part 2. Modeling of stress-strain state. Strength Mater. **38**(5), 481–491 (2006). https://doi.org/10.1007/s11223-006-0067-9

Algorithmic Support of the System of Automatic Control of Longitudinal Movement of the Small Unmanned Aerial Vehicle Vertigo

Artem Nikitin(✉), Sergii Kochuk(✉), and Sergii Firsov(✉)

National Aerospace University "Kharkiv Aviation Institute",
Kharkiv 61070, Ukraine
vampal9999@gmail.com, sergei.kochuk@gmail.com,
sn.firsof@gmail.com

Abstract. Rapid development of unmanned aerial vehicles (UAV) is currently being observed. Expansion of both their applications and the tasks they solve lead to the growth of requirements to on-board control systems. This paper presents various options for constructing an automatic control system (ACS) for the longitudinal motion of the Vertigo small unmanned aerial vehicle (SUAV). A linear mathematical model of the SUAV movement was developed and its time characteristics were obtained. The ACS was synthesized on the basis of a traditional PID regulator, a nonlinear fuzzy PD type regulator and a neural regulator. As a training device of the neural regulator there was a reference model of the aircraft with a training block, which was switched off automatically when the required parameters of quality of transient processes were reached by the control object. The purpose of the work is to synthesize various algorithms of UAV ACS operation, as well as to analyze their influence on the dynamic properties of the SUAV "Vertigo".

Keywords: Unmanned Aerial Vehicle (UAV) · PID controller · Non-linear fuzzy PD type controller · Neural controller

1 Introduction

The development of unmanned aviation has become possible, primarily due to the miniaturization and increase in computing performance. An excellent example is the development of Christophersen, Pickell and other control boards of the SUAV Flight Control System 20 [1]. With dimensions of only 55 mm × 85 mm, this device has a sufficient set of measuring devices, software and data storage/processing facilities that allow SUAV to solve a wide range of problems.

The increasing use of UAVs in both civilian and military applications is increasing the demands on on-board equipment and traffic control systems. In particular, the ability of modern SUAVs to adapt to changing flight conditions and external disturbances. The availability of such systems and their modern algorithmic support make it possible to significantly increase the efficiency, reliability and service life of the SUAV.

M. Nechyporuk et al. (Eds.): *Integrated Computer Technologies in Mechanical Engineering*, AISC 1113, pp. 86–97, 2020.
https://doi.org/10.1007/978-3-030-37618-5_8

2 The Mathematical Model of the Control Object

The object of control is the SUAV "Vertigo". This aircraft is made according to the aerodynamic scheme of the "flying wing" type, and is able to fly in the "airplane" and "helicopter" modes. 3D model of the aircraft is built in SolidWorks environment and presented in Fig. 1.

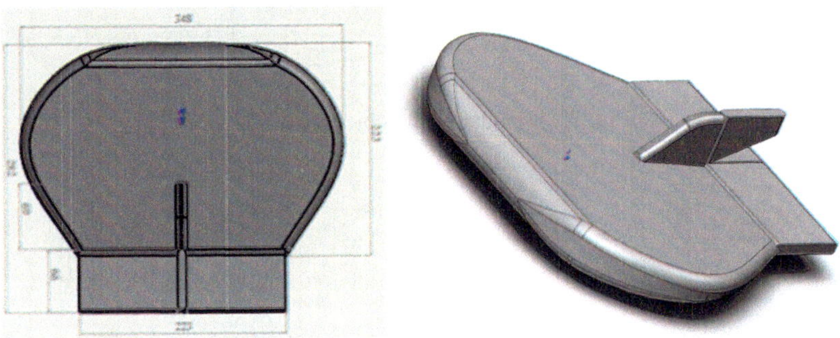

Fig. 1. Appearance of 3D UAV model in SolidWorks environment.

To simplify the task of automatic control system (ACS) synthesis, it was decided to study the dynamics of the control object (CO) at longitudinal short-period motion (LSPM) [2], the structural scheme of which is shown in Fig. 2.

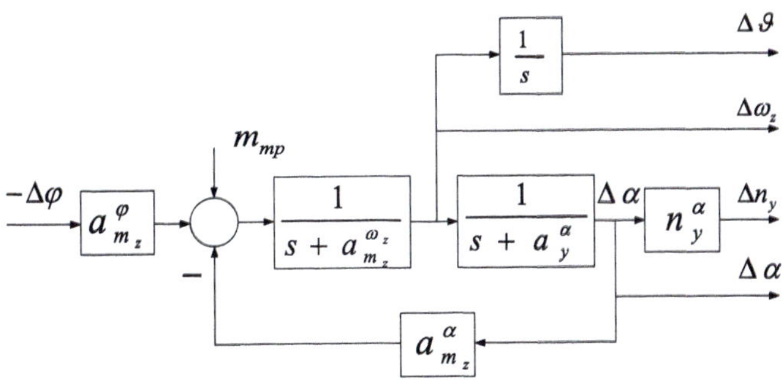

Fig. 2. The structure of the LSPM SUAV scheme.

To determine the dynamic characteristics of the aircraft in Matlab Simulink, the corresponding s-model was assembled, presented in Fig. 3. The s-model coefficients were obtained by purging the 3D model of the aircraft in FlowSimulation environment.

The driving action for the s-model (Fig. 3) was formed as a step signal $\Delta\phi = -1$.

Modeling of longitudinal uncontrollable efficiency of LSM shows (Figs. 4 and 5) that free movement on the basic parameters ω_z, ϑ, n_y is unstable. Analytically, this position can be confirmed by the nature of the roots (small actual parts) of the drone transfer function in the longitudinal efficiency of the pitch rate or vertical overload.

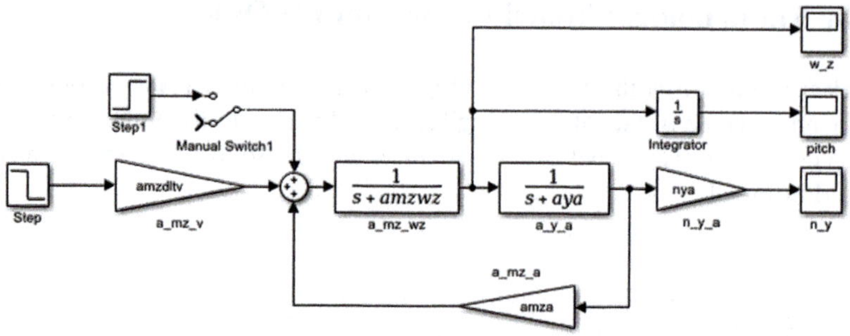

Fig. 3. S-model of the object of study.

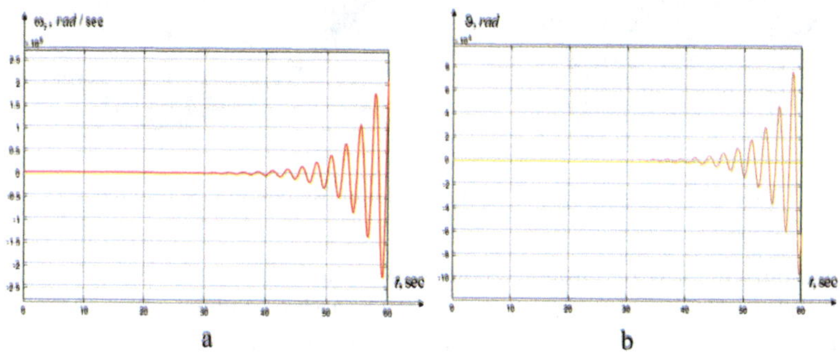

Fig. 4. System transient in terms of angular velocity (a) and pitch angle (b).

Write down the vertical overload transfer function of the UAV and find the roots of the characteristic equation to check the above.

$$W_{n_y}^{\delta_e}(s) = a_{m_z}^{\delta_e} \cdot \frac{\frac{1}{s+a_{m_z}^{\omega_z}} \cdot \frac{1}{s+a_y^{\alpha}}}{1+a_{m_z}^{\alpha}} \cdot n_y^{\alpha} = \frac{0.7962}{(s+0.4284)\cdot(s-0.8387)} \tag{1}$$

Find the roots of the characteristic equation.

$$\begin{gathered} A(s) = (s+0.4284)\cdot(s-0.8387) = s^2 - 0.4103 \cdot s - 0.3593; \\ D = b^2 - 4ac = 0.1683 + 1.4372 = 1.6055; \\ s_1 = \frac{-b \pm \sqrt{D}}{2a} = \frac{0.4103 + \sqrt{1.6055}}{2} = 0.8387; \\ s_2 = \frac{-b \pm \sqrt{D}}{2a} = \frac{0.4103 - \sqrt{1.6055}}{2} = -0.4284. \end{gathered} \tag{2}$$

One of the roots of the characteristic equation (s_1) is positive, which is a confirmation of the instability of the longitudinal efficiency.

3 Automatic Control System Based on PID Controller

As can be seen from the transient graphs (Fig. 4), which have a divergent oscillatory character, the control system of the SUAV "Vertigo" is unstable when open. To perform the control task, the controller must be entered into the control loop. The traditional approach to the solution of the control problem is the use of PID-regulator and the presence of negative feedback, which can be observed in the work of Konovalov and Burakov [3]. Thus, the S-model of the ACS of the "Vertigo" SUAV will take the following form (Fig. 5).

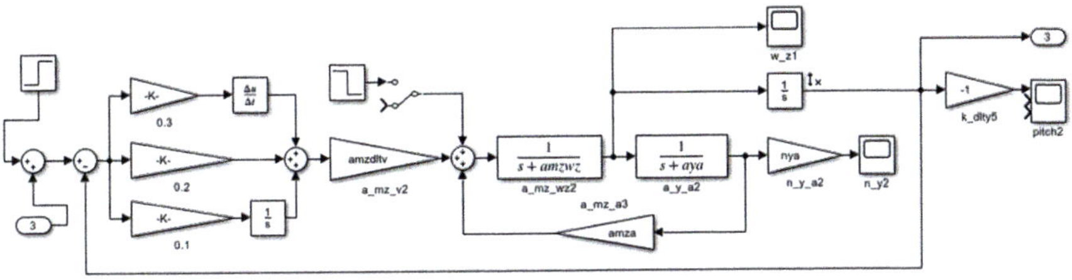

Fig. 5. S-model "Vertigo" UAV with PID regulator.

To find optimal parameters of the automatic control device, the Check Step Response Characteristics model settings block was used. Let's set the desired pitch transient parameters (the transition time up to 6 s, the value of over-regulation up to 20% and the established error of ±5%), as well as enter the necessary variables into the Check Step Response Characteristics setup block. The result of the Check Step Response Characteristics block is shown in Fig. 6.

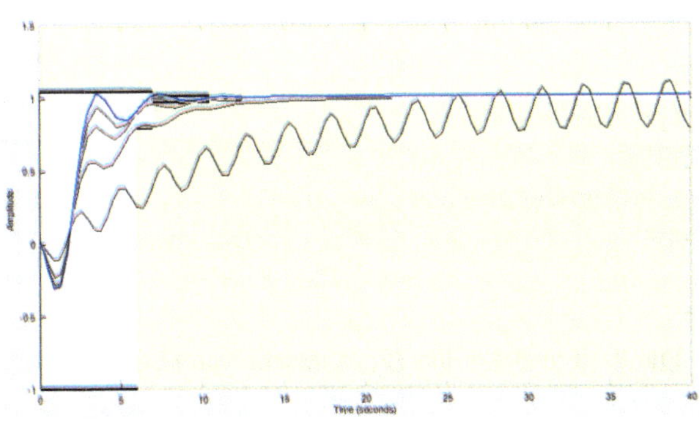

Fig. 6. Changing the transient process of the ACS Vertigo when tuning the PID controller.

When the Check Step Response Characteristics block is finished, the parameters of the PID controller became equal: P = 0.4694, I = 0.0028, D = −0.6417. Figure 7 shows that the control object has become controllable, and the quality of the transition process meets the requirements set earlier: the transition time is 4 s, there is no overshoot, and the established error does not exceed 1.6%.

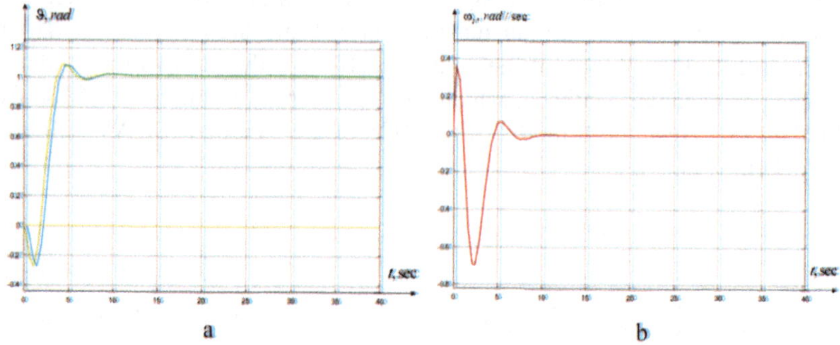

Fig. 7. Transient process of the system in pitch (a) and angular velocity (b) after setting the PID controller parameters.

4 Automatic Control System Based on Fuzzy PD Controller

Linear PID controllers are widely used in control systems when the parameters of the control object are known in advance. However, in practice, the dynamic characteristics of the control object may change, or be partially unknown, which complicates the use of linear PID controllers. To solve such problems, fuzzy logic controllers (FLC) are used. In the work of Burakov [3], synthesis and research of controllers with fuzzy logic is carried out, their advantages are determined in comparison with traditional PID controllers. Using the recommendations of the article by Burakov [3], the synthesis of ACS on the basis of linear FLC PD type (Fig. 8) was performed.

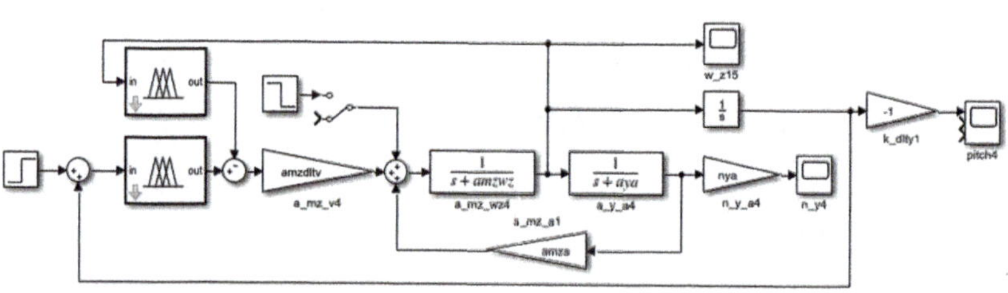

Fig. 8. S-model of the SPLM system with FLC PD-type.

For the proportional and differential component, the laws of distribution of the input and output signal were chosen, as shown in Figs. 9 and 10. The 3 simplest rules have been chosen as the rules for both components of the FLC. The transition graph for pitch angle is shown in Fig. 11. Figures 7 and 11 show that the use of FLC has improved the quality of management. The steady-state error is zero.

Transient time is about 4 s and there is no overshoot. Let's carry out further studies of the obtained regulators for stability in case of perturbations in Sect. 5.

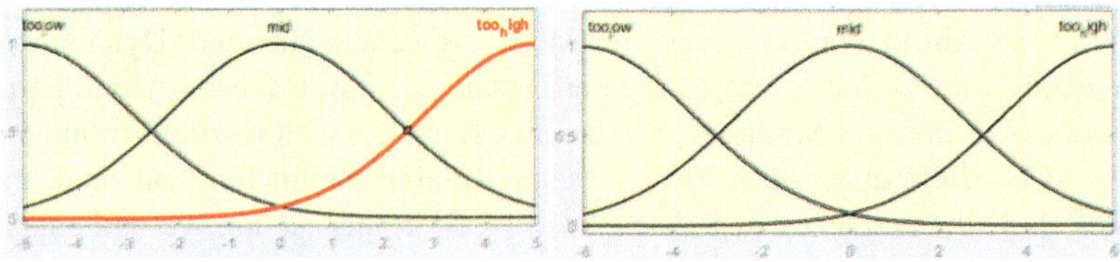

a b

Fig. 9. Input signal distribution law for FLC for P-component (a) and D-component (b).

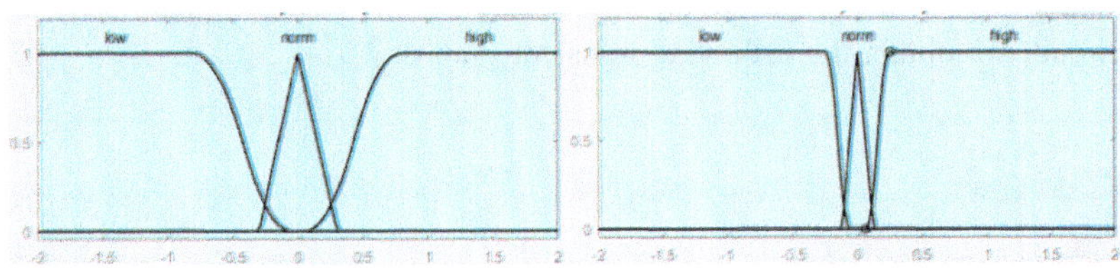

a b

Fig. 10. FLC output distribution laws for P-component (a) and D-component (b).

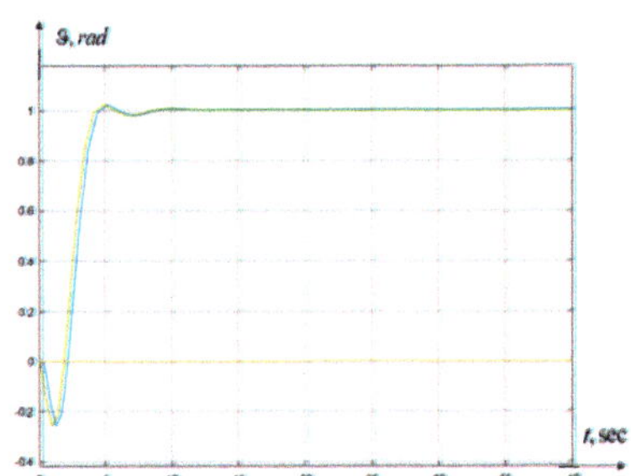

Fig. 11. Transient process of the LSPM system on pitch angle with FLC PD type.

5 Investigation of Resistance to Disturbing Effects of a Linear PID Controller and PD Type FLC

In the process of drone exlutation, there are often disturbances of various nature: from a sudden gust of wind to malfunction in the aircraft control circuit. Occurrence of such factors can influence both the setting influence and the parameters of the control object, which leads to deterioration of the quality of control or its loss. When developing modern ACS for manned and unmanned aircraft, the automatic control device is based on the principles of adaptation. They allow the regulator's parameters to be changed in such a way as to minimize the impact of disturbances on the control quality. The work of Chikasha and Dube [4] considers the application of the predictive control method to the implementation of adaptive control system of quadrocopter height. In assessing the response of the control system to the stepwise disturbance, it was concluded that the use of this algorithm compared to the linear PID controller and Dynamic Matrix Control (DMC) shows better adaptation performance and noise resistance in particular. In the work of Pedro and Mathe [5] use a fuzzy regulator as a regulator implementing adaptive ACS of the quadcopter. The presented tuning parameters and final transient schedules show the efficiency of its use. As in [4, 5] in this section we will analyze the previously obtained ACS for resistance to perturbations and adaptive properties.

Let us consider the case of changing the parameters of the mathematical model of the control object as a disturbing influence. Let us change the coefficient in the range $a^{\alpha}_{mz} = -7.262$ of ±20% of its true value. Thus, the extreme values of the range of changes $a^{\alpha}_{mz} = [-8.7144 : -5.8096]$. By substituting the changed parameters into the ACS circuit, we obtain the following transient graphs:

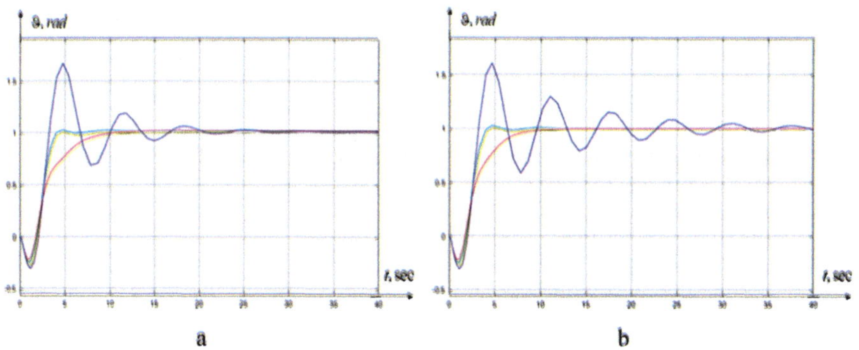

Fig. 12. Pitch transient processes using a PID controller (a) and PD type FLC (b).

Where the initial system transient is marked in green, red – at $a^{\alpha}_{mz} = -8.7144$, blue – at $a^{\alpha}_{mz} = -5.8096$. It can be seen from Fig. 12 (a) that there is no overshoot when the coefficient decreases, the transient process time is 8 s, the steady-state error does not exceed 2.3%. The increase of the coefficient leads to oscillation and overshooting of $\sigma = 66.5\%$. The transient time is increased to 19 s. The value of the steady-state error tends to 0.

The transient process in Fig. 12 (b) shows that there is no overshoot when the coefficient decreases, the transient process time is 7.5 s, the steady-state error does not exceed 0.1%. The increase of the coefficient leads to oscillation and overshooting of $\sigma = 60.5\%$. The transient time is increased to 28.3 s. The value of the steady-state error tends to 0. We will carry out research of the influence of disturbing stepwise effect on this ACS. To simulate the perturbation effect, we will introduce a step signal at the 20 s simulation time into the control loop. The results of modeling are shown in Fig. 13. From the Figs. 12 and 13, the following conclusions can be drawn: the fuzzy logic PD type controller and the PID controller are able to process disturbances in the form of changes in the parameters of the mathematical model of the control object, although the quality of control remains unsatisfactory. In the event of disturbances of an external nature (Fig. 13), the adaptive nature of the transition process is observed when using a PID controller. This is due to the presence of an integrating component, in contrast to the PD type FLC.

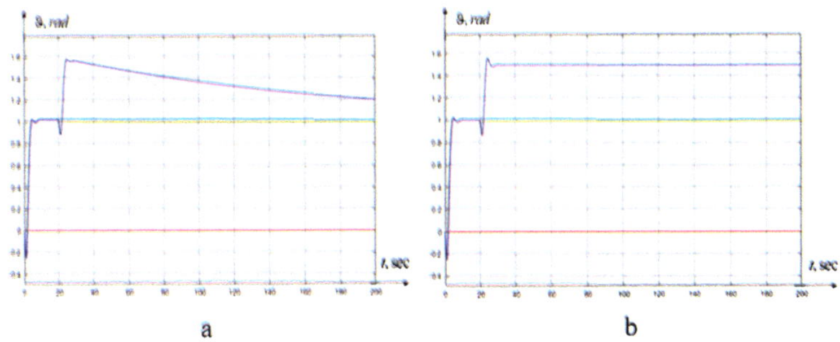

Fig. 13. Pitch transient process when using PID controller (a) and PD type FLC (b).

Thus, the use of a fuzzy logic PD type controller allows us to provide the best quality indicators of the ACS transient process in the absence of disturbances, while the presence of an integrating component of the PID controller allows it to exhibit the properties of an adaptive controller.

6 Synthesis and Analysis of a Simple Neural Regulator

In recent years, automatic control systems based on neural networks have become widely used. The essence of this method is to use a neural network and a reference model of the control object in order to change the parameters of the ACS regulator. Thus, the obtained regulator implements the principles of adaptive and optimal control. Patel and Bhandari [6] implemented a non-linear adaptive controller based on a two-layer neural network for a fixed wing UAV. The resulting regulator based on 12 neurons was able to perform self-training both independently and online, which significantly increased the learning speed of the neural network and its adaptive properties to the disturbing stochastic processes, as compared to the use of a traditional PID regulator.

In the work of Matassini and Innocenti [7], a decision was made to synthesize a control system based on a neural network with an observer, which is able to track undefined disturbing effects without knowing their boundaries and suppress them. This assumption was confirmed by the results of numerical modeling. Positive results of modeling with the use of a neural network were also obtained by researchers Mikhailenko [8], Shi, Cheng [9] and Wang, Zhang [10], experimentally confirming the prospects of development of this research area.

Black, Haghi [11] and Mo and Farid [12] carried out a detailed analysis of existing models and methods of synthesis of adaptive control systems. Their advantages and disadvantages were determined, as well as possible directions for further research.

The use of the reference model is also possible in adaptive systems that do not use neural networks [13]. The work of Boudiba O. and Firsov S. demonstrated the work of an adaptive automatic control system using a reference model and a custom PID controller. Controller parameters vary depending on the results of comparing the output values of the control object and the reference model.

The following structural scheme of an adaptive control system based on a simple two-layer neural network with one input and one output consisting of five neurons using a reference model (Fig. 14).

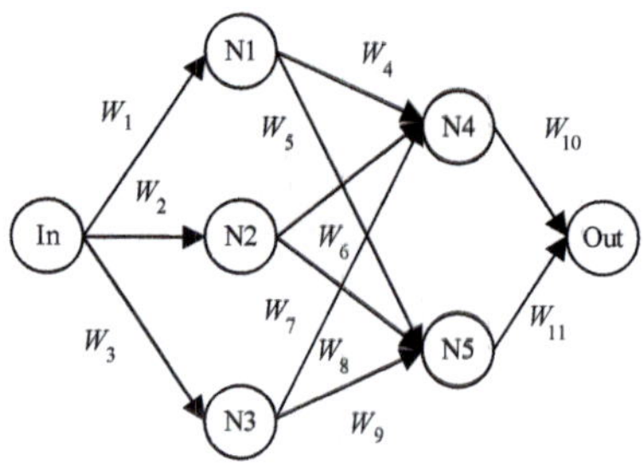

Fig. 14. Neural network structure.

The S-model of an adaptive neural controller in the SUAV control loop in Matlab Simulink has the form (Fig. 15). The simulation duration was 2100 s. The PID controller parameter to be set is the P-component. As a perturbation, we use the same effect on the control object as in Sect. 5. Figure 16 shows the graphs of transient processes.

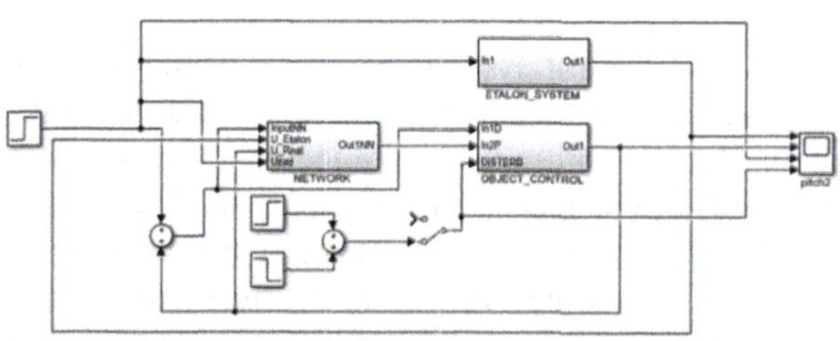

Fig. 15. S-model of ACS based on neural network.

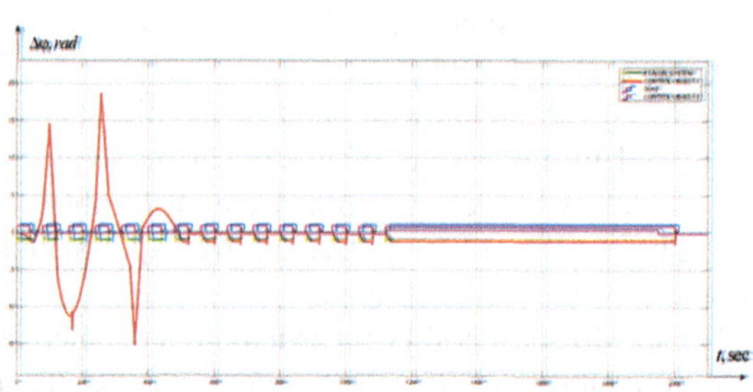

Fig. 16. Transient process of the ACS LSPM “Vertigo” based on a neural regulator.

Figure 16 shows the general view of the transition process of the adaptive control system LSPM "Vertigo" based on a neural regulator. This transient process consists of two main parts: the first one is identification of the control object by the neural network using the output data of the reference model and the plant (training of the neural network); the second one is the reaction of the trained neural network to an input signal in case of disturbance. Let us consider each of them in more detail below.

Figure 17-a shows that from about 500 s of modeling the neural network recognizes the control object and increases the convergence of transients in the pitch angle of the control object and the reference model. Approximately at 1080 s of modeling there is a decrease in the steady-state error up to 5% (Fig. 17-b). Thus, it can be stated that from about 1080–1100 s of modeling the neural network is trained and ready for further research.

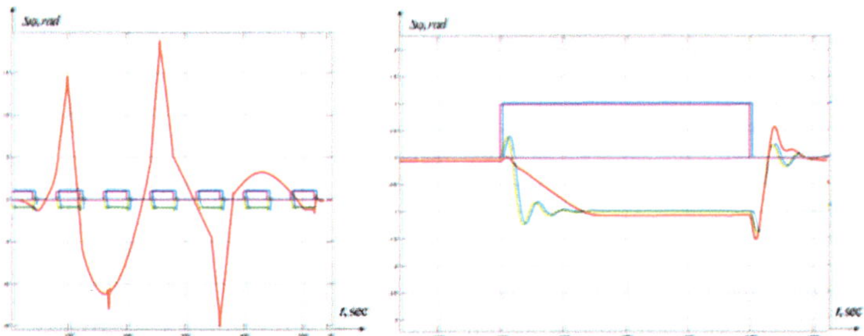

a b

Fig. 17. Transient in pitch angle of the reference model (green), control object with neural controller (red) and input signal (blue) first part (a) and the second part (b).

The second stage of modeling is the study of the reaction of the adaptive control system based on the neural network to the disturbing impact. The results of the experiment are presented in Fig. 18 (a,b).

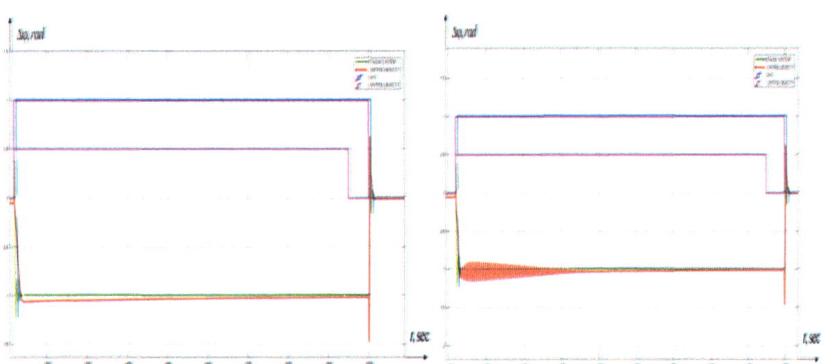

a b

Fig. 18. The transient process in the pitch angle of the reference model (green), the control object with a neural controller (red), the input signal (blue) and the disturbance (purple) at $a_{mz}^{\alpha} = -8.7144$ (a) and at $a_{mz}^{\alpha} = -5.8096$ (b).

Figure 18-a shows that with $a_{mz}^{\alpha} = -8.7144$ the transient process lasts approximately 100 s. At 1220 s, the steady-state error is 5%. At $a_{mz}^{\alpha} = -5.8096$ (Fig. 18-b) the transient process takes about 296 s, and the steady-state error takes 5% for 1416 s of modeling. After the perturbation at 1950 has disappeared and the initial impact at 2000 s, the system remains stable and controllable.

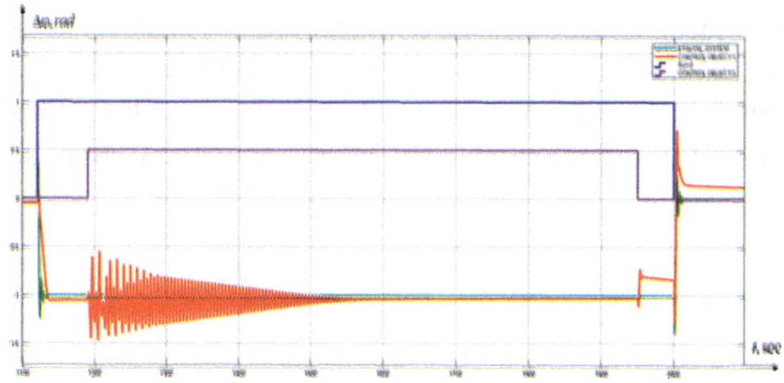

Fig. 19. The transient process in the pitch angle of the reference model (green), control object with neural regulator (red), input signal (blue) and disturbance (purple).

Figure 19 shows that when a pulsed type of disturbance is applied to the system, the control object's pitch-angle transient process has a damping oscillatory character. At 1560 s, after 370 s from the moment of disturbance, the steady-state error is about 5%. After the disturbance has stopped, the system also remains stable and controllable.

7 Conclusion

This article examined the synthesis of automatic control systems based on a traditional PID controller, a non-linear PD type controller and a simple neural controller, as well as their analysis for the presence of adaptive properties. The simulation results showed that the presence of an integrating link in the control loop allows one to accumulate a steady-state error and bring the system output signal to the required value, but for a sufficiently large period of time. The absence of this component, as in the case of PD type FLC (Fig. 13), does not allow ACS to work out the disturbing effect. In this article synthesis of automatic control systems on the basis of traditional PID regulator, non-linear PD type regulator and simple neural regulator and their analysis on the subject of adaptive properties were considered. The results of modeling showed that the presence of an integrating link in the control loop allows to accumulate the established error and bring the output signal of the system to the required value, but for a sufficiently long period of time. The absence of this component, as in the case of the FLC PD type (Fig. 13), does not allow ACS to work out the disturbing effect. It should be noted that the obtained results of modeling show the best general indicators of the system quality in the case of using the regulator with fuzzy logic, as well as greater stability reserves in comparison with the traditional PID regulator.

The use of a simple two-layer neural network in the control loop allowed to suppress the disturbing influence of both internal (change of the parameter a_{mz}^{α}) and external (impulse perturbation) character. This indicates that the neural regulator has adaptive properties. Further research will be aimed at expanding the input and output parameters of the regulator based on the neural network, as well as changing its internal structure in order to improve its adaptive properties and overall control quality.

References

1. Christophersen, H.B., Pickell, W.J., Koller, A.A., Kannan, S.K., Johnson, E.N.: Small adaptive flight control systems for UAVs using FPGA/DSP technology. In: AIAA 3rd "Unmanned Unlimited" Technical Conference, AIAA-2004, Chicago, p. 6556 (2004)
2. Aslanian, A.E.: Aircraft Automatic Flight Control Systems. Kiev Higher Military Aviation Engineering School Printing House, Kiev (1984)
3. Konovalov, A.S., Burakov, M.V.: Synthesis of fuzzy logic controllers. Inf. Control Syst. **1**, 22–27 (2011)
4. Chikasha, P.N., Chioniso, D.: Adaptive model predictive control of a quadrotor. IFAC PapersOnLine **50**(2), 157–162 (2017)
5. Pedro, J.O., Mathe, C.: Nonlinear direct adaptive control of quadrotor UAV using fuzzy logic technique. In: 10th Asian Control Conference (ASCC). IEEE, Kota Kinabalu (2015)
6. Patel, N., Bhandari, S.: Robust nonlinear adaptive control of a fixed-wing UAV using multilayer perceptrons. In: AIAA Guidance, Navigation, and Control Conference, AIAA-2017, p. 1524. Grapevine (2017)
7. Matassini, T., Shin, H.-S., Tsourdos, A., Innocenti, M.: Adaptive control with neural networks-based disturbance observer for a spherical UAV. IFAC PapersOnLine **47**(17), 308–313 (2016)
8. Mikhailenko, V.S.: Algoritm of adjusting the adaptive neuro-fuzzy PI-controller. Odes'kyi Politechnichnyi Universytet. Pratsi **2**(36), 149–154 (2011)
9. Shi, X., Cheng, Y., Yin, C., Huang, X., Zhong, S.: Design of adaptive backstepping dynamic surface control method with RBF neural network for uncertain nonlinear system. In: Wang, Z., Hoi, S. (eds.) Neurocomputing 2018, vol. 330, pp. 490–503. Elsevier, The Netherlands (2018)
10. Wang, Y., Zhang, H., Han, D.: Neural network adaptive inverse model control method for quadrotor UAV. In: Jie, C. (eds.), Proceedings of the 35th Chinese Control Conference, vol. 35, pp. 3653–3658. Shanghai xi tong ke xue chu ban she, Chengdu (2016)
11. Black, W.S., Haghi, P., Ariyur, K.B.: Adaptive systems: history, techniques, problems, and perspectives. Systems **2**(4), 606–660 (2014)
12. Mo, H., Farid, G.: Nonlinear and adaptive intelligent control techniques for quadrotor UAV – a survey. Asian J. Control **21**(2), 989–1008 (2019)
13. Boudiba, O., Firsov, S.: Designing adaptive PID-controller non-sensitive to changes in aerodynamic characteristics of an unmanned aerial vehicle. East.-Eur. J. Enterp. Technol. **97**(1/9), 69–75 (2019)

Using the Theory of Similarity in the Formation of the Shape of Ground Launch Devices

Vladyslav Sereda(✉) and Aleksey Kornev

Inter-branch Scientific Research Institute of Physical Simulation Problems at National Aerospace University «Kharkiv Aviation Institute», Chkalova Street 17, Kharkiv 61070, Ukraine
v.sereda@khai.edu

Abstract. Based on the theory of dimensions and similarity, a method is proposed for obtaining the criteria for the energy perfection of ground launch devices for launching unmanned aerial vehicles. The principle of rationing the appearance of ground launch devices using special similarity criteria is proposed. A criterion space has been built for assessing the energy perfection of pneumatic and pyrotechnic ground catapults depending on the mass of the launched aircraft. The principle of deterministic design decision-making in the formation of the appearance of ground catapults according to the tactical and technical characteristics of the launched unmanned vehicle is proposed.

Keywords: Dimension and similarity theory · Criterion of energy perfection · Ground launch device

1 Getting Similarity Criteria Based on π-Theorem

1.1 Introduction

The apparatus of the theory of dimension and similarity [1, 2] serves as the traditional basis for the design of various technical object. In the presence of an array of tactical and technical characteristics, a limited set of special similarity criteria makes it possible to evaluate the development of any technical object [3]. However, in order to avoid dead-end development paths on the basis of the theory of dimensions and similarity, earlier stages of development can be systematically adjusted. Using this approach makes it possible to impart a systemic character to the process of formation the shape of an object, which provides a radical improvement in the quality of design and a reduction in the resource intensity of the R&D cycle [4].

In contrast to the "big" aviation, the catapult start of light unmanned aerial vehicles (UAV) is most expedient, since it provides high autonomy and mobility of the complex, and also allows to increase the range. At the moment, there are many technical solutions of ground launch devices (GLD) for light UAVs [5], which require appropriate systematization and comparative analysis. Such an analysis can be carried out on the basis of a limited universal set of similarity criteria characterizing the mechanical properties of the transport system.

M. Nechyporuk et al. (Eds.): *Integrated Computer Technologies in Mechanical Engineering*, AISC 1113, pp. 98–104, 2020.
https://doi.org/10.1007/978-3-030-37618-5_9

1.2 UAV Launch Model

It is obvious that the energy capabilities of the GLD should be proportional to the weight and size characteristics of the launched UAV, however the endless increase in the energy of GLD is impossible due to the difficulty of implementation, as well as economic inexpediency. Analysis of the existing launch systems indicates that some known technical solutions are spontaneous and are not the result of system design.

For example, the use of a pyrotechnic type of start to small-sized UAVs (HESA Ababil-2 UAV weighing 83 kg), as well as pneumatic systems – to launch large-sized UAVs (EADS Atlante UAV weighing 420 kg) looks controversial in soft form. The main argument here is, in the first case, high energy and mobility of the launch system, and in the second – the secrecy of the launch and the absence of the need for licensing equipment.

Thus, a formal answer is required to the question of the energy efficiency of the application of one or another type of launch to UAVs of various weight and size characteristics. It is proposed to reduce the comparative analysis of the energy efficiency of GLDs to the formal use of the rules of norms and use for these purposes the criteria of energy perfection (CEP) in the form of minimal homogeneous factor matrixes of the GLD.

The similarity criteria will be formed on the basis of the π-theorem, taking into account the algorithm for finding the dimension proposed by prof. Ryzhenko [6]. To reduce the number of possible similarity criteria, we consider a simplified model of a UAV start, but fairly fully reflecting the physics of the start process and including the equations of conservation of momentum (1) and energy (2), as well as initial conditions (3). The lifting force of the wing, the force of aerodynamic drag and the thrust force of the engine acting on the UAV while driving along the guideway, is neglected.

$$m_{UAV}\frac{dV}{dt} = F - m_{UAV}g(f\cos\theta + \sin\theta); \tag{1}$$

$$m_F H_U \xi = m_{UAV}\frac{V^2}{2} + m_{UAV}gH\sin\theta; \tag{2}$$

$$V|_{t=0} = 0; \quad x|_{t=0} = 0; \quad \left.\frac{dV}{dt}\right|_{t=0} = 0, \tag{3}$$

where x is the coordinate of the UAV on the guide; V is UAV movement speed; g is the gravitational constant; m_{UAV} is mass UAV; m_F is fuel mass of the starting accelerator; ξ is the coefficient of completeness of fuel combustion; θ is the angle of inclination of the guide; f is the coefficient of sliding friction; H is guide height.

1.3 Dimension Search Method

Based on the physical formulation of problem (1)–(3), there follows a list of dimensional and dimensionless physical quantities (Table 1). As the main physical quantities that characterize the UAV launch process without any connection with other quantities, it is rational to assign (as is customary in gas dynamics in the study of self-similar processes) dimensions [M], length [L] and time [T]. The dimensions of the secondary physical quantities are expressed symbolically in terms of the basic ones in the form of a power formula:

$$[p_i] = [M]^{\mu_i}[L]^{\lambda_i}[T]^{\tau_i}, \quad \mathrm{i} = 1\ldots n, \tag{4}$$

where p is the secondary unit of measurement;

μ, λ, r is the degree of dimension of the basic units of measurement;

n = 7 is the number of dimensional values in the model.

Table 1. List of dimensional and dimensionless physical quantities.

Parameter designation	Dimensional parameters	Dimension in SI	Dimension		
			μ	λ	τ
p_1	x	m	0	1	0
p_2	t	s	0	0	1
p_3	V	m/s	0	1	–1
p_4	g	m/s^2	0	1	–2
p_5	m	kg	1	0	0
p_6	F	kg m/s^2	1	1	–2
p_7	Hu	m^2/s^2	0	2	–2
c_1	θ	–	–	–	–
c_2	ξ	–	–	–	–
c_3	f	–	–	–	–

We define a general expression that allows us to obtain a similarity criterion for the problem under study. The similarity criterion is a defining dimensionless variable composed of secondary physical quantities that are not a function of independent variables [6]. To do this, we represent the dimensions of the basic parameters $p_{1\ldots n}$ in terms of the degrees of the dimensions of the basic units of measurement [M], [L], [T].

$$\begin{aligned} \Pi_m &= c[p_1]^{z_1}[p_2]^{z_2}\ldots[p_3]^{z_n} \\ &= c[l]^{\mu_1 z_1+\mu_2 z_2+\ldots+\mu_n z_n}[V]^{\lambda_1 z_1+\lambda_2 z_2+\ldots+\lambda_n z_n}[\rho]^{\tau_1 z_1+\tau_2 z_2+\ldots+\tau_n z_n}, \end{aligned} \tag{5}$$

where

c is a dimensionless constant;

$z_{1\ldots n}$ is some real numbers; m is the number of independent similarity criteria.

To calculate the number of independent similarity criteria, we find the rank of the matrix of coefficients, for which we arbitrarily choose a minor composed of degrees of the dimensions of the columns $p_{1...n}$. The rank of the matrix is 3, then the number of independent similarity criteria is m = 7 − 3 = 4. Write in general the system of equations for determining the primary system of similarity criteria and substitute the dimensions of the values into it.

$$\begin{cases} \mu_1 z_1 + \mu_2 z_2 + \ldots + \mu_n z_n = 0; \\ \lambda_1 z_1 + \lambda_2 z_2 + \ldots + \lambda_n z_n = 0; \\ \tau_1 z_1 + \tau_2 z_2 + \ldots + \tau_n z_n = 0; \end{cases} \tag{6}$$

$$\begin{cases} 0+0+0+0+z_5+z_6+0=0; \\ z_1+0+z_3+z_4+0+z_6+2z_7=0; \\ 0+z_2-z_3-2z_4+0-2z_6-2z_7=0. \end{cases} \tag{7}$$

Criteria $\Pi_{1...4}$ together with dimensionless values form the primary system of similarity criteria (Table 2). Based on the assertion that any combination of criteria is a similarity criterion, and therefore, a solution to the system of equations, we obtain all the similarity criteria for evaluating the energy perfection of the GLD. From the criterion series $\Pi_{1...4}$, the criteria for energy perfection are formulated as an analogue of efficiency, which expresses the ratio of the beneficial effect to the cost of obtaining it in equivalent terms.

Table 2. Getting the primary system of similarity criteria.

Numbers z_i that are assumed to be equal		Solving a system of equations for z_i			Similarity criteria
Zero	One	1	2	5	
4, 6, 7	3	–1	1	0	$\Pi_1 = \frac{p_3 p_2}{p_1} = \frac{Vt}{x}$
3, 6, 7	4	–1	2	0	$\Pi_2 = \frac{p_4 p_2^2}{p_1} = \frac{gt^2}{x}$
3, 4, 7	6	–1	2	–1	$\Pi_3 = \frac{p_6 p_2^2}{p_1 p_5} = \frac{Ft^2}{xm}$
3, 4, 6	7	–2	2	0	$\Pi_4 = \frac{p_7 p_2^2}{p_1^2} = \frac{H_U t^2}{x^2}$

The generality of the estimates of the energy perfection of GLD is expressed by the universal applied logic contained in them: the similarity criterion should demonstrate the ratio of the beneficial effect to the cost of obtaining it in equivalent terms. Thus, in the numerator, there should be an effective work performed by the GLD, and in the denominator – the equivalent work expended:

$$\kappa_{CEP} = \frac{A_{use}}{A_{exp}}; \tag{8}$$

$$\kappa_{PnC} = 2^{-1}\Pi_1^2\Pi_3^{-1} = \frac{1}{2}\frac{V^2t^2}{x^2}\frac{xm}{Ft^2} = \frac{m_{UAV}V^2}{2FL} \tag{9}$$

$$\kappa_{PyC} = 2^{-1}\xi^{-1}\Pi_1^2\Pi_4^{-1} = \frac{1}{2}\frac{V^2t^2}{x^2}\frac{x^2}{H_U\xi t^2} = \frac{m_{UAV}V^2}{2H_U\xi m_F}, \tag{10}$$

where

A_{use} is the effective realizable work;
A_{exp} is the equivalent expendable work;
κ_{PnC} is the criterion of energy perfection pneumatic GLD;
κ_{PyC} is criterion of energy perfection of pyrotechnic GLD.

During the operation of the starting accelerator, the internal energy of solid rocket fuel is expended. The internal energy of solid fuel is determined by its calorific value and mass $H_U\xi m_T$. In turn, the required mass of fuel depends on the flow through the nozzle and the accelerator operation time. Then the work expended during the combustion of a charge of solid rocket fuel:

$$A_{PyC} = \xi H_U m_T = \xi H_U \frac{F}{J\varphi}\tau, \tag{11}$$

where J is the specific impulse of the solid propellant solid propellant; φ is the specific loss coefficient; τ is the total engine running time.

The main costs in the preparation of pneumatic GLD is the operation of the compressor for the compression of gas in the balloon. We accept the assumption that the filling of the container is quite slow. Then we will consider the process as equilibrium and proceeding at a constant temperature. Then work under isothermal compression of an ideal gas.

$$A_{PnC} = l_I m_B = p_0 W_B \ln\left(\frac{p_B}{p_0}\right)\frac{p_B W_B}{RT_0}, \tag{12}$$

where

l_I is the specific work of gas compression in an isothermal process;
m_B is the mass of the compressible gas in the balloon;
p_0 is the initial pressure level in the balloon;
p_B is the pressure in the balloon after gas filling;
T_0 is the initial gas temperature in the balloon;
R is the gas constant.

2 The Dependence of the Scale Factor on the Energy Capabilities of Ground Launch Devices

The dependence of the Consider the task of finding the criteria of energy perfection for a variety of geometrically similar UAVs of various masses in the range of 50...250 kg. As such values for all types of UAVs, we take the maximum permissible starting overload at 15 g, and for GLD, these will be: the maximum pressure in the balloon is 12 bar, the volume of the balloon relative to the cylinder is 4, the transmission ratio is 4. Also we will apply composite fuel with the same physicochemical characteristics (Table 3).criteria of energy perfection κ_{PnC} and κ_{PyC} on the mass of the launched UAV indicates that with increasing mass of the UAV, the efficiency of using pneumatic ejection systems drops sharply, and pyrotechnic starting accelerators, on the contrary, slightly increase. The intersection of the criterion diagrams allows us to obtain the ranges of the most efficient way to start, depending on the mass of the UAV (Fig. 1).

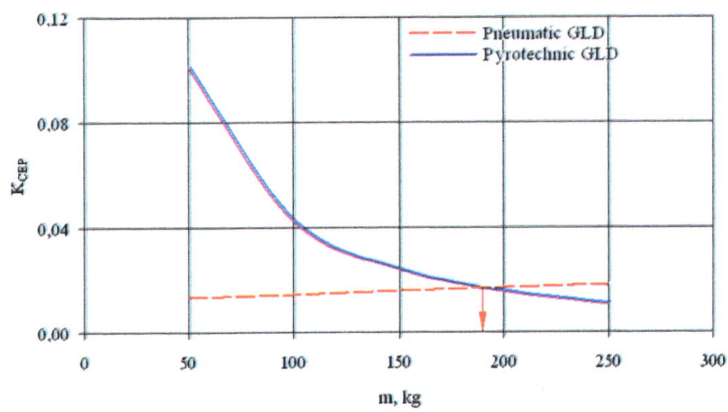

Fig. 1. Energy efficiency of GLD using depending on the mass of the launched UAV.

Table 3. Baseline data for solving a theoretical problem.

Parameter name	Numerical value				
	1	2	3	4	5
Mass launched UAV, kg	50	100	150	200	250
Minimum speed UAV, m/s	55	60	65	70	75
Permissible overload, g	15	15	15	15	15
CEP pneumatic GLD	0,101	0,043	0,024	0,016	0,011
CEP pyrotechnic GLD	0,013	0,015	0,016	0,017	0,018

When starting a UAV weighing up to 190 kg, it is advisable to use a pneumatic catapult, and over it – pyrotechnic systems. CEP (9) and (10) also adequately reflects the scale factor of the GLD itself: with the growth of GLD energy (to launch a heavy UAV with a high specific load on the wing), GLD of “cyclopean” dimensions is required, while the dimensions of the launch accelerator will grow slightly.

It is obvious that when a UAV is launched with mass limits (for example, up to 5 kg and more than 5 tons), the criteria degenerate and the energy efficiency inclines towards other methods of launch (respectively, manual and airfield). However, these assessments require a more detailed formalized study and are beyond the scope of this study, for the reason that, strictly speaking, do not apply to the catapult start.

3 Conclusion

The apparatus of the criterial complexes [3, 4], by virtue of its general physical base, is universally applicable to various classes of transport systems, including pneumatic and pyrotechnic GLD. Given the kinetic energy of the GLD and the starting overload of the UAV, we can determine the CEP, on the basis of which it is possible to unambiguously solve the problem of selecting the type of GLD in the early stages of the design of the complex. The proposed method of searching for the similarity criterion and the subsequent functional analysis can be extended to other methods of launching a UAV in flight, for example, an airfield or air launch.

The proposed CEP (9) and (10) show that high energy perfection of pneumatic GLD is achieved when launching light UAV (weighing 50–100 kg), and pyrotechnic GLD – when launching heavy UAV (weighing over 300 kg). Thus, the use of starting accelerators allows you to fill an extensive niche of UAVs launched from the surface in a wide range of masses. In this sense, pyrotechnic GLD look more promising than pneumatic catapults, losing only in terms of the visibility of the start.

The developed criterion assessments provide a general analysis of the development of GLD classes based on quantitative assessments of functional perfection, and can be used to identify promising areas and trends in the development of unmanned aircraft. The proposed method provides a comprehensive intensification of the R&D process at the stage of forming the GLD look through the achievement of the required level of energy excellence, subject to minimizing technical risk and resource-intensive work.

References

1. Kirpichev, M.V.: Theory of Similarity, 96 p. PH of the Academy of Sciences USSR, Moscow (1953)
2. Sedov, L.I.: Methods of Similarity and Dimension in Mechanics, 440 p. Science, Moscow (1977)
3. Kartashev, A.S.: A criterion method for selecting tactical and technical characteristics and formation the shape of a small-sized aircraft. Automob. Transp. **21**, 82–86 (2007)
4. Ambrozhevich, A.V., Borisyuk, M.D., Dolzhenko, I.Yu.: Method of formation of the aerogasdynamic shape of a promising armor-piercing feathered subcaliber projectile. Artillery and small arms, № 3/36, pp. 3–7 (2010)
5. Vasilin, N.Ya.: Unmanned Aerial Vehicles, 272 p. Popurri LLC, Minsk (2003)
6. Ryzhenko, A.I.: Definition of a System of Criteria and Similarity Scales When Designing Free-Flying Dynamically Similar Aircraft Models, 100 p. National Aerospace University N. Ye. Zhukovsky, Kharkov (1992)

Brushless Direct Current Propulsion System Identification

Rafael Trujillo Torres(✉) and Firsov Sergii(✉)

National Aerospace University "Kharkiv Aviation Institute", Kharkiv, Ukraine
rafael.trujillo.torres@gmail.com, sn.firsof@gmail.com

Abstract. This work reports Parameter estimation of Propulsion System driven by Brushless Direct Current (BLDC) motor used in multirotor unmanned vehicles, parameters was obtained by experimental process. In order to obtaining the parameters in mathematical model given response to different excitation inputs, the analysis method for system identification that is used is the reaction curve. To establish the structure and parameters that mathematically describe the motor, the motor is stimulated with a step signal, the response to this signal is observed. Response in speed was observed graphically and from the experimental analysis the parameters that define his transitory response were determined then is used numerical approximations using optimization tools as objective function linear regression using least square method through objective function in Python programing language. Finally, the model found was replicated by computer simulation and compared, the relevant parameters were adjusted and validated by comparison between the experimental results and what was theoretically stated.

Keywords: Multirotor · Propulsion system · Systems identification · Simulation · First order systems

1 Introduction

Control systems are currently a fundamental part in the promotion of new technologies, since they allow automation of processes and the development of intelligent systems [1]. Propulsion systems in multicopter unmanned systems creates forces and moments than ensures the aircraft flight (in hover, forward and transition flight). Those are achieved due the changing the magnitude and the direction of the resultant forces and moments generated by the engines. Suitable mathematical representation of the system to be controlled is necessary in order to control system [2], this implies, establish the mathematical model that represents it properly. In classical control, the Transfer Function (T.F.) [1], is the mathematical representation that describes the model of the system and that allows to establish its output value before a certain input or disturbance to it. Through the T.F are determined stability, the errors present before external disturbances and the parameters to be controller. However, it is not always possible to establish the T.F., unless the physical parameters of the plant to be controlled are known a priori. Obtaining acceptable results and they have also been raised methods that resort to the use of computational tools as support for the identification process. In

M. Nechyporuk et al. (Eds.): *Integrated Computer Technologies in Mechanical Engineering*, AISC 1113, pp. 105–113, 2020.
https://doi.org/10.1007/978-3-030-37618-5_10

presented work is proposed to obtain the T.F. describing the dynamic behavior of the F450 DJI quadrotor Fig. 1 propulsion system, given excited step signal response in speed was observed graphically and from the experimental analysis the parameters that define his transitory response were determined. Finally, the model found was replicated by computer simulation, the relevant parameters were adjusted and validated by comparison between the experimental results and what was theoretically stated. The proposal presented here is characterized by being accessible in the experimental design as well as in the mathematics that sustains it.

Fig. 1. F450 DJI quadrotor [4]

2 Systems Identification Method

2.1 Systems Identification in Mechanical Systems

Systems identification is a method used for accurate characterizing the dynamic response behavior of dynamic systems [3], can be applied to a complete aircraft, subsystem, or individual components from measured data [4]. According to [5] Systems Identification method can be described as in Fig. 2.

The identification methods can be classified [7] in: "offline" when the data is taken first and subsequently processed and "online" methods, when data capture and processing are performed simultaneously. In the case of linear models, the "identification offline" is sufficient, since valid models are achieved. In addition, identification can be done either in "open loop" or in "closed loop". Benefits from applying these techniques include the reduction of flight-test time required for control system optimization and handling-qualities evaluation, especially for complex control-law architectures, and improvements in the final system performance. Frequency-domain methods offer a transparent understanding of component and end-to-end response characteristics that can be critical in solving system integration problems.

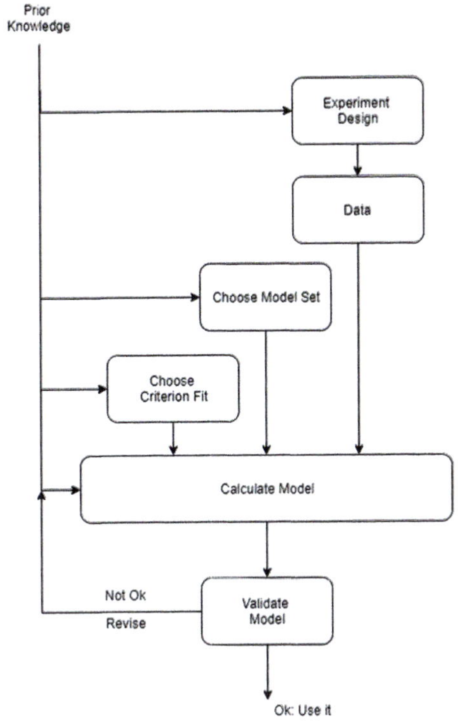

Fig. 2. Systems Identification Method.

2.2 Propulsion System Description

Systems architecture of a propulsion system in F450 Drone is show in Fig. 3. In this work all these elements are summarized as a block and analyzed the overall performance of the system.

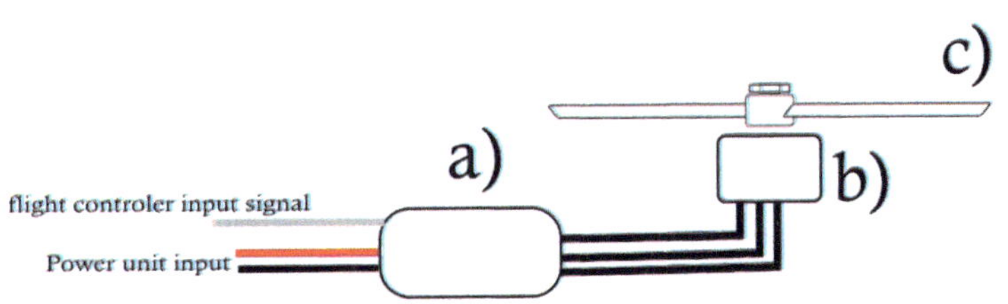

Fig. 3. Architecture f450 Drone (a) ESC, (b) BLCD engine, (c) Propeller 1045.

Brushless Direct Current (BLDC) motors are one of the motor types rapidly gaining popularity. BLDC motors are used in industries such as Appliances, Automotive, Aerospace, Consumer, Medical, Industrial Automation Equipment and Instrumentation. As the name implies, BLDC motors do not use brushes for commutation; instead, they are electronically commutated [8]. BLDC motors have many advantages over brushed DC motors and induction motors. A few of these are:

- Better speed versus torque characteristics
- High dynamic response
- High efficiency
- Long operating life
- Noiseless operation
- Higher speed ranges

The control of BLCD on F450 Propulsion system is realized by PWM signal carried out by ESC element. In order to characterize system signal input is consider in range from 1000 ms to 2000 ms.

2.3 Mechanical Model Parameter

Unmanned aircraft propulsion system can be represented as a Linear Time Invariant System [6]. First order systems are, by definition, systems whose input-output relationship is a first order differential equation. In this work is used First Order Linear system with time delay, generally, the First Order Plus Dead Time (FOPDT) [3] model is used to obtain initial controller tuning constants, given the nature of this work and the dynamic system the formulation for carrying out the experiment FOPDT is more convenient. Mathematical model of FODT can be written as:

$$\tau_p \frac{dy(t)}{dt} = -y(t) + K_p u\left(t - \theta_p\right) \tag{1}$$

Where K_p is the proportional constant defined as the ratio between steady state output and the input signal, τ_p is time constant defined as a delay in seconds where the 63.2% of the steady-state output is reached by the system and θ_p is process dead time.

Figure 4 show graphical representation of the first order differential equation with step input.

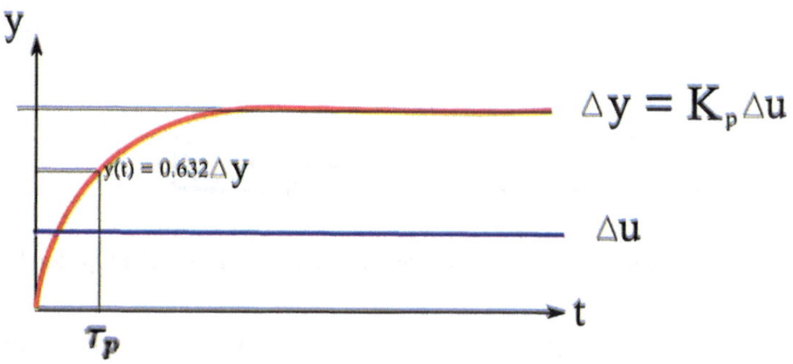

Fig. 4. First order system step response.

First order differential equation can be rewritten as a transfer function:

$$G(s) = \frac{Y(s)}{U(s)} = \frac{K_p}{\tau_p s + 1} \tag{2}$$

3 Experiment Design

Experiment is based on reaction curve method, this method consists of introducing a disturbance system, and verifying the response y(t). Subsequently, it is enough in the behaviors in the previous section, we propose a transfer function that represents the system that is in mode and the output is close to the real response of the system [6] defines this parameter. In order to gathering data is used dynamometer where is possible to measure input ESC control signal and Thrust force. This is a very suitable method to characterize systems in the industry where there are no precise references of the system and a design of Proportional-Integral-Derivative (PID) controllers or advanced controllers is required. The method is heuristic and easy to implement in various contexts where there are no sophisticated identification tools.

3.1 Testbed Description

In order to gathering data is used dynamometer where is possible to measure input ESC control signal and Thrust force, was chosen the dynamometer series 1580 from RCBechmark company Fig. 5, where is possible to program the routine for signal input data and sample time data recorder using Java Programing language. Due data gathering sensors sample is no possible to reach step response in fisical dynamic system, but is possible to approximate using ramp input with very small ratio input-time response compared to the overall experiment time.

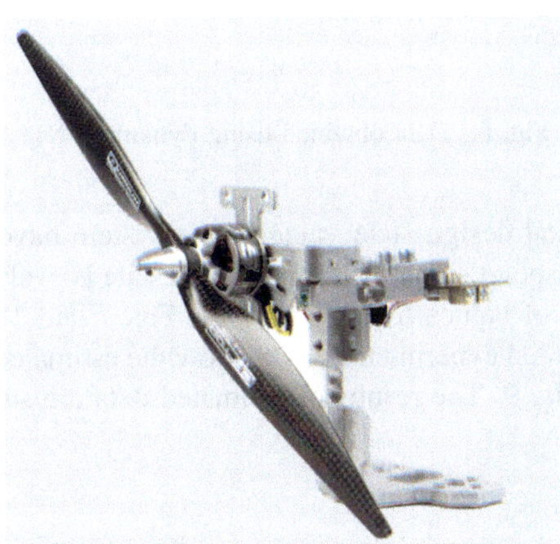

Fig. 5. RCBenchmar dynamometer [8]

As mentioned before ESC signal operates in range 1000 to 2000 ms, by experience is observed than for engine drive initial inertial breakout is after 1200 ms, therefore initial input is driven in 1200 ms in 10 s time and to reach to 1800 ms in 20 s.

3.2 Data Gathering

Due data gathering sensors sample is no possible to reach step response in physical dynamic system [7], however is possible to approximate using ramp input with very small ratio input-time response compared to the overall experiment time. Data is obtained from sensors given the sensor characteristics [8] is written program and obtained data in sample time of 50 ms, information is saved in .csv document. Python program is used to visualize data using information as Data frame, Fig. 6 shows the data obtained.

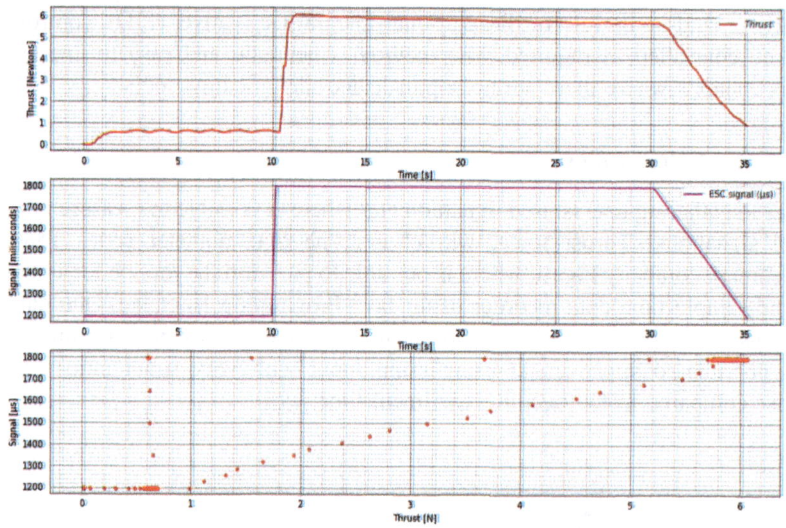

Fig. 6. Data obtained using dynamometer.

Due the experimental design state value of the system have to be estimated previous and father step respond signal, 1200 ms signal state is evaluated and estimated in simple mean approach of values from 3 to 10 ms Fig. 7 and 1800 ms signal state is evaluated after 10 s of total experiment, Fig. 8 show the estimated data. Step respond is analyzed as show in Fig. 9. The results of estimated data are summarized in Table 1.

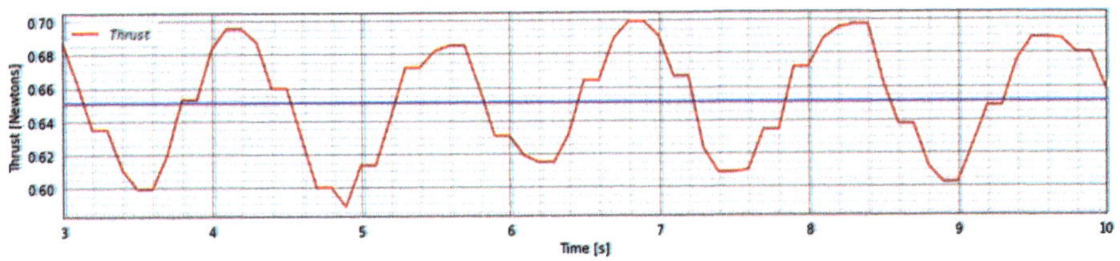

Fig. 7. Data from 3 to 10 s and mean value data

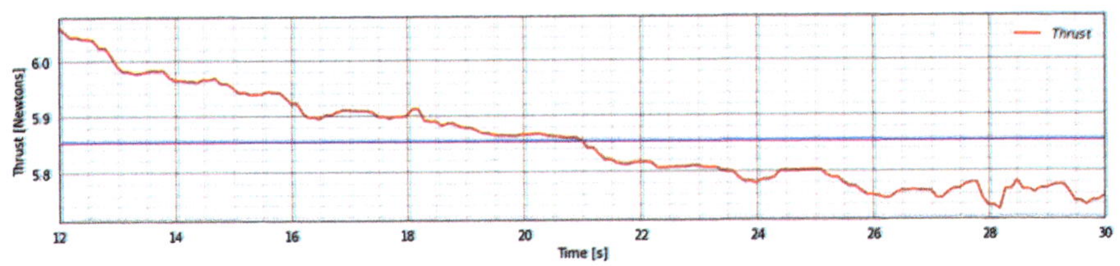

Fig. 8. Data from 12 to 30 s and mean value data

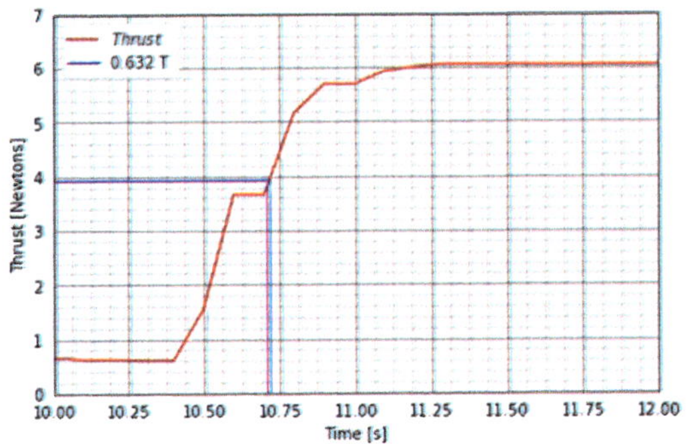

Fig. 9. τ_p time constant parameter graphical estimation

Given the data estimated the resulting estimated transfer function of the system is showed in Eq. 3.

$$G(s) = \frac{Y(s)}{U(s)} = \frac{0.00866}{0.71s + 1} \tag{3}$$

4 Optimization Method

In order to estimate the parameters in Eq. 1 is used optimization method implemented in Python programming language. Optimization is carried out by objective function, in this case is the minimum value of least squares method given the iteration of parameters FOPDT, the least squares criterion is one of the foundation of estimation theory, concerning extracting the true value of signals form noisy measurements, values in Table 1 are used as initial guess (Noisy parameter) derived form the data obtained graphical estimation. Curve fitting by least squares method concerns combining a set of measurements to derive estimates of measurements which specify the curve that best fits the data [9]. By the least squares criterion, given a set of N (noisy) measurements $f_i, i \in 1, N$, which are to be fitted to a curve $f(\mathbf{a})$, where a is vector of parameter values, seeking to minimize the square of the difference between the measurements and the values of the curve to give and estimated of parameters a according to Eq. 4:

$$\hat{a} = \min \sum_{i=1}^{N} (f_i - f(x_i, \mathrm{a}))^2 \tag{4}$$

The Estimated first order system is implemented in system defined in Python and plotting; Resulting trust given step input is show in Fig. 10. In order to verify the parameter system proposed, is compared stated response in experiment and system simulated in Fig. 10 show comparison data.

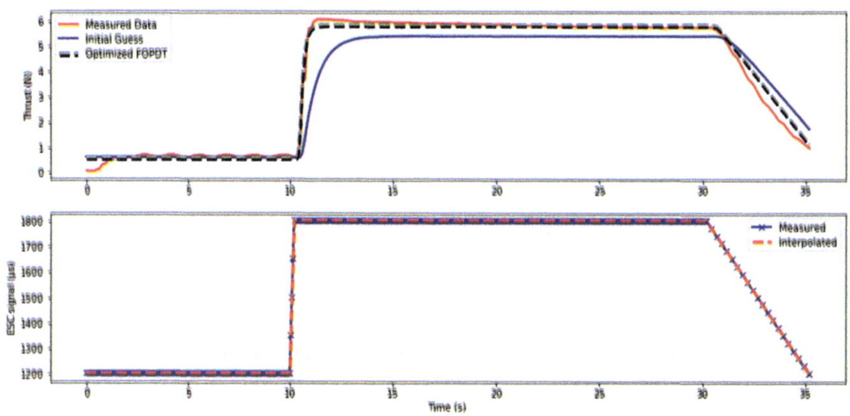

Fig. 10. Comparison between model computed, initial guess and data gathering.

Table 1. First order plus dead time parameters.

Parameter	Initial guess	Optimized value
K_p	0.0866	0.00874
τ_p	0.71 s	0.157 s
θ_p	0.5 s	0.36 s

Given the computed results differential Eq. 1 can be rewritten using parameters as:

$$0.157\frac{dy(t)}{dt} = -y(t) + 0.00874\,u(t - 0.36) \tag{5}$$

5 Conclusions

In The results that have been presented and the methodology that has been proposed represent an experimental modeling of systems for obtaining models of the dynamics of unknown systems. The methodology used through the reaction curve has been sufficient to calculate the parameters that allow to completely establish the first Order Differential equation and thereby simulate the response of the system to different conditions and/or external disturbances. The results allow to conclude that the BLDC motor can be described and represented by a first order Differential equation with delay, which presents responses with sufficient approximation to the real response of the system, has been demonstrated that through relatively simple procedures, an experiment can be designed using "open source" tools as Python and Data frame structures for the identification of systems through analysis of the transitory response of the system and with results that represent an approximation quite close to what was observed experimentally.

References

1. Ogata, K.: Modern Control Engineering, 4th edn. Prentice Hall, New Jersey (2002)
2. Lao, S., Ye, C., Identification of parameters for a DC-motor by LabView, February 2012
3. AP Motnitor Homepage. http://apmonitor.com. Accessed 12 Oct 2019
4. Tischler, M.: System identification methods for aircraft flight control development and validation. In: Advances in Aircraft Flight Control, pp. 35–69 (1995)
5. Ljung, L.: System Identification: Theory For The User, 2nd edn. Prentice Hall PTR, Upper Saddle River (1999)
6. Lee, B.-K., Ehsani, M.: Advanced simulation model for brushless DC motor drives. Electr. Power Compon. Syst. **31**, 841–868 (2003)
7. Ezeta, R.F.: Análisis y diseño de sistemas de control digital, 1st edn. Mcgraw-Hill. México City (2013)
8. RCBenchmark Homepage. https://www.rcbenchmark.com/pages/series-1580-thrust-stand-dynamometer
9. Nixon, M.S., Aguado, A.S.: Feature Extraction & Image Processing for Computer Vision, pp. 519–523, 3rd edn. Academic Press (2012)

The Method of Triads in the Aircraft Design

Oleksandr Karatanov(✉) and Viktoriia Chetverykova

National Aerospace University «KhAI», Kharkiv, Ukraine
a.karatanov@khai.edu

Abstract. The article describes the features of the triad method application at the initial stages of aircraft design for the selection of fundamental decisions in aircraft design. The possibility of using the triad method to determine the initial appearance at the stage of conceptual design is analyzed. A brief overview and features of the conceptual design are given in this article. We are considering the general application algorithm and the basic concepts of the triad method. There are examples of the triad method used for the airfoil, aerodynamic design, and the selection of landing gear type. The formulated main postulates which allowed you to transfer this experience to the design of engineering products.

Keywords: Method of triads · Design · Minimal context method · Aviation technology · Conceptual design · Development of technical specifications · Airfoils · Aerodynamic configuration · Wing position · Landing gear

1 Introduction

Formulation of the Problem. In the entire lengthy process of creating an aviation technology, from the conception to the serial production start and operation start, design is worth highlighting. This is a complex process that requires huge amounts of computation. The new technical systems design is one of the most complex types of engineering creative activity.

The stage preceding the direct design of the requirements development for aviation technology, which carries out jointly by the customer and the development design office. This stage is called external or conceptual design.

A new aircraft design begins with the concept development – the general concept of its creation. The concept determines in what ways and means, what parameters will ensure the high efficiency and competitiveness of the designed aircraft, his superiority compared to aircraft in operation or development [1].

At this stage, based on parametric studies of promising aviation techniques, analysis of their interaction with the components of the complex in which they will operate, the required general characteristics of the future aircraft are forecasted.

At the same time, multi-variate calculations are carried out to determine and optimize the feasibility indicators of the operation of the proposed aircraft on the intended network of air routes. The necessary technical, economic and tactical, and technical characteristics of the aircraft are determined, allowing to formulate the requirements for its design, as a result of this work. It should be noted the reasonable task of requirements largely determines the success of the program of creating a new aircraft.

M. Nechyporuk et al. (Eds.): *Integrated Computer Technologies in Mechanical Engineering*,
AISC 1113, pp. 114–125, 2020.
https://doi.org/10.1007/978-3-030-37618-5_11

Analysis of Recent Research and Publications. The main feature of the study an issue is the absence of publications directly related to the study subject, and the existing publications are out of date and indirectly related to the issue:

- publications related to conceptual design – focus on preliminary design, in fact [1, 2];
- publications on the method of triads focus on issues of its use in psychology and only consider [3] its use in design;

The Aim of the Study. The main feature of conceptual design is the need to make many decisions when it's insufficient or, conversely, excessive information. This publication outlines the way the triad method is used in the conceptual design of aviation equipment, which has not been used in this field before. Use will potentially help to shape the appearance of aircraft more in line with customer expectations.

2 Method of Triads

The method of triads is a design technique that allows you to identify the features of the research object based on comparison.

The method of triads is a powerful technique that effectively manifests itself in the initial stages of design, with which you can identify the concepts underlying the projected object.

George Kelly was the first to propose this technique. The basics of this method are described in his book "Theories of Personal Constructs". However, these and subsequent publications consider the use of the triad method in psychological studies, and only at later works are it considered the possibility of its application for engineering and design.

Initially, the method was considered within the framework of the method of repertory lattice.

The repertory grid technique was designed specifically to identify the individual categorical structure of the personality, to minimize the bias and any impact of the researcher on the respondent during the interview.

The minimal context method or triad method is most often used to identify constructs.

The construct is a speculative construction introduced hypothetically (theoretical) or created about observable events or objects (empirical) according to the rules of logic with well-defined boundaries and precisely expressed in a particular language [4].

According to George Kelly's Personal Constructs Theory, the people create subjective classifications – personal constructs to understand the world around us.

In general, the method of triads can be represented as an iterative sequence. (see Fig. 1).

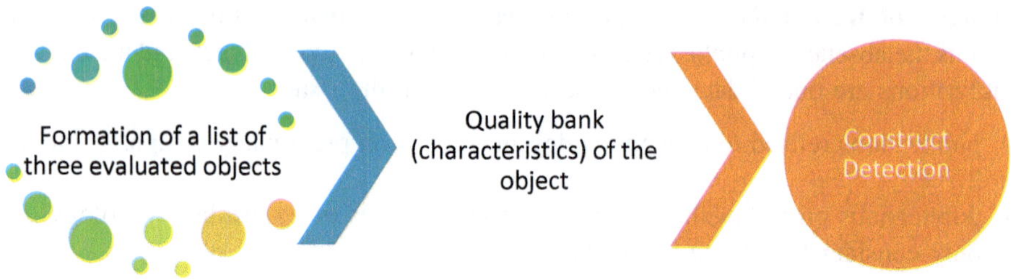

Fig. 1. Three steps of the triad method.

Step 1. Items are presented in groups of three. This is the minimum number that allows us to determine the similarities and differences.

The groups are formed by some pre-selected list. With each of the elements, which the test subject should be familiar in advance.

Step 2. The test subject was imposed three elements from the entire list and proposed to name some important quality by the two of them are similar and therefore different from the third.

Step 3. The test subject is asked to name what exactly is the difference between the third element and the other two after the experimenter writes down the answer (if the test subject doesn't indicate which two elements have been assessed as similar to each other, he's asked to do so). The answer to this question is the opposite pole of the construct.

Repetition. The test subject is presented with as many triads of elements as the experimenter sees fit. There are no specific rules. It all depends on the size of the sample, which is the number of constructs to be investigated.

An example. There is a list of fruit names. "The apple-pear-orange" triad is taken. The respondent identifies two similar objects — "the apple and pear"; similarity quality — "the lack of an allergic reaction in the respondent", the difference between the third object is "allergic". So identified personality construct "allergy/lack thereof".

At the beginning of the triad study, a researcher or respondent selects six to ten specific related brands, products or services from a particular area. These examples of research incentives should be of a certain variety. Ideally, the respondents will be introduced to each of them before the session, because the purpose of the study is to identify what is important and meaningful to the respondent.

The researcher asks the respondent to select three examples (triad) for discussion, after identifying six to 10 examples/incentives from a particular area, and then invites him to explain how, according to his feelings; two of them differ from the third. This process can be repeated repeatedly, each time concentrating on a new triad to identify as many mental structures available to the respondent related to this area, if it's necessary

When many respondents participate in this study, we get a large amount of data in a specific area. As a rule, the constructs and their rating are different for different people.

The results of such a study are often surprising and noncorrelative and suggest that they may not have come to this study before the study.

At first glance, the triad study procedure may seem too simple, but it's actually about building an interview, which allows revealing deep-rooted opinions and perceptions, requires researchers to work very seriously and scrupulously.

The method of triads can be successfully used to analyze competitors and their products, as well as to compare different interface design options.

The method of triads, as a whole, is a powerful interviewing technique that helps researchers and designers understand how personal constructs relate to specific products and services.

3 Applying the Method of Triads in Conceptual Design

The object image or its constituent parts can create in a person's imagination as a result of the creative process in the early stages of design. A significant help in this action is the use of a method, an approach that could simplify, normalize and introduce into some framework the process of creative search.

In the conceptual design, process decisions are made that determine the subsequent appearance of the aircraft. There is no place for specific numbers – they will be obtained later, there is an operation with qualitative characteristics – better/more/stronger…

The aircraft design usually begins with the choice of an airfoil (Fig. 2).

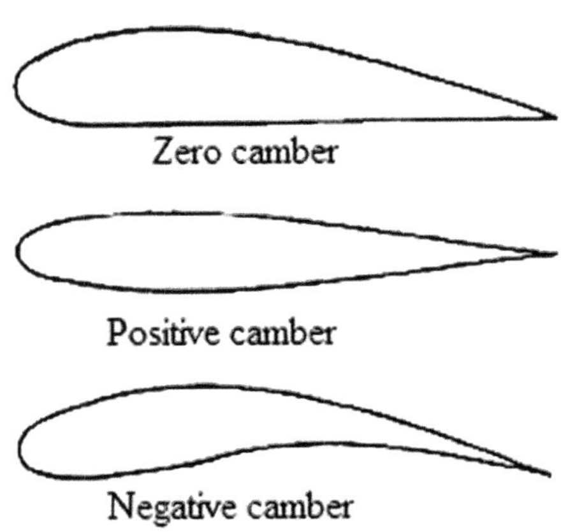

Fig. 2. Airfoils

Much depends on the profile shape; profiles have different maximum lift factor $C_{y\ max}$:

- for symmetrical (positive camber), the wing lift force coefficient is 1.2 to 1.4;
- ordinary asymmetrical with a convex lower surface (zero camber) – till 1.8;

- with a strong concavity of the lower surface (negative camber), it sometimes reaches 2.

However, need to remember that profiles with very high $C_{y\ max}$ usually have high C_x и m_z is pitching-moment coefficient. The aircraft with this profile, the tail plumage must develop greater strength to balance. As a result, its aerodynamic resistance is increasing, and the overall gain from the high-carrier profile is significantly reduced.

The following Table 1 gives a triad of an airfoil comparison. The arrow up shows the increase in the parameter, and down, respectively, a decrease.

Table 1. Triad of an airfoil comparison.

	Symmetrical wing	Asymmetric profile with convex bottom surface	Asymmetric profile with strong concavity of the lower surface
C_y max	↓	↑	↑
C_x	↓	↑	↑
m_z	↓	↑	↑

In this table:

$C_{y\ max}$ – maximum lift force coefficient;
C_x – drag coefficient;
m_z – longitudinal moment or pitching moment.

Also undoubtedly important is the choice of the location of the tail (aerodynamic configuration):

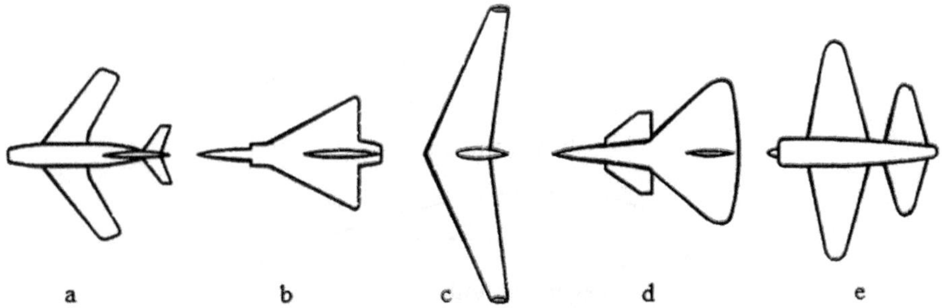

Fig. 3. Aerodynamic Configuration.

- Aft-Tail Configuration (Tail Plumage, see Fig. 3a);
- Tailless Aircraft (see Fig. 3b).
- Flying Wing (see Fig. 3c);
- «Canard» (Front Plumage, see Fig. 3d);
- Tandem-Seat Configuration (see Fig. 3e)

The normal aerodynamic scheme of the aircraft is characterized by the arrangement of horizontal plumage behind the wing. The wing is covered with an unperturbed stream. The plumage is in the worst conditions, but provides good longitudinal stability of the aircraft and control it.

In aircraft of the "Canard" type, the horizontal tail is located in front of the aircraft and is the bearing, which allows to reduce the wing area and weight of the aircraft (but makes horizontal tail more loaded). The front location of the horizontal tail increases its effectiveness (but worsens its road stability), which reduces the required angles of surface deviation and resistance when balancing an aircraft. The bearing horizontal tail radically changes the structural strength of the structure. In this case, the fuselage in flight "rests" on the wing and plumage; as a result of the loading and strength of it have the best performance.

The "Tailless" aircraft has a smaller mass and frontal resistance but requires measures to ensure the necessary in-flight stability of controllability.

As you can see, such a text isn't very convenient for perception. The presentation of this information in the table according to the triad method will be much more visual. Let's summarize the data on aerodynamic circuits for their comparison by the triad method in the Table 2.

Table 2. Comparison of aerodynamic circuits by method of triads.

	Aft-tail configuration	Tailless aircraft	«Canard»
S_{wing}	↑	↑	↓
C_x	↑	↓	↓
m	↑	↓	↓
Pitching effectiveness	↑	↓	↓

In this table:

S_{wing} – wing area;
m – aircraft mass.

As much as important is the choice of the wing position.

Monoplanes are divided by wing height relative to the fuselage into the following types:

- Low-winged Aircraft;
- Mid-winged Aircraft;
- High-winged Aircraft;
- Parasol wing Aircraft (the wing is located above the fuselage).

The construction of the parasol wing aircraft isn't widespread due to the low aerodynamic characteristics, therefore, we consider the first three according to the triad method (Fig. 4).

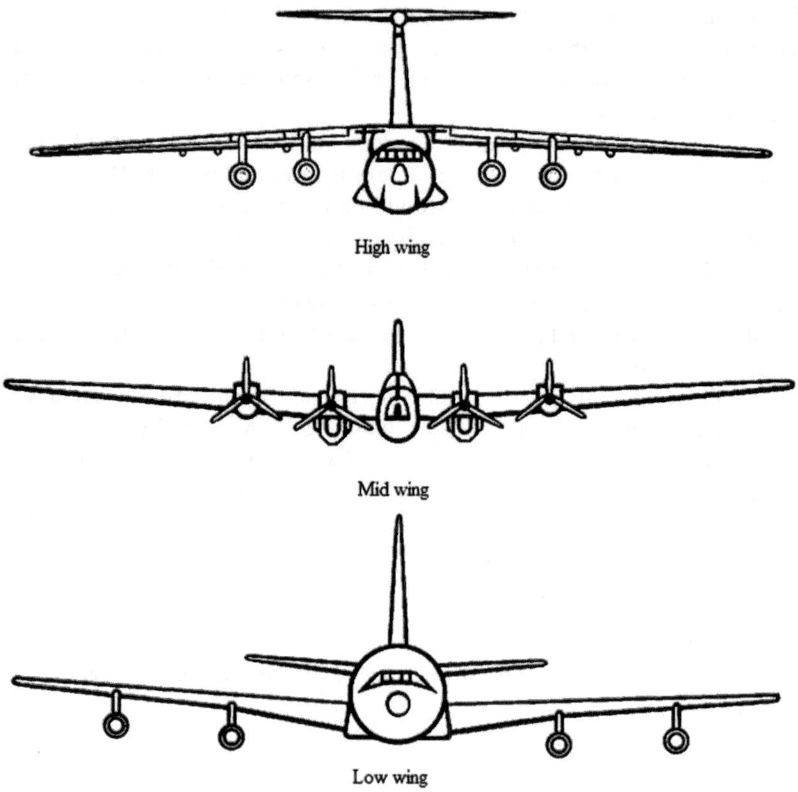

Fig. 4. Wing position

Let's considering their features by the triad method (Table 3).

Table 3. Comparison of wing position by method of triads.

	Low wing	Mid wing	Shoulder wing
The speed and simplicity of a loading/unloading aircraft	↓	↓	↑
The runway requirements	↑	↑	↓
The simplicity access to engines and wing	↑	↑	↓
The fuselage protection	↑	↓	↓
The fuselage free space	↑	↓	↑
The simplicity of the wing/body intersection	↓	↑	↓
The wing-body interference	↑	↓	–
The dihedral (lateral stability) stability	↓	–	↑-
The landing gear length	↓	–	↑

We deliberately allowed deviations from the rules of the triad method in the last two compared qualities, whereby we will highlight only signs common for two of the three compared. This suggests that you need to allow such a departure or to discard characteristics, which don't fit this rule.

The chassis configuration extremely conveniently fits into consideration using the triad method.

The aircraft landing gear layouts divide into three main types depending on the main supports location and auxiliary supports location relative to the aircraft center of gravity (Fig. 5):

- Tailwheel-Type Landing Gear;
- Tricycle-Type Landing Gear;
- Bicycle Landing Gear.

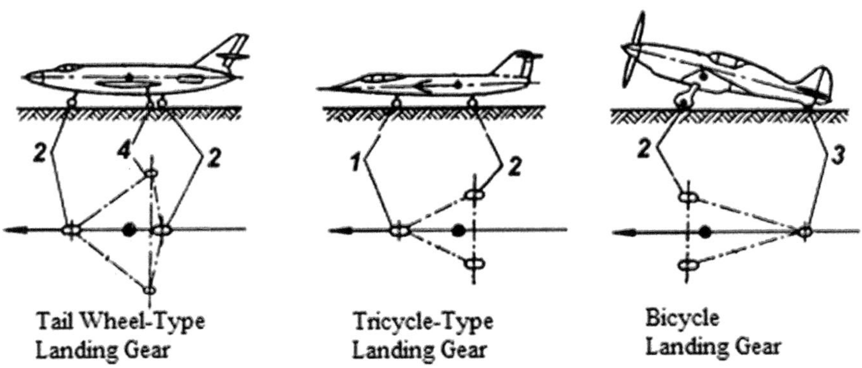

Fig. 5. Landing Gear: 1 – Nose Landing Gear; 2 – Main Landing Gear; 3 – Tail Landing Gear; 4 – Outrigger Undercarriage; Black Spot – Center of Gravity.

Let's summarize the data on Landing Gear for their comparison by the triad method in the Table 4.

Table 4. Comparison of Landing Gear by method of triads.

	Tailwheel-type landing gear	Tricycle-type landing gear	Bicycle landing gear
The pilot skill requirements	↑	↓	↑
The chassis rack breakage risk	↓	↑	↓
The aerodynamic characteristics of aircraft	↑	↓	↑
The cockpit view	↓	↑	

4 Level Detection of the Designer Competencies

The conceptual and sketch design processes are the most important stages of aircraft development, which at a maximum cost from 20% to 25% of the time of all work and as many as 5% to 10% of funds accept from 75% to 80% of the key project decisions (technical and organizational).

The fate of the project often depends on how correct decisions are made in the early stages, and not only because mistakes made in the early stages of project development lead to too much cost and time for its completion in the process of detailed design and construction, but and because the possibility of project implementation may even depend on them.

As seen, the conceptual design phase is one of the most important stages. Therefore, all decisions are so important at this stage.

Accordingly, there are increased requirements for the designer employed at this stage. He has to be an expert in his field. A team of experts should be used to reduce subjectivity.

There are different approaches to selecting the number of experts (m). The source [5] states that the number of experts should be no less than the number of alternatives (n), which are subject to ranking (m $\geq$ n). According to the sources [6, 7], the number of experts is recommended to be determined by the following formula:

$$m \geq 0,5\left(\frac{0,33}{b}+5\right),$$

where b – the level of the permissible error of the result of expert analysis, usually b from 0 to 0.1.

In general, there are recommendations to include from 6 to 15 (from 7 to 20 [8]) people in the expert group.

It's also important to determine their quality in addition to determining the number of experts. The following methods of the performance indices of experts are the highlights: a heuristic, a statistical, a test, a documentary and a combine. The heuristic evaluation methods are based on the idea that has developed about this expert in others (or himself), and correctly reflecting its true quality.

Many expert assessment methods now propose as an indicator of expert competence (C):

$$C_i = \frac{Cn_i + ar_i}{2},$$

where Cn_i – the coefficient of the level of familiarity of the expert with the discussed problem; the expert himself assesses the level of his familiarity with the issue;

$Cn_i =$
- 0 – the expert isn't familiar with the question;
- 0.1..0.3 – the expert is new to the question, but the question falls within the scope of his interests;
- 0.4..0.6 – the expert is satisfactorily familiar with the issue, doesn't directly participate in the practical solution of the issue;
- 0.7..0.9 – the expert is well acquainted with the issue, participates in the practical solution of the issue;
- 1 – the question is within the expert's narrow specialization.

ar_i – the argumentation coefficient, which takes into account the structure of the arguments that served as the basis for the expert for a certain assessment, is determined by Table 5 by summing up the values selected by the expert.

Table 5. The values of the argumentation coefficient

Sources of argument	The influence quantity of the source of the argument on the expert opinion		
	High	Medium	Lower
Theoretical analysis	0.3	0.2	0.1
Production experience	0.5	0.4	0.2
Generalizations of works by domestic authors	0.05	0.05	0.05
Generalizations of works by foreign authors	0.05	0.05	0.05
Personal acquaintance with the state of affairs abroad	0.05	0.05	0.05

The coefficient of consistency or competency (CC) of the expert group is calculated according to the source [5] according to the following formula:

$$CC = \sum_{i=1}^{m} C_i,$$

where C_i – the competency coefficient of the i-th expert.

The working group of experts formed is competent and able to correctly solve the tasks set before it, if the level of its competence meets the following condition [9]:

$$0,67 \leq CC \leq 1.$$

The issues of determining the number of experts and assessing their competence consider in more detail in the source [10]. A sufficient level of the competency coefficient allows us to confirm the reliability of decisions made by experts.

5 Conclusions

Certainly, the triad method isn't without drawbacks.

Firstly, it is the subjectivity of the results obtained - they are based on the experience of the designer and reflect only his point of view. The use of experience, relying on opinion, and not on accurate scientific results, leads to the fact that we can get incorrect results, distorted through the prism of perception of the designer.

Secondly, obviously, to achieve more accurate results, a corresponding extremely high degree of competence of the designer will be required. A deep understanding of all aspects of aircraft engineering design, a vision of the general concept, design experience supported by practice, and an understanding of the interconnectedness of various design decisions are needed.

Thirdly, the considered example of the application of the triad method in conceptual design doesn't take into account the mutual influence of one or another constructive solution on each other. But even a slight change in the appearance of the aircraft, for example, can radically change the flow pattern of an aircraft.

Fourth, the failure to use this method for the subsequent design stages is obvious due to its specificity.

However, despite these shortcomings, the triad method is quite applicable for the conceptual design of both aviation equipment and complex engineering products.

The subjectivity of the result can be reduced using the knowledge of several design experts. The final result can be obtained in several ways, for example, using the "Kemeny median".

The issues of the design competency definition, as well as a group of design experts, discusses in Sect. 4 of this article.

As for taking into account the mutual influence of constructive decisions on each other, at this development stage, this method doesn't take this into account and relies only on the designer experience and knowledge. However, it's quite possible to improve this method, using decision theory achievements in the field of solving multicriteria problems. The fundamental inapplicability of the stages method at other than conceptual design doesn't reduce the significance at the initial stage of design.

Moreover, in addition to the already described method of applying this method for the aircraft design, it's also possible to use the developments of this method in other development-related processes. This method can help to simplify (providing) the process of determining the main structural solutions for a customer who doesn't have sufficient knowledge in the aviation field. So, for example, in the design bureau engaged in the development of unmanned aerial vehicles, an order comes to create an original model with a look that must be approved by the customer. The application of the triad method, in this case, will allow you to quickly and effectively determine the preferences of the customer.

References

1. Komarov, V.A., et al.: Conceptual Aircraft Design: Electronic Textbook. The Ministry of Education and Science of the Russian Federation, Samara State Aerospace University, Samara (2013)
2. Nawar, J., Probha, N., Shahriar, A., Wahid, A., Bakaul, R.: Conceptual Design of a Business Jet Aircraft (2014). https://www.researchgate.net/publication/298791385_CONCEPTUAL_DESIGN_OF_A_BUSINESS_JET_AIRCRAFT
3. Martin, B., Huntington, B.: Universal Design Methods, St. Petersburg, p. 208 (2014)
4. Abushenko, V.L., Shvyrev, V.S.: Construct/Humanitarian Encyclopedia: Concepts. Center for Humanitarian Technologies (2019). https://gtmarket.ru/concepts/6889
5. Margolin, E.: Methodology for Processing Expert Survey Data. Polygraphy 5, 14–16 (2006). (in Russian)
6. Lukicheva, L.I., Yegorychev, D.N.: Management Decisions. Omega-L, 383 p. (2009). (in Russian)
7. Petrov, A.Y.: Integral Methodology for Assessing the Commercial Potential of an Investment Product. Moscow Printer, 23 p. (2010). (in Russian)
8. Zerny, Y.V., Polyvany, A.G., Yakushin, A.A.: Quality Management in the Instrumentation. New center, 479 p. (2011). (in Russian)
9. Mikhnenko, P.: Secrets of Effective Business Solutions. NT Press, 288 p. (2007). (in Russian)
10. Postnikov, V.M.: Analysis of approaches to the formation of the composition of an expert group focused on the preparation and making decisions. Sci. Educ. Publ. House MSTU **3** (27), 333–346 (2012). Bauman, Edition. (in Russian)

Automation of the Manipulator

Dmitriy Kritskiy, Olha Pohudina(✉), Serhii Koba, Olha Kritskaya, and Andrii Pohudin

National Aerospace University, Kharkiv Aviation Institute, Kharkiv, Ukraine
d.krickiy@khai.edu, o.pogudina@gmail.com

Abstract. The subject of study in the article is the process of implementing the search for an object, its recognition (its geometric shape) and moving the manipulator to the corresponding coordinates to capture the object using a probe (ticks) and moving it to the given coordinates. The goal is to improve the quality of the robot manipulator when recognizing complex images of objects. Problems: study of human-robot interaction (HRI), gesture recognition for HRI, development of a generalized information processing algorithm, automation of collaboration between the working hand and the person, which should be guided by the movement of the hand using the camera. The models used: gesture recognition model for human-robot interaction. The following results are obtained: gesture recognition algorithm obtained during human-robot interaction, the functional structure of the system is obtained. The scientific novelty of the obtained results is as follows: improved model for recognizing objects of complex shape using the Kinect camera, proposed a special algorithm for gesture recognition in HRI, formed a series of gesture recognition tasks in HRI.

Keywords: Industrial robot · Human-robot interaction · Gesture recognition · The algorithm of information processing

1 Introduction

Nowadays robots are an indispensable attribute of modern human society. Despite the diversity of robots, their use in industry has determined the rapid growth and development of the technical direction, therefore, the park of industrial robots is the most widespread and in demand today.

Industrial robot (IR) or industrial manipulator (IM) is a mechanical arm that has an actuating mechanism at the end. Often, the manipulator copies the human arm, which has a shoulder, elbow, wrist, and hand, but for more complex technological operations, a few more links may be added.

Industrial robots are used to replace human labor in the process of performing various major and minor technological operations. It also eliminates the possibility of injury of varying severity, if the robot performs operations in hazardous working conditions for humans. Manipulators do an excellent job with monotone types of work, and surpass human labor in the quality. Automated production in which robots are used, allows solving problems with a wide range of products by reprogramming [1].

M. Nechyporuk et al. (Eds.): *Integrated Computer Technologies in Mechanical Engineering*, AISC 1113, pp. 126–138, 2020.
https://doi.org/10.1007/978-3-030-37618-5_12

The history of mechanics is rich in examples, which testify to the constant desire of man to create mechanisms and devices similar to living beings.

The process of controlling the actions of the automatic manipulation robots can take place with the participation and without the direct participation of a person. The manipulation robot consists of a manipulator, actuators, devices intended to replace feelings, communication devices with the operator and the computer. The manipulator simulates the movements of the human hand and is a multi-faceted open mechanism with one moving rotary and translational kinematic pairs. The number of degrees of the mobility of manipulators varies in the range from 3 to 10. The function of the brush in the manipulator performs the so-called probe, the design of which involves performing operations with a certain type of objects of manipulation.

The work is devoted to the realization of finding the object, recognizing it (its geometric shape) and moving the manipulator to the corresponding coordinates for capturing the object with a probe (pincers) and moving it to the given coordinates. The solution of this problem allows implementing a system of automatic operation of industrial robotic manipulators [2].

Also based on operator recognition, the Kinect camera is used to implement a manipulator control; it stops the programmed actions and waits for the instructions from the operator. The instructions from the operator are transmitted to the manipulator using the principles of gesture recognition.

2 Main Focus of the Chapter

Human-Robot Interaction. Nowadays, industrial manipulation works are used in various fields of industry, which administrate a wide range of technological tasks. The main type of robot manipulation systems are mechanical manipulators.
Human-robot interaction is the study of interactions between humans and robots. It is often referred as HRI by researchers.

Despite the fact that human workers have sufficiently advanced problem-solving skills and sensory-motor abilities, they have limited strength and accuracy [1]. Nevertheless, robotic systems have high resistance to fatigue, speed, repeatability and better performance, but they are limited in flexibility [2]. HRI can free human workers from difficult tasks and will enable communication channels between people and robots to increase their overall effectiveness [4].

Ideally, a bundle (team) HRI should work just like a human collaborative team [3]. To create an effective HRI team, we can analyze human-to-human groups as examples. In human teamwork and collaboration, there are two theories: joint intention theory [4] and situated learning theory [3, 5].

Gesture Recognition for HRI. Gesture is one type of communication methods. Facial expressions, hand gestures and body postures are effective communication channels in human-human collaboration [6]. Gestures can be divided into three types [7]:

- body gestures: actions or movements of the whole body;
- hand and arm gestures: arm posture, hand gestures;

- head and facial gestures: nodding or shaking head, winking.

To recognize gestures in the context of HRI, we need to develop an information processing algorithm for our system [4].

Imagine a generalized information processing algorithm consisting of four stages [8]:

Stage 1: Data collection.
Stage 2: Data analysis.
Stage 3: Decision making.
Stage 4: Response.

A fairly comprehensive review and analysis of data collection technologies was given in [9], Table 1 represents the result:

Table 1. Advantages and disadvantages of different sensor technologies

Sensor technology	Advantages	Disadvantages
Marker	Low computational workload	Markers on user body
Single camera	Easy setup	Low robustness
Stereo camera	Robust	Computational complexity, calibration difficulties
ToF camera	High frame rate	Resolution depends on light power and reflection
Microsoft Kinect	Fast emerging, software support for body gesture recognition	Cannot be used for hand gesture recognition over 2 m
Sensor technology	Advantages	Disadvantages
Glove	Fast response, precise tracking	Cumbersome device with a load of cables
Band sensor	Fast response, large sensing area	Band needs to contact with human body
Non-wearable	Avoid contact with human body	Low resolution, technology not mature enough

Based on the generalized algorithm, a specific gesture recognition algorithm in HRI was proposed. As shown in Fig. 1, there are six basic steps involved in the process of gesture recognition for HRI (besides checking the camera functioning):

- receiving data from the camera.
- identification of found objects.
- finding a gesture among them.
- gesture classification.
- determination of the distance to the object and its position.
- the reaction of the robot manipulator.

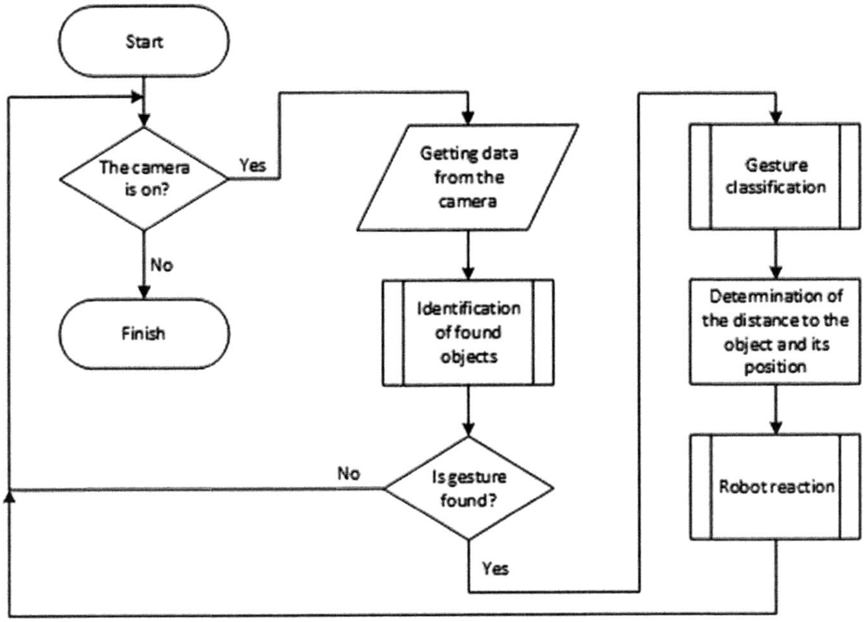

Fig. 1. The algorithm of the gesture recognition in the human-robot interaction

Robots have been successfully used in different stages of production for decades. Robots replace a person when performing routine, energy-intensive, dangerous operations. Machines do not get tired, they do not need pauses on rest, water and food. The robots do not require wage increases and are not members of trade unions. As a rule, industrial robots do not possess artificial intelligence. Typical is the repetition of the same moves of the manipulator on the hard program. Great success has been achieved, for example, in the use of robots on conveyors of automobile factories. Already there are plans for the automotive industry, where all the processes of assembling cars and transporting semi-finished products will be carried out by robots, and people will only control them. In 2016 1.8 million industrial robots were used all over the world, it is forecasted that by 2020 their number will exceed 3.5 million units [5].

Before writing a software for a physical robot, it is recommended to try it on simulators such as Gazebo, V-Rep ROS, etc., regardless of whether there is a physical robot in stock [10, 11]. This allows to write the final version of the program, as well as set it up offline, before trying it on a physical robot. One of the most popular robotics applications is the 3D modeling of the robot itself, as well as its surroundings. Such programs allow the simulator to program a virtual robot that is capable of reproducing the movements and the logic of physical robot in its working environment.

Functional Structure of the System. The main task of this work is to automate the joint work between the work-arm and the person, which should be guided by the movement of the hand with the help of the camera.

Figure 2 describes the functional structure of the system considering a risk-oriented design approach as described in [12], which consists of two levels of decomposition.

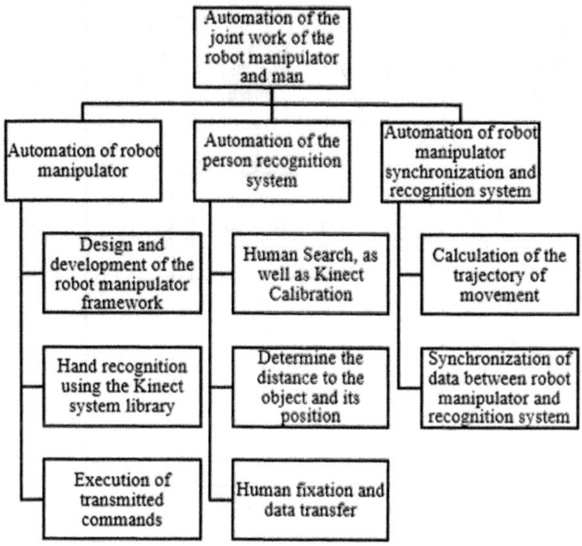

Fig. 2. Functional structure of the system

The functional structure of the system is considered in the general plan, from the moment when the system enters the indicators from the camera, until the system performs calculations and synchronization of data between the robot manipulator.

Automation of Manipulator Work. Figure 3 describes the "Automation of robot manipulator" block.

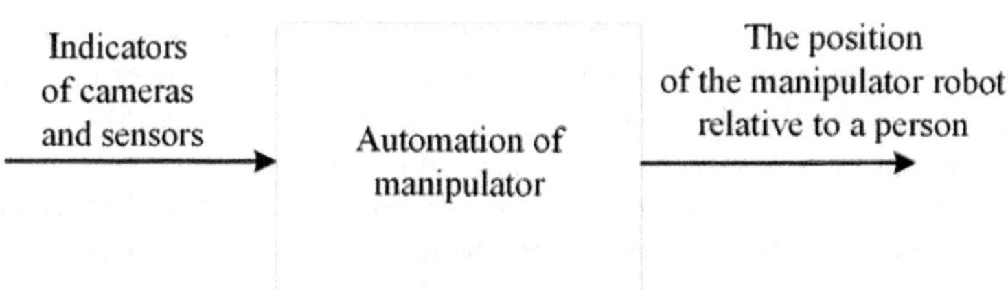

Fig. 3. The "Automation of robot manipulator" block

In the "Automation of robot manipulator" subtask the developer must analyze the configuration of the machine, manipulator or recognition system, as well as the motherboard that will be responsible for processing the data. The developer analyzes the possibility of implementing any other program element. If the developer can not programmatically implement any element, he communicates with the project manager.

The project manager decides whether it is possible to change the item and how it is. The optimization of software for one or another motherboard is often a problematic element [6].

As a result of solving this problem, it is necessary to identify the main elements that affect the automation of the manipulator:

- design and development of the manipulator construction;
- recognition of the human hand movement;
- adjustment of the manipulator position.

For the "Automation of robot manipulator" subtask the input data is obtained indicators from the sensors, the camera and the reduction to the required format. The initial data will be executed by the command and coordinates of the operator (person) movement.

Automation of the Wheel Robot. Figure 4 describes the "Automation of the recognition system of the operator's hand position in space" block.

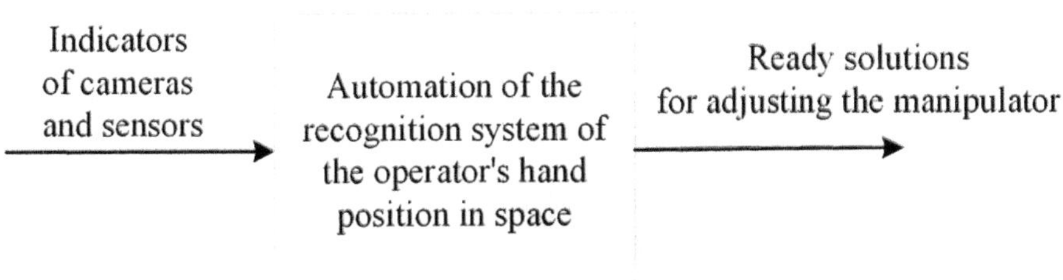

Fig. 4. The "Automation of the recognition system operation" block

In the subtask "Automation of the recognition system operation" the developer must determine: what type of robot manipulator and what functionality will perform in pairs with the recognition system.

When the task, that the recognition system will perform, should be formed, the developer is engaged in calculating the choice of algorithm that will be used by the system when recognizing a person and his position in the space, as well as the algorithm of behavior in extreme situations (when a signal is lost with a manipulator, several objects (people) are found). It is also necessary to determine how the recognition system will calculate the distance to the objects, as well as the algorithm of behavior.

As a result of the solution to this problem, it is necessary to analyze the functionality and type of the manipulator, and perform the following steps:

- movement of the manipulator;
- definition of the distance to the object.

For the "Automation of the recognition system operation" subtask the input data will be derived from sensors, cameras and then it will be reduced to the required format. Output data will be the coordinates of the manipulator position.

2.1 The Process of Robot Manipulator Assembling

To begin with, the materials for the assembly were purchased, as well as a frame for the robot manipulator was printed on a 3D printer. First the details, that will be the basis of the turning mechanism, have been taken. A bearing has been put in one of the designated holes to ensure the rotation of two parts relatively to each other with the correct load distribution and maximum reduction in friction. Next, we needed to consolidate these two details. Then the base plate should be fixed to in the rotary mechanism [7]. This base plate serves as a base for fixing the stepper motor. As the platform for fixing the stepper motor is ready, it should be fastened on a rotary base.

We have performed similar operations with other platforms and stepping motors and received the following result, which is shown in Fig. 5.

Fig. 5. Ready-made base with fixed stepping motors

After the base has been assembled, the pipe with prepared metal tubes should be fastened and fixed to the swivel mechanism. That leads to the following result (Fig. 6).

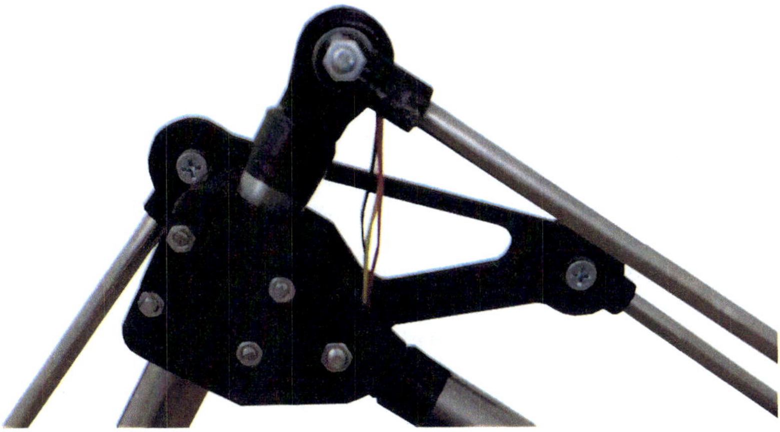

Fig. 6. Swivel mechanism with fixed metal tubes

Next the probes for robot manipulator should be assembled, the final result can be seen in Fig. 7.

Fig. 7. Assembly probe for robot manipulator

When the probe is assembled, it can be fixed on the manipulator. After this step, the assembly can be called finished, the robot manipulator is assembled.

3 The Interaction of the Manipulator and Kinect

The way of interaction of the manipulator and Kinect recognition camera will be described next.

An important part of adaptive robots is their developed software, designed for processing information coming from external and internal sensors and operational changes to the motion program. Due to the ability to perceive changes in the external environment and adapt to existing conditions, adaptive work can be manipulated with non-oriented parts of arbitrary shape and produce assembly operations [8].

The copying control of biotechnical manipulation robots is carried out with the help of a human operator who moves the master device, and the manipulator repeats these movements simultaneously in all degrees of mobility of the device.

Kinect's touchscreen gaming controller will be used to get information about human hand motion. In order to transfer these movements to the manipulator, a software product will be developed in the Python programming language, because this language has the ability to interact with the ROS library, and is compatible with Kinect.

As an electronic computer, the minicomputer Raspberry Pi 3 is used. As a robot arm of the manipulator, Robot Arm MK2 is used with some developments and modifications.

The manipulator layout management system is developed based on Arduino, which connects to a personal computer via the COM port, through which the board receives power and control signals for controlling the servomotors of the manipulator [9].

Servo Library is a library of the Arduino Controller and provides a set of functions for managing servo drives. Standard servo drives can rotate the drive to a certain angle from 0 to 180° normally. Some servo drives enable full rotation at a given speed. The Servo library lets simultaneously manage 12 servo drives on most Arduino boards and the 48 on Arduino Mega. If the library is used for controllers other than Mega, the 9-th and 10-th outputs are in PWM (pulse-width modulation) mode, even if the drive is not connected to these pins. With the Mega board, up to 12 servo drives can be used without losing the PWM functionality. When using Mega for controlling from 12 to 23 servo drives it will not be possible to use 11-th and 12-th outputs for PWM.

The Stepper library provides a convenient interface for managing bipolar and unipolar stepper motors. In order to control the stepper motor, depending on its type (bipolar or unipolar) and the chosen connection method, some additional electronic components will be needed.

In order to realize all the features of the proposed construction, it is necessary to solve a number of tasks, such as:

1. Synthesis and analysis of the kinematic model of the manipulator, considering the peculiarities of the kinematic pairs formed by the pairs of segments.
2. Synthesis and analysis of dynamic, complete and explicit models of the manipulator.
3. Development of a multichannel data collection and processing system in real time for the information provision of manipulator management tasks and experimental study of the dynamic properties of the manipulator.

4. Development of engineering techniques for selecting the optimal parameters of the design of the manipulator, taking into account the solvable problems.
5. The synthesis of the control system of the manipulator.

Consider the first task of the synthesis and analysis of the kinematic model of the manipulator.

The manipulator has a complex geometry, so we will link the reference frame with different mechanical components of the manipulator and describe the relationship between these frameworks.

To study the processes in the manipulator, at first kinematic model should be made. This model would relate the movement of its links with the position of the center of the probe in space.

In a three-dimensional space, to indicate the location of a point in space, it is enough to accurately determine its coordinates in a fixed coordinate system. When describing the position of a solid body, an associated coordinate system is associated with it. Three parameters that specify the orientation of the axes of the connected coordinate system in relation to the motionless (Euler's corners) and the three coordinates of the beginning of the associated coordinate system. Thus, the main task of describing the kinematic characteristics of a manipulator is the transformation between its own and the chosen working coordinate systems.

The calculation of direct problems of kinematics is an important first step to using a new robot, a kind of familiarity with it. The direct problem of kinematics of manipulators is formulated as follows: the given kinematic scheme of the manipulator and at a certain moment of time are known values of general coordinates, which determine the position of all parts of the manipulator in relation to one. It is necessary to determine the position and orientation of capturing the last link of the manipulator in the reference frame associated with the base.

The investigated two-stage mechanism (Fig. 8) is a two-leg mechanical manipulator with two rotating pairs.

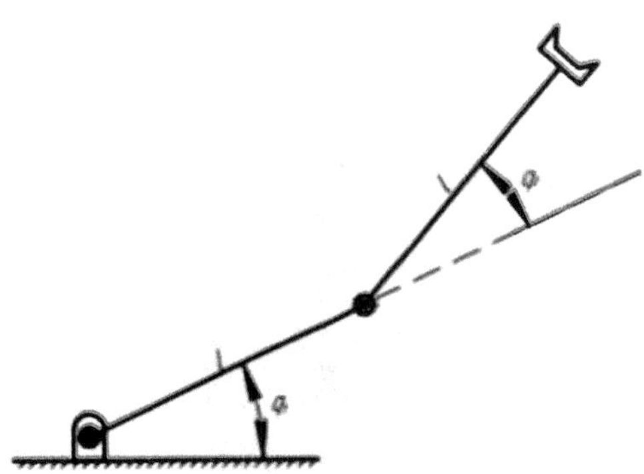

Fig. 8. Two-stage mechanism

The result of solving all the above tasks is the system presented in Fig. 9, which includes a manipulator control system (the program code is implemented with the Python programming language), robot arm of the Robot Arm MK2 manipulator, as well as the Kinect touch proximity gaming controller, as a microcomputer Arduino Duo is used.

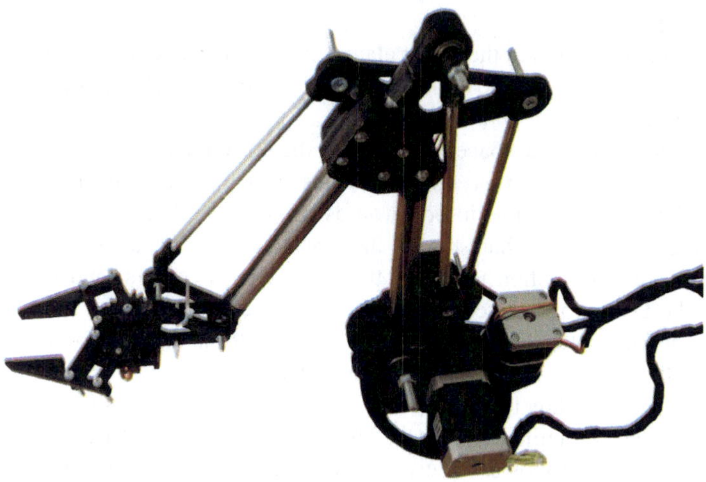

Fig. 9. The manipulator

The functionality of the developed robot manipulator includes the movement in the automatic mode, as well as in the mode of receiving commands: the coordinates of the manipulator position will come from the recognition system of the operator's hand movement. For the wheel robot it is possible to send signals to move forward, back, turn left and right.

Fig. 10. The calibrated manipulator operator

The motion recognition system will use the coordinates of the operator's hand who passed the first calibration. Other operators will not be able to influence on the position calculation (Fig. 10).

4 Conclusion

As a result of the work done, the main problems encountered when controlling the robot using the Kinect sensor were identified, such as:

- overlapping of operator's controls during operation;
- loss of productivity when sending control actions;
- inaccuracy of control due to the limited number of degrees of mobility;
- the difficulty of determining the angle corresponding to the standard position of the servomotor.

The advantages of this type of control are:

- the possibility of forming control commands for the robot in real time;
- expansion of the possibilities of interaction of the robot with the external environment (manipulation with objects);
- the ability to record certain movements and call them with voice commands;
- low cost of necessary equipment for management.

References

1. Krüger, J., Lien, T., Verl, A.: Cooperation of human and machines in assembly lines. CIRP Ann.-Manuf. Technol. **58**, 628–646 (2009)
2. Green, S.A., Billinghurst, M., Chen, X., Chase, G.: Human-robot collaboration: a literature view and augmented reality approach in design. Int. J. Adv. Robot. Syst. **5**, 1–18 (2008)
3. Vygotsky, L.S.: Mind in Society: The Development of Higher Psychological Processes. Harvard University Press, Cambridge (1980)
4. Cohen, P.R., Levesque, H.J.: Persistence, intention, and commitment. In: Reasoning About Actions and Plans, pp. 297–340 (1990)
5. Breazeal, C., Brooks, A., Gray, J., Hoffman, G., Kidd, C., Lee, H., Lieberman, J., Lockerd, A., Mulanda, D.: Humanoid robots as cooperative partners for people. Int. J. Humanoid Robots **1**(2004), 1–34 (2004)
6. Khatib, O.: Real-time obstacle avoidance for manipulators and mobile robots. In: Autonomous Robot Vehicles, pp. 396–404. Springer, New York (1986)
7. Saracen, M.J., Carrano, A.W., Henderson, T.D.: U.S. Patent No. 8,160,205. U.S. Patent and Trademark Office, Washington, DC (2012)
8. Fumagalli, M., Naldi, R., Macchelli, A., Forte, F., Keemink, A.Q., Stramigioli, S., Carloni, R., Marconi, L.: Developing an aerial manipulator prototype: physical interaction with the environment. IEEE Robot. Autom. Mag. **21**(3), 41–50 (2014)
9. Yamamoto, Y., Yun, X.: Modeling and compensation of the dynamic interaction of a mobile manipulator. In: Proceedings of the 1994 IEEE International Conference on Robotics and Automation, pp. 2187–2192. IEEE, May 1994

10. Kritsky, D.N., Ovsiannik, V.M., Pogudina, O.K., Shevel, V.V., Druzhinin, E.A.: Model for intercepting targets by the unmanned aerial vehicle. In: Palagin, A., Anisimov, A., Morozov, A., Shkarlet, S. (eds.) Mathematical Modeling and Simulation of Systems, MODS 2019. Advances in Intelligent Systems and Computing, vol. 1019. Springer, Cham (2020)
11. Kritskiy, D., Alexander, K., Koba, S., Druzhinin, E.: Increasing the reliability of drones due to the use of quaternions in motion. In: Proceedings of 2018 IEEE 9th International Conference on Dependable Systems, Services and Technologies, DESSERT 2018 (2018)
12. Kritsky, D.N., Druzhinin, E.A., Pogudina, O.K., Kritskaya, O.S.: A method for assessing the impact of technical risks on the aerospace product development projects. Advances in Intelligent Systems and Computing, vol. 871, pp. 504–521. Springer, Heidelberg (2019)

Influence of Electromagnetic Radiation of Different Quantum Energy on Dielectric Properties of Composites Based on Crystals CdZnTe and ZnSe

O. M. Chugai[1], S. V. Oliyník[1(✉)], O. O. Voloshin[1], S. M. Galkin[2], L. Iu. Sidelnikova[2], and O. O. Sosnytska[2]

[1] National Aerospace University "Kharkiv Aviation Institute", Ministry of Education and Science of Ukraine, Chkalova 17, Kharkiv 61070, Ukraine
oleg.chugai@khai.edu, oleynick1981@gmail.com
[2] Institute for Scintillation Materials, National Academy of Sciences of Ukraine, Nauky ave. 60, Kharkiv 61072, Ukraine

Abstract. The features of changes in the real and imaginary parts of the complex permittivity of the samples with CdZnTe crystallites in the frequency range 10–10^3 Hz of the electric field in the absence of exposure to electromagnetic radiation are obtained. It has been shown that composites without these crystals have a higher diffuse reflection coefficient. Under the influence of non-monochromatic light, an increase in the changes in the complex permittivity of the samples with increasing frequency of the measuring field was detected. Moreover, the changes are larger for the samples with the largest crystallite size. The low inertia of changes in the dielectric parameters of the samples under the influence of non-monochromatic illumination is established. Samples with ZnSe (Al) crystallites are characterized by significant changes in the complex permittivity under the influence of non-monochromatic illumination in the frequency range 10^3–10^4 Hz. Negative changes of this parameter are especially unusual. The peculiarities of the spectral dependences of the constituents $\Delta\varepsilon'_{ef}$ and $\Delta\varepsilon''_{ef}$ complex permittivity of the structures, measured at a light intensity of 10 mW/cm^2, have been established. The changes in the magnitudes of the increments of the real and imaginary components of the complex permittivity under the action of X-Ray radiation with a dose rate of 200 mR/h at different frequencies of the electric field are investigated and their characteristic features are determined. An analysis of the results from the point of view of non-equilibrium electronic processes in semiconductors is presented.

Keywords: Dielectric properties · Composite materials · Complex permittivity

1 Introduction

Despite the fact that composite materials (CM) have long been known and have been widely used in various fields of technology, these materials still attract the attention of scientists and practitioners. Moreover, in recent years, there has been a rapid expansion

M. Nechyporuk et al. (Eds.): *Integrated Computer Technologies in Mechanical Engineering*, AISC 1113, pp. 139–149, 2020.
https://doi.org/10.1007/978-3-030-37618-5_13

of types of composites and areas of their practical application. This is due to the ability to control a wide range of physical properties of the compound-filler system by selecting both the composition of its components and the geometric characteristics of the filler. In this regard, of particular interest are the CdZnTe and ZnSe crystals, which are widely used in ionizing radiation detectors. It should be noted that the scope of these crystals can be significantly expanded by creating composites based on them. Thus, it is promising to develop and manufacture photosensors based on composite materials using CdZnTe and ZnSe crystals to convert light characteristics into an electrical signal. Note that these sensors do not require high voltage or high current. In view of the above, it is relevant to study the electronic processes associated with changes in the real and imaginary parts of the complex dielectric permittivity of composites based on CdZnTe and ZnSe crystals and an insulating matrix under the influence of light or ionizing radiation.

2 Sample Preparation

The technology of manufacturing the samples [1] includes: obtaining fractions of $A^{II}B^{VI}$ (CdZnTe), crystals acting as filler, preparing the equipment to provide the desired shape and size CM, preparation of filler epoxy and polymerization of composites.

Photosensitive crystals were obtained from predetermined fragments of ZnSe and CdZnTe crystalline ingots, which exhibit high photoconductivity.

Used crystallites, the size of which was d_1 = 130 mkm and d_2 = 370 mkm. To obtain crystallites of the desired size, the fragments of the ingot were ground in a porcelain mortar and then separated using specially made straps, the parameters of which were determined by a PЭM – 106 raster electron microscope.

The equipment was made of fluoroplast and allowed to produce several samples simultaneously in the form of flat parallel plates. Remains of contamination were removed from the equipment before use, all surfaces were degreased with acetone, and then a layer of grease was applied to prevent adhesion of the samples to the equipment.

Cold filler epoxy resin LH288, was used as a filler, which has low flammability but high chemical resistance. Such a filler does not need to be heated for polymerization, which eliminates the thermal degradation of the photoelectric properties of ZnSe and CdZnTe, crystals, and the contamination of crystallites by such a filler is minimal.

After polymerization of the mixture, plates were obtained from which rectangular specimens were produced with a side size up to 10 mm and a thickness up to 5 mm.

3 Results of the Research of Dielectric, Photodielectric and Optical Properties of CM on the Basis of CdZnTe Crystals and Epoxide Resin

Epoxy resin as a matrix for photosensitive composites is of great interest because it allows these composites to be easily integrated into a variety of technical structures, including aerospace.

The Fig. 1 shows the frequency dependences of the real and imaginary parts of the complex dielectric constant of composites based on CdZnTe crystals of different characteristic sizes. It is seen that the introduction into the matrix of epoxy resin crystallites of the specified composition significantly affects the studied values. Noteworthy is the characteristic for the region of rapid decrease of this value in the frequency range 10 ... 100 Hz, which is probably related to the contribution to the dielectric loss of ohmic conductivity.

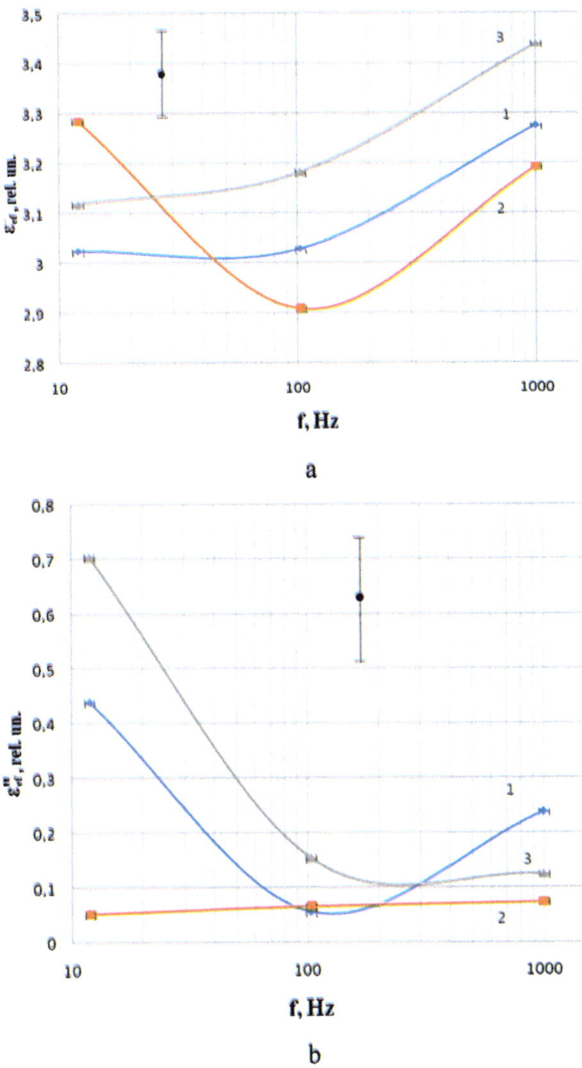

Fig. 1. Frequency dependences of the real (a) and imaginary (b) parts of the dielectric permittivity of samples containing no CdZnTe crystallites and with crystallites of sizes d_1 and d_2 (1–3 respectively).

The specular and diffuse reflection spectrum by matrix and composites were obtained. These dependencies for composites are characterized by a higher diffuse reflection coefficient than for the matrix. However, such a reflection coefficient is more unevenly distributed across the spectrum. This is obviously due to the multiple reflection of light rays from the crystallites, which increases the proportion of light energy absorbed by the composite.

The measurement of the change under the influence of non-monochromatic illumination of the dark values of the imaginary and the real part of the complex dielectric permittivity of the samples for different frequencies of the electric field gave the following results (see Table 1). For epoxy resins without filler (hereinafter referred to as the matrix), these changes are minor. As the frequency of the electric field increases, the changes in these parameters increase. This is most pronounced in composites with a larger crystallite size.

Table 1. Change in the complex dielectric permittivity of the samples based on CdZnTe crystallites under the influence of non-monochromatic light.

Samples	Frequency, Hz							
	10^2		10^3		10^4		10^5	
	$\Delta\varepsilon'_{ef}$	$\Delta\varepsilon''_{ef}$	$\Delta\varepsilon'_{ef}$	$\Delta\varepsilon''_{ef}$	$\Delta\varepsilon'_{ef}$	$\Delta\varepsilon''_{ef}$	$\Delta\varepsilon'_{ef}$	$\Delta\varepsilon''_{ef}$
Without filler	≈0	≈0	0.03	−0.0001	0.001	−0.0009	−0.03	−0.0004
Fine fractions	≈0	≈0	0.13	−0.0044	0.26	−0.0027	0.51	−0.0015
Large fractions	≈0	≈0	0.23	−0.0134	0.55	−0.0039	0.55	−0.0005

Measurements of the inertia of changes ε' and $tg\delta$ in both the matrix and the composites found that these parameters of the matrix do not change significantly over time under the influence of short-term illumination of the sample with non-monochromatic light. In Fig. 2 shows the time dependences of the dielectric parameters for a matrix without a CdZnTe filler. Gray circles represent dielectric parameters measured without exposure to light. The red squares indicate the permittivity and the dielectric loss tangent of the samples when exposed to a step of non-monochromatic light. We see that the changes and are not significant in the case of the matrix, which is consistent with the data in Table 1.

In the case of composites, such changes not only exceed the level of fluctuations in magnitudes, but also occur over a period of about one second. Such a small inertia of changes in the investigated values indicates the prospect of creating sensors based on such composites.

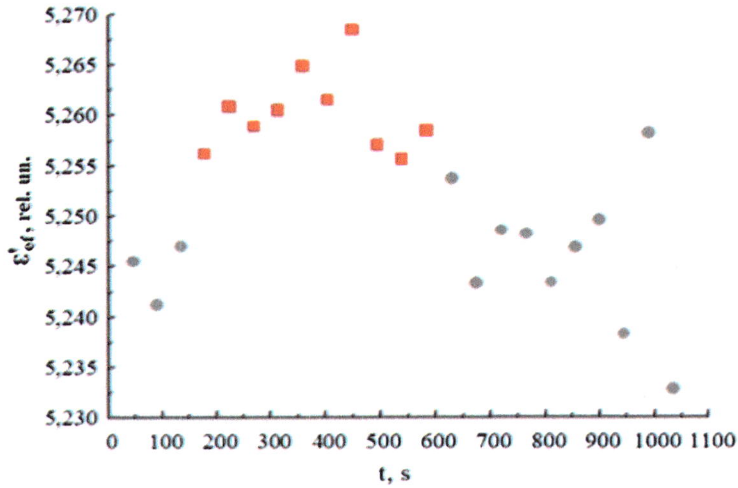

a

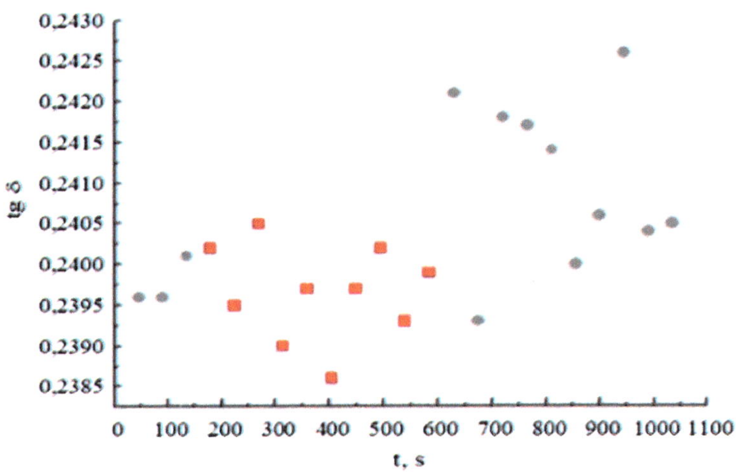

b

Fig. 2. Time dependences of the dielectric constant (a) and dielectric loss tangent (b) of samples containing no CdZnTe crystallites under the influence of non-monochromatic step of light.

4 Results of the Investigation of the Influence of Non-monochromatic Light and Ionizing Radiation on the Dielectric Properties of CM on the Basis of ZnSe and Caoutchouc

Composites based on caoutchouc and ZnSe crystals doped with aluminum were investigated. The concentration of alloying atoms was 0.005 at. %. It should be noted that these composites are characterized by intense luminescence [1], especially under the influence of X-ray radiation. Therefore, in view of the disclosure of the mechanism of influence of electromagnetic radiation on electronic processes in CM, it was advisable to investigate the changes of complex dielectric permittivity not only under the influence of non-monochromatic light but also under the influence of X-ray radiation.

The changes of the real and imaginary components of the complex dielectric permittivity of these composites under the influence of non-monochromatic light for different frequencies of the electric field are investigated (see Table 2). The structures were created using ZnSe crystallites of different conductivity. The crystallites of ZnSe (Al) – 3 samples had the lowest electrical conductivity, and the crystallites of ZnSe (Al) – 2 samples had the highest. Non-monochromatic intensity photo excitation was used 100 mW/cm2. The field strength in the sample was 30 mV/cm. The effect of light caused a significant change in the complex dielectric permittivity in all samples tested. Moreover, the greatest absolute values of these changes reached in the frequency range of the electric field 103 – 104 Hz for the sample ZnSe (Al) – 2. Depending on the frequency of the field, the changes have both positive and negative sign. In this regard, we note that the negative values of changes in the constituents of the dielectric parameters when exposed to light – an unusual and poorly studied phenomenon [2, 3].

Table 2. Change in the complex dielectric permittivity of the samples based on ZnSe crystallites under the influence of non-monochromatic light.

Samples	Frequency, Hz					
	10^3		10^4		10^5	
	$\Delta\varepsilon'_{ef}$	$\Delta\varepsilon''_{ef}$	$\Delta\varepsilon'_{ef}$	$\Delta\varepsilon''_{ef}$	$\Delta\varepsilon'_{ef}$	$\Delta\varepsilon''_{ef}$
ZnSe(Al) – 1	0.010	0	0.099	0.041	0.30	0.21
ZnSe(Al) – 2	2.58	0.67	−2.00	−1.30	0.47	0.27
ZnSe(Al) – 3	0.044	−0.0017	0.010	0.003	0.032	0.120

It is established that in the range of white light intensity 10–100 mW/cm^2 changes in values $\Delta\varepsilon'_{ef}$ and $\Delta\varepsilon''_{ef}$ are proportional to the light intensity.

The peculiarities of the spectral dependences of the constituents $\Delta\varepsilon'_{ef}$ and $\Delta\varepsilon''_{ef}$ dielectric permittivity of the structures, measured at a light intensity of 10 mW/cm^2, have been established (Fig. 3). The spectral dependences of these quantities are proportional to each other. In these dependences, a maximum of 700 nm is observed. We emphasize that, according to the literature data [4], for the ZnSe crystals in this region of the spectrum, the maximum photoconductivity or photodielectric effect was not observed previously. Therefore, it can be argued that the creation of composites based on these crystals and caoutchouc allows you to change the spectral distribution of photosensitivity of materials.

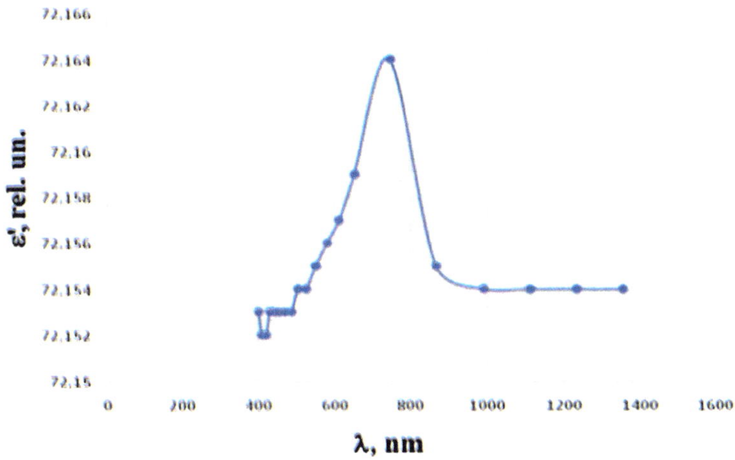

a

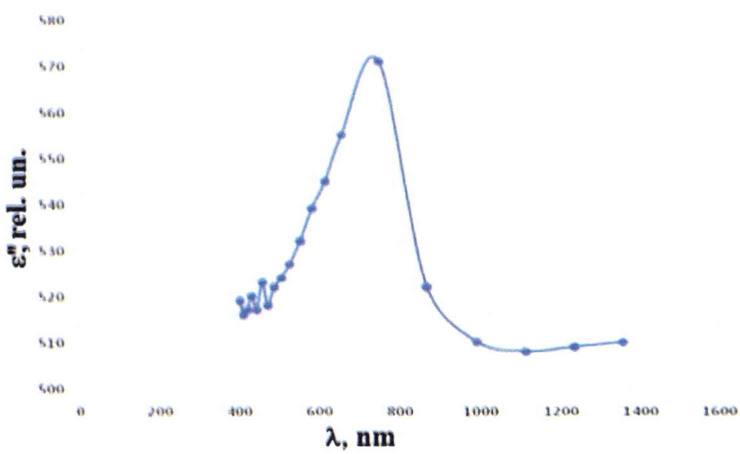

b

Fig. 3. Spectral dependences of the real (a) and imaginary (b) parts of the dielectric permittivity of samples with crystallites of ZnSe(Al), f = 1 kHz.

Changes in the magnitudes of the increments of the real and imaginary components of the complex dielectric permittivity of the studied structures under the action of X-ray radiation with a dose rate of 200 mR/h at different frequencies of the electric field are investigated and their characteristic features are determined (see Table 3).

Table 3. Change in the complex dielectric permittivity of the samples based on ZnSe(Al) crystallites under the influence of X-ray radiation.

Samples	Frequency, Hz					
	10^2		10^3		10^4	
	$\Delta\varepsilon_{ef}^{'}$	$\Delta\varepsilon_{ef}^{''}$	$\Delta\varepsilon_{ef}^{'}$	$\Delta\varepsilon_{ef}^{''}$	$\Delta\varepsilon_{ef}^{'}$	$\Delta\varepsilon_{ef}^{''}$
ZnSe(Al) – 1	−0.008	−0.004	0.00995	−0.0001	−0.0015	0
ZnSe(Al) – 2	−0.0032	−0.002	0.40	0.0050	0.0016	0.0001
ZnSe(Al) – 3	0	0	0.27	0.013	0.015	0.0004

In our view, the comparative analysis of non-monochromatic light and X-ray exposure data is of most interest, since the conditions for obtaining these data differ only in the energy of the quantum that influenced the sample. The comparison made it possible to find the following features:

1. maximum values of $\Delta\varepsilon_{ef}^{'}$ and $\Delta\varepsilon_{ef}^{''}$ for both types of radiation are observed on the sample ZnSe (Al) – 2 at an electric field frequency of 1 kHz;
2. for light and X-ray quantum, these changes have the opposite character – the changes are negative in the first but positive in the second case;
3. when exposed to the X-ray pattern, small changes in the studied values, as a rule, have a negative sign.

5 Analysis of the Results of the Experiment

The investigated composites can be considered as a set of dielectric – semiconductor – dielectric structures randomly distributed in the sample volume. Therefore, it is important to dwell on the dielectric and photoelectric properties of the structure of the type.

In general, for semiconductor-based structures containing deep impurity levels distributed in the normal direction to the structure, electron concentrations n(x), holes p (x), and electric field strengths are described by a system of continuity equations and Poisson equations [5]

$$\frac{\partial n}{\partial t} + \frac{\partial q_n}{\partial x} = G - R_n. \tag{1}$$

$$\frac{\partial p}{\partial t} + \frac{\partial q_p}{\partial x} = G - R_p. \tag{2}$$

$$\frac{\partial E}{\partial x} = \frac{4\pi e}{\varepsilon}[p - p_* - n + n_* - N_t(f - f_*)]. \tag{3}$$

These equations are usually solved in a diffuse-drift approximation. In this case, the density of carrier flows q_n and q_p have a standard appearance

$$q_n = -D_n \frac{\partial n}{\partial x} - \mu_n En. \tag{4}$$

$$q_p = -D_p \frac{\partial p}{\partial x} - \mu_p Ep. \tag{5}$$

The rate of generation of electrons and holes under the action of external radiation is expressed by the ratio

$$G = \alpha I_0 \exp(-\alpha x), \tag{6}$$

in which α – is the light absorption coefficient.

When solving Eqs. (1)–(3), taking into account relations (4)–(6), they use some or other assumptions for simplification. Thus, in [6] it was assumed that the semiconductor is high-ohmic and does not contain localized (usually impurity) states of carriers in the band gap. Note that in reality crystals always contain these states. Moreover, their energy spectrum is quite wide. Another significant assumption in solving Eqs. (1)–(3) is the neglect of the electric field and, as a consequence, the localized states of the carriers near the semiconductor-dielectric contact [7]. The reason for this is the use in the study of the structure of the voltage, which exceeds the difference between the values of the electronic affinity of the contact materials. Obviously, this assumption is unfair to the materials we have studied, which can be considered as a set of contacts of a dielectric – semiconductor in an insulating matrix. In this regard, it should be noted that the analysis of charges and surface states in the area of these contacts is a separate task [8]. Finally, another condition that has a decisive influence on the behavior of non-equilibrium charge carriers in an external electric field is the relationship between the lifetime τ_r and the spatial distribution of carriers of opposite signs $\tau_s \sim e\ell^2/(\mu E)$ [9]. In the latter ratio ℓ – the characteristic length depends on the dominant scattering mechanism of the carriers. If $\tau_r \le \tau_s$, then the charge carriers of opposite signs are separated in space slightly.

In view of the above, we consider it expedient to explain, in explaining the features of the photo-dielectric effect in the investigated composites, the idea of the existence in the sample volume of the many electric dipoles formed in the volume of the crystallites themselves and at the boundaries between the crystallites and the insulating matrix. If we denote the total dipole moment in unit volume of the sample of dipoles of the first grade as $\sum_{i=1}^{N} \vec{P}_{vi}$, the second grade as $\sum_{j=1}^{2N} \vec{P}_{sj}$, then for $\Delta\varepsilon'$ there will be a fair ratio $\Delta\varepsilon\prime_{ef} \sim \frac{1}{\varepsilon_o E}[\sum_{i=1}^{N} \vec{P}_{Vi} + \sum_{J=1}^{2N} \vec{P}_{sj}$, in which - electric steel. In turn, the dipole electric moment created by the non-equilibrium charge carriers in the volume of single crystallites is expressed as

$$\sum_{i=1}^{N} \vec{P}_{vi} = \Delta q'_{+} (\vec{r}_{c+} - \vec{r}_{c-}), \tag{7}$$

where $\Delta q'_+$ – is the charge of the positive sign carriers defined by the expression

$$\Delta q'_+ = \int \rho'_+ dV, \tag{8}$$

in which ρ'_+ – the bulk density of the charge, the integration is performed throughout the crystallite except the contact layer. Similarly, the charge of negative sign carriers is also described. In expression (7), the parameters $\vec{r}_{c+}$ and $\vec{r}_{c-}$ – are the radius vectors of the centers of positive and negative charges determined by the relations

$$\vec{r}_{c+} = \frac{1}{\Delta q'_+} \int \rho'_+ \vec{r}_+ dV$$

and

$$\vec{r}_{c-} = \frac{1}{\Delta q'_-} \int \rho'_- \vec{r}_- dV,$$

in which $\vec{r}_+$ and $\vec{r}_-$ – are the radii-vectors of point charges in the elementary volume. In other words, we represent the electric dipole moment of a bulk charge as a macrodipole with the baseline $(\vec{r}_{c+} - \vec{r}_{c-})$.

Relatively, $\vec{P}_{sj}$ we can only assume that this value depends on the density of the surface states, their energy spectrum, as well as the difference in the affinity values of the contacting materials to the electron. Capturing these states of nonequilibrium electrons will naturally cause a change $\vec{P}_{sj}$ and, accordingly, $\Delta\varepsilon'_{ef}$.

Thus, in the case of composites the value $\Delta\varepsilon'_{ef}$ is determined by the set of parameters and characteristics on which the generation, recombination, drift, and diffusion of charge carriers in a semiconductor depend. Various combinations of these quantities and characteristics are manifested in the peculiarities of the change of magnitude $\Delta\varepsilon'$ with the frequency of the electric field in certain situations. In particular, the decrease $\Delta\varepsilon'_{ef}$ in frequency based on CdZnTe crystals with decreasing frequency can be explained by the capture of some of the non-equilibrium carriers from the volume of crystallites to the surface states and, as a consequence, by the preferential decrease in size $\sum_{i=1}^{N} \vec{P}_{vi}$. With this assumption, the data for composites based on ZnSe (Al) – 1 crystals are in good agreement. However, in the case of composites based on annealed and therefore more conductive crystals of the same composition, the nature of the frequency dependence is quite different. Moreover, at frequencies of 104 Hz, significant negative values of this value are observed. This feature of the photodielectric effect in composites can be attributed to the preferential capture of non-equilibrium carriers on "slow" surface states [10], which were formed due to the high-temperature annealing ZnSe (Al) crystals. The special role of x-rays in the change of $\Delta\varepsilon'_{ef}$ composites, we believe, is to change the localized states both in volume and on the surface of crystallites due to the effect of radiation "shaking" [11].

References

1. Litichevskyi, V., Galkin, S., Lalaiants, O., Voronkin, E., Breslavskiy, I., Tretiak, S., Kosinov, N.: Scintillation panels based on zinc selenide and oxide scintillators. Funct. Mater. **18**(3), 391–397 (2011)
2. Penin, N.A.: Negative capacitance in semiconductor structures. Fiz. Tekh. Poluprovodn. **30** (4), 626–633 (1996)
3. Penin, N.A.: Photo-capacitive effect in a monopolar metal-dielectric-semiconductor capacitor at low temperatures. Fiz. Tekh. Poluprovodn. **34**(5), 562–566 (2000)
4. Chugay, O.N., Komar, V.K., Puzikov, V.M.: Modification of the physical properties of AIIBVI crystals by the formation of a naturally ordered defective structure. ISMA, Kharkov (2008)
5. Zi, S.: Physics of Semiconductor Devices, vol. 1. Mir, Moscow (1984)
6. Reznikov, B.I.: Non-stationary photoelectric effect in high-resistance pure strongly biased metal-semiconductor and metal-insulator-semiconductor structures. Fiz. Tekh. Poluprovodn. **31**(8), 1003–1010 (1997)
7. Kovtonyuk, N.F., Misnik, V.P., Sokolov, A.V.: Sensitivity of dielectric-semiconductor structures to non-stationary light fluxes. Fiz. Tekh. Poluprovodn. **39**(11), 1336–1339 (2005)
8. Berman, L.S., Belyakova, E.I., Kostina, L.S., Kim, E.D., Kim, S.C.: Analysis of charges and surface states at the semiconductor-insulator-semiconductor interface. Fiz. Tekh. Poluprovodn. **34**(7), 814–817 (2000)
9. Shik, A., Ruda, H., Sargent, E.H.: Photoelectric phenomena in polymer-based composites. J. Appl. Phys. **88**(6), 3448–3453 (2000)
10. Karas, N.I.: The effect of oxide coating on negative photoconductivity in macroporous silicon structures. Optoelektron. Poluprovodn. Tekhn. **48**, 136–139 (2013)
11. Gavrilov, V.V., Kulikov, V.D., Chernov, S.A.: Radiation shaking and macroacoustic waves in alkali halide crystals. Fiz. Tverd. Tela **32**(4), 1124–1127 (1990)

Cutting Irregular Objects from the Rectangular Metal Sheet

Sergiy Plankovskyy[1], Yevgen Tsegelnyk[1(✉)], Olga Shypul[1], Alexander Pankratov[2], and Tatiana Romanova[2]

[1] National Aerospace University "Kharkiv Aviation Institute", 17 Chkalova Str., Kharkiv 61070, Ukraine
y.tsegelnyk@khai.edu

[2] A. Pidgorny Institute of Mechanical Engineering Problems of the National Academy of Sciences of Ukraine, 2/10 Pozharskogo Str., Kharkiv 61046, Ukraine

Abstract. A problem of cutting irregular objects from the domain (rectangular sheet) taking into account technological requirements is studied. As mathematical models of real objects two-dimensional phi-objects, bounded by circular arcs and line segments, are considered. Our objects allow free translations and rotations within a domain. We reduce a collection of technological requirements to geometrical constraints (such as minimal allowable distances, prohibited areas, range of the possible object rotations, changing shapes of objects by adding auxiliary circular zones). A nonlinear programming model of the irregular cutting problem, employing the phi-function technique, is provided. We develop an efficient optimization algorithm, which involves a fast feasible starting point and local optimization procedures. To show the benefits of our methodology we present some computational results.

Keywords: Cutting problem · Irregular objects · Phi-function technique · Mathematical model · Nonlinear optimization

1 Introduction

Laser cutting technology is gradually replacing conventional methods of manufacturing thin-sheet parts in the aircraft industry [1, 2]. General problems and criteria for transition to laser technology of manufacturing include maintaining the quality of parts, reducing of the process cycle, computer aided design, reducing production costs. Manufacturing application of laser technologies requires solving a lot of problems, among which the search for optimal parameters for laser processing is only the first stage. Ordinarily, is not simply a transition from an operation to laser processing, but the whole process chain subject to revision, and it is often possible to reduce and simplify [3].

The modern CNC system of laser cutting machine must not only control the treatment process, but also have the tools to organize optimal nesting and tool path plan [4, 5].

M. Nechyporuk et al. (Eds.): *Integrated Computer Technologies in Mechanical Engineering*, AISC 1113, pp. 150–157, 2020.
https://doi.org/10.1007/978-3-030-37618-5_14

The nesting is design in order to reduce material consumption. However, the laser cutting is characterized by some processing limitations, such as piercing [6]. In order to reduce the processing time to solve the problem of design an optimal cutting path should take into account the location of the processing limitations [5, 6].

In addition, when designing a nesting card, the size of allowable distances between contours should be taken into account. Its choice is determined by minimizing of overheating and thermal influence zones on adjacent parts during processing, especially in the braking areas [7, 8], which may be achieved through the use of special S-curve acceleration/deceleration algorithm, as implemented, e.g. in the machining processes [9].

Cutting problem is a part of operational research and computational geometry that has a lot of applications, e.g., in sheet metal cutting, selective laser sintering (SLS) technology, furniture making, garment industry, shoe manufacturing, glass industry, shipbuilding industry and aircraft manufacturing. The problems are NP-hard [10], and as a result solution methodologies predominantly utilize heuristics and nearly all practical algorithms deal with shapes, which are approximated by polygons (see [11, 12] and reference therein). Several efficient methods have been proposed to solve irregular nesting problems which operate with polygonal object approximations (see, e.g., [12–15]). In the present paper we further develop the methodology for solving irregular cutting problem based on the phi-function technique [16–21].

In [16] authors consider a convex domain of variable sizes and two irregular objects bounded by circular arcs and line segments, that can be continuously translated and rotated. Minimal allowable distances between objects may be imposed. The objects should be arranged within a domain such that a given objective will reach its minimal value. The paper presents a mathematical model and a solution strategy, and provide new benchmark instances of finding the containing region that has either minimal area, perimeter or homothetic coefficient of a given domain, as well as finding the convex polygonal hull (or its approximation) of a pair of objects.

Paper [17] provides mathematical models and practical algorithms for solving the cutting and packing problem. The paper reviews and further enhances phi-functions. It also demonstrates that in many realistic cases the phi-functions can be described by quite simple formulas without radicals and other complications. A general solution strategy using the phi-functions is outlined and illustrated by several examples.

Ready-to-use free radical phi-functions are derived in [18], which allow to model relations between arbitrary shaped objects bounded by arcs and line segments with continuous rotations. We consider the cutting problem in the following formulation.

Paper [19] review the concept of phi-functions and describe its application for modeling and solving 2D&3D irregular packing problems that consider continuous object rotations.

Paper [20] formulates a basic placement problem for irregular objects. Each object is presented as a union of four types of basic objects. It provides an exact nonlinear programming model of the problem. An efficient solution algorithm to search for local optimal solutions for the problem in a reasonable time is proposed. The algorithm reduces the basic problem to a sequence of nonlinear programming subproblems and employs optimization procedures to generate starting feasible points and feasible subregions.

In [21] the methodology for solving NP-hard irregular placement problems is presented. The irregular placement problem, which covers a wide spectrum of practical packing, cutting, nesting, clustering, and layout problems is formulated. Authors provide a nonlinear programming model of the problem, employing the phi-function technique. The model involves a large number of inequalities with nonsmooth functions. A solution tree for the placement problem is constructed and evaluation of the number of its terminal nodes is provided. The solution strategy is based on a combination of discrete and continuous optimization methods.

Irregular Cutting Problem Cut a collection of irregular objects T_i, $i = 1, 2, \ldots, N$, from a rectangular domain Ω of minimal length, taking into account the given technological requirements, such as the cutting thickness, positioning error of the sheet, edge defects of the sheet, lead in and lead out technological zones, defected zones, technological zones for avoiding sharp angles, restrictions on orientation of objects (indifference ranges).

2 Mathematical Model and Solution Algorithm

We assume that each object T_i is a two-dimensional multi-connected phi-object [17], bounded by line segments and circular arcs [16].

Without loss of generality we assume that each object involve its technological zones.

Each technological zone related to the particular object is described by a circle of the given radius depending on the specific cutting tool and metal width (see Fig. 1).

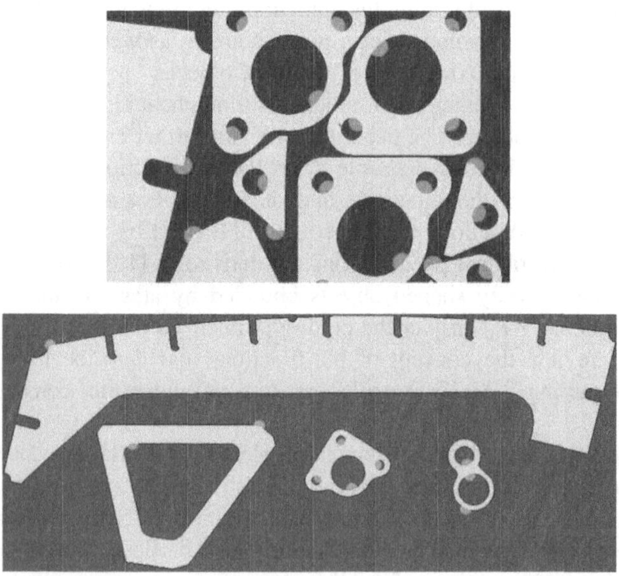

Fig. 1. Examples of technological zones (shown as small circles) related to the appropriate objects.

The location and orientation of an object T is defined by a variable vector of its placement parameters $u_T = (v_T, \theta_T)$, where θ_T is a rotation angle and $v_T = (x_T, y_T) \in R^2$ is a translation vector.

The rotation of T by angle θ_T and the translation of object T by vector v_T define $T(v_T, \theta_T) = \{t \in R^2 : t = v_T + M(\theta_T)t^0, \forall t^0 \in T^0\}$, where T^0 denotes the non-translated and non-rotated object T, $M(\theta_T)$ is a standard rotation matrix.

The cutting conditions include: containment of objects into a domain, non-overlapping of objects, given minimal allowable distances between objects, as well as between objects and the boundary of the domain, a given allowable ranges of the object rotation angles, prohibited areas.

In order to describe relations between objects (such as continuous object rotations, distance constraints) we use the phi-function technique [16–21].

We denote a vector of all variables by $u = (l, u_1, u_2, \ldots, u_N) \in R^{3N+1}$, where $u_i = (v_i, \theta_i) \in R^3$ is a vector of variable placement parameters (motion vector), $v_i = (x_i, y_i)$ is a translation vector, θ_i is a rotation parameter of T_i, $i \in I_N$, $\Omega = \{(x, y) \in R^2 : 0 \le x \le l, 0 \le y \le w\}$ is a rectangle of variable length l and the given width w.

A mathematical model of the irregular cutting problem can be presented in the form

$$\min l, \text{ s.t. } u \in W \subset R^{3N+1}, \tag{1}$$

$$\begin{aligned} W = \{u \in R^{3N+1}&: \widehat{\Phi}_{ij}(u_i, u_j) \ge 0, i > j = 1, \ldots, N, \\ &\widehat{\Phi}_i(u_i) \ge 0, i = 1, \ldots, N, \omega_t(u) \ge 0, t = 1, \ldots, m\} \end{aligned} \tag{2}$$

where

$\widehat{\Phi}_{ij}(u_i, u_j)$ is an adjusted phi-function [17] for describing the distance constraint $\text{dist}(T_i(u_i), T_j(u_j)) \ge \rho$,

$\text{dist}(T_i(u_i), T_j(u_j)) = \min\limits_{a \in T_i(u_i), b \in T_j(u_j)} dist(a, b)$, $\rho > 0$ is a given minimal allowable distance between objects $T_i(u_i)$ and $T_j(u_j)$,

$\widehat{\Phi}_i(u_i)$ is an adjusted phi-function for describing the distance constraint $\text{dist}(T_i(u_i), \Omega^*) \ge \rho_0$, $\Omega^* = R^2 \backslash \text{int}\Omega$, $\text{dist}(T_i(u_i), \Omega^*) = \min\limits_{a \in T_i(u_i), b \in \Omega^*} dist(a, b)$, $\rho_0 > 0$ is a given minimal allowable distance between objects $T_i(u_i)$ and Ω^*,

$dist(a, b)$ is the Euclidean distance between points a and b,

$\omega_t(u) \ge 0$, $t = 1, 2 \ldots, m$, is a system of additional restrictions on values of components of the vector u.

We propose here an approach that combines heuristic and local optimization procedures. The proposed algorithm works sufficiently fast and uses multistart strategy. The approach allows us to search for "good" local optimal solutions of the problem (1)–(2).

Our solution strategy involves the following procedures.

(1) Generation of a feasible starting point $u^0 \in W$ for problem (1)–(2). The algorithm employs the strip approximation of objects [21]. It is especially efficient for cutting multi-connected objects from a rectangular domain of minimal length.
(2) Search for a local minimum of the problem (1)–(2), starting from each feasible point $u^0 \in W$.

We develop a special solver for the irregular cutting problem. The solver allows us: to operate with irregular shaped objects, using the DXF format; to generate systems of inequalities; to provide an exact calculation of elements of Jacobian and Hessian matrixes for the generated inequality systems.

The search for local extrema of nonlinear programming problems in our procedures is performed by IPOPT [22].

(3) Choice the best of local minima obtained at the second step for the set of feasible starting points as a solution of problem (1)–(2).

Searching for Starting Feasible Points. We employ the optimization method by groups of variables with respect to the given object sequence. Our objects, as well as, a domain are approximated by small rectangles of the same width. The algorithm allows us to fill holes of multi-connected objects by smaller objects. Assuming that each smaller object fixed within the appropriate hole of composed object, we further deal with one-connected irregular objects. The algorithm works very fast and deals with domains, bounded by arcs and line segments, prohibited areas, distance constraints and discrete object rotations.

We detect isolated subdomains of our domain on each step. In most cases such subdomains appear after arranging each object. If the subdomain area is too small to contain one of the rest objects this subdomain is removed. Otherwise the subdomain is used to arrange objects. The complexity of the feasible starting point algorithm has estimation $O(N)$ for the practical problems.

Local optimization algorithm

For each starting point $u^0 \in W$ the following iterative procedure is applied. First we construct the system of inequalities for each object $T_i(u_i^0)$, $u_i^0 = (v_i^0, \theta_i^0) = (x_i^0, y_i^0, \theta_i^0)$, $i \in I_N$. It provides such arrangement of object $T_i(u_i)$ that any point $t = (x, y) \in T_i(u_i^0)$ can vary within the appropriate square with the diagonal points $(x - \frac{\delta}{2}, y - \frac{\delta}{2})$ and $(x + \frac{\delta}{2}, y + \frac{\delta}{2})$. Here δ is a decomposition step of the procedure.

Further, we denote the inequality system that provides an arrangement of all objects $T_i(u_i)$, $i = 1, 2, .., N$ within the appropriate squares at the point u^0 by $\Delta(u^0) \geq 0$.

The key idea of the procedure lies in the following. We add system $\Delta(u^0) \geq 0$ to (2) and reduce the number of inequalities, which describe the feasible region W, eliminating adjusted phi-functions for pairs of objects located "far" from each other, provided that $\Delta(u^0) \geq 0$.

Then we search for a point of local minimum u^{0*} of the subproblem starting from the feasible starting point $u^0 \in W^0 \subset W$.

$$\min l \text{ s.t. } u \in W^0.$$

We take the point $u^1 = u^{0*}$ as a new feasible starting point, and form subregion W^1. Then our algorithm searches for a point of local minimum u^{1*} of the subproblem

$$\min l \text{ s.t. } u \in W^1.$$

We take the point $u^2 = u^{1*}$ as a starting point for further local optimization until $l(u^{t*}) = l(u^{(t+1)*})$, $t = 1, 2, \ldots$.

The point u^{t*} is taken as a point u^* of local minimum of problem (1)–(2).

Our algorithm monitors O(N) pairs of objects (this depends on the value of δ), instead of O(N^2) pairs of objects, because for each object only its "δ-neighbors" have to be controlled.

3 Computational Results

Figure 2 shows some instances demonstrating the efficiency of our methodology. The proposed algorithm of automatic arrangements of objects allows increasing packing factor up to 5% in comparison with heuristic arrangements of irregular objects found for benchmark instances [20].

Fig. 2. Instances of the optimized sheet cuttings.

4 Conclusions

We consider the problem of cutting irregular objects from a rectangular domain of minimal length, taking into account the given technological requirements including the cutting thickness, positioning error of the sheet, edge defects of the sheet, lead in and lead out technological zones, defected zones, technological zones for avoiding sharp angles, restrictions on orientation of objects (indifference ranges).

A mathematical model of the irregular cutting problem, using the phi-function technique, is provided. The model takes into account the object continuous translations and rotations, containment of objects into a domain, non-overlapping of objects, given minimal allowable distances between objects, as well as between objects and the boundary of the domain, a given allowable ranges of the object rotation angles, prohibited areas. The model may be applied for metal cutting problems and can be realized by the current state-of-the art local or global solvers.

The proposed solution strategy employs a fast algorithm to generate feasible starting arrangements of objects and an efficient local optimization algorithm, which allows reducing the number of nonlinear inequalities in nonlinear programming subproblems.

References

1. Tam, S.C., Williams, R., Yang, L.J., et al.: A review of the laser processing of aircraft components. J. Mater. Process. Technol. **23**(2), 177–194 (1990). https://doi.org/10.1016/0924-0136(90)90156-o
2. Barbosa, G., Cordeiro, E., Costa, F.: Systems and methods for manufacturing aircraft furniture parts using an integrated automated cell. SAE Technical Paper 2015-01-2600 (2015). https://doi.org/10.4271/2015-01-2600
3. Maiorov, V.S.: Technological feature of laser cutting by manufacturing parts for aircraft plants. In: Panchenko, V.Ya., Golubev, V.S. (eds.) 6th International Conference on Industrial Lasers and Laser Applications 1998, vol. 3688, pp. 201–205. SPIE (1999). https://doi.org/10.1117/12.337508
4. Dewil, R., Vansteenwegen, P., Cattrysse, D.: A review of cutting path algorithms for laser cutters. Int. J. Adv. Manuf. Technol. **87**(5–8), 1865–1884 (2016). https://doi.org/10.1007/s00170-016-8609-1
5. Petunin, A.A., Stylios, C.: Optimization models of tool path problem for CNC sheet metal cutting machines. IFAC-PapersOnLine **49**(12), 23–28 (2016). https://doi.org/10.1016/j.ifacol.2016.07.544
6. Pocorni, J., Powell, J., Frostevarg, J., Kaplan, A.F.H.: Dynamic laser piercing of thick section metals. Opt. Lasers Eng. **100**, 82–89 (2018). https://doi.org/10.1016/j.optlaseng.2017.07.012
7. Sharma, A., Yadava, V.: Experimental analysis of Nd-YAG laser cutting of sheet materials – a review. Opt. Laser Technol. **98**, 264–280 (2018). https://doi.org/10.1016/j.optlastec.2017.08.002
8. Kim, Y., Gotoh, K., Toyosada, M.: Global cutting-path optimization considering the minimum heat effect with microgenetic algorithms. J. Marine Sci. Technol. **9**(2), 70–79 (2004). https://doi.org/10.1007/s00773-004-0176-8

9. Kombarov, V., Sorokin, V., Fojtů, O., et al.: S-curve algorithm of acceleration/deceleration with smoothly-limited jerk in high-speed equipment control tasks. MM Sci. J. **2019**(04), 3264–3270 (2019). https://doi.org/10.17973/MMSJ.2019_11_2019080
10. Chazelle, B., Edelsbrunner, H., Guibas, L.J.: The complexity of cutting complexes. Discrete Comput. Geom. **4**(2), 139–181 (1989). https://doi.org/10.1007/bf02187720
11. Bennell, J.A., Oliveira, J.F.: A tutorial in irregular shape packing problem. J. Oper. Res. Soc. **60**, 93–105 (2009). https://doi.org/10.1057/jors.2008.169
12. Leao, A.A.S., Toledo, F.M.B., Oliveira, J.F., et al.: Irregular packing problems: a review of mathematical models. Eur. J. Oper. Res. (2019). https://doi.org/10.1016/j.ejor.2019.04.045
13. Alvarez-Valdes, R., Martinez, A., Tamarit, J.M.: A branch & bound algorithm for cutting and packing irregularly shaped pieces. Int. J. Prod. Econ. **145**(2), 463–477 (2013). https://doi.org/10.1016/j.ijpe.2013.04.007
14. Baldacci, R., Boschetti, M.A., Ganovelli, M., Maniezzo, V.: Algorithms for nesting with defects. Discrete Appl. Math. **163**(1), 17–33 (2014). https://doi.org/10.1016/j.dam.2012.03.026
15. Fadel, G., Wiecek, M.: Packing optimization of free-form objects in engineering design. In: Fasano, G., Pintér, J. (eds.) Optimized Packings with Applications, vol. 105, pp. 37–66. Springer, Cham (2015). https://doi.org/10.1007/978-3-319-18899-7_3
16. Bennell, J.A., Scheithauer, G., Stoyan, Y., et al.: Optimal clustering of a pair of irregular objects. J. Global Optim. **61**(3), 497–524 (2015). https://doi.org/10.1007/s10898-014-0192-0
17. Chernov, N., Stoyan, Y., Romanova, T.: Mathematical model and efficient algorithms for object packing problem. Comput. Geom.: Theory Appl. **43**(5), 535–553 (2010). https://doi.org/10.1016/j.comgeo.2009.12.003
18. Chernov, N., Stoyan, Y., Romanova, T., Pankratov, A.: Phi-functions for 2D objects formed by line segments and circular arcs. Adv. Oper. Res. **2012**, Article ID 346358 (2012). https://doi.org/10.1155/2012/346358
19. Stoyan, Yu., Romanova, T.: Mathematical models of placement optimization: two- and three-dimensional problems and applications. In: Fasano, G., Pintér, J. (eds.) Modeling and Optimization in Space Engineering, vol. 73, pp. 363–388. Springer, Cham (2013). https://doi.org/10.1007/978-1-4614-4469-5_15
20. Stoyan, Y., Pankratov, A., Romanova, T.: Cutting and packing problems for irregular objects with continuous rotations: mathematical modelling and non-linear optimization. J. Oper. Res. Soc. **67**, 786–800 (2016). https://doi.org/10.1057/jors.2015.94
21. Stoyan, Y., Pankratov, A., Romanova, T.: Placement problems for irregular objects: Mathematical modeling, optimization and applications. In: Butenko, S., Pardalos, P.M., Shylo, V. (eds.) Optimization Methods and Applications, vol. 130, pp. 521–559. Springer, Cham (2017). https://doi.org/10.1007/978-3-319-68640-0_25
22. Wachter, A., Biegler, L.T.: On the implementation of an interior-point filter line-search algorithm for large-scale nonlinear programming. Math. Program. **106**(1), 25–57 (2006). https://doi.org/10.1007/s10107-004-0559-y

Artificial Intelligence and Smart Systems

Method for Identifying and Counting Objects

Olha Pohudina(✉), Dmitriy Kritskiy, A. N. Bykov, and T. Szalay

National Aerospace University «KhAI», Kharkiv, Ukraine
o.pogudina@gmail.com, d.krickiy@khai.edu

Abstract. The article presents an implementation of object recognition methods using the OpenCV library. The analysis of existing approaches, as well as recognition algorithms. The best methods were selected for the quality and number of recognizable objects in the image, taking into account the time to search. A recognition algorithm based on the analysis of brightness histogram (three color components of objects is proposed); a recognition algorithm based on the analysis of the contour length has been developed; programs for counting objects of the same type in the image are developed, taking into account their separate location on the image and the possibility of overlapping (merging contours). The above algorithms are implemented in the programming language C++ using the OpenCV library. The results obtained in the course of using the implemented methods, both individually and in combination, are presented.

Keywords: OpenCV library · RGB · Histogram of color code · Contour analysis · Compare contours · Count of objects

1 Introduction

Traditionally, an image is considered as a combination of individual characteristics and data sets, such as size, brightness, contrast, etc. However, upon closer examination of the image, it is an abstract "object" that has multidimensional characteristics and consists of elements that also have many parameters. This means that the methods and concepts of analytical data processing are applicable to image processing and analysis. Descriptive and attribute data of objects in the image represent some indicators or numerical values. These values are the assessment of a particular item on a predetermined scale of signs. Having estimates of the object on different grounds, it can be compared with other objects that have similar features. Thus, it is possible to establish the relationship of objects and their correlations with each other.

There are several ways to search by image: search by content (find a photo of a person or an image, for example, of an airplane), search by a visual pattern (find images similar to a given one), search by descriptions (find an image marked as "Flowers") and etc. Each of the search directions has its own characteristics and spheres of application.

Systems that search by visual pattern actively use various signs of images. These attributes differ a great variety.

Analysis of Recent Research and Publications. The use of color characteristics is discussed in [1]. Such image attributes as the overall color, the most frequent colors,

M. Nechyporuk et al. (Eds.): *Integrated Computer Technologies in Mechanical Engineering*, AISC 1113, pp. 161–172, 2020.
https://doi.org/10.1007/978-3-030-37618-5_15

the regularity of color placement, the complexity of patterns, etc. are highlighted. In [2], it is proposed to use color attributes. For this purpose, a color histogram is constructed, considering the spatial arrangement of colors, and color moments are calculated. To estimate the similarity of two images, the distribution parameters of their colors are compared: the expectation for each of the color channels and the pairwise covariance of the channel distributions.

The work [3] is devoted to the use of textural attributes. The authors identify signs of contrast, roughness, directionality, linear images, regularity and texture roughness. In [4], to search for images by content, it is proposed to use the homogeneous texture descriptor (HTD), the edge histogram descriptor (EHD), the scalable color descriptor (SCD), the dominant color descriptor (DCD).

The work [5] is devoted to the study of the object form characteristics in the image. It is emphasized that the outlines of images are areas with a high information concentration, which barely depends on color and brightness. The image analysis usually involves obtaining the outer contour of the depicted objects and recording the coordinates of the points of this contour. Most often the outer contour is needed in the form of a closed curve or a set of arc segments. Three general approaches to representing the boundaries of an object are considered: curve approximation; contour tracking; linking drop points.

Another significant attribute for searching objects in image collections is point features [6]. A point is the simplest and most obvious geometric element that can be considered as an element of the function discrete representation. If a number of special points in the scene are selected and their position is determined in the stream of images, the obtained data will be able to give a huge amount of information about the structure of the scene, as well as about the parameters of the camera from which the images were taken.

Thus, there is a fairly large number of attributes that could be used in image search tasks on a visual sample. However, most of the works include only a part of the existing image features, combining them and developing signatures based on them. Due to the fact that many signs can be interconnected, their rational selection is a very important task.

The Aim of the Study. The aim of the work is to consider the issues of building a search system for a visual sample in image collections. This method will allow, using the experience in the management of drones [7, 8], to recognize objects (structures, cattle, insects, etc.) and their calculation using information obtained from cameras installed on drones.

2 Identification of Objects with the Statistical Characteristics of Color

In the process of identifying objects from a digital image, the main parameter is often the color of the object. Color identification algorithms are an integral part of optical sorters. They are in demand in the identification of agricultural products (fruits, vegetables, flowers, etc.), and they sort salts, minerals, glass, plastic and waste during recycling.

As a rule, the tasks of identifying objects by color are performed through preliminary training of the system with examples. The input of the identification algorithm is dichromatic, trichromatic or RGB — red-green-blue images, in which there is a background of a certain color (often white, black or gray). Further, in the task set, it is necessary to solve three problems: the transition from the space of points to the space of signs, the creation of a database of signs, the implementation of a solution choice based on the set of signs. Consider solving these problems using the OpenCV library.

Transition from the Space of Points to the Space of Signs. The source information is read from the image file. Suppose an input is an image of the RGB model. In this model, each image point is described by three input signals - Red, Green and Blue. To solve the problem, first of all, we need to cut off the background points. For example, for gray, the values of RGB signals remain almost the same, ranging from values of 0, 0, 0 (black) to 255, 255, 255 (white). Analysis have shown that the mismatch between the RGB signals for gray does not exceed 15%. For example, a background point can be determined by signals with red, green and blue values of 175, 150 and 167, respectively. This feature allows to simply separate the points of the object from the points of the background.

The color of most objects is heterogeneous and, therefore, it cannot be determined by any particular code number. It can only be described by a histogram of the repeatability of the colors of each of the 256 codes (Fig. 1). When constructing a histogram, the repeatability of color n should be correlated to the total number of points N of the object.

In the case of an image of the RGB model, it is necessary to construct three histograms for each color component. To get a separate color component in OpenCV, the following combination of commands can be used:

cvSetImageCOI (img, nCan); cvCopy (img, channel); cvResetImageCOI (img);

where img and channel are pointers to an IplImage * structure that store the input image (img) and single-channel image (channel), and nCan is an integer type variable that stores the channel number in the selected color model.

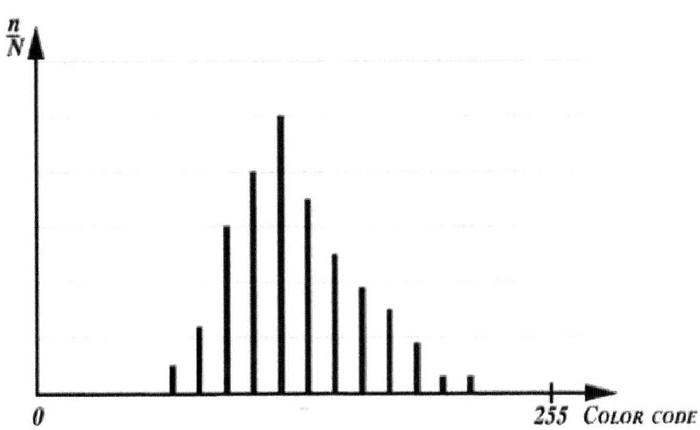

Fig. 1. Histogram of color code repeatability

For example, in the RGB model, nCan = 3 corresponds to the color component of Blue. The cvSetImageCOI and cvResetImageCOI commands set and remove the channel of interest, and cvCopy copies part of the data of the three-channel image (img) into the single-channel image (channel).

To get the histogram, the cvCreateHist() function can be used, that creates the histogram object. After that, the histogram itself is calculated using the cvCalcHist() function. The calculation results can be saved to a database file in the form of an array for further analysis.

As formal features, that help to distinguish one type of object from another, the static characteristics of the histograms can be taken: average value; dispersion; asymmetry; excess.

Let N – the number of points on the object – the total sample size, xi – the observed colors (0 … .255) – options, ni – the number of points per color – frequency, ni/N – normalized frequency. Then the formula for calculating indicators will be as follows:

– the average value

$$M = \bar{X} = \frac{1}{N}\sum_{i=0}^{255} x_i n_i = \sum_{i=0}^{255} x_i \left(\frac{n_i}{N}\right),$$

– the dispersion

$$D = \frac{1}{N}\sum_{i=0}^{255} (x_i - \bar{X})^2 n_i = \sum_{i=0}^{255} (x_i - \bar{X})^2 \frac{n_i}{N},$$

– the standard deviation

$$\sigma = \sqrt{D},$$

– the initial moment of the k-th order

$$\mu_k = \sum_{i=0}^{255} (x_i - \bar{X})^k \frac{n_i}{N},$$

– the asymmetry of distribution

$$A = \frac{\mu_3}{\sigma^3},$$

characterizes histogram skewness. If $A = 0$, then the histogram is symmetric about the central value, $A > 0$ – the right end is extended, $A < 0$ – the left end is extended;

– the excess of the distribution

$$E = \frac{\mu_4}{\sigma^3 - 3},$$

characterizes the degree of sharpness of the peak of the curve compared to the normal law. If $E > 0$ – a sharper peak, and $E < 0$ – less.

Creating a Database of Signs of Objects. Images of objects are provided for learning the system (for example, apples, cucumbers, etc.). They are matched with the attributes of the identified object. During the training, the signs are stored in the database, and when they are identified, they are compared with the signs of the object being identified.

In the learning mode, when the system processes the next image for each type of object, there is a process of recalculating the signs and saving their values. In this case, a set of signs of the object is identified by the name of the product.

Multiple Choice Decision. The problem of choosing a solution based on a set of signs can be solved by adding estimates for all signs for each type of object, information about which is available in the database.

In this case, the coincidence of the sign of the test image with the corresponding sign of any type of object from the database for each feature can be estimated by 1 (one), while the mismatch is 0 (zero). The selection is carried out at the highest overall rating. It is considered that the signs coincide if the value of the sign being tested is within the confidence interval defined by the formula:

$$\Delta = 2.25\sigma.$$

Another way to choose the solution is to compare histograms using the cvCompareHist() function of the OpenCV library. Two histograms and a comparison method are transmitted to the function, which determines the measure of proximity of one histogram to another:

CV_COMP_CORREL – the correlation method. The return value lies in the interval [−1,1]: 1 – the maximum match, −1 – the maximum discrepancy, 0 – there is no correlation;

CV_COMP_CHISQR – the chi-square. The return value lies in the interval [0, unlimited): 0 – the maximum match, and the extreme discrepancy depends on the number of elements of the histogram;

CV_COMP_INTERSECT – the intersection. If the histograms are normalized to 1, then the perfect correspondence of the histograms is 1, and the perfect discrepancy is 0.

CV_COMP_BHATTACHARYYA – the Bhatacharyya's distance. If the histograms are normalized to 1, then the perfect match is 0, the mismatch is 1.

Figure 2 shows the result of comparing two images by histograms of three image channels. The source images are shown in separate windows, then histograms are plotted. And the result of their comparison is obtained (displayed in the console window).

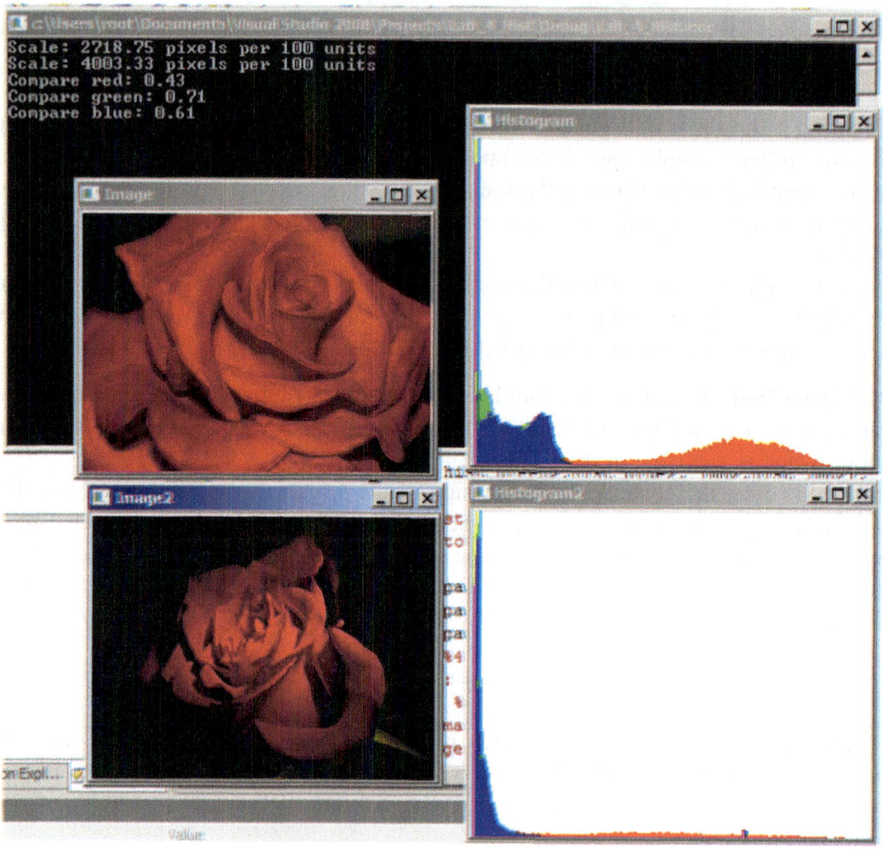

Fig. 2. Example of histogram comparison

3 Methods of the Contours Recognition

Contour analysis is one of the most important and very useful methods for describing, storing, comparing and searching for images/objects.

When conducting a contour analysis: it is assumed that the contour contains sufficient information about the shape of the object; internal points of the object are not considered.

The above provisions impose significant restrictions on the scope of contour analysis, which are associated with the problems of contour selection on images: due to the same brightness with the background, the object may not have a clear border, or may be noisy, which makes it impossible to select the contour; overlapping of objects or their grouping leads to the fact that the contour is selected incorrectly and does not correspond to the boundary of the object.

However, the transition to the consideration of only the contours of objects allows to move away from the image space to the space of contours, which significantly reduces the complexity of algorithms and calculations.

The OpenCV library implements convenient methods for detecting and manipulating image contours.

To search for contours, the function is used to find contours on the binary image cvFindContours():

```
CVAPI(int) cvFindContours(
CvArr * image,
CvMemStorage * storage,
CvSeq ** first_contour,
int header_size CV_DEFAULT(sizeof (CvContour)),
int mode CV_DEFAULT(CV_RETR_LIST),
int method CV_DEFAULT(CV_CHAIN_APPROX_SIMPLE),
CvPoint offset CV_DEFAULT(cvPoint (0,0)));
```

where image is the original 8-bit single-channel image (non-zero pixels are processed as 1 and zero are 0), storage is the storage for storing the data of the detected contours, first_contour is a pointer that points to the first element of the sequence containing the data of the contours found, header_size is a size of the element header of the sequence, mode is a search mode (Table 1), method is an approximation method (Table 2), offset is an offset by which the contour points are shifted (useful if the contours are extracted from ROI and then must be analyzed in the context of the whole image).

Table 1. The search contour mode of the cvFindContours() function

№	Title	Explanation
0	CV_RETR_EXTERNAL	Find only extreme outer contours
1	CV_RETR_LIST	Find all contours and list them
2	CV_RETR_CCOMP	Find all contours and place them in a 2-level hierarchy
3	CV_RETR_TREE	Find all contours and place them in the hierarchy of nested contours

Table 2. The approximation method of the cvFindContours() function

№	Title	Explanation
1	CV_CHAIN_APPROX_NONE	All points of the chain code are converted to points
2	CV_CHAIN_APPROX_SIMPLE	Compresses horizontal, vertical, and diagonal segments and leaves only their end points
3	CV_CHAIN_APPROX_TC89_L1 CV_CHAIN_APPROX_TC89_KCOS	The Teh-Chin approximation algorithm is used

The usual sequence of actions when recognizing objects with the method of contour analysis is as follows:

Step 1: Image preprocessing (smoothing, noise filtering, contrast enhancement);
Step 2: Image binarization;
Step 3: Selection of the contours of objects;
Step 4: Primary filtering of the contours (using a perimeter value, an area value, etc.);
Step 5: Equalization of contours (reduction to a single length, smoothing) – allows to achieve scale invariance;
Step 6: Enumeration of all found contours and search for a template that is as close as possible to the given contour (or the sorting of contours according to some value, for example, an area).

Consider a software application that searches for an object with clear boundaries in the image-scene, in which there are other objects (Fig. 3).

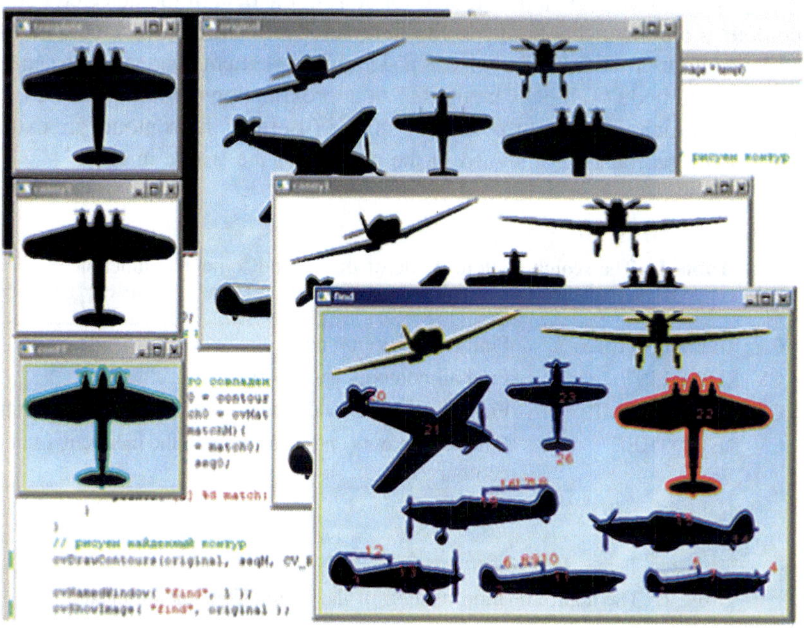

Fig. 3. Search for the contour of an object using the cvMatchShapes() function

The program receives two images as the input, which are converted into a binary form. Then, using the cvFindContours() function, the external and internal borders are searched (on the sample using the cvDrawContours() function, the external borders are marked in yellow, the internal ones are blue). Then a cycle is organized in which all contours in the image-scene are compared with the contour of the template and the matching contour is highlighted in a different color (in the sample it is highlighted in red). The following parameters can be used to compare contours: the moments, the contour length, the contour size in bytes, etc.

For example, in the following snippet, a search is performed using the comparison of two contours by their moments using the cvMatchShapes() function. The function uses a linear combination of normalized central moments, that makes it possible to create an invariant representation of the contours, independent of the scale, rotation and reflection of the contour.

If the result of the comparison is displayed on the screen, the cvMatchShapes() function for comparing an aircraft object with an image of a scene in a contour highlighted in red will give 0.00, that corresponds to the complete coincidence of the contours. As you can see, the two top aircraft in the image-scene are highlighted in yellow, as they are external contours. This is due to the fact that several pixels of these objects are in contact with the edge of the image. This feature should be considered when shaping the scene. In addition, there are 10 aircraft objects in the scene image, but when detecting the contours, 26 objects were found (the object number was displayed on the resulting image of the scene). To eliminate this disadvantage after binarization of an object, morphological operations such as building, erosion, and their combination should be applied.

Consider the following examples using the methods of recognition of the contours in images using the OpenCV library:

- the character recognition in Japanese language learning systems;
- the recognition of road signs;
- the object counting in the image.

Each of the examples has its own characteristics: in the case of character recognition, a clear outline can be got, because there is no background, but the order of writing strokes should be considered, therefore we need to compare not the resulting outline, but all the outline of individual strokes created in the writing process. The second task is characterized with the background and the size of the searched object. The third task is complicated with the possibility of overlapping of the set of objects.

4 Subsystem of Object Counting in the Image with Specified Characteristics of Their Contours

There are tasks related to the rapid and effective counting of many similar objects in the image (grains, vegetables, fruits), as well as the evaluation of their parameters. One of the areas requiring the use of such methods, for example, is the selection of wheat and potatoes.

Consider the program of counting the number of potatoes in a given image (Fig. 4). The algorithm includes the following steps:

1. Creation of variables for image storage: IplImage structure is used;
2. Declaration and initialization of a variable pointer to the name of the file with the image;

Fig. 4. Image to count the number of potatoes

3. Image loading: cvLoadImage function is used;
4. Creation of storage variables (container for storing all contours, type CvMemStorage), first_contour (a pointer to the first contour in the CvSeq structure);
5. Pre-processing of the image for the correct detection of the necessary objects, the correct translation into a black and white image. Therefore, when finding a contour in OpenCV, it is important to consider what will be black or white in the image, since the black spots are recognized as an object, the contour is located on these objects, and the white area of the image is considered as a "hole" (nothing). For such transformations, a number of standard functions can be used, as well as splitting the image into channels and converting each channel using different functions. The type of transformation is determined experimentally on several images with different locations of the same object;
6. Use of the contour search function cvFindContours;
7. Removal of the unnecessary contours, if the size of the object has a certain value (considering that the length of the object contour cannot be more or less than some value). To do this, it is possible to use a loop that is organized according to the list stored in the cvSeq structure and access to each contour can be obtained using the links to the elements h_next, h_prev. The length of the contour is determined by the parameter total;
8. Use the contour drawing function cvDrawContours() to confirm the correct operation of the contour search function;
9. Creation of windows in which images will be displayed using the cvNamedWindow function,
10. Loading images using the cvShowImage() function, displaying the number of objects counted.

During the research of contour analysis, an algorithm for finding objects in an image was developed, as well as an information subsystem that allows to find and count objects. This subsystem uses an image of 980 * 550 pixels in size, the average length of the contour is 100 pixels, and the length of the contours to be deleted is 28 pixels. Thus, the program in total counted 1949 contours, however, using the cleaning of the contours received the result – 94 objects. The resulting value is equal to the number of objects in the image.

5 Development of the Subsystem of Counting the Overlapping Objects in the Image

It is relatively easy to calculate simple and detached objects. Their outlines can be got from a binary image. But for connected or overlapping objects this is quite difficult to do. The Distance transform method should be used for it, the result of which is a binary image, where the value of each pixel is replaced by its distance to the nearest background pixel.

The OpenCV function of the Distance transform algorithm is as follows:

```
distanceTransform(
    InputArray src, // 8-bit single-channel image
    OutputArray dst, // 32-bit floating point single channel image
    int distanceType, // One of: CV_DIST_L1, CV_DIST_L2, CV_DIST_L3
    int maskSize // Size of distance mask
)
```

For example, the loaded image was converted to a binary form. Then the algorithm Distance transform is executed, the result of which is shown in Fig. 5.

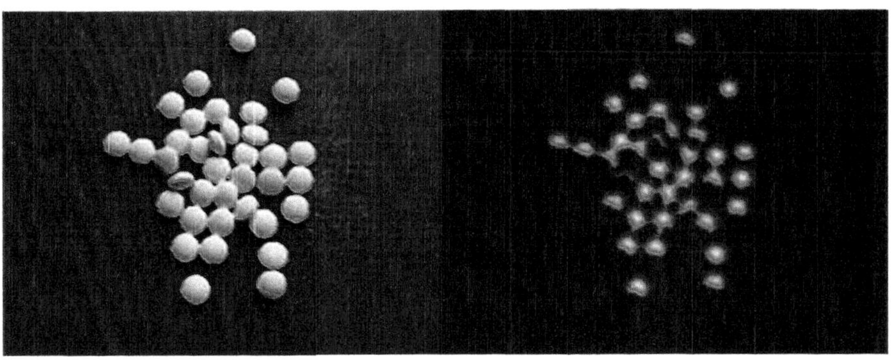

Fig. 5. Result transformation using the distanceTransform function

In the image stored in the variable dist, the result [0,1] is normalized:
normalize (dist, dist, 0, 1., cv :: NORM_MINMAX).
Threshold conversion was performed for peaks:
threshold (dist, dist, .5, 1., CV_THRESH_BINARY).
Next, a search for contours and the number of objects was processed. For the image shown in Fig. 5 - 34 out of 36 objects were found.

6 Conclusions

Modern recognition methods combine classical algorithms for video and image analysis and new approaches that show high results in related computer vision tasks, such as identifying objects (buildings, livestock, insects) and counting their number.

The implemented methods show a high percentage of correct recognition, but in order to increase the likelihood of correct recognition, these methods can be used simultaneously. In this case, the time spent will increase slightly, but the probability of correct recognition increases and reaches almost 100% (individually, the probability of recognition is 0.94, when used together, 0.99). The obtained quality of object recognition corresponds to modern trends in this area [9, 10].

In the future, the implementation of these methods will be adapted to the recognition of objects and counting their number using a camera mounted on unmanned aerial vehicles.

References

1. Mojsilovic, A., Kovacevic, J., Hu, J., Safranek, R.J., Ganapathy, S.K.: Matching and retrieval based on the vocabulary and grammar of color patterns. IEEE Trans. Image Process. **9**(1), 38–54 (2000)
2. Fleyeh, H.: Color detection and segmentation for road and traffic signs. In: IEEE Conference on Cybernetics and Intelligent Systems, vol. 2, pp. 809–814. IEEE (2004)
3. Tamura, H., Mori, S., Yamawaki, T.: Textural features corresponding to visual perception. IEEE Trans. Syst. Man Cybern. **8**(6), 460–473 (1978)
4. Quack, T., Mönich, U., Thiele, L., Manjunath, B.S.: Cortina: a system for large-scale, content-based web image retrieval. In: Proceedings of the 12th Annual ACM International Conference on Multimedia, pp. 508–511. ACM, October 2004
5. Cohen, L.D.: On active contour models and balloons. CVGIP: Image Understand. **53**(2), 211–218 (1991). https://doi.org/10.1016/1049-9660(91)90028-n
6. Barinova, O., Konushin, V., Yakubenko, A., Lee, K., Lim, H., Konushin, A.: Fast automatic single-view 3-D reconstruction of urban scenes. In: European Conference on Computer Vision, pp. 100–113. Springer, Heidelberg (2008)
7. Kritsky, D.N., Druzhinin, E.A., Pogudina, O.K., Kritskaya, O.S.: A method for assessing the impact of technical risks on the aerospace product development projects. Advances in Intelligent Systems and Computing, vol. 871, pp. 504–521. Springer, Heidelberg (2019)
8. Kritskiy, D., Alexander, K., Koba, S., Druzhinin, E.: Increasing the reliability of drones due to the use of quaternions in motion. In: 2018 Proceedings of 2018 IEEE 9th International Conference on Dependable Systems, Services and Technologies, DESSERT 2018 (2018)
9. Kuplyakov, D.A., Shalnov, E.V., Konushin, V.S., Konushin, A.S.: A distributed tracking algorithm for counting people in video. Program. Comput. Softw. **45**(4), 163–170 (2019)
10. Sofiiuk, K., Barinova, O., Konushin, A.: AdaptIS: Adaptive Instance Selection Network. arXiv preprint arXiv:1909.07829 (2019)

Application of Artificial Neural Networks in the Problems of the Patient's Condition Diagnosis in Medical Monitoring Systems

Viktoriia Strilets[1(✉)], Nina Bakumenko[1], Serhii Chernysh[2], Mykhaylo Ugryumov[1], and Volodymyr Donets[1]

[1] V. N. Karazin Kharkiv National University, Kharkiv, Ukraine
striletsvictoria@gmail.com, n.bakumenko@karazin.ua, ugryumov.mykhaylo52@gmail.com, vovan.s.marsa@gmail.com

[2] National Aerospace University "Kharkiv Aviation Institute", Kharkiv, Ukraine
91sergey@gmail.com

Abstract. The problem of accurate medical diagnosis is always urgent for any person. Existing methods for solving the problem of classification of the state of a complex system are considered. The paper proposes a method of classification of patients' status in medical monitoring systems using artificial neural networks. The artificial neural networks training method uses bee colonies to simulate less training error. The research purpose is to determine the patient's belonging to a particular class according to the variables of his condition, which are recorded. Examples of using the method to determine the status of patients with urological diseases and liver disease are given. The classification accuracy was more than 80%.

Keywords: Classification problem · Artificial neural networks · ROC-curve · Medicine monitoring

1 Introduction

The defects appearance and defects development in complex systems is a complex dynamic process. Experts cannot always predict their appearance and behavior of the system in this case. It is not always possible to come to one thought at what stage of development is the defect and in what state of the system, what methods must be applied to eliminate the defect and normalize the state of the system. Controlling and predicting the dynamic process of a complex system helps experts make decisions that lead to better performance.

As a complex system was considered medical and biological system, which includes a doctor, patients and a system for diagnosing the condition of patients. Each treatment cycle is characterized by a set of final patient's states. The number of states under consideration is determined by the expert based on the results of the cluster analysis. The method for solving the problem of patient clustering is presented in [1].

M. Nechyporuk et al. (Eds.): *Integrated Computer Technologies in Mechanical Engineering*, AISC 1113, pp. 173–185, 2020.
https://doi.org/10.1007/978-3-030-37618-5_16

We accept the hypothesis that the patient's condition is uniquely determined by the system of variables. The problem of recognizing the patient's current state is reduced to the problem of classifying the values of the patient's state variables.

Much attention is paid to the classification problems of complex systems. Many papers have been published today that describe methods for solving the classification problem for technical and biomedical systems.

A naive Bayes classifier is a simple probabilistic classifier based on the Bayes theorem with a "naive" assumption of independence. Bayesian classifiers can be very effective [2] or minor modifications of algorithms can greatly increase its efficiency [3].

Despite the apparent simplicity, the naive Bayesian classifier is capable of producing excellent results even in complex life problems [2].

The advantage of this classifier is a small amount of data necessary for training, parameter evaluation and classification [4]. Also, the naive Bayesian method is often in demand in the classification of text because of the implementation simplicity and the very high efficiency of the classification of text or entities, the nature of which is tied to a static difference [3, 4].

K-Nearest Neighbor Classifier is a metric algorithm for automatic classification of objects or regression, it is also one of the most commonly used and well-known methods for performing recognition tasks and solving data mining problems [4]. Among the advantages of this method, one can single out the simplicity of implementation and high performance and not a high probability of error [5]. Among the shortcomings are high requirements for the computer memory, intolerance to noise and, as a consequence, poor efficiency, the solution to these problems is considered in [6].

Random Forest Classifier is a machine learning algorithm proposed by Braiman [12], the essence of which is to use a certain set of decision trees. This algorithm combines the Breiman bagging method and the random space method proposed by Ho [7]. Due to the fact that the decision tree is the basis of the algorithm, the algorithm provides a comparison of the importance of parameters and helps to identify extremely important properties for determining the final class or state [8–10]. The works in which this algorithm was used to determine the condition of patients, speech recognition and handwriting are given [8]. Modifications of this algorithm help determine the sensitivity to drugs and select candidates for medical research [9]. Random Forest Classifier also helps determine the type of cancer, highlighting important parameters that doctors can pay attention to on their own classification, which can increase the likelihood of diagnosing cancer even in the absence of access to this system [10]. Another area of use of modified Random Forest algorithms is the prediction of molecular interactions [11].

Logistic Regression is a statistical model that in its basic form uses a logistic function to model a binary dependent variable, although there are many more complex extensions. The binary model of logistic regression has extensions to more than two levels of the dependent variable: categorical output data with more than two values are modeled using polynomial logistic regression, and if several categories are ordered using ordinal logistic regression, for example, an ordinal proportional coefficient logistic model. The model itself simply simulates the probability of an output in terms of input data and does not perform statistical classification (this is not a classifier), although it can be used to create a classifier, for example, by choosing a limit value and classifying the input data with a higher probability than a cutoff as one class, lower cutoffs like the other; This is a common way to make a binary classifier. Coefficients are usually not computed by a closed-form expression, in contrast to linear least squares. Also, due to the nature of this model, it is possible to obtain a graph with the importance of properties [12, 13].

Decision Tree Classifier is a decision support tool used in machine learning, data analysis and statistics. This tool uses a tree-like model of decisions and their possible consequences, including random outcomes of events, resource costs, and utility. This is one way to display an algorithm consisting only of conditional statements. Decision trees are commonly used in operations research, especially in decision analysis, to determine a strategy with the highest probability of achieving a goal [14, 15].

AdaBoost, short for Adaptive Boosting, is a machine learning meta-algorithm formulated by Yoav Freind and Robert Shapira [20]. This algorithm can be used in combination with several classification algorithms to improve their efficiency. The algorithm strengthens the classifiers by uniting them into a "committee". AdaBoost is adaptive in the sense that each subsequent committee of classifiers is built on objects incorrectly classified by previous committees. AdaBoost is sensitive to data noise and outliers. However, it is less prone to retraining compared to other machine learning algorithms [17, 18].

AdaBoost calls weak classifiers in the cycle t = 1, …, T. After each call, the distribution of weights Dt is updated, which correspond to the importance of each of the objects in the training set for classification. At each iteration, the weights of each incorrectly classified object increase, thus the new committee of classifiers "focuses its attention" on these objects. Among the advantages, it is possible to single out the simplicity of implementation and high work efficiency, as the algorithm tends to increase the margin [17–19].

Publications [20, 21] detail the general principles of the theory of artificial neural networks that are being widely used to construct diagnostic models in the form of regression equations.

The research purpose is to determine the patient's belonging to a particular class according to the variables of his condition, which are recorded, using artificial neural networks.

2 Research Problem Statement

On the basis of the systematic analysis of the process of diagnosing the patient's condition, a hierarchy of stages of diagnosis was determined: laboratory diagnosis (blood tests, etc.), visual diagnosis (ultrasound, MRI, etc.) and doctor's examination. At each stage, the corresponding patient status variables are logged. An experimental sample of the variables being recorded was formed, characterizing the condition of the patients being monitored.

Suppose there is a multidimensional state matrix $X = \{x_{i,j}\}$ (i = 1 … I, j = 1 … J), where I is the number of patients in the sample, J is the number of measured state variables. Traditionally, the rows of this matrix are called precedents. Centering and normalization of data is performed by the formula

$$\overset{\circ}{x}_{ij} = \left(x_{ij} - \langle X_j \rangle\right)/\sigma_j, \tag{1}$$

where $\langle X_j \rangle$ is the mean value of the jth state variable, σ_j is its mean square deviation.

The problem of a diagnostic model building: a vector function is given by the set of training pairs $\left(\vec{X}^{(0)}, \vec{d}\right)_p$, $p = 1 \ldots P$, where $\vec{X}^{(0)}, \vec{d}$ are the input vectors with dimensions H_0 and the output with dimensions H_{k+1}. This sample should be approximated. The result of the solution of the problem must be some mathematical mechanism that would obtain any value of the vector function $\vec{Y}^{(K+1)}\left(\vec{X}^{(0)}\right)$, which is given in the form of a training sample, by a given input vector in the range that is limited by the input.

The classification problem statement is formulated. Let $\vec{X}^*$ is the vector of variables that describe the state of precedents, and let M is the set of class numbers (scenarios). The number of possible scenarios is known, as well as for each scenario (class), subsets of observed state variables (symptoms) are formed. According to the values of the projections of the vector $\vec{X}^*$, the precedent is assigned to one of the possible sets R_m, where $m = 0 \ldots M - 1$. It is necessary to find such an m-th scenario for which the maximum density of the conditional probability distribution of the occurrence of $\vec{X}^*$ in the precedent under the m-th scenario is

$$\exists!\ m^* \in C_m\left(\rho\left(\vec{X}_m^* | R_m\right)\right)(m = 0\ldots M - 1) : \rho\left(\vec{X}_m^* | R_m\right) \rightarrow max, \tag{2}$$

here $C_m\left(\rho\left(\vec{X}_m^* | R_m\right)\right)$ – the set of m-th indices of the density distribution of the conditional probability of the occurrence of $\vec{X}^*$ in the precedent in the m-th scenario.

The logistic function argument was used as the target function values. The use of logistic regression makes it possible to predict the likelihood of an event occurring by the values of a certain number of attributes. To do this, we introduce the so-called dependent variable y, taking only one of the two values - as a rule, numbers 0 (event did not occur) and 1 (event occurred), and independent variables (signs) - $\vec{X}^*$, based on the values of which you need to calculate the probability of accepting one or the other value of the dependent variable.

It is assumed that the probability of an event $y = 1$ is:

$$Pr\left\{y = 1 | \vec{X}^*\right\} = f(z),$$

here $z = \theta^T x = \theta_1 x_1 + \ldots + \theta_n x_n$, $\theta_i x$ – he vectors of the independent variables values $x_1, \ldots, x_n$ i and the parameters (regression coefficients) are real numbers $\theta_1, \ldots, \theta_n$, respectively, and $f(z)$ is the so-called logistic function (sometimes also called sigmoid or logit function):

$$f(z) = \frac{1}{1 + e^{-z}}.$$

Since y takes only values 0 and 1, then the probability of the second possible value is:

$$\Pr\{y=0|x\} = 1-f(z) = 1-f(\theta^T x).$$

For brevity, the distribution function y for a given x can be written as follows:

$$\Pr\{y|x\} = f(\theta^T x)^y \left(1-f(\theta^T x)\right)^{1-y}, y \in \{0,1\}.$$

To solve the object state classification problem, we used a unidirectional multilayer artificial neural network (UMN) to study and a radial basis artificial neural network (RBN). Let's look at their architecture and the specificities of learning methods to solve the problem under consideration.

3 Structure and Training Method of Unidirectional Multilayer Artificial Neural Networks

The simplest UMN with one hidden layer ($K = 1$) is shown in Fig. 1. Here $\left\{Y_{ph}^{(0)}\right\}$ – is the set of input data, $\left\{Y_i^{(k)}\right\}$ – is the set of output data of the k-th layer; k is the layer number, $k = 1 \ldots (K + 1)$, K is the number of hidden layers; $p = 1 \ldots P$, P is the number of analogues; $\left\{w_{ij}^{(k)}\right\}$ is the set of weights of the k-th layer; i is an element of the k-th layer; j is the element of the (k–1)-th layer; H_0 is number of network inputs; H_1 is number of hidden layer neurons; H_2 is the number of network outputs.

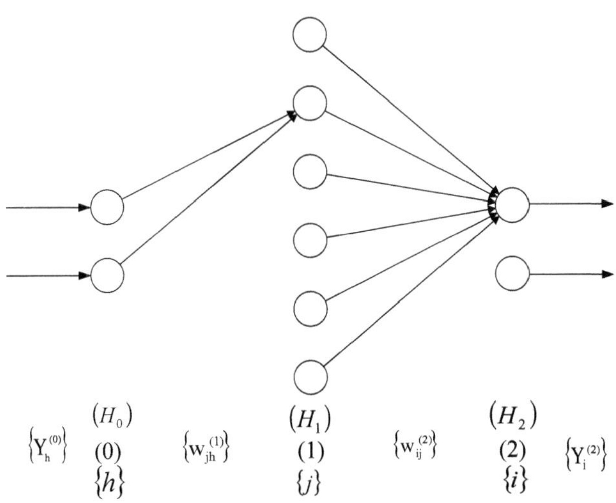

Fig. 1. UMN structure

Analytical representation of the required functions for a unidirectional multilayer network, according to Fig. 1, has the following structure:

$$Y_i^{(2)} = f\left(s_i^{(2)}\right), s_i^{(2)} = w_{i0}^{(2)} + \sum_{j=1}^{H_1} w_{ij}^{(2)} Y_j^{(1)}, i = 1 \ldots H_2,\ j = 1 \ldots H_1$$

$$Y_j^{(1)} = f\left(s_j^{(1)}\right), s_j^{(1)} = w_{j0}^{(1)} + \sum_{h=1}^{H_0} w_{ih}^{(1)} Y_h^{(0)}, h = 1 \ldots H_0,$$

here $f(s) = th(\beta s) = \frac{e^{\beta s} - 1/e^{\beta s}}{e^{\beta s} + 1/e^{\beta s}}$ is the chosen activation function,

$f_s^{'} = \beta[1 - f^2(s)]$ is the derivative of the activation function.

Stable (robust) statistical methods of UMN parameters estimation were used as methods of UMN training, detailed description of which was published in the monograph [20]. These methods provide robustness and informative parameters of systems (processes) statistical models with a priori uncertainty of the input data.

The model parameters were denoted by the vector M – the vector of random numbers of dimension m, and the data of measurements by the vector D – the vector of random numbers of dimension H0 + HK + 1. If p measurements of vector D were observed, then the posterior probability density distribution of vector M after P (p = 1 … P) measurements would be equal to

$$\rho(M|D_P) = \frac{\rho(D_P|M)\rho(M)}{\rho(D_P)},$$

here $\rho(D_P|M) = \prod_{p=1}^{P} \rho(D_p|M)$, $\rho(M) = \prod_{n=1}^{m} \rho(M_n)$, $\rho(D_P) = \prod_{p=1}^{P} \rho(D_p)$ is constant, $m = \sum_{k=1}^{K+1} (H_{k-1} + 1)H_k$ – for UMN ($m = H_0 H_1 + (H_1 + 1)H_2$ – for radial basis neural networks), $m \leq PH_{k+1}$.

The vector $\widehat{M}$ was found as a solution to the problem

$$\widehat{M}_{t+1} = \arg\ \inf_{M \in D_M} E(M|D_P),$$

where function E is a scalar convolution of objective functions by $f_i \equiv Y_i^{(2)}$, $x_h = Y_h^{(0)}$. has the form

$$E = \frac{1}{2PI} \sum_{p=1}^{P} \gamma^{P-p} \sum_{i=1}^{I} \left\{ f_{fit} \left[4 \left(\frac{\Delta_{f,p}}{f_i^*} \right)^2 \left(1 + \sigma_{f,p}^0 \right)^{-2} \right] + \beta_{t+1} f_{fit} \left[\left(\sigma_{f,p}^0 \right)^2 - 1 \right] \right\},$$

for which $\beta_{t+1} = \frac{\sigma_D^2}{\sigma_M^2}$, σ_D^2, σ_M^2 – the variance of random numbers, $t = 1 \ldots T$ – the number of the epoch of training, $I = H_k$, γ – the level of significance (γ = [0.95, 0.99]), $f_{fit}(d_i) = 1 - exp\left[\left(-\frac{L_{fit}}{4}\right) d_i\right]$, $L_{fit} \geq 4 (d_i > 0)$, $\left(\sigma_{f_i}^{\circ}\right)^2 = \frac{\left(\sigma_{f_i}\right)^2}{\left(\sigma_{f_i}^*\right)^2}$, $\sigma_{f_i}^2 = \beta^4 \left[1 - f^2\left(s_i^{(2)}\right)\right]^2$

$\sum_{j=1}^{H_1}\left\{\left(w_{ij}^{(2)}\right)^2\left[1-f^2 s_j^{(1)}\right]^2\sum_{h=1}^{H_0}\left[\left(w_{ih}^{(1)}\right)^2\left(\sigma^*_{F_h^{(0)}}\right)^2\right]\right\}$, $\left(\sigma^*_{f_i}\right)^2=\left(\frac{2l_f}{f_{i,max}-f_{i,min}}\right)^2\left[\frac{\Delta^\circ_{f_i}}{300}f_{i,max}\right]^2 n_\alpha$ $(n_\alpha=1)$, $\Delta^\circ_{f_i}=\frac{\Delta_{f_i}}{f_{i,max}}100\%$.

The solution – approximate functions of the form $Y_i^{(k+1)}\left(\vec{Y}^{(0)}\right)$ – is found by stochastic approximation on the basis of the tiered gradient conjunction method [20]. The training used a regularization algorithm that implements interrupts in the iterative process in cases of accumulation of calculation errors [20].

Training and moment coefficients were dependent on the following:

$$\eta_{ij}^{(k)}=\rho_{ij}^{(k)}(t)\eta_{ij}^{(k)}(t-1), \eta_{ij}^{(k)}(0)=\eta_{max},$$

here $\rho_{ij}^{(k)}(t)=-\frac{1}{\left.\frac{\partial s_{ij}^{(k)}}{\partial w_{ij}^{(k)}}\right|_t}\left(1-exp\left(h\left.\frac{\partial S_{ij}^{(k)}}{\partial w_{ij}^{(k)}}\right|_t\right)\right)$, $\alpha_{ij}^{(k)}(t)=-\frac{S_{ij}^{(k)}(t)}{S_{ij}^{(k)}(t)-S_{ij}^{(k)}(t-1)}$, the implementation of the condition $E_{\tau+1}>E_\tau, E_\tau<E_{\tau-1}$ is given in the form:

(a) if $E^\circ_\tau\le K_w$ and $S_{ij}^{(k)}(t)S_{ij}^{(k)}(t-1)>0$ then $\alpha_{ij}^{(k)}(t)=\alpha_{mid}$;

(b) if $E^\circ_\tau\le K_w$ and $S_{ij}^{(k)}(t)\left(w_{ij}^{(k)}(t)-w_{ij}^{(k)}(t-1)\right)\alpha_{ij}^{(k)}(t)<0$ then

$$w_{ij}^{(k)}(t+1)=w_{ij}^{(k)}(t)+\mu(t)\left\{\eta_{ij}^{(k)}(t)r_{ij}^{(k)}(t)-\upsilon(t)\alpha_{ij}^{(k)}(t)\left[w_{ij}^{(k)}(t)-\widehat{w}_{ij}^{(k)}\right]\right\}+\widetilde{w}_{ij}^{(k)}(t+1),$$

where the quantities $\widehat{w}_{ij}^{(k)}$ were determined according to the method of simulating the movement of bee colonies by the formula

$$\widehat{W}=\arg\inf_{\substack{W\in D_w,\\ \tau\in[1,t-1]}} E(W,\tau), W=\left\{w_{i,j}^{(k)}\right\}$$

The following parameter values were accepted h = 0.95, K_w = 1.04, $\rho_{ij}^{(k)}(t)=[0.7,1.05]$, $\eta_{ij}^{(k)}(t)=[0.01,0.4]$, $\alpha_{ij}^{(k)}(t)=[-0.8,0.8]$, α_{mid} = 0.005, α_{opt} = 0.5.

When pairwise comparison of formal mathematical models, the variance of the signal variance, which characterizes the robustness of a particular model, is estimated:

$$D_{Y_i,dB}=10\lg\left(\frac{D_{Y_i}^{(\lambda)}}{D_{Y_i}^{(0)}}\right), \text{ decibel; } \lambda=1,2.$$

The residual variance values for each of the compared formal mathematical models were then used as estimates of the signal dispersions.

4 Structure and Training Method of Radial Basis Artificial Neural Networks

The structure of the simplest RBN with one hidden layer ($K = 1$) is similar to that shown in Fig. 1. We have the following designations:

$\vec{Y}^{(k)} = \left[Y_1^{(k)}, \ldots, Y_{H_1}^{(k)}\right]^T$, $k = 0, 1, 2$, is vector of the input data of the k-th layer;

$\vec{c}_j = \left[c_{j1}, c_{j2}, \ldots, c_{jH_0}\right]^T, j = 1 \ldots H$, is vector of coordinates of centers of activation function for hidden layer neurons;

$\vec{\sigma}_j = \left[\sigma_{j1}, \sigma_{j2}, \ldots, \sigma_{jH_0}\right]^T$, $j = 1 \ldots H_1$, is vector that specifies the width of the window of the activation function of the j-th neuron of the hidden layer;

$\varphi_j = \left(\vec{Y}_p^{(0)}, \vec{c}_j, \vec{\sigma}_j\right) = exp\left(-\frac{1}{2}\sum_{h=1}^{H_0} Z_{pjh}^2\right) \equiv \varphi_{pj}$ is radial-base activation function of the hidden layer neuron, $Z_{pjh} = \frac{Y_{ph}^{(0)} - c_{jh}}{\sigma_{jh}}$;

w_{ij} is the weight of the connection between the i-th neuron of the source layer and the j-th neuron of the hidden layer (here, according to the notation in Fig. 1, it is implied that $w_{jh}^{(1)} = e_{jh} = 1, w_{ij}^{(2)} \equiv w_{ij}$).

The analytical representation of the desired functions for the radial basis network in accordance with Fig. 1 has the following structure:

$$Y_i^{(2)} = f\left(s_i^{(2)}\right), s_i^{(2)} = w_{i0}^{(2)} + \sum\nolimits_{j=1}^{H_1} w_{ij}^{(2)} Y_j^{(1)}, i = 1\ldots H_2,\ j = 1\ldots H_1;$$

$$Y_j^{(1)} = \varphi\left(s_j^{(1)}\right), s_j^{(1)} = \frac{1}{2}\sum\nolimits_{h=1}^{H_0} \left(z_{jh}\right)^2, h = 1\ldots H_0,$$

where $\varphi(s) = \exp(-s)$ – is the activation function selected, $\varphi_s'(s) = -\varphi(s)$ – s a derivative of the activation function.

A hybrid algorithm was used to train RBNs when the number of training pairs significantly exceeds the number of neurons in the hidden layer $P \gg H_1$. The learning process is divided into two stages:

- selection of linear network parameters (weights of the output layer) based on the use of the pseudo inversion method;
- adaptation of nonlinear parameters of activation functions (centers $\vec{c}_j$ i width $\vec{\sigma}_j$ for these functions).

Both stages are closely intertwined. In the presence of P training pairs $\left(\vec{Y}_p^{(0)}, \vec{d}_p\right)$, $p = 1\ldots P$, and fixation of specific values of centers and width of activation functions (in the first moment these will be initial values) we get a system of equations ($P \gg H_1$):

$$\Phi \vec{w}_i = \vec{d}_i, i = 1\,..\,H_2,$$

where $\Phi = \left[\varphi_{pj}\right], p = 1\ldots P, j = 0\ldots H_1;$

$$\varphi_{p0} = 1, \vec{w}_i = \left[w_{i0}, w_{i1}, \ldots, w_{iH_1}\right]^T, \ \vec{d}_i = \left[d_{0i}, d_{1i}, \ldots, d_{pi}\right]^T.$$

Weight vector $\vec{w}_i$ can be defined in one step by a pseudo inversion of the matrix Φ, i.e. $\vec{w}_i = \Phi^+ \vec{d}_i$, where Φ^+ – pseudo-inverted to Φ matrix. In practice, the pseudo inversion is calculated using eigenvalue decomposition. (SVD - decomposition), according to which $\Phi = USV^T$ [17].

In the second stage of the algorithm, when the values of the output weights are fixed, the excitation signals are passed over the network to the output layer, which allows to calculate the magnitude of the error for the sequence of vectors $\left\{\vec{Y}_p^{(0)}\right\}$. There is a return to the hidden layer (back propagation) after that. The gradient vector of the choice function with respect to specific centers $\vec{c}_j$ i ширини $\vec{\sigma}_j$ is determined by the error value $\left\|\vec{Y}^{(2)} - \vec{d}\right\|_{L_2}$.

The algorithm of forming the "coverage area" by radial basis functions with consideration of K-neighbors $\sigma_{jh}^2 = \sum_j = \frac{1}{K}\sum_{k=1}^{K}\sum_{h=1}^{H_0}\left(c_{jh} - c_{kh}\right)^2, k = 1\ldots K \ \ K \in [3, 5]$ was applied to determine the values of $\sigma = \left\{\sigma_{jh}\right\}$, which will reduce the duration of training RBN [17].

Refining the values of the covariance matrix $\sigma = \left\{\sigma_{jh}\right\}$ and the coordinates of the centers $C = \left\{c_{jh}\right\}$ completes the next training cycle. Correction of the elements of the covariance matrix is carried out according to the formula (a recurrent learning algorithm is used, which corresponds to the stochastic approximation method based on the ravine conjugate gradient method and ensures convergence $\sigma_{jh}(t) \underset{t\to\infty}{\rightarrow} \widehat{\sigma}_{jh}$ with probability $P = 1$):

$$\sigma_{jh}(t+1) = \sigma_{jh}(t) + \mu(t)\left\{\eta_{jh}(t)r_{jh}(t) + v(t)\alpha_{jh}(t)\left[\sigma_{jh}(t) - \widehat{\sigma}_{jh}\right]\right\} + \widetilde{\sigma}_{jh}(t+1),$$

where the projections of the search direction vector $r_{jh}(t)$ were in accordance with the conjugate gradient method.

Repeated repetition of both steps leads to complete and fast learning of the network, especially if the initial values of the parameters of the radial basis functions are close to optimal values.

In practice, the selected steps affect the adaptation of parameters to varying degrees. As a rule, the SVD algorithm functions faster (it finds the local minimum of the function in one step). To balance this disproportion, one refinement of linear parameters is usually accompanied by several cycles of nonlinear parameter adaptation.

5 An Example of the Use of Artificial Neural Networks for the Classification of Patients' Conditions in Medical Monitoring Systems

Two samples were considered, which represent the laboratory data of patients on two types of diseases: liver disease and urological diseases.

The Urological Data Sample contains 47 state variables, whose values were of three types: real numbers, booleans, and enumerators. The training sample consisted of data on 30 patients, the test sample contained data on 10 patients. The result of the classification is to determine the patient's condition (healthy or ill). For classification, the RBN with logistic function was used. RBN structure contains 47 neurons of the input layer corresponding to the number of state variables, 26 neurons of the hidden layer, and 2 neurons of the output layer corresponding to the two classified states. As a result of training, the error was 0.122 and the standard deviation of 0.0024. The recognition result on the control sample is 100%.

As a result of training and network testing of the Urological Data Sample, the confusion matrix and ROC curves were constructed to analyze the quality of the classification (Figs. 2 and 3).

Redraw confusion matrix for 2 classes.

Truth data

Classifier results		Class 1	Class 2	Classification overall	Producer Accuracy (Precision)
	Class 1	11	0	11	100%
	Class 2	0	29	29	100%
	Truth overall	11	29	40	
	User Accuracy (Recall)	100%	100%		

Overall accuracy (OA): 100%

Kappa[1]: 1

Fig. 2. Confusion matrix of the Urological Data Sample classification problem

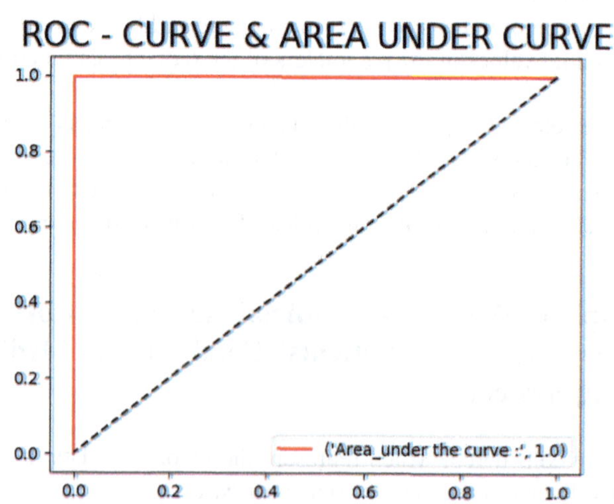

Fig. 3. ROC-curve of the Urological Data Sample classification problem

The liver disease sample contains 10 state variables, the values of which were of three types: real numbers, booleans, and enumerators. The study sample consisted of data on 550 patients, and the test sample contained data on 40 patients. RBN structure contains 10 neurons of the input layer corresponding to the number of state variables, 540 neurons of the hidden layer, and 2 neurons of the output layer corresponding to the two classified states. As a result of training, the error was 0.0107 and the standard deviation of 0.0185. The recognition result on the control sample is 80%.

As a result of training and network testing of the liver disease sample, the confusion matrix and ROC curves were constructed to analyze the quality of the classification (Figs. 4 and 5).

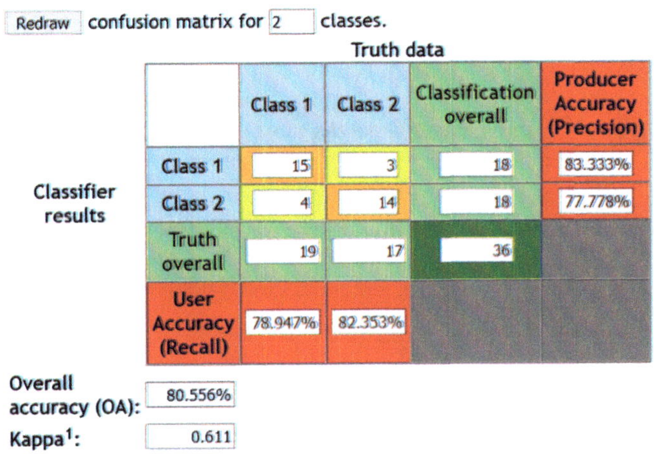

Redraw confusion matrix for 2 classes.

Truth data

Classifier results		Class 1	Class 2	Classification overall	Producer Accuracy (Precision)
	Class 1	15	3	18	83.333%
	Class 2	4	14	18	77.778%
	Truth overall	19	17	36	
	User Accuracy (Recall)	78.947%	82.353%		

Overall accuracy (OA): 80.556%

Kappa[1]: 0.611

Fig. 4. Confusion matrix of the liver disease sample classification problem

A confusion matrix, also known as an error matrix, [25] is a specific table layout that allows visualization of the performance of an algorithm, typically a supervised learning one (in unsupervised learning it is usually called a matching matrix). ROC-curve (receiver operating characteristic) – a graph that allows you to evaluate the quality of binary classification displays the relationship between the share of objects from the total number of characteristic carriers correctly classified as bearing the attribute (called the sensitivity of the classification algorithm) and the share of objects from the total number of objects that do not carry the attribute, erroneously classified as bearing a trait [22–24].

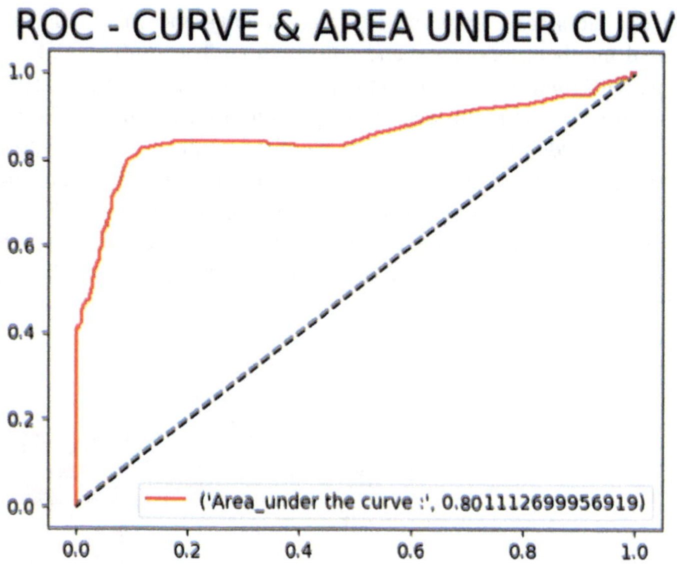

Fig. 5. ROC-curve of the liver disease sample classification problem

References

1. Bakumenko, N.: Application of the c-means fuzzy clustering method for the patient's state recognition problems in the medicine monitoring systems. In: Bakumenko, N., Strilets, V., Ugriumov, M. (eds.) CEUR Workshop Proceedings of 3rd International Conference on Computational Linguistics and Intelligent Systems, COLINS 2019, vol. 2362, pp. 218–227 (2019)
2. Zhang, H.: The Optimality of Naive Bayes. American Association for Artificial Intelligence (2004)
3. Metsis, V., Androutsopoulos, I., Paliouras, G.: Spam filtering with Naive Bayes—which Naive Bayes? In: Third Conference on Email and Anti-Spam, 27–28 July 2006, Mountain View, California USA (2006)
4. Rennie, J., Shih, L., Teevan, J., Karger, D.: Tackling the poor assumptions of Naive Bayes classifiers. In: Proceedings of the Twentieth International Conference on Machine Learning, Washington D.C. (2003)
5. Cover, T.M., Hart, P.E.: Nearest neighbor pattern classification. IEEE Trans. Inf. Theory **IT-13**(1), 21–27 (1967)
6. Garcia, S., Derrac, J., Cano, J., Herrera, F.: Prototype selection for nearest neighbor classification: taxonomy and empirical study. IEEE Trans. Pattern Anal. Mach. Intell. **34**(3), 417–435 (2012)
7. Ho, T.K.: Random decision forests. In: Proceedings of the 3rd International Conference on Document Analysis and Recognition, Montreal, QC, pp. 278–282, 14–16 August 1995
8. Denisko, D., Hoffman, M.M.: Classification and interaction in random forests. Proc. Natl. Acad. Sci. U.S.A. **115**(8), 1690–1692 (2018)
9. Riddick, G., Song, H., Ahn, S., Walling, J., Borges-Rivera, D., Zhang, W., Fine, H.A.: Predicting in vitro drug sensitivity using Random Forests. Bioinformatics (Oxford, England) **27**(2), 220–224 (2011)

10. Touw, W.G., Bayjanov, J.R., Overmars, L., Backus, L., Boekhorst, J., Wels, M., van Hijum, S.A.: Data mining in the Life Sciences with Random Forest: a walk in the park or lost in the jungle? Brief Bioinform. **14**(3), 315–326 (2013)
11. Basu, S., Kumbier, K., Brown, J.B., Yu, B.: Iterative random forests to discover predictive and stable high-order interactions. Proc. Natl. Acad. Sci. U.S.A. **115**(8), 1943–1948 (2018)
12. Breiman, L.: Random forest, machine learning. In: Proceedings of the Thirteenth International Conference, vol. 45, pp. 5–32 (2001)
13. Strano, M., Colosimo, B.M.: Logistic regression analysis for experimental determination of forming limit diagrams. Int. J. Mach. Tools Manuf **46**(6), 673–682 (2006)
14. Cramer, J.S.: The origins of logistic regression. Tinbergen Inst. **119**, 167–178 (2002)
15. Kaminski, B., Jakubczyk, M., Szufel, P.: A framework for sensitivity analysis of decision trees. Central Eur. J. Oper. Res. **26**(1), 135–159 (2017)
16. Karimi, K., Hamilton, H.J.: Generation and interpretation of temporal decision rules. Int. J. Comput. Inf. Syst. Ind. Manag. Appl. **3** (2011)
17. Gao, W., Zhou, Z.-H.: On the doubt about margin explanation of boosting. Artif. Intell. J. **203**, 1–18 (2013)
18. Freund, Y., Schapire, R.E.: A short introduction to boosting. J. Jpn. Soc. Artif. Intell. **14**(5), 771–780 (1999)
19. Kegl, B.: The return of AdaBoost. MH: multi-class Hamming trees. In International Conference on Learning Representations, 2014
20. Системное совершенствование элементов сложных технических систем на основе концепции обратных задач [Текст]: монография/ В. Е. Стрелец, А. А. Трончук, Е. М. Угрюмова и др.; под общ. ред. М. Л. Угрюмова. – Х. : Нац. аэрокосм. ун-т им. Н. Е. Жуковского « Харьк. авиац. ин-т » , 148 с (2013)
21. Угрюмова Е.М. Обучаемые искусственные нейронные сети в построении формальных математических моделей систем при априорной неопределенности данных/Е.М. Угрюмова// Вісник Харківського національного університету: зб. наук. пр. Сер. Мате матичне моделювання. Інформаційні технології. Автоматизовані системи управління. – 2010. – Випуск 13 (№890). – С. 237–253 (2010)
22. Powers, D.M.W.: Evaluation: from precision, recall and f-measure to ROC, informedness, markedness & correlation. J. Mach. Learn. Technol. **2**(1), 37–63 (2011)
23. Flach, P., Hernandez-Orallo, J., Ferri, C.: A coherent interpretation of AUC as a measure of aggregated classification performance. In: Appearing in Proceedings of the 28th International Conference on Machine Learning, Bellevue, WA, USA (2011)
24. Bi, J., Bennett, K.P.: Regression error characteristic curves. In: Proceedings of the Twentieth International Conference on Machine Learning (ICML-2003), Washington DC (2003)
25. Stehman, S.V.: Selecting and interpreting measures of thematic classification accuracy. Remote Sens. Environ. **62**(1), 77–89 (1997)

Development and Analysis of Intelligent Recommendation System Using Machine Learning Approach

Pavlo Piletskiy, Dmytro Chumachenko(✉), and Ievgen Meniailov

National Aerospace University "Kharkiv Aviation Institute", Chkalova Street 17, Kharkiv 61070, Ukraine
pavlo1997pileckii@gmail.com, dichumachenko@gmail.com, j.menyailov@khai.edu

Abstract. Recommendation systems have changed the way inanimate websites interact with their users. Instead of providing static information when users search and possibly buy products, recommendation systems increase the degree of interactivity to expand the possibilities provided to the user. Recommendation systems generate recommendations independently for each specific user based on his past purchases and searches, as well as on the basis of the behavior of other users. This article describes recommender systems and their algorithms. Recommendation system is developed using SVD algorithm. The architectural pattern MVC was used. The implementation represents a three-tier architecture (DAL, BLL, WEB). Program realization of recommendation system was made using .Net language.

Keywords: Recommendation system · Artificial intelligence · SVD · Recommender · Machine learning

1 Introduction

The modern works of many scientists are devoted to the development of intellectual problem-oriented systems and their application to population dynamics [1–7]. The most important tool for studying these systems is adequate mathematical models for predicting the processes to make adequate management decisions [8–11]. To date, a significant number of such theoretically valid models have been created. They rely on the mathematical apparatus of statistics and probability theory [12–16]. A common drawback of existing models is the low accuracy of forecasting, as well as its short-term.

Recommendation systems are programs that try to predict which objects (movies, music, books, news, websites) will be of interest to the user, having certain information about his profile.

Two main strategies for creating recommendation systems are content-based filtering and collaborative filtering [17, 18]. When filtering based on the content, user and object profiles are created, user profiles can include demographic information or

M. Nechyporuk et al. (Eds.): *Integrated Computer Technologies in Mechanical Engineering*, AISC 1113, pp. 186–197, 2020.
https://doi.org/10.1007/978-3-030-37618-5_17

answers to a specific set of questions, object profiles can include genre names, actor names, artist names and other attributive information depending on the type of object. For example, in the Music Genome Project, a music analyst evaluates each track by hundreds of different musical characteristics that can be used to identify a user's musical preferences. Collaborative filtering uses information about past user behavior - for example, information about purchases or ratings. In this case, it does not matter with what types of objects the work is carried out, but at the same time implicit characteristics can be considered, which would be difficult to consider when creating the profile. The main problem of this type of recommender systems is the "cold start": lack of data on users or objects that have recently appeared in the system.

In the process, recommender systems collect user data using a combination of explicit and implicit methods. Explicit Data Collection examples:

- request from the user to evaluate the object on a differentiated scale;
- request from the user to rank a group of objects from best to worst;
- presenting to the user two objects with a question about which one is better;
- a proposal to create a list of objects that the user likes.

Examples of implicit data collection:

- monitoring what the user is viewing in online stores or other types of databases;
- keeping records of user behavior online;
- tracking the contents of a user's computer.

Recommender systems compare the same type of data from different people and calculate a list of recommendations for a particular user. Some examples of their commercial and non-commercial use are given in the article on collaborative filtering. To calculate the recommendations, the graph of interests is used [19]. Recommender systems are a convenient alternative to search algorithms, as they allow you to detect objects that cannot be found last. Curiously, recommender systems often use search engines to index unusual data.

2 Content Filtering

Content filtering is based on creating a user profile and an object profile. It is necessary to consider the parameters of the objects and their compliance with user preferences. For this purpose, recommender systems use tags (keywords) to describe objects, and the user profile reflects the assessment of certain tags or their combination.

In recommendation systems with content filtering, the satisfaction function $h(u, s)$ of user u by some object s is determined based on information about the user's satisfaction with objects $s_i \in S$, which are like s. That is, if the recommendation system is used in a web application, then in order to recommend video games to the user u, the system must determine what connects different games to which the user has previously praised. Then the system will recommend to the user the games with the maximum match to those that were highly rated by the user earlier.

In a way, the user profile in the system is formed from the content consumed by the user in the form of a set of parameters that define the object s. These parameters are usually keywords and their corresponding weights for each object. Therefore, it is necessary to determine the weights of these parameters. In information retrieval, one of the most well-known methods for determining keyword weights is the TF-IDF measure [20].

Suppose that N is the total number of objects that are likely to be recommended to the user and that the keyword k_j occurs in n_i objects, and $f_{i,j}$ is the number of occurrences of this word in the object d_j. Then $TF_{i,j}$ (term frequency) is the ratio of the number of occurrences of the tag to the total number of tags of the object, that is:

$$TF_{i,j} = \frac{f_{i,j}}{max_z f_{z,j}} \tag{1}$$

But if we consider only the frequency of occurrence of the tag, then in most objects the most common tags will have maximum weight, which, most likely, will lead to an incorrect assessment of user preferences. To avoid this, IDFi (inverse document frequency) is used - the reciprocal of the frequency the tag entered the collection object. We define it as:

$$IDF_i = log\frac{N}{n_i} \tag{2}$$

Thus, the weight $w_{i,j}$ of the keyword k_i in the object d_j is denoted as the product of the frequency of occurrence of the tag on the inverse frequency of the object:

$$w_{i,j} = TF_{i,j} \times IDF_i \tag{3}$$

As stated above, content filtering recommender systems offer objects based on those that the user praised earlier. Various objects are compared with the ones that the user liked and recommended, those that have maximum similarity.

The set of objects that the user rated earlier forms the user profile ContentBasedProfile(u) or, in other words, the weight vector ($w_{u,1},\ldots, w_{u,k}$), where each weight $\underline{w_{u,i}}$ determines the importance of the k_i tag for user u. Consequently, ContentBasedProfile (u) and Content (s) can be represented as TF-IDF vectors w_u and w_s, while the user satisfaction function $h\ (u,\ s)$ can be represented as the cosine coefficient of the vectors w_u and w_s:

$$(u, s) = \cos\left(\overrightarrow{w_u}, \overrightarrow{w_s}\right) = \frac{\overrightarrow{w_u}\,\overrightarrow{w_s}}{\left\|\overrightarrow{w_u}\right\|\left\|\overrightarrow{w_s}\right\|} = \frac{\sum_{i=1}^{K} w_{i,u} w_{i,s}}{\sqrt{\sum_{i=1}^{K} w_{i,u}^2}\sqrt{\sum_{i=1}^{K} w_{i,s}^2}} \tag{4}$$

where K is the total number of tags in the system.

In addition to heuristics based mainly on information retrieval methods, quite often other techniques of recommendations with content filtering are used, for example, a naive Bayes classifier, various machine learning techniques, including neural networks, decision trees and clustering [21]. But unlike the methods of information retrieval, these techniques try to predict user satisfaction using heuristics as the basis of the cosine coefficient, rather than a consistent model that was derived from the data using mathematical statistics and machine learning. For example, in musical audio streaming, you can select information that you haven't heard through the naive Bayesian classifier using information about favorite and non-liked musical compositions and evaluate the likelihood that composition 9 of p_j games to a specific C_i class (liked and not liked).

3 Collaborative Filtering

Collaborative filtering uses the well-known preferences of the user group in which the user for whom you need to predict recommendations is a member. The main reason for this approach is that users who have rated the same object in the past equally are likely to rate the other object in the future.

In contrast to content filtering, collaborative filtering methods try to predict whether an object will match user preferences based on the ratings of other users of the same group, i.e. we obtain a satisfaction estimate *h(u, s)* with some object s for user u, which is calculated depending on satisfaction $h(u_j, s)$ with the same object *s* of users $u_j \in U$, which have similar characteristics with user *u*.

It is worth noting an important feature of this method: signs of similarity of one user to others are not always part of the system under consideration. That is, in system 10, user groups can be determined not only based on personal preferences of users, but also based on more general characteristics, such as geographical location, profiles on social networks and other parameters. This kind of information can serve as an initial assessment of the similarity of a new user to those already registered in the system and can be used to solve the cold start problem common for collaborative filtering.

There are a fairly large number of user-based recommendation systems developed with the participation of both scientists and commercial structures of various fields of activity. Relatively the first system of this kind is the "Grundy system" [22] - the librarian program asks users questions on various topics, thus creating a user profile and recommending books that could please the user. If the user replied that the proposed book was not interesting to him, the system began to find out the reason, thereby deepening the profile information. Thus, the "Grundy system" was a personal model for each user, thanks to which she could very accurately advise interesting literature. At the same time, another system called "Tapestry" [23], instead of determining similarity through questions, suggested the user to list users with the appropriate choice of preferences.

Also, there is an opinion [24] that collaborative filtering methods for content are divided into two main categories: memory-based and model-based. The first category of algorithms tries to predict the rating of an object using knowledge of all objects that the user has managed to evaluate earlier. The rating value $r_{u,s}$ for user u and object s is calculated as the aggregated value of ratings of other most similar *N* users for the same object *s*:

$$r_{u,s} = aggr_{u \in \widehat{U}} r_{u',s'} \tag{5}$$

where $\widehat{U}$ denotes the set N of users who rated object s and are most like user u. The simplest example of an aggregation function is the calculation of the mean:

$$r_{u,s} = \frac{1}{N}\sum\nolimits_{u\in\widehat{U}} r_{u',s} \tag{6}$$

The most common case of aggregation is:

$$r_{u,s} = k\sum\nolimits_{u\in\widehat{U}} sim(u,u') \times r_{u',s'} \tag{7}$$

$$r_{u,s} = \overline{r_u} + k\sum\nolimits_{u\in\widehat{U}} sim(u,u') \times \left(r_{u',s} - \overline{r_{u'}}\right) \tag{8}$$

where k is used to normalize and, as a rule,

$$\mathrm{k} = \frac{1}{\sum\nolimits_{u'\in\widehat{U}} sim|u,u'|} \tag{9}$$

$$\overline{r_u} = \left(\frac{1}{S_u}\right)\sum\nolimits_{s\in S_u} r_{u,s'} \tag{10}$$

and sim (u, u') is a measure of "similarity", which is the reciprocal of the distance and in most cases is taken as the weight. Thus, the more users u and u' are similar, the more weight will be considered when calculating $r_{u,s}$. From formula (9) we can see that in different recommendation systems, quite possibly, different similarity measures can be applied. Let's look at the two most common of them. In the case of weighted sums (9), there are several difficulties, for example, the inability to consider that different users do not always evaluate objects with the same level of severity. To avoid such a problem, approach (10) is applied. This formula considers differences from the average rating for a user, instead of using absolute ratings.

There are also many different methods for finding similarities between different *sim (c, c')* users in collaborative filtering recommendation systems. In most cases, the basis for the similarity of users are the ratings assigned by them to the same system objects.

4 Hybrid Filtration Methods

Some recommendation systems use combinations of content filtering and collaborative filtering approaches, called hybrid methods. They allow one way or another to avoid the disadvantages of both approaches. There are several basic options for combining different methods in hybrid recommendation systems:

1. Building a single model using the characteristics of both methods.
2. The introduction of item-based features in user-based systems.
3. The introduction of user-based features in item-based systems.
4. Implementing item-based and user-based approaches separately and using combinations of ratings obtained.

5 SVD Algorithm

The SVD (Singular Value Decomposition) algorithm, translated as a singular matrix decomposition, is designed to implement a recommendation system.

Singular Value Decomposition (SVD) - decomposition of a real matrix with the aim of its reduction to canonical form. Singular decomposition is a convenient method for working with matrices. It shows the geometric structure of the matrix and allows you to visualize the available data. The singular decomposition is used to solve a variety of problems - from the least squares approximation and solving systems of equations to image compression. In this case, various properties of the singular decomposition are used, for example, the ability to show the rank of a matrix and approximate matrices of a given rank. SVD allows you to calculate large inverse and pseudo-inverse matrices, which makes it a useful tool for solving regression analysis problems.

For any real *(n* x *n)* - matrix *A*, there are two reals orthogonal *(n* x *n)* - matrices *U* and *V* such that $U^T AV$ is the diagonal matrix Λ,

$$U^T \cdot AV = \Lambda \tag{11}$$

The matrices *U* and *V* are chosen so that the diagonal elements of the matrix Λ have the form

$$\lambda_1 \geq \lambda_2 \geq \ldots \geq \lambda_r > \lambda_{r+1} = \ldots = \lambda_n = 0 \tag{12}$$

Where *r*—matrix rank *A*. In particular, if *A* is non-degenerate,

$$\lambda_1 \geq \lambda_2 \geq \ldots \geq \lambda_n > 0 \tag{13}$$

Index *r* of element is λ_r actual dimension of matrix eigenspace *A*.

The columns of the matrices *U* and *V* are called the left and right singular vectors, respectively, and the diagonal values of the matrix Λ are called the singular numbers.

Equivalent notation of singular decomposition—$A = U\Lambda V^T$.

It is easy to see that the matrices *U* and *V* are orthogonal,

$U^T U = UU^T = I$ and $V^T V = VV^T = I$, and the sum of the squared values from the columns is equal to one.

In addition to the usual decomposition, it can also be truncated when, from lambdas, only the first numbers remain, and we assume the rest to be zero.

$$\lambda_{d+1}, \ldots, \lambda_{\min(n,m)} := 0 \tag{14}$$

This is equivalent to the fact that for the matrices *U* and *V* we leave only the first columns and cut the matrix Λ to a square $d \times d$.

$$A' = \underset{n\times d}{U} \times \underset{d\times d}{\sum} \times \underset{d\times m}{V} T \tag{15}$$

The resulting matrix A' well approximates the original matrix A and, moreover, is the best low-ranking approximation in terms of standard deviation.

Simplify the decomposition of matrices by denoting the product of the first two matrices in one matrix (Fig. 1):

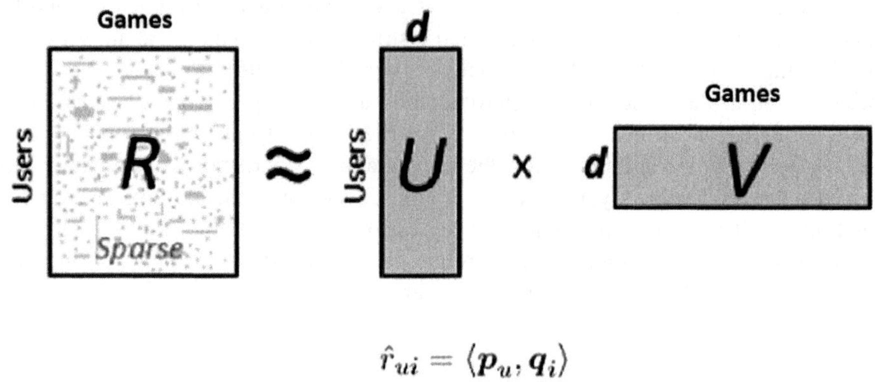

Fig. 1. Simplified version of matrix decomposition

To predict the user's rating U for the film I, we take some vector p_u (a set of parameters) for this user and a vector for this film q_i. Their scalar product will be the prediction we need: $\hat{r}ui = \langle pu, qi \rangle$.

The algorithm is quite simple but gives amazing results. It does not just allow us to predict estimates. With its help, we can reveal the hidden signs of objects and the interests of users by the history of users. For example, it may happen that on the first coordinate of the vector each user will have a number indicating whether the user is more like a boy or a girl, on the second coordinate - a number reflecting the approximate age of the user. In the film, the first coordinate will show whether it is more interesting to boys or girls, and the second - to what age group of users it is interesting.

But the matrix R is unknown to us, so we will use machine learning to determine it. So, we cannot find the SVD decomposition of the matrix, because we do not know the matrix itself. But we want to use this idea and come up with a prediction model that will work in a manner like SVD. Our model will depend on many parameters - user vectors and films. For the given parameters, in order to predict the estimate, we take the user vector, the film vector and obtain their scalar product:

$$\widehat{r_{ui}}(\Theta) = p_u^T q_i \tag{16}$$

$$\Theta = \{p_u, q_i | u \in U, i \in I\} \tag{17}$$

But since we do not know the vectors, they still need to be obtained. The idea is that we have user ratings, with which we can find optimal parameters for which our model would predict these ratings as best as possible:

$$E_{(u,i)}(\widehat{r_{ui}}(\Theta) - r_{ui})^2 \rightarrow \min_{\Theta}. \quad (18)$$

So, we want to find such parameters θ so that the squared error is as small as possible. But there is a paradox: we want to make less mistakes in the future, but we don't know what estimates we will ask. Accordingly, we cannot optimize this. But we know the ratings that users have already made. Let's try to choose the parameters so that the error that we already have is as small as possible. In addition, we add one more term - the regularizer.

$$\sum_{(u,i)\in D}(\widehat{r_{ui}}(\Theta) - r_{ui})^2 + \lambda \sum_{\theta\in\Theta}\theta^2 \rightarrow \min_{\Theta}. \quad (19)$$

How do we find the optimal parameters? We need to optimize such a functional:

$$J(\Theta) = \sum_{(u,i)\in D}(p_u^T q_i - r_{ui})^2 + \lambda\left(\sum_u \|p_u\|^2 + \sum_i \|q_i\|^2\right) \quad (20)$$

There are many parameters: for each user, for each object, we have our own vector, which we want to optimize. We have a function that depends on many variables. How to find her minimum? Here we need a gradient - a vector of partial derivatives for each parameter.

$$\nabla J(\Theta) = (\frac{\partial J}{\partial \theta_1}, \frac{\partial J}{\partial \theta_2}, \ldots, \frac{\partial J}{\partial \theta_n})^T \quad (21)$$

The gradient is very convenient to visualize. In the illustration, we have a surface: a function of two variables. For example, altitude. Then the gradient at any point is such a vector directed in the direction where our function grows most. And if you let water from this point, it will flow in the direction opposite to the gradient (Fig. 2).

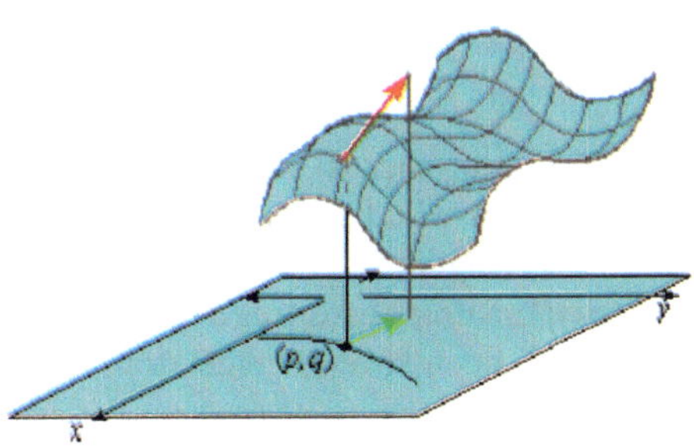

Fig. 2. Gradient direction of our model

The most famous function optimization method is gradient descent. Suppose we have a function of many variables; we want to optimize it. We take some initial value, and then we look where it is possible to move to minimize this value. The gradient descent method is an iterative algorithm: it repeatedly takes the parameters of a certain point, looks at the gradient and steps against its direction:

$$\Theta_{t+1} = \Theta_t - \eta \nabla J(\Theta) \tag{22}$$

The problems of this method are that, firstly, in our case it works very slowly and, secondly, it finds local rather than global minima. The second problem is not so scary for us, because in our case, the value of the functional at local minima is close to the global optimum.

If we want to improve the quality of recommendations, we need to learn how to measure it. For this, an algorithm trained on one sample - training, is checked on another - test. Netflix suggests measuring the quality of RMSE metrics recommendations:

$$RMSE = \sqrt{\frac{1}{|D|} \sum_{(u,i) \in D} \left(\widehat{r_{ui}} - r_{ui}\right)^2}$$

Today it is the standard metric for predicting scores. However, it has its drawbacks:

- Each user has their own idea of the rating scale.
- Users with a wider range of ratings will have a greater influence on the value of the metric than others.
- An error in predicting a high score has the same weight as an error in predicting a low score. At the same time, predicting a score of 9 instead of a true score of 7 is worse than predicting 4 instead of 2 (on a ten-point scale).
- You can have an almost perfect RMSE metric, but have a very poor quality of ranking, and vice versa.

6 Implementation

For this study, a game store based on the.Net Framwork platform was developed. The architectural pattern MVC was used. The implementation represents a three-tier architecture (DAL, BLL, WEB) (Fig. 3):

Before the algorithm starts, the user needs to go to the site, log in and simulate the purchase of several games. Also open several games for familiarization with detailed information.

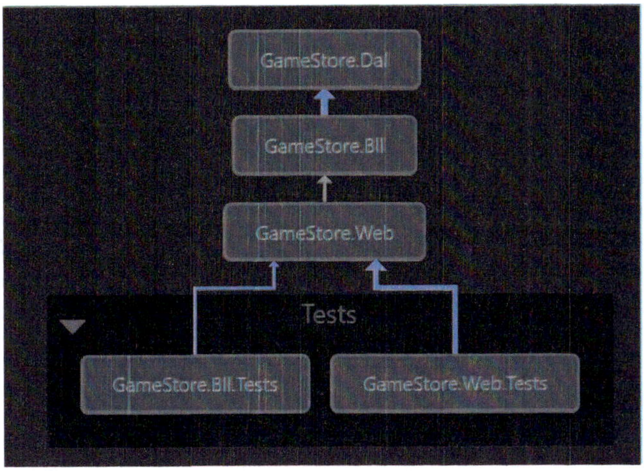

Fig. 3. Application architecture

As a result of the SVD algorithm, the user will be provided with a list of recommending games for him, based on the collected statistics (Fig. 4).

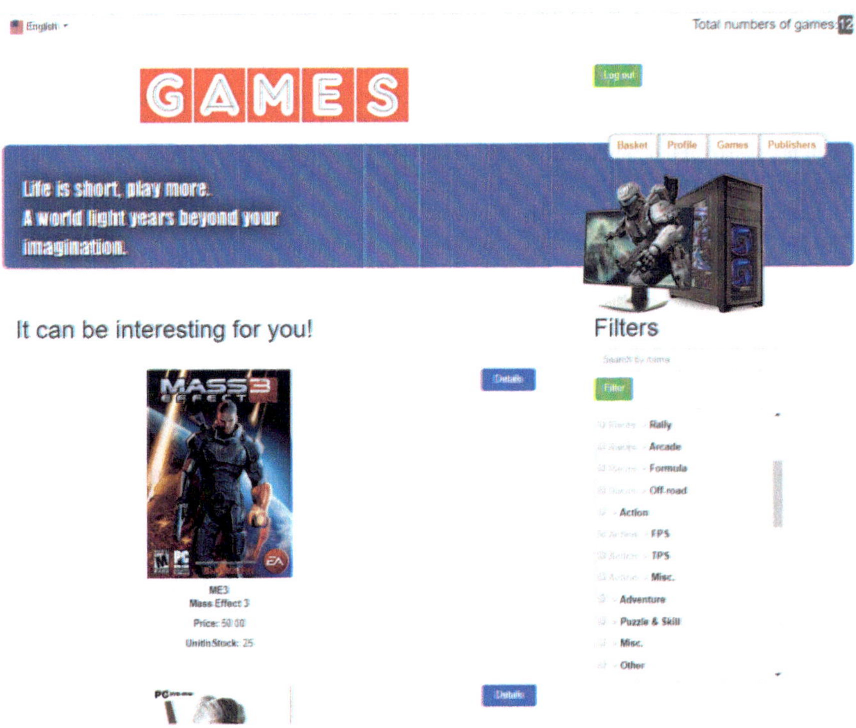

Fig. 4. Games which were recommended by svd algorithm

7 Conclusions and Prospects for Further Development

As a result of the analysis of the types of recommendation systems for use in a Web-based video game selection system, a hybrid method was selected that gives the most accurate results. The method includes collaborative filtering and its own development of a solution to the problem of selecting games for new users of the system.

After collaborative filtering, the SVD algorithm developed for rating matrices was chosen. A solution was chosen for the problem of the "cold start" of new users, consisting in the selection of games for users based on their activity on the site, and in the selection of the most correlated by genre with the user's preferred genres.

References

1. Meniailov, I., et al.: Using the K-means method for diagnosing cancer stage using the Pandas library. In: CEUR Workshop Proceedings, vol. 2386, pp. 107–116 (2019)
2. Chumachenko, D., et. al.: Development of an intelligent agent-based model of the epidemic process of syphilis. In: 2019 IEEE 14th International Scientific and Technical Conference on Computer Sciences and Information Technologies (CSIT), Lviv, vol. 1, pp. 42–45 (2019)
3. Chumachenko, D., et al.: On agent-based approach to influenza and acute respiratory virus infection simulation. In: Proceedings of 14th International Conference on Advanced Trends in Radioelectronics, Telecommunications and Computer Engineering, TCSET 2018, pp. 192–195 (2018)
4. Polyvianna, Y., et al.: Computer aided system of time series analysis methods for forecasting the epidemics outbreaks. In: 2019 15th International Conference on the Experience of Designing and Application of CAD Systems (CADSM), pp. 7.1–7.4 (2019)
5. Chumachenko, D., Chumachenko, T.: Intelligent agent-based simulation of HIV epidemic process. Adv. Intell. Syst. Comput. **1020**, 175–188 (2019)
6. Chumachenko, D., Chumachenko, K., Yakovlev, S.: Intelligent simulation of network worm propagation using the code red as an example. Telecommun. Radio Eng. **78**(5), 443–464 (2019)
7. Chumachenko, D., Yakovlev, S.: On intelligent agent-based simulation of network worms propagation. In: 2019 15th International Conference on the Experience of Designing and Application of CAD Systems (CADSM), pp. 3.11–3.13 (2019)
8. Dotsenko, N., Chumachenko, D., Chumachenko, I.: Modeling of the processes of stakeholder involvement in command management in a multi-project environment. In: 2018 IEEE 13th International Scientific and Technical Conference on Computer Sciences and Information Technologies, CSIT 2018 – Proceedings, pp. 29–32 (2018)
9. Dotsenko, N., Chumachenko, D., Chumachenko, I.: Project-oriented management of adaptive teams' formation resources in multi-project environment. In: CEUR Workshop Proceedings, vol. 2353, pp. 911–923 (2019)
10. Dotsenko, N., Chumachenko, D., Chumachenko, I.: Management of critical competencies in a multi-project environment. In: CEUR Workshop Proceedings, vol. 2387, pp. 495–500 (2019)
11. Dotsenko, N., Chumachenko, D., Chumachenko, I.: Modeling of the process of critical competencies management in the multi-project environment. In: 2019 IEEE 13th International Scientific and Technical Conference on Computer Sciences and Information Technologies, CSIT 2019 – Proceedings, vol. 3, pp. 89–93 (2019)

12. Bazilevych, K., et al.: Stochastic modelling of cash flow for personal insurance fund using the cloud data storage. Int. J. Comput. **17**(3), 153–162 (2018)
13. Chumachenko, D., et al.: Intelligent expert system of knowledge examination of medical staff regarding infections associated with the provision of medical care. In: CEUR Workshop Proceedings, vol. 2386, pp. 321–330 (2019)
14. Chumachenko, D.: On intelligent multiagent approach to viral hepatitis B epidemic processes simulation. In: Proceedings of the 2018 IEEE 2nd International Conference on Data Stream Mining and Processing, DSMP 2018, pp. 415–419 (2018)
15. Mazorchuck, M., Dobriak, V., Chumachenko, D.: Web-application development for tasks of prediction in medical domain. In: 2018 IEEE 13th International Scientific and Technical Conference on Computer Sciences and Information Technologies (CSIT), Lviv, pp. 5–8 (2018)
16. Chumachenko, D., Yakovlev, S.: Development of deterministic models of malicious software distribution in heterogeneous networks. In: 2019 3rd International Conference on Advanced Information and Communications Technologies (AICT), Lviv, Ukraine, pp. 439–442 (2019)
17. Melville, P., Mooney, R., Nagarajan, R.: Content-Boosted Collaborative Filtering for Improved Recommendations, pp. 187–192. University of Texas, USA (2002)
18. Zhernakova, O.: System recommendations and search for video content. Telemedia (2012)
19. Hossain, N.: Why the Interest Graph Is a Marketer's Best Friend. Mashable (2013)
20. Fleder, D., Hosanagar, K.: Blockbuster culture's next rise or fall: the impact of recommender systems on sales diversity. Manag. Sci. **55**(5), 1–49 (2009)
21. Xiaoyuan, S., Taghi, M.: A Survey of Collaborative Filtering Techniques A Survey of Collaborative Filtering Techniques. Hindawi Publishing Corporation, Advances in Artificial Intelligence archive, USA, pp. 1–19 (2009)
22. Yehuda, K.: Factor in the neighbors: scalable and accurate collaborative filtering. Yahoo! Res. Haifa: Mag. 1–11 (2009)
23. Linden, G., Smith, B., York, J.: Item-to-Item Collaborative Filtering, pp. 76–80. IEEE Internet Computing, Los Alamitos (2003)
24. Sarwar, B., et al.: Item-Based Collaborative Filtering Recommendation Algorithms, pp. 285–295. University of Minnesota, Minneapolis (2001)
25. Melville, P., Mooney, R., Nagarajan, R.: Content-Boosted Collaborative Filtering for Improved Recommendations, pp. 187–192. University of Texas, USA, Conf. Materials (2002)
26. Zan, H., Xin, L., Hsinchun, C.: Link prediction approach to collaborative filtering. In: JCDL 2005, Denver, Colorado, USA (2005)
27. Ponizovkin, D.M.: The construction of an optimal graph of relations in collaborative filtering systems. Softw. Syst.: Theory Appl.: J. **4**(8), 107–114 (2011)
28. Sammut, C., Webb, J.: Encyclopedia of Machine Learning, pp. 829–838. IBM T. J.Watson Research Center, NY, USA (2010)
29. Jacobson, A.: Unified software development process software, 496 p. (2010)
30. Adomavicius, G.: Toward the next generation of recommender systems: a survey of the state-of-the-art and possible extensions. IEEE Trans. Knowl. Data Eng. **17**(6), 734 (2005)
31. Salton, G.: Automatic Text Processing. Addison-Wesley, Boston (1989)
32. Pazzani, M.: Learning and revising user profiles: the identification of interesting web sites. Mach. Learn. **27**, 313–331 (1997)
33. Rich, E.: User modeling via stereotypes. Cogn. Sci. **3**, 34 (1979)
34. Goldberg, D.: Using collaborative filtering to weave an information tapestry. Commun. ACM **35**(12), 61–70 (1992)

Non-linear Estimation Methods in Multi-objective Problems of Robust Optimal Design and Diagnostics of Systems Under Uncertainties

Ievgen Meniailov[1(✉)], Mykhaylo Ugryumov[2], Dmytro Chumachenko[1], Kseniia Bazilevych[1], Sergiy Chernysh[1], and Iryna Trofymova[1]

[1] National Aerospace University "Kharkiv Aviation Institute", Chkalova str., 17, Kharkiv 61070, Ukraine
{j.menyailov,k.bazilevych}@khai.edu, dichumachenko@gmail.com, 91sergey@gmail.com, irina.trofymova@gmail.com

[2] V.N. Karazin Kharkiv National University, 4, Independence sq., Kharkiv 61022, Ukraine
ugryumov.mykhaylo52@gmail.com

Abstract. The advanced methodology of solution synthesis for multi-objective stochastic optimization problems is offered and realized. To validate the methodology regarding to some particular object one solved the problem of robust optimal designing of centrifugal impeller fitted with backward curved blades in the conditions of stochastic nature of the input data.

Keywords: Design optimization · Sensitivity analysis · Robust meta-models · Neural networks · Memetic algorithm

1 Introduction

The consistent set of system units design parameters and questions of strength are needed to be taken into account for serial production of new technical systems at enterprises [1]. These problems are basis for manufacturing process formation.

Today the engineering defects percent is about 5% for every 100 items, and the defects percent of blades proper frequency is about 20% under verification. Decrease of large costs risk of complex technical system development is possible by introducing the methods of robust optimal design under serial production conditions (rejection of selective assembly of products). These designing methods are based on computing intelligence methods [2].

Their imperfections are high cost and lack of possibility of solving multi-objective stochastic optimization problems at MV-statement [3].

M. Nechyporuk et al. (Eds.): *Integrated Computer Technologies in Mechanical Engineering*, AISC 1113, pp. 198–207, 2020.
https://doi.org/10.1007/978-3-030-37618-5_18

2 Rationale and Purpose of the Research

We will consider the direct problem of calculating the design dimensional chains: it is necessary to determine the nominal values (mathematical expectations) and confidence intervals of the values of parameters, state variables, selection criteria for the constituent links of the dimensional chain, based on the specified nominal values and confidence intervals of the state variables, selection criteria for closing link.

At present, there are a number of methods for constructing confidence intervals for parameters, state variables, and criteria for selecting the solutions of system elements based on the calculation of design dimensional chains: limit values, interval calculations, probabilistic.

If we are talking about assembling mechanical parts, then the tolerance of the closing link is a separable function. In general, the tolerance of the closing link is a non-linear function of its variables, which are the tolerances of the constituent links of the chain.

Reducing the risks of high costs for fine-tuning complex technical systems in the conditions of mass production is possible due to the introduction of robust optimal design methods based on the use of computational intelligence methods. In turn, from the point of view of the need to develop methods for robust optimal design, the problem of improving existing and developing new mathematical models and methods for solving one-criterion problems of stochastic optimization

As is well known, when structuring multicriteria stochastic optimization problems, there arises an uncertainty in the choice of metrics when evaluating objective functions and unknown quantities (parameters, control variables, or state variables) in the case when the data are random variables. In order to solve ill-posed problems of this type, regularizing algorithms should be used, which will ensure the obtaining of stable (robust) estimates of the unknown quantities, and the mathematical models synthesized on their basis will have the property of robustness.

To date, a lot of works have been published devoted to the description of methods for estimating target functions and unknown quantities in multicriteria problems of identification of mathematical models, optimization and decision-making in the design, improvement and diagnosis of technical as well as medical and biological systems.

When structuring the classification problems of multidimensional observations, the application of the concept of power averages was used Kolmogorov, according to which it is proposed to use a number of functionals to estimate the quality of partitions: the distance between classes (for example, the Mahalanobis-type distance); The measure of the concentration of points corresponding to the partition; measure of intraclass scattering (for example, generalized intraclass dispersion).

As a computational method for synthesizing quasisolutions of ill-posed problems, the method of regularizing the solution of ill-posed problems was used. Tikhonov, in which the smoothing functional is used as the scalar convolution of the objective functions.

As an additional term, the stabilizing functional (membership function) is included in the smoothing functional, which allows, on the one hand, to take into account the system of preferences of the person making the decision, and, on the other hand, to ensure the correctness of the method for synthesizing quasisolutions. In this case, the choice of the stabilizing functional should be carried out according to a number of conditions, in particular, it must be a continuous nonnegative convex function.

Analysis of existing literature sources shows that when developing methods for estimating objective functions and unknown quantities in multicriteria problems of identification of mathematical models, optimization and decision making, especially in cases of a priori uncertainty of data, a number of mathematical problems arise: the formation of a system of preferences for the decision-maker: generalized (scalar convolutions) of objective functions, a system of restrictions, a set of correctness; structurization of regularizing algorithms for synthesis of quasisolutions; High computational complexity of the described methods.

As computational methods for synthesis of solutions to stochastic optimization problems, locally stochastic methods are used (including on the basis of self-organization): stochastic quasigradient algorithms; evolutionary (genetic algorithms, immune); population (imitation movement: flocks of migratory birds, ant, bee colonies).

Some mathematical problems arise in designing a robust optimal design and intellectual diagnostic methods. These problems are uncertainty estimation, regularization algorithms structuring and high computational complexity of quasi-solution synthesis methods for practical problems under uncertainties.

Today these problems are solved by participants of some scientific programs, e.g., EU FP6: NODESIM-CFD, EU H2020: UMRIDA. The researches resulted in designing the solution methods for M-, V-, P-problems and stochastic optimization with mixed condition.

M-problem: it is required to find such the x_c, which could ensure the minimal value of mathematical expectation of decision selection criteria (DSC) f_c when σ_x is given. In order to solve modification problem it is expedient to reduce this problem to the problem of minimization of decision selection criterion such as $|f_c - f^*|$, where f^* is given DSC value. This problem is the nearest one to optimization problems of classical type. If stochastic nature of the variable x and of DSC itself is ignored, the problem is reduced to classical problem.

V-problem: it is required to find such the x_c, which ensures the minimal value of σ_f when σ_x is given. The optimization problem conversation into the modification problem can be carried out similarity to M-problem conversation.

P-problem: it is required to find such an x_c, under which $P(f_{min} \leq f(x) \leq f_{max})$ probability could attain its maximum value when σ_x is given.

There are only a few software packages in the world, which afford an opportunity of robust optimal design (e.g., «IOSO Technology, Robust design optimization»[4], «ESTECO, modeFRONTIER» [5], «DYNARDO, optiSLang» [6], «NUMECA International, FineDesign3D» [7], «DASSAULT SYSTÈMES, Isight» [8], «VANDERPLAATS R&D, VisualDOC» [9]).

System model for decision-making process of information-analytical support of the formation of technical appearance of the complex technical system functional elements based on the original mathematical models is shown in Fig. 1.

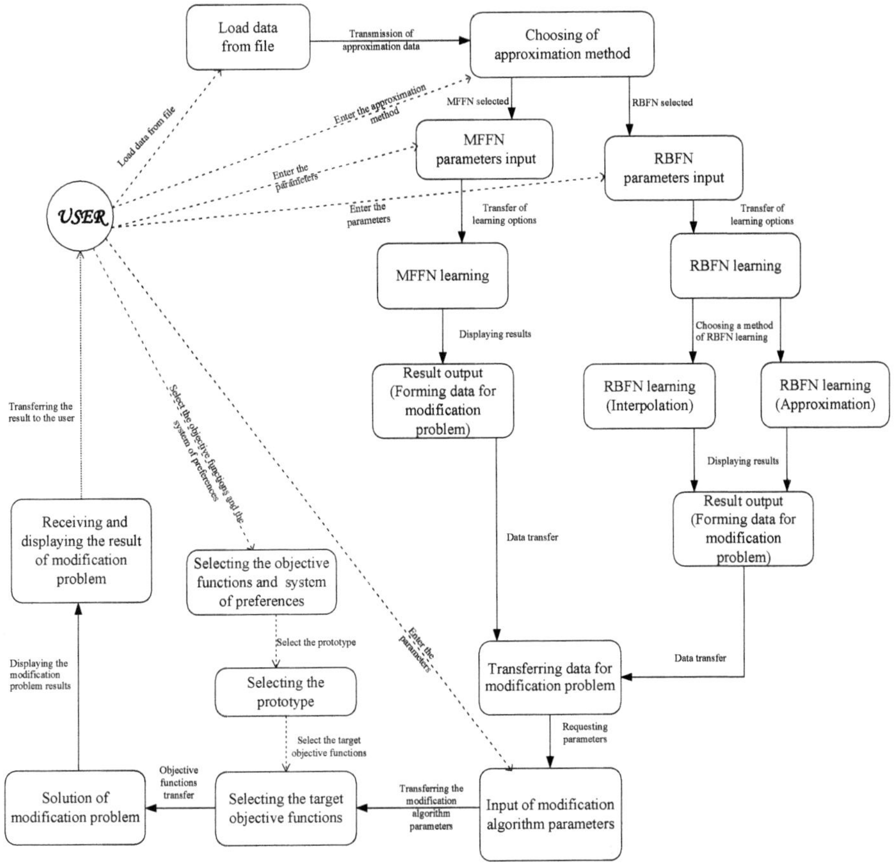

Fig. 1. System model for decision-making process of information-analytical support of the formation of technical appearance of the complex technical system functional elements based on the original mathematical models

Purpose of this research is methodology, computational methods and realization in interactive decision support computer system development. The advanced methodology of solution synthesis for multi-objective stochastic optimization problems is offered and realized at first time, which contrary to the known models allows searching for rational solutions of multi-objective MV-problem by hierarchical double-level solution synthesis scheme making.

The scheme contains robust surrogate models of system and process and effective robust desired quantities estimation under parametric data uncertainties. Thus the solution of engineering dimensional chains calculation direct problem is searched by probability method: by given values of mathematical expectations and confidence intervals of decision criterion (objective function) values or by phase variables of system (subsystem) or process under consideration the mathematical expectations and confidence intervals of control subsystem (functional unit) variables are searched.

Sensitivity analysis – the novel estimation method of multidimensional model variables informativity (importance) for system and process is developed. This method takes into account an accuracy of state variables measurement and pair correlation presence between them. This feature allows using the estimation method for model completeness analysis and control informative variables set synthesis.

The novel regularization method was developed. The method allows searching Tikhonov's normal solutions by of MV-problems modification. The efficient memetic algorithm with the consistent application of advanced real coded evolutionary method, decremental neighborhoods method, randomized path relinking method of these problems solutions synthesis is proposed. Application memetic algorithm provides a reduction of the computing time expenditures for the solution of real-life problems at several times.

The interactive decision making support software system «ROD&IDS®» was developed. The software system performs the following functions:

- preparation of input data: methods of preliminary rationing of input data (Fig. 2). As input data, alternatives data (test samples) are used - a set of design and operational parameters, control and phase variables, decision criteria (target functions). Test samples are formed using either solutions in the deterministic formulation of the direct analysis problem, or the results of experimental studies of analogs;

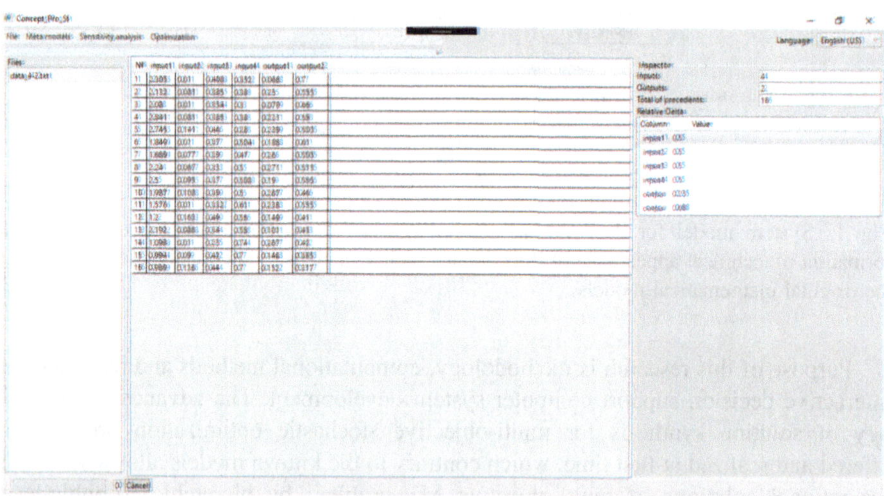

Fig. 2. A set of industrial centrifugal fan alternatives (analogues)

- methods for constructing robust metamodels of the systems and processes under consideration: methods for approximating vector functions of vector variables based on the use of trained artificial neural networks (ANN) - unidirectional multilayer and radial-based ANN (Fig. 3). Education considered ANN is carried out by the method of stochastic approximation based on the ravine conjugate gradient method. The application of the proposed development allows to obtain stable (robust) estimates of the parameters of neural network models under conditions of uncertainty of input data, which ensures the synthesis of robust metamodels;

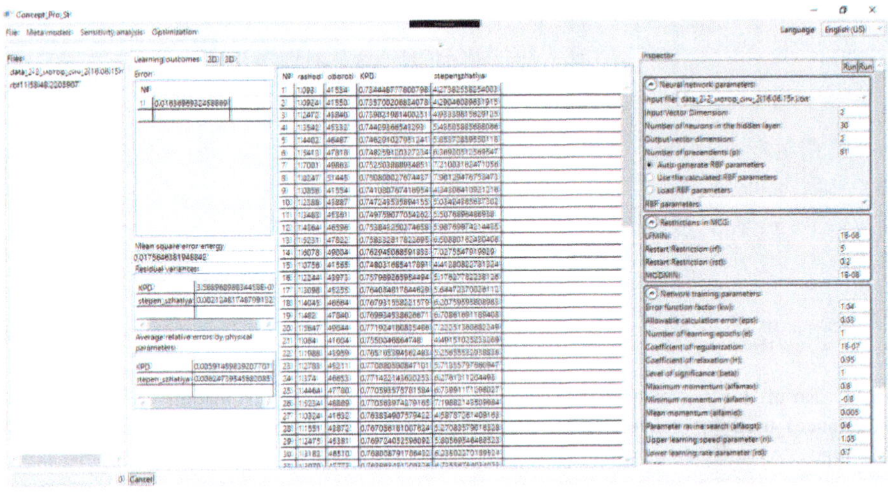

Fig. 3. Pressure characteristics robust approximations result for multistage axial flow compressor (MAFC) via a RBFN

- graphic means of three-dimensional representation of metamodels;
- methods for evaluating informativeness (significance) of variable metamodels, taking into account pair correlation and accuracy of measurement of variables (Fig. 4). On the basis of the obtained results, in particular, a solution to the problem of synthesizing a set of controlled variables can be obtained for diagnosing system failures under conditions of uncertainty of input data;
- methods for solving the problem of classifying systems and processes using monitoring data of monitored variables based on the use of unidirectional multilayer and radial basis ANN;
- method of synthesis modification of the solutions.

At the first stage, for the prototype, the solution of the inverse problem of calculating design dimensional chains is found (Monte-Carlo analysis): evaluation of confidence intervals for the values of decision criteria (target functions) for given confidence intervals for the values of control variable subsystems (functional elements). Thus, a comparative analysis of various systems manufacturing techniques can be carried out.

At the second stage, the quasi-solutions of multi-criteria MV-problems are synthesized by the regularization method. The use of the proposed development allows to obtain stable (robust) estimates of the unknown quantities in the conditions of parametric uncertainty of the input data. As a computational method, an effective memetic algorithm was used, based on the joint use of an evolutionary method with parameters

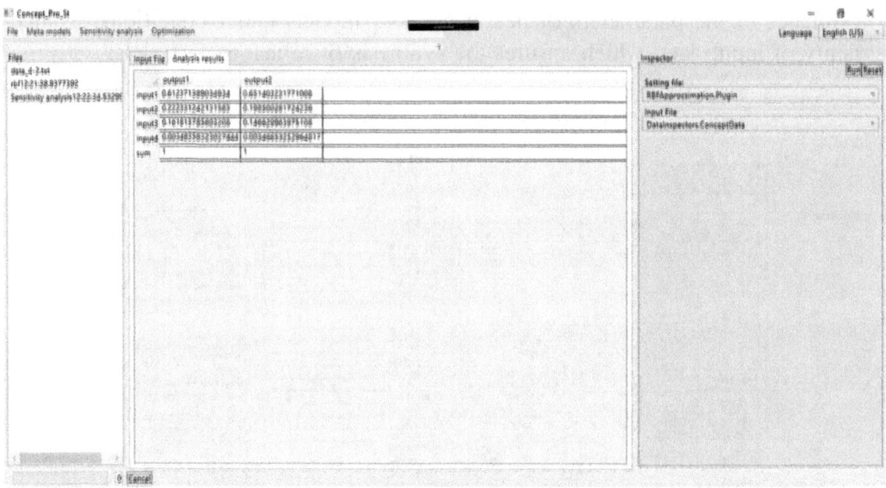

Fig. 4. Result for neural network variables sensitivity analysis (estimation of informatively, importance) used for calculation of the aerodynamic characteristics of industrial radial fan prototype

varying from epoch to epoch: material coding operators, fitness and relaxation functions, the number of individuals (the number of mini-populations) and the randomized method of laying paths in deterministic and stochastic (MV-task) formulations.

Thus, the probabilistic method finds the solution of the direct problem of calculating design dimensional chains: according to given values of mathematical expectations and confidence intervals of the values of decision criteria (target functions) or phase variables of the considered systems (subsystems) or processes, the mathematical expectations and confidence intervals of the values of the control variable subsystems are found (functional elements).

3 Experiments and Results of the Modeling

Let us consider an example of robust optimal design of radial fan fitted with backward curved blades in the conditions of the stochastic nature of the input data. The input data is a set of options – there are 16 variants of analogues. Values of geometrical parameters of radial fan were specified: chord pitch ratio of cascade (x_1), relative maximum deviation of camber line from chord line (x_2), dimensionless chord of the profile (x_3), dimensionless inlet radial fan blades tip diameter (x_4). There were assigned as a decision selection criteria (objective functions): dimensionless parameters of flow rate (f_1) and total pressure (f_2) for the maximum efficiency rating. Further the robust neural network models in the form of radial-basis ANN function were used as the meta-models of systems under consideration.

Prototype data given in the table. As the target values of the selection criteria used solutions nominal values and their confidence intervals (normalized flow parameter f1 = 0,2390 ± 0,0080, normalized total pressure setting f2 = 0,5600 ± 0,0075). The general results of the synthesis of the solution of the modification problem in deterministic (Fig. 5) and stochastic (Fig. 6) form are presented in the Table 1.

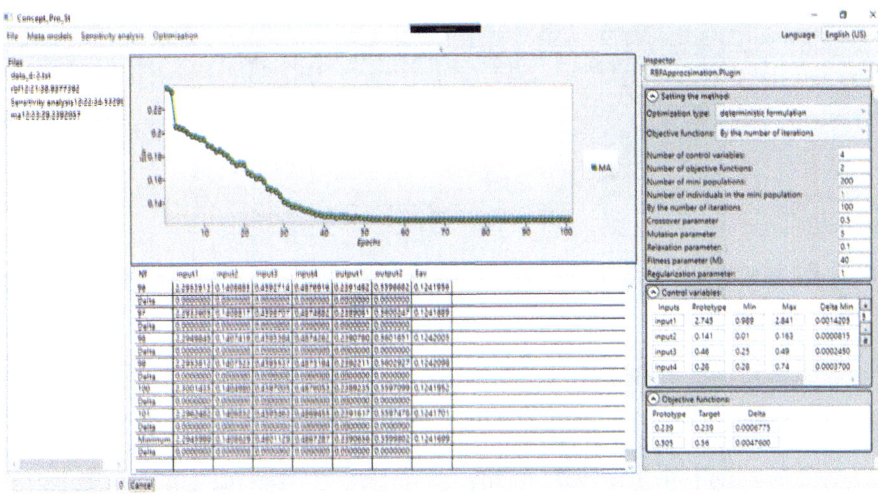

Fig. 5. An example of the solution for the formation of industrial centrifugal fan shape at the deterministic formulation of problem

The solution for the formation of industrial centrifugal fan shape at the stochastic formulation of problem was obtained using «ROD&IDS®».

Based on the comparison of calculation results it should be noted:

- the maximum difference between the nominal values of the unknown quantities obtained as problem solution in different formulations runs up to 3–18%;
- unknown quantity estimates obtained as result of modification problem solution at stochastic formulation are effective and stable (robust). It is a result of confidence intervals decreasing for decision selection criteria. The confidence intervals are less than 1.8–3.6 times compared to the results of modification problem solution at deterministic formulation. The confidence intervals of unknown quantity were changed insignificantly.

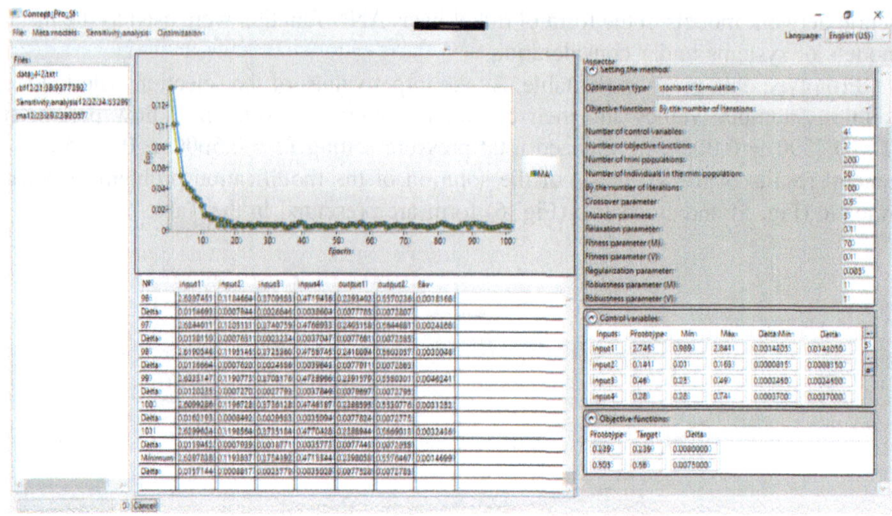

Fig. 6. An example of the solution for the formation of industrial centrifugal fan shape at the stochastic formulation

Table 1. I – Overall results of calculations

Input data	Prototype	Deterministic formulation	Stochastic formulation
x_1	2.7450 ± 0.0142	2.2946	2.6298 ± 0.0157
x_2	0.1410 ± 0.0008	0.1409	0.1194 ± 0.0009
x_3	0.4600 ± 0.0024	0.4601	0.3754 ± 0.0026
x_4	0.2800 ± 0.0037	0.4867	0.4713 ± 0.0035
f_1	0.2390 ± 0.0120*	0.2391 ± 0.0141*	0.2398 ± 0.0078
f_2	0.5050 ± 0.0115*	0.5600 ± 0.0264*	0.5576 ± 0.0073

* – the solution results of engineering dimensional chains calculation inverse problem.

4 Conclusion

The authors proposed a methodology for the synthesis of solutions of multi-purpose problems of stochastic optimization. It includes: the creation of reliable metamodels of systems and processes, and then an effective robust estimate of unknown values occurs with parametric uncertainty. To solve optimization problems, a computational method is used based on a memetic algorithm that implements the joint use of an evolutionary method with parameters varying from epoch to epoch: real-time coding operators, fitness and relaxation functions, as well as the method of narrowing surroundings and the randomized method of tracing. This ensures a reduction of the information and time complexity of the proposed computational method, as compared with the classical genetic algorithm, by at least several times.

A solution is obtained for the robust optimal design of a radial fan with backward curved impeller blades, a two-stage axial compressor in the stochastic nature of the input data. The result was obtained using the "ROD&IDS®" computer-aided decision-making system developed by the authors.

The "ROD&IDS®" is invariant with regard to the field of application: mechanical engineering, including control using project monitoring data, production for ensuring the quality of products manufactured by enterprises.

References

1. Bazilevych, K., et al.: Stochastic modelling of cash flow for personal insurance fund using the cloud data storage. Int. J. Comput. **17**(3), 153–162 (2018)
2. Meniailov, I., et al.: Using the K-means method for diagnosing cancer stage using the Pandas library. In: Proceedings of CEUR Workshop, vol. 2386, pp. 107–116 (2019)
3. Chumachenko, D., et al.: Development of an intelligent agent-based model of the epidemic process of syphilis. In: 2019 IEEE 14th International Scientific and Technical Conference on Computer Sciences and Information Technologies (CSIT), vol. 1, pp. 42–45 (2019)
4. Sigma technology. Robust design optimization and robust optimal control (2017). http://www.iosotech.com/robust.htm
5. Robust design and reliability. ESTECO's integration platform for multi-objective and multi-disciplinary optimization (2017). http://www.esteco.com/modefrontier/robust-design-reliability
6. Dynamic Software and Engineering. OptiSLang: Software for sensitivity analysis, multiobjective and multidisciplinary optimization, robustness evaluation, reliability analysis and Robust Design Optimization (2017). www.dynardo.de/en/software/optislang.html
7. NUMECA. FINE™/Design3D: an integrated environment for the design and optimization of turbomachinery channels and blades (2017). http://www.numeca.com/product/finedesign3d
8. Isight and the simulia execution engine. Process automation and design exploration (2017). http://www.3ds.com/products-services/simulia/products/isight-simulia-execution-engine/
9. Vanderplaats R&D. VisualDOC: Software for Process Integration and Multidiscipline Design Optimization (2017). http://www.vrand.com/products/visualdoc/

Japanese Text Recognition

Olha Pohudina(✉), Dmitriy Kritskiy, A. N. Bykov, and A. D. Morikova

National Aerospace University «KhAI», Kharkiv, Ukraine
o.pogudina@gmail.com, d.krickiy@khai.edu

Abstract. The article describes the method of image recognition on the example of recognition of Japanese characters. Recognition is carried out by determining the lengths of the contours. After that, the moments of each circuit are compared. Also, taking into account the peculiarities of Japanese characters, the comparison occurs line by line. This approach can also be used to teach writing hieroglyphs. The same principle has been applied to the recognition of road signs.

Keywords: Contour comparison · OpenCV · Automated knowledge testing system · Moments of contours · Recognition of the road signs

1 Introduction

Currently, the development of digital technology allows people to transmit and receive information with great speed. The availability and development of modern technology has made this type of data, such as digital images, especially prevalent. In this regard, programs that allow the average user to receive and work with data obtained from the image are in great demand.

Already today, one can encounter the fruits of scientific discoveries in everyday life: text recognition software, face recognition systems, numbers of cars driving along the highway, and a system for recognizing a person's emotional state by facial expression [1, 2]. These tasks can be solved using digital processing methods [3, 4]. Initially, these methods were developed and studied by specialists working in the field of applied mathematics. Now there are several public libraries [5, 6], for example OpenCV, that allow you to work with digital data [7]. Such computer vision libraries do not contain ready-made solutions and are only a tool for creating working algorithms. Computer vision focuses on processing three-dimensional scenes projected onto one or more images. Using one or more images, you can restore the structure or other information about a three-dimensional scene. The scope of computer vision is unusually wide: process control systems (industrial robots, autonomous vehicles), video surveillance systems, information management systems (for example, for indexing image databases), and systems for modeling objects or the environment (analysis of medical images, topographic modeling), interaction systems (for example, input devices for a human-machine interaction system) [8–10].

The purpose of this article is to recognize the Chinese language. To recognize the font of the text, the methods of contour analysis are used. The outline of the letters or

M. Nechyporuk et al. (Eds.): *Integrated Computer Technologies in Mechanical Engineering*, AISC 1113, pp. 208–215, 2020.
https://doi.org/10.1007/978-3-030-37618-5_19

numbers from the input image is compared with the outline of the same character in different fonts. For this work, a demo font database was originally created that contains reference images of letters and numbers.

In the process of text recognition, the main stages can be distinguished: (1) image filtering. The image that will be recognized should be as clear as possible to avoid errors. Also, black and white to simplify the recognition process. Therefore, there is a need to minimize the noise of the original image, as well as the use of a monochrome filter; (2) image segmentation. It is assumed that the sentences in the text are written clearly horizontally, the letters do not overlap and are located at the same distance, different from the distance between words; (3) highlighting the characteristics of characters and their classification. At this stage, character recognition using Hamming metrics; (4) text writing—merging separately recognized characters into text.

2 Tools for Recognition of Symbols in the Systems of the Japanese Language Learning

The process of hieroglyphs memorizing is quite monotonous and time consuming, that makes it inappropriate for the teacher to participate in it. For effective memorization, it is necessary to use an automated knowledge testing system, including the ability to control the writing of hieroglyphs (Fig. 1).

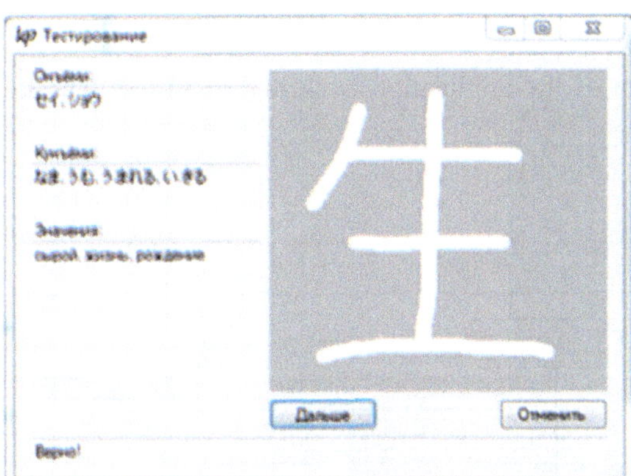

Fig. 1. Subsystem of the hieroglyphs writing control

Consider the information subsystem of character recognition for learning Japanese. We used the capabilities of the OpenCV library for working with contours when solving the problem of comparing the test user response with a sample. The space of contours made it possible to check not only the external similarity of a character written by a user with a sample, but also such important and often overlooked character characteristics as the number and order of strokes. To solve this problem, not only the final result was controlled, but also the writing process itself.

The generalized algorithm for solving the problem of character recognition is as follows:

Step 1: The user in the specified window writes the current feature of the hieroglyph by moving the mouse with the button pressed,
Step 2: After the user releases the mouse button, the written stroke is saved, and the image workspace is updated.
Step 3: If the written stroke satisfies the user and the user does not consider this stroke last, go to step 1.

Otherwise: - the user presses the "cancel" button, deletes the last written stroke and returns to step 1.
– the user presses the "next" button to compare the hieroglyph with the template.

Step 4: Comparison of the number of hieroglyph strokes a pattern is carried out.
Step 5: If the number of strokes coincides, go to Step 6, otherwise, a message about the incorrect number of strokes is invoked
Step 6: The next stroke of the hieroglyph and the next stroke of the pattern are searched.
Step 7: Checking of the contours matching, if a match is found, go to step 8, otherwise a message about the incorrect spelling will be displayed.
Step 8: Checking the end of the list of stroke contours of the hieroglyph and the template. If the end of the list is reached, the contour of the finished hieroglyph is checked, and a message about the correct spelling is displayed, otherwise, return to step 6.

The architecture of the application is shown in Fig. 2. The key functions are as follows:

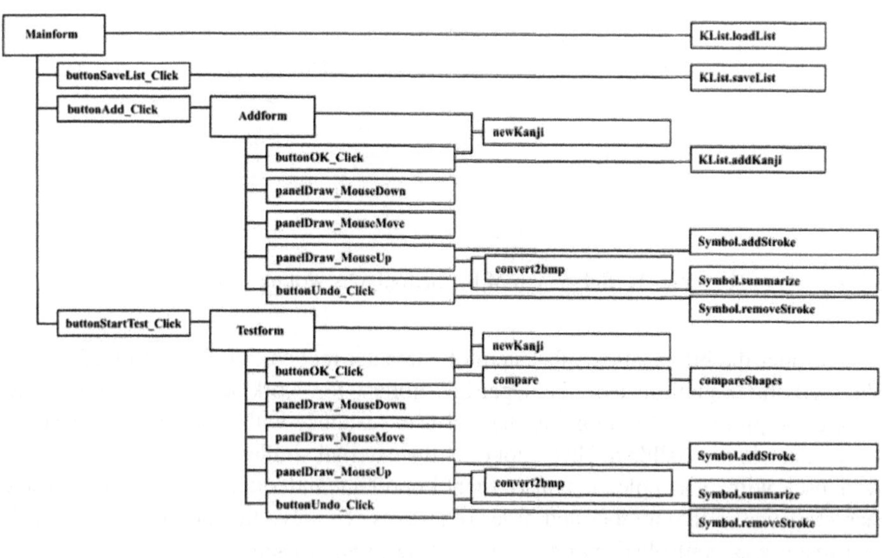

Fig. 2. The architecture of the subsystem that controls writing of hieroglyphs

- `int compare (Symbol templ)` – compares the hieroglyph written by the user with the template. Returns -1 if the number of strokes does not match, the number of the incorrectly written stroke, if there is one, or 0, if the hieroglyph is written correctly,
- `double compareShapes (IplImage img1, IplImage img2, int method)` – compares two images (strokes) with the moments of their contours.

There is also a Symbol class in the application that is designed to work with symbols. It contains the following fields:

- `IplImage [] strokes` – an array of images, each of which is a stroke,
- `int strokeCount` – the number of strokes in the hieroglyph.
- And also the class has the following methods:
- `void addStroke(IplImage stroke)` – adds a stroke to the symbol.
- `void removeStroke(int number)` – removes a stroke from the hieroglyph.
- `IplImage summarize()` – combines all strokes into one image.

Another class `KList` is designed to work with a list of hieroglyphs. This class is static. It contains the `KanjiData` class, which includes three fields of the string type: `onYomi` (it contains go-on readings), `kunYomi` (it contains kan-on readings) and `meanings` (it contains the meanings of the hieroglyph). In addition, the `KList` class contains a Kanji structure consisting of two fields: `KanjiData data` (this field stores data about a hieroglyph: readings and values) and `Symbol symbol` (this contains the spelling of a hieroglyph), as well as a *static* `static` field `Kanji [] kanjiList` – the list which contains all the hieroglyphs entered by the user.

Despite the fact that the solved problem may seem rather trivial (single contours are simple, the background and the object in the image are rather contrasting to each other), there is a danger of a false recognition of the correct sign. For example, if we compare

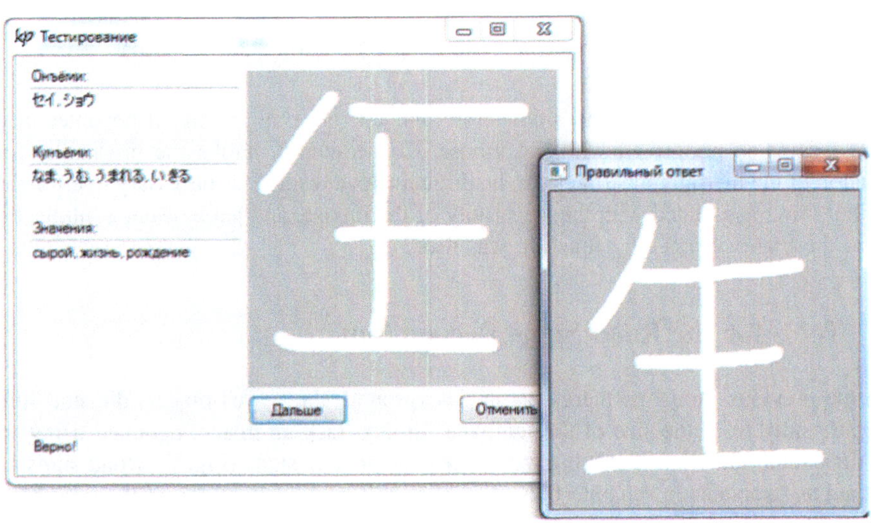

Fig. 3. False recognition in case of comparing only individual traits of a hieroglyph

only the contours of individual strokes, then there is a probability of incorrect interposition of strokes in a hieroglyph that will be recognized with the sign "true". For example, the symbol 生 is taken (Fig. 3).

In addition, the `cvMatchShapes` library function method may not always be useful in this case (Fig. 4). The fact is that this method is a linear combination of central moments, which allows to create an invariant representation of contours, independent of scale, rotation, and reflection. Therefore, using the `cvMoments` function, all the necessary parameters can be got.

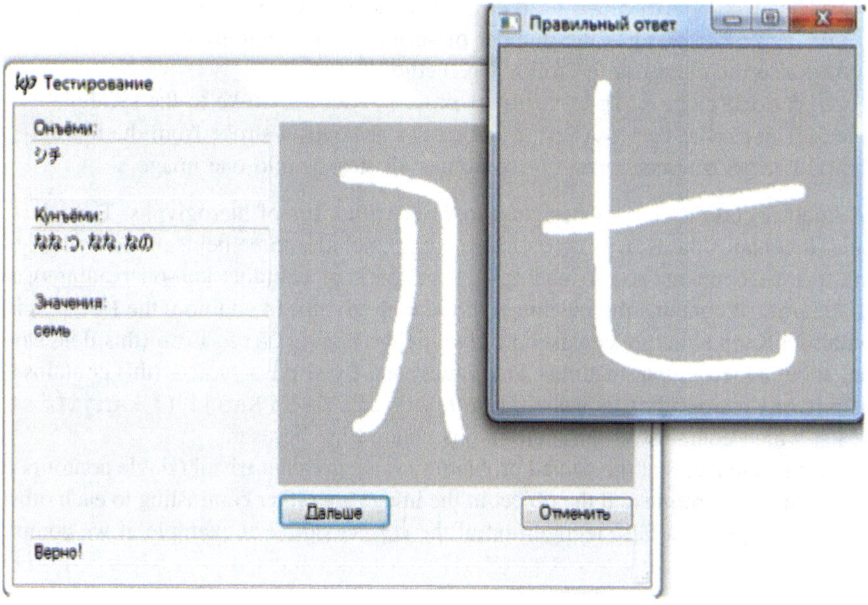

Fig. 4. False recognition when using the cvMatchShapes function

In addition, analyzing the architecture and operation algorithm, it becomes clear that writing is performed with the mouse. Consequently, unlike the brush, the formation of the hieroglyph strokes can be difficult. As a result, it is necessary to eliminate minor flaws associated with the sensitivity of the mouse, and get a neater contour. For this, the `cvApproxPoly` function was used.

3 Tools for the Road Signs Recognition

In the previous problems there was no background, the search objects did not differ significantly from the size of the template. In next task an image, obtained using the DVR behind the windshield, is given to search for road signs (Fig. 5). Road signs are stored as templates in separate files. As, there will be a background on the image, the signs can be located at different distances from the received frame, therefore, they will differ significantly in size.

Fig. 5. Image taken with a DVR

Most of the existing methods of recognition of the road signs can be described as the following three-step scheme: the detection of a sign in the image, the clarification of the position of the sign and the segmentation of the background, the recognition of the class of the sign. The clarification of the position can be useful if the results of the detector, that often produces several detections around an object of interest, are processed. The background removal can significantly improve the recognition accuracy of the sign class.

To implement the software subsystem, the original image and the patterns of road signs in gradations of gray are loaded into the `IplImage` type variables. Images are further processed using the `cvCanny()` – Canny's bounds detection function. This function allows to select the edges of the image – the curves along which there is a sharp change in brightness or other types of heterogeneities. This will remove noise and unnecessary details from the image, make the edges thin (edge thinning), link the edges into the contour (edge linking).

The next step is to create a data warehouse in which the contours, found in the previous step, are stored, and the function of searching for contours `cvFindContours()` is called.

Then the contours, that satisfy the specified conditions, are selected. The first thing to be analyzed was the area of the contour, since it fluctuates within previously known boundaries. It can be got with calling the `cvContourArea` function.

`cvContourPerimeter` returns the length of the contour, which is necessary to calculate the second parameter – compactness. This is the ratio of the area to the perimeter squared. So, it characterizes the similarity of an object with a circle, since the circle is the most compact figure (compactness factor is about 0.79).

A check was also made for the coincidence of contour moments using the `cvMatchShapes` function. Experiments have shown that almost complete similarity is expressed by a value less than 0.06–0.08. Thus, according to the result of the `cvMatchShapes` function, it is impossible to say for sure whether the contour is suitable or not, therefore the last of the methods for difference establishing is the analysis of a color in the contour.

The proposed approach allows recognition of approximately 70% of the objects. Therefore, it needs modification.

4 Conclusion

This paper presents an agglomeration algorithm for recognizing Japanese characters using the OpenCV library. That provides the ability to read the contours of the object. Since the analysis of the correct spelling of a character is complicated, it is necessary to compare not a whole character with a pattern, but each line, then you have to save each line separately.

In this regard, problems arise:

- due to the same brightness with the background, the object may not have a clear border or may be noisy with noise, which makes it impossible to highlight the contour;
- overlapping objects or their grouping leads to the fact that the contour is chosen incorrectly and does not correspond to the boundary of the object.

However, due to the specifics of the processed images (since they are created manually, there is no noise on them), these restrictions are insignificant, which makes the contour analysis an optimal method for recognizing hieroglyphs.

These problems will be resolved in future work.

References

1. He, W., Zhang, X.-Y., Yin, F., Luo, Z., Ogier, J.-M., Liu, C.-L.: Realtime multi-scale scene text detection with scale-based region proposal network. Pattern Recogn. 107026 (2019). https://doi.org/10.1016/j.patcog.2019.107026
2. Tu, W.-C., He, S., Yang, Q., Chien, S.-Y.: Real-time salient object detection with a minimum spanning tree. In: 2016 IEEE Conference on Computer Vision and Pattern Recognition (CVPR) (2016). https://doi.org/10.1109/cvpr.2016.256
3. Buczel, K., Wrzuszczak-Noga, J.: Prefiltration analysis for image recognition algorithms for the android mobile platform (2019). https://doi.org/10.1007/978-3-030-30604-5_30

4. Deshpande, S., Shriram, R.: Real time text detection and recognition on hand held objects to assist blind people. In: 2016 International Conference on Automatic Control and Dynamic Optimization Techniques (ICACDOT) (2016). https://doi.org/10.1109/icacdot.2016.7877741
5. Tsai, C.-J., Huang, P.-H.: Keyword-based approach for recognizing fraudulent messages by keystroke dynamics. Pattern Recogn. **98**, 107067 (2020). https://doi.org/10.1016/j.patcog.2019.107067
6. Johns, E., Leutenegger, S., Davison, A.J.: Pairwise decomposition of image sequences for active multi-view recognition. In: 2016 IEEE Conference on Computer Vision and Pattern Recognition (CVPR) (2016). https://doi.org/10.1109/cvpr.2016.414
7. Tareen, S.A.K., Saleem, Z.: A comparative analysis of SIFT, SURF, KAZE, AKAZE, ORB, and BRISK. In: 2018 International Conference on Computing, Mathematics and Engineering Technologies (iCoMET) (2018). https://doi.org/10.1109/icomet.2018.8346440
8. Kritsky, D.N., Ovsiannik, V.M., Pogudina, O.K., Shevel, V.V., Druzhinin, E.A.: Model for intercepting targets by the unmanned aerial vehicle. In: Palagin, A., Anisimov, A., Morozov, A., Shkarlet, S. (eds.) Mathematical Modeling and Simulation of Systems. MODS 2019. Advances in Intelligent Systems and Computing, vol. 1019. Springer, Cham (2020)
9. Kritskiy, D., Alexander, K., Koba, S., Druzhinin, E.: Increasing the reliability of drones due to the use of quaternions in motion. In: 2018 Proceedings of 2018 IEEE 9th International Conference on Dependable Systems, Services and Technologies, DESSERT 2018 (2018)
10. Kritsky, D.N., Druzhinin, E.A., Pogudina, O.K., Kritskaya, O.S.: A method for assessing the impact of technical risks on the aerospace product development projects. Advances in Intelligent Systems and Computing, vol. 871, pp. 504–521. Springer, Heidelberg (2019)

4. [illegible]
5. [illegible]
6. [illegible]
7. [illegible]
8. [illegible]
9. [illegible]
10. [illegible]

Big Data and Data Science

Proximate Objects Probabilistic Searching Method

Andrey Chukhray and Olena Havrylenko(✉)

National Aerospace University “Kharkiv Aviation Institute”, Kharkiv, Ukraine
achukhray@gmail.com, o.havrylenko@khai.edu

Abstract. The method of objects probabilistic search is developed based on the necessary proximity conditions in a Euclidean space, which were previously proved for the Levenshtein’s metric. The method is based on a random selection of k pivots in Euclidean space among the original objects, projecting all source objects in a k-dimensional Euclidean space, filling special hash data structures, and fast search facilities, similar to the desired, based on proven necessary conditions for the objects proximity in Euclidean space. Experimental studies of the proposed method show the higher speed in comparison with the known method.

Keywords: Probabilistic search · Euclidean space · Necessary proximity conditions

1 Problem Statement

Recent decades in the artificial intelligence sphere there are developed and used a set of “nearest neighbor” search methods for objects arrangement and clustering [1–3]. Nevertheless, the problem of the high-performance methods development in circumstances when calculation of distances between objects still take the certain search time is still actual. The method based upon the necessary proximity conditions, which is a generalization of the conditions previously proved for Levenshtein’s metric, is developed, experimentally tested and described in [4, 5].

In common, the problem statement is the following. Assume that an edit distance δ between objects of a certain class Cl satisfies the conditions:

$$\begin{cases} \delta(X,Y) \geq 0; \\ \delta(X,X) = 0; \\ \delta(X,Y) = \delta(Y,X); \\ \delta(X,Z) \leq \delta(X,Y) + \delta(Y,Z). \end{cases} \quad (1)$$

A certain object rt of a class Cl and a set of objects $ET = (et_1, et_2, \ldots, et_n)$ of the same class are given. It is required to find all et_i of the set ET, such that the distance δ between et_i and rt is not greater than a given positive integer λ. Formally required to find $ET_s = \{et_{s1}, et_{s2}, \ldots, et_{sl}\}$, such that $\forall et_{si} \in ET_s \subseteq ET : \delta(et_{si}, rt) \leq \lambda$, $\lambda \in N$, $l \leq n$.

The proposed method consists of two steps. 1st step. k elements $o_1, o_2, \ldots, o_k$, $k \leq n$ are randomly selected from the ET set. These elements are

M. Nechyporuk et al. (Eds.): *Integrated Computer Technologies in Mechanical Engineering*, AISC 1113, pp. 219–227, 2020.
https://doi.org/10.1007/978-3-030-37618-5_20

considered as k pivots in a k-dimensional Euclidean space E^k. After that each element et_i of the ET set is associated with E^k point coordinates which are equal to the distances to the pivots, i.e. $P(et_i)_j = \delta(et_i, o_j)$, $i = \overline{1, n}$, $j = \overline{1, k}$. 2nd step. The rt object is associated in space E^k with the point coordinates $P(rt)_j = \delta(rt, o_j)$, $j = \overline{1, k}$. Distances are calculated only between the object rt and those objects, whose corresponding points in space E^k are located closest to the point $P(rt)$.

To determine points closeness in Euclidean space it is necessary introduce the proximity conditions for a given objects X, Y and Z of class *Cl*.

Proposition 1. For a given object X, Y and Z of a class *Cl* the distance δ between them satisfies the conditions (1) and the following inequality $\forall X, Y, Z\ \delta(X, Y) \geq |\delta(X, Z) - \delta(Z, Y)|$ is true.

Proof. Consider two of the triangle inequality: $\delta(X, Z) \leq \delta(X, Y) + \delta(Y, Z)$; $\delta(Y, Z) \leq \delta(Y, X) + \delta(X, Z)$. From the first inequality it is followed $\delta(X, Z) - \delta(Y, Z) \leq \delta(X, Y)$, from the second - $\delta(Y, Z) - \delta(X, Z) \leq \delta(Y, X)$. By combining both expressions and using the symmetry property, we obtain the system of inequalities: $\begin{cases} \delta(X, Y) \geq \delta(X, Z) - \delta(Z, Y); \\ \delta(X, Y) \geq \delta(Z, Y) - \delta(X, Z), \end{cases}$ or, as follows, $\delta(X, Y) \geq |\delta(X, Z) - \delta(Z, Y)|$ Q.E.D.

Proposition 2. For a given objects et_i and et_j of a class *Cl*, distance δ between which satisfies conditions (1) and does not exceed a threshold λ, corresponding points $P(et_i)$ and $P(et_j)$ it space E^k are situated on a distance of no more than $\lambda\sqrt{k}$, i.e. $\forall i \forall j \neq i\ \delta$ $(et_i, et_j) \leq \lambda$: $\rho(P(et_i), P(et_j)) \leq \lambda\sqrt{k}$.

Proof. From the definition of the metric E^k follows: $\rho(P(et_i), P(et_j)) = \sqrt{(P(et_i)_1 - P(et_j)_1)^2 + (P(et_i)_2 - P(et_j)_2)^2 + \ldots + (P(et_i)_k - P(et_j)_k)^2}$. According to Proposition 1: $|\delta(et_i, o_1) - \delta(et_j, o_1)| \leq \delta(et_i, et_j) \ldots, |\delta(et_i, o_k) - \delta(et_j, o_k)| \leq \delta(et_i, et_j)$. Hence, transitively: $|P(et_i)_1 - P(et_j)_1| \leq \lambda \ldots, |P(et_i)_k - P(et_j)_k| \leq \lambda$ and, therefore: $\sqrt{(P(et_i)_1 - P(et_j)_1)^2 + \ldots + (P(et_i)_k - P(et_j)_k)^2} \leq \sqrt{\lambda^2 k} = \lambda\sqrt{k}$, Q.E.D.

Proposition 3. For a given objects et_i and et_j of a class *Cl* the distance δ between which satisfies conditions (1) and does not exceed a threshold λ, point $P(et_j)$ located in the space E^k within a hypercube centered at $P(et_i)$ with the side equal to 2λ.

Proof. According to Proposition 1 the following systems of inequalities can be consequently obtained:

$$\left\{\begin{array}{l} |\delta(et_i,o_1)-\delta(et_j,o_1)|\leq\delta(et_i,et_j); \\ \\ |\delta(et_i,o_2)-\delta(et_j,o_2)|\leq\delta(et_i,et_j); \\ \\ \dots \\ |\delta(et_i,o_k)-\delta(et_j,o_k)|\leq\delta(et_i,et_j), \end{array}\right. = \left\{\begin{array}{l} P(et_j)_1\geq P(et_i)_1-\lambda; \\ P(et_j)_1\leq P(et_i)_1+\lambda; \\ P(et_j)_2\geq P(et_i)_2-\lambda; \\ P(et_j)_2\leq P(et_i)_2+\lambda; \\ \dots \\ P(et_j)_k\geq P(et_i)_k-\lambda; \\ P(et_j)_k\leq P(et_i)_k+\lambda. \end{array}\right. \quad (2)$$

The geometric meaning of the system of inequalities (2) is a hypercube centered at $P(et_i)=(\delta(et_i,o_1),\delta(et_i,o_2),\ldots,\delta(et_i,o_k))$ with the side length 2λ, Q.E.D.

Proposition 4. For a given objects et_i and et_j of a class Cl the distance between which δ satisfies conditions (1) and does not exceed a threshold λ, absolute value of the difference of the distances from points $P(et_i)$ and $P(et_j)$ to the origin in the space E^k does not exceed $\lambda\sqrt{k}$, i.e. $|\rho(P(et_i),0)-\rho(P(et_j),0)|\leq\lambda\sqrt{k}$.

Proof. According to the property of the Euclidean metric space (triangle inequality) $|\rho(P(et_i),0)-\rho(P(et_j),0)|\leq\rho(P(et_i),P(et_j))$. On the other hand, according to Proposition 2: $\rho(P(et_i),P(et_j))\leq\lambda\sqrt{k}$. From the above it can be obtained: $|\rho(P(et_i),0)-\rho(P(et_j),0)|\leq\rho(P(et_i),P(et_j))\leq\lambda\sqrt{k}$ and hence $|\rho(P(et_i),0)-\rho(P(et_j),0)|\leq\lambda\sqrt{k}$, Q.E.D.

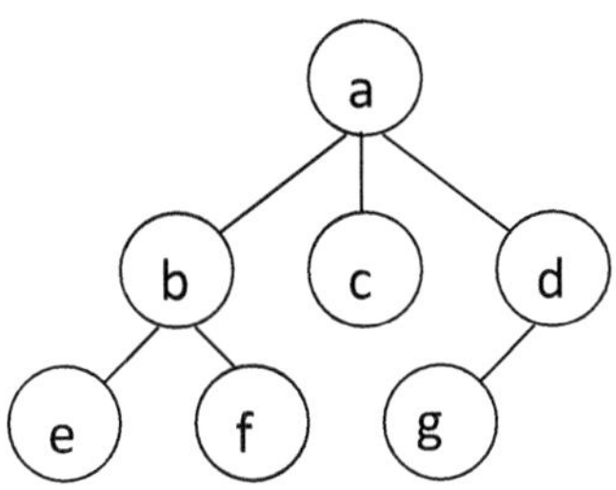

Fig. 1. An example of a tree as an object

Proposition 5. Assume that $u,w\in R$ and $u,w>0$. Then from $[u]\leq w$ it is followed: $[u]\leq[w]$, where $[u]$, $[w]$ are the whole parts of the numbers u and w respectively.

Proof. There are two cases when $[u]\leq w$ is true: $[u]=[w]$ and $[u]<[w]$. Generalizing both of them, we can obtained condition $[u]\leq[w]$. It is also obvious that assumption $[u]>[w]$ could not be true for $[u]\leq w$, therefore what was required to prove $[u]\leq[w]$ is true.

Proposition 6. Assume that $u,v,w\in R$ and $u,v,w>0$. Then from $|u-v|\leq w$ it is followed: $|[u]-[v]|\leq[w]+1$, where $[u]$, $[v]$, $[w]$ - the whole parts of the numbers u, v, w respectively.

Proof. First consider the case, when $u \geq v$. Then from $u \leq w + v$ it could be obtained $[u] \leq w + v$, considering $u \geq [u]$, furthermore $[u] \leq w + [v] + 1$, considering $v < [v] + 1$. Further, according to the Proposition 5 and the fact that any positive number greater than any negative one, if $[u] - [v] \leq w + 1$, then $[u] - [v] \leq [w] + 1$. Second case $v > u$ could be considered the same way and obtained the next inequity:$[v] - [u] \leq [w] + 1$. Summarizing both cases: $|[u] - [v]| \leq [w] + 1$, Q.E.D.

Definition 1. A size of an object et_i of a class Cl is the number of its elements and marked as $\overline{et_i}$. For example, if et_i is the string "home", then $\overline{et_i} = 4$ (line length); if et_i is the tree shown on Fig. 1, then $\overline{et_i} = 7$ (the number of vertices).

Proposition 7. If the absolute value of the difference between the sizes of objects et_i and et_j of a class Cl, distance δ between which satisfies conditions (1), is greater than λ, then the distance between these objects is also greater than λ, i.e. $(|\overline{et_i} - \overline{et_j}| > \lambda) \Rightarrow (\delta(et_i, et_j) > \lambda)$.

Proof. This proposition is obvious and follows from the fact that if the objects sizes differ on λ, then to convert an object et_i into object et_j or vice versa, it is should be completed at least λ element deletions in the best case and more – in the other cases.

After necessary objects proximity conditions have been proved, the essence of the proposed method could be described more detailed.

On the first step, after the random k pivots from a set ET selection and the points coordinates $P(et_i)$ calculation, distance by the points $P(et_i)$ to the origin in space E^k could be obtained as: $\rho(P(et_i), 0) = \sqrt{P(et_i)_1^2 + P(et_i)_2^2 + \ldots + P(et_i)_k^2}$.

In addition, we form a matrix D, which is distribution of distances in space E^k by points $P(et_i)$ to the origin. To do this, the set $\Psi = \{[\rho(P(et_1), 0)], [\rho(P(et_2), 0)], \ldots, [\rho(P(et_n), 0)]\} = \{\psi_1, \psi_2, \ldots, \psi_z\}$, $z \leq n$ is introduced, where $[\rho(P(et_i), 0)]$ means the integer part of $\rho(P(et_i), 0)$.

Each index $\psi_i \in \Psi$ deals with a set of integers $IND_i = \{ind_{i1}, ind_{i2}, \ldots, ind_{iw}\}$, which are the objects indexes with the distance from the origin equal to ψ_i. In this case, the following condition is true: $\forall q \in \{1, \ldots, w\}, ind_{iq} \in \{1, \ldots, n\}, \exists [\rho(P(et_{ind_{iq}}), 0)] = \psi_i$. Considered matrix D has dimension $(\max(\Psi) - \min(\Psi) + 1) \times (\max\{|IND_1|, \ldots, |IND_z|\})$. Using auxiliary set IND_i, there is assigned: $D_{\psi_i q} = ind_{iq}$, what means, that raw ψ_i of the matrix D contains the indices of the objects, for which an integer part of the distance in space E^k to the origin is equal to ψ_i.

For the target object rt it is required to find row with index $[\rho(P(rt), 0)]$ in the matrix D. After that, according to Propositions 4 and 6, it is need to review the neighbor rows of the matrix D with the indices from the set $\Psi 1 = \{[\rho(P(rt), 0)] - [\lambda\sqrt{k}] - 1, [\rho(P(rt), 0)] - [\lambda\sqrt{k}], \ldots, [\rho(P(rt), 0)] - 1, [\rho(P(rt), 0)], [\rho(P(rt), 0)] + 1, \ldots, [\rho(P(rt), 0)] + [\lambda\sqrt{k}], [\rho(P(rt), 0)] + [\lambda\sqrt{k}] + 1\} = \{\psi 1_1, \psi 1_2, \ldots, \psi 1_v\}$, and $\Psi 1 \subset \Psi$. $v \leq z$.

Then, when viewing the element rows of the matrix D with index $\psi 1_t$, i.e. $D_{\psi 1_{tq}}$, further screening "candidates" can be find out among the proximate objects: first, by checking the conditions from the Proposition 7: $|\overline{et_i} - \overline{et_j}| > \lambda$, and second, by checking the condition of the Proposition 3: does the $P(et_{D_{\psi 1_{tq}}})$ lie in a hypercube centered at the point $P(rt)$ and side of 2λ. Finally, if $P(et_{D_{\psi 1tq}})$ is within the hypercube, then the distance $\delta(rt, et_{D_{\psi 1tq}})$ between the objects rt and $et_{D_{\psi 1tq}}$ is calculated.

2 Method Instantiation

There are considered two cases: (1) Class Cl objects are ordered m-ary trees, $m \in N$; (2) class Cl objects are strings.

1st Case. Objects of class Cl are ordered m-ary trees. Then the statement of the problem is as follows. A tree rt and a set of trees $ET = (et_1, et_2, \ldots, et_n)$, $i = \overline{1, n}$ are given. It is required to find all the trees et_i, such that the distance $\delta(rt, et_i)$ is not greater than a given positive integer λ.

In this case, one of the metrics for δ can be chosen. The metric used in [7] and the metric from [234, 235] can be considered as reasonable alternatives. Differences between these two metrics are in the set of valid tree editing operations: each metric permits rename operation, removal and insertion of tree nodes, but the first one consider the last two operations as applicable to the tree leaves only, i.e. to the vertices with no descendants, whereas the second one applies them to any tree node.

Further the essence of the insertion and deletion of tree nodes in a second metric will be review in details. As a result of the insert operation, some or all descendants of the parent node for the inserted node transform into inserted node descendants. After the delete operation all deleted node descendants transform into the descendants of its parent node. Consider concrete examples.

Example 1. Two ordered binary trees X and Y are represented on Figs. 2 and 3. The edit distance between X and Y in the Selkow metric [7] is 6.

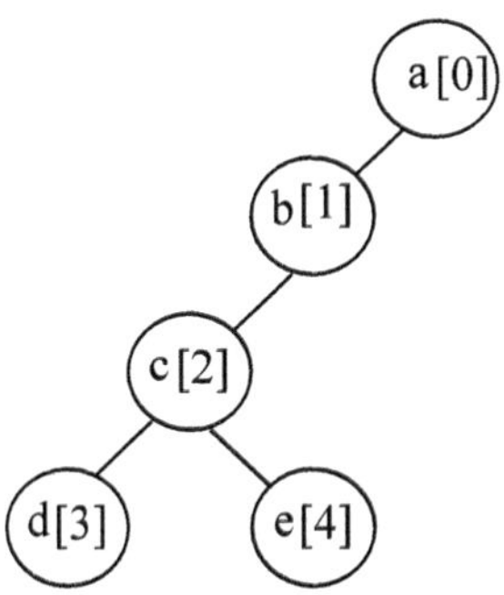

Fig. 2. Ordered binary tree X

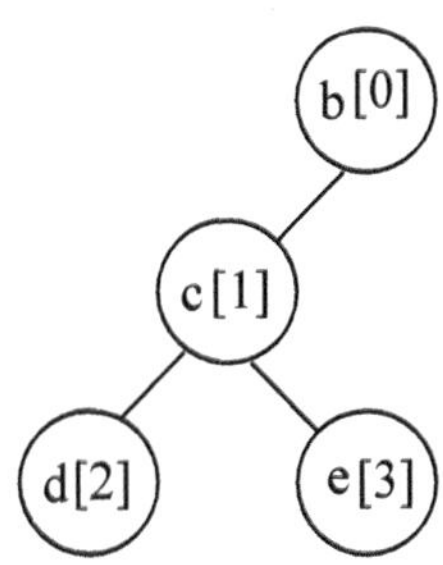

Fig. 3. Ordered binary tree Y

The minimal set of editing operations which convert X to Y includes operations:

(1) replace the node “a” with index 0 with the name «b»;
(2) replace the node “b” with index 1 with the name “c”;
(3) delete the leaf node “e” with index 4;

(4) delete the leaf node “d” with index 3;
(5) replace the name node “c” with index 2 with the name “d”;
(6) insert a right child “e” to the node “b” with index 1;

For metric used in [7, 8], the edit distance between X and Y will be equal to 1, as for converting Y to X it is necessary to delete node “a” with index 0.

Example 2. Two ordered ternary tree X1 and Y1 are represented on Figs. 4 and 5.

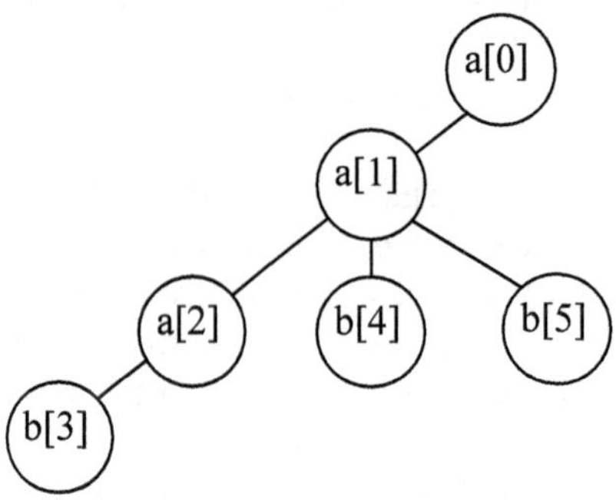

Fig. 4. Ordered ternary tree X1

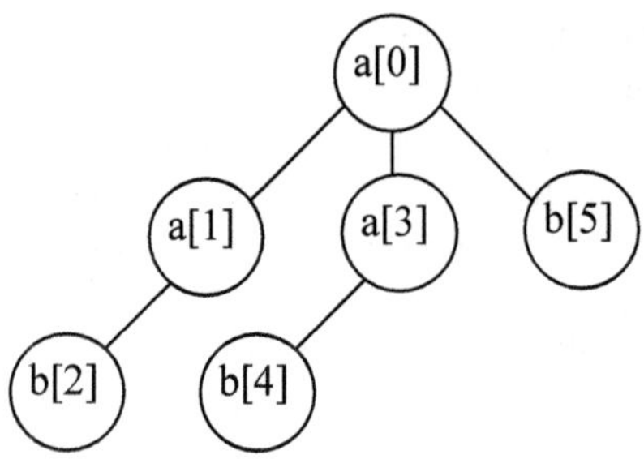

Fig. 5. Ordered ternary trees Y1

According to Selkow metric [7] edit distance between X1 and Y1 is equal to 7. The minimal set of editing operations which convert X to Y includes operations:

(1) insert a child “a” to node “a” with index 0;
(2) insert a child “b” to node “a” with index 0;
(3) insert a child “b” to the newly inserted node «a»;
(4) delete the leaf “b” with index 3;
(5) delete the leaf “b” with index 4;
(6) delete the leaf “b” with index 5;
(7) replace the name node “a” with index 2 with the name “b”.

For the second metric the distance between X1 and Y1 is equal to 3, as to convert X1 to Y1 it is necessary to perform operations:

(1) remove the node “a” with index 0;
(2) insert a child “b” to the node “b” with the index 4;
(3) replace the name node “b” with index 4 with the name “a”.

As follow from the above examples, trees are more proximate if the second metric is used. In common, the metric selection criteria should be determined depending on the specific practical problem.

2nd Case. Class *Cl* objects of are the strings. For this case using the Levenshtein distance, which is a minimal number of string edit operations for its conversion into another, is the best choice. That was proved theoretically and empirically in [4, 5], where the problem of finding similar strings is considered.

3 Experimental Research of the Method

Experimental studies of this method were carried out for the metric Zhang and Shasha [8]. To compare the results method described in [1] was chosen, where the condition formulated in Proposition 7 was also embedded to exclude a series of extra "expensive" tree edit distance calculations.

For randomly generated trees in every method time spent on the edit distance calculation was excluded from the overall time of the second search stage. The results of experimental research of the lists of 10, 100 and 1000 trees are shown on Figs. 6, 7 and 8.

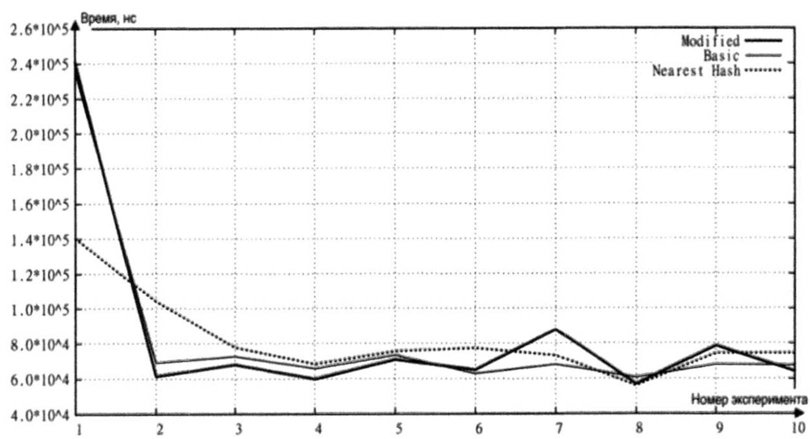

Fig. 6. The experimental results for the list of 10 trees

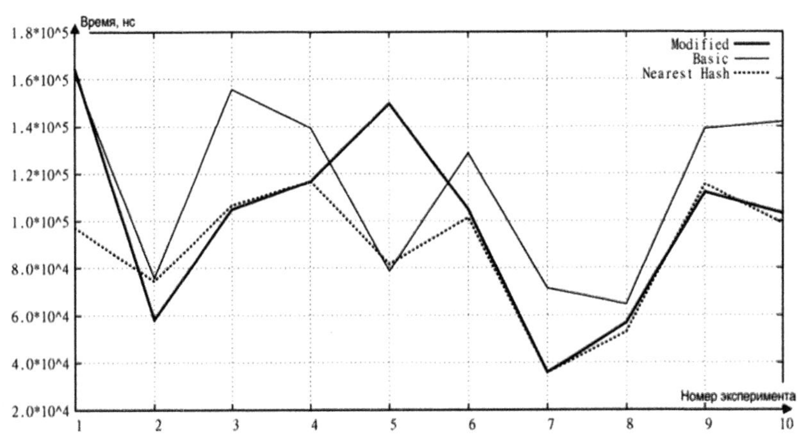

Fig. 7. The experimental results for the list of 100 trees

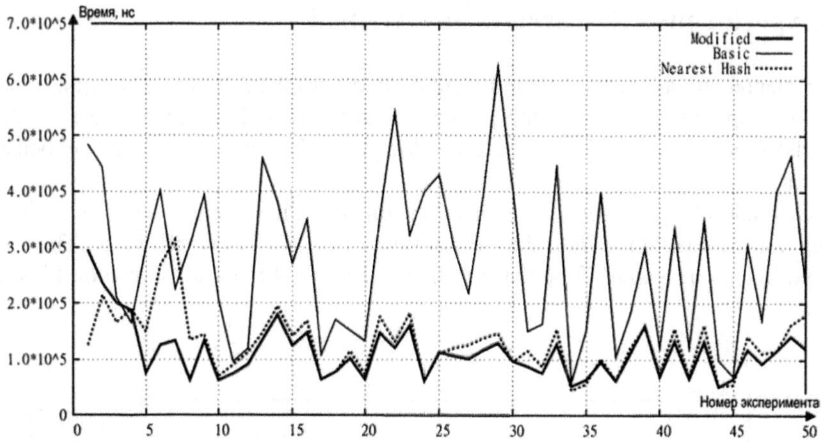

Fig. 8. The experimental results for the list of 1,000 trees

In the experiments shown on Figs. 6 and 7, the first 10 trees from the original list were target, and in experiment on Fig. 8 search target were first 50 trees, that are noted on the horizontal axis. As following from the figures, the proposed search method (Nearest Hash) is better than known ones (Basic and Modified) by performance 1.71 and 1.67 times – in the first experiment, 1.66 and 1.69 times – in the second experiment and 2.30 and 3.80 times – in the third experiment. Thus theoretical performance increase due to exclude extra objects on the first step and hash matrix constructing on the second is proved empirically.

4 Summary

Concluded, the improved method of proximity objects probabilistic search is represented with the introduction of the necessary theoretically proved the proximity objects conditions, which allow improving search performance. Proposed method can be used for the fast and precise search in Intelligent Search Systems, Big Data technology as well as in Intelligent Tutoring Systems to solve the problems of knowledge clustering and trainee answers analysis for effective pedagogical feedback.

Further research will focus on new experimental research obtained by the method: comparison with other known methods, the choice of the pivots number, the rational selection of specific pivots, as well as various applications of the method in practical tasks.

References

1. Bustos, B., Navarro, G., Chavez, E.: Pivot selection techniques for proximity searching in metric spaces. Pattern Recogn. Lett. **24**(14), 2357–2366 (2003)
2. Batko, M., Falchi, F., Lucchese, C., et al.: Building a web-scale image similarity search system. Multimed. Tools Appl. **3**(47), 599–629 (2010)
3. Zezula, P., Amato, G., Dohnal, V., Batko, M.: Similarity Search: The Metric Space Approach. Springer, New York (2006)
4. Chukhray, A.G.: Quick search method "similar" relational tuples relations. Radioelectron. Comput. Syst. **2**(2), 64–69 (2003)
5. Kulik, A., Chukhray, A., Zavgorodniy, A.: Similar strings detecting methods. In: Proceedings of the East-West Fuzzy Colloquium, pp. 38–47. IPM, Zittau (2005)
6. Selkow, S.M.: The tree-to-tree editing problem. Inf. Process. Lett. **6**(6), 184–186 (1977)
7. Tai, K.C.: The tree-to-tree correction problem. J. ACM **26**(3), 422–433 (1979)
8. Zhang, K., Shasha, D.: Simple fast algorithms for the editing distance between trees and related problems. Soc. Ind. Appl. Math. J. Comput. **18**(6), 1245–1262 (1989)

Towards the Technology of Employers' Requirements Collection Development

Olga Cherednichenko[1(✉)], Maryna Vovk[1], Olha Yanholenko[1], and Olena Yakovleva[2]

[1] National Technical University "Kharkiv Polytechnic Institute", Kharkiv, Ukraine
olha.cherednichenko@gmail.com, marihavovk@gmail.com, olga.yan26@gmail.com
[2] Kharkiv National University of Radio Electronics, Kharkiv, Ukraine
olena.yakovleva@nure.ua

Abstract. The article demonstrates the results of the research, which studied the effectiveness of organizing a proper connection between companies and universities. A methodology of employers' requirements collection and evaluation is given. To get information about the requirements, it is proposed to analyze the open vacancies of companies. The technology for reviewing, extracting and systematizing requirements from the description of employers' vacancies was created in order to use the gathered information in university Learning Management Systems. It is suggested comparator identification method as a basis for mechanism of vacancies description search, data retrieval, an indicator value transformation, and comprehensive assessment performing. It is proposed the model of requirements collection based on the method of comparator identification. The ontology of employers' requirements can be used for requirements collection, analysis and estimation. However, there is still a problem of requirements thesaurus creation, vacancies, and features for different specialties in different languages. Description of vacancies in IT sphere is unified enough worldwide. But we cannot say the same about all other areas of industry. That's why thesaurus of equivalent terms and their translations have to be created. Future work will be directed on the implementation of the suggested approach in the domain of IT specialists.

Keywords: Information system · Employers' requirements · Data collection · Informational retrieval · Competency · Learning management system · Model

1 Introduction

The labour market is changing due to total globalization nowadays. Often education systems do not meet the requirements of the labor market. One of the peculiarities of the modern labor market and world education area is high dynamics. Unfortunately, today a gap between educational output and labor market requirements exists. Cooperation with the representatives of the labor market is an essential factor of education quality provided by universities. Therefore, university curricula should take into account employers' requirements. Moreover, curricula should be adapted to new

M. Nechyporuk et al. (Eds.): *Integrated Computer Technologies in Mechanical Engineering*, AISC 1113, pp. 228–239, 2020.
https://doi.org/10.1007/978-3-030-37618-5_21

requirements and environment. Universities should know what they are preparing their students for, what kind of labor market they are oriented on, and what sort of employers they want to cooperate with. Curricula should respond to employers' needs.

Quality of education is the most important indicator of university success. It combines various aspects including performance level of students, correspondence to requirements of labour market, research activity, resources quality, etc. Our previous research was devoted to development of comprehensive assessment framework of students' satisfaction with education they obtain in university [1]. Also it is considered the problems of university resources evaluation which surely influences education quality. Another direction of our research was represented by evaluation of the level of research activity of university's employees [2]. Also, it is developed the web-based quality monitoring information system, which allows measuring and analysing the outcomes of university functioning based on the data available on the web [3]. Now we are concentrated on information retrieval. The aim of the paper is developing an approach for retrieving from advertising description the employer demands to an applicant.

The rest of the paper is organized as follows: in Sect. 2 the problem statement is given. Section 3 studies related works and analyses information systems (IS), which are used in university for effective management. In Sect. 4 the model for the construction of vacancies patterns for some majors is described. Experiment results and technology of data collection from vacancies description that is located on job search sites are presented in Sect. 5. In Sects. 6 and 7 discussions and conclusion are presented.

2 Problem Statement

The problem of discrepancy between the competencies of alumnus and employers' demands is considered. The issue is to receive the labor market requirements and use them as learning goals. During recruiting, employers mold vacancy description. So the vacancy description contains the main applicant characteristics, which are crucial during the selection. Vacancy descriptions are published on employers' sites or sites for job searching. The problem of requirements gathering is quite complicated. The experts in such case can't serve, because of the huge amount of data from different companies.

Descriptions of vacancies are a source of data for the requirements of employers. Analysis of sites for searching work and jobs matching shows that the job description contains certain information, such as the candidate's basic knowledge, experience, personal qualities and characteristics, working conditions, etc. From this volume of information, it is possible to single out the requirements, which in the future will be used in the process of learning management in the university.

During the processing such information, there are 2 problems that need to be solved. The first problem is the processing of semistructured text of job advertisements. Vacancy description can contain different languages (English, Ukrainian, etc.), and there is difference in the structure of sites' pages. The solution to this problem can be the use of NLP methods. The second problem is the automation of arranging data from vacancies description. It is necessary to automate the process of recognizing similar vacancies, identifying common requirements, highlight the most important of them in order to complete the education problem in a proper way. An actual task, in which the search and processing of vacancies are conducted, is the inclusion of this kind of information use within the information systems of university management.

The purpose of this work is to create a framework for reviewing, extracting and arranging requirements from the description of employers' vacancies.

3 Literature Review

Issues related to the search and retrieval of data from vacancies were considered in work [4], but received data was used mainly to compare with the resume data and to find the most suitable candidate. In [5] was created the intelligent management system based on text mining methods for supporting recruitment services. But this system collects data into the separate data storage from the University IS, and there is no connection between these two components.

Effective management in university requires the application of various IS. These IS support different areas of university activities [6]. The following types of IS in university are distinguished:

- Systems of administrative and financial management;
- Learning management systems (LMS);
- Learning content management systems (LCMS);
- Research management systems;
- Information resources management systems.

Systems of administrative and financial management usually are present in all universities and they are similar to ordinary enterprise resource planning systems [7]. LMS support the development of studying curricula and learning programs. Also, they help to construct timetables of lessons and exams [8]. Accounting of students' academic progress is one of the functions of LMS as well. LCMS are oriented on the educational contents, courseware, handbooks and other materials. The online access to learning contents, construction of individual learning courses, students' testing, e-learning, library management are all the functions of LCMS [9]. Research management systems are not so widespread. Still, universities use conference management IS, publishing IS and systems which support the work of academic boards [10]. Systems of information resources management combine the university websites, e-mail, forums, blogs, workflow control and databases [11].

Traditionally the existing IS process data from internal sources. It could be official reports and documents. However, the external data sources stay out of consideration. Employers' needs must be taken into account while developing learning curricula since they form the demand for particular specialists in different industrial areas. Such cooperation between universities and employers definitely exists now. The discussions about the content of learning programs are organized by departments and faculties in the form of seminars, conferences, and roundtables. Also, companies can be invited to participate in some surveys concerning their vision of educational content in the university and desirable directions of learning.

However, the information collected from the employer's side in such a way is not processed automatically, except for the survey results. And it is not incorporated to the university IS, which means that it is not used in the automatic procedures of intelligent decision-making and curricula development. So, the existing IS don't formalize employers' requirements and don't realize the models of their accounting during the management of the studying process. Actually, now universities can only control the results of students studying while analyzing the success of graduates' job placement. The information about the first place of a job is interesting for a university, but it is not enough and it doesn't provide flexible management of the studying process.

The requirements of employers should be analyzed and taken into account in the process of curricula development. Therefore, problems of vacancies data parsing, extracting, collection and analysis arise. And all that data should be actual and retrieved. Thus, the models of employers' requirements collection in order to use the gathered information in university LMS should be developed.

4 Materials and Methods

The core idea of this research is that data about employers' requirements can be collected from the description of vacancies (Fig. 1). The degree of requirements satisfaction can be evaluated based on the estimates of graduates' competencies.

The main process of requirements evaluation comparison is underlined. The basic stages of the process of requirements evaluation include: Search of vacancies' descriptions; Collection of data related to employers' requirements; Evaluation of employers' requirements; Collection of data related to graduates' competencies; Evaluation of graduates' competencies; Comparison of requirements with competencies; Construction (or correction) of curricula.

Thus, in order to evaluate requirements, it is needed a mechanism for vacancies description search, data retrieval, an indicator value transformation, and comprehensive assessment performing. Each of the listed points is a separate task. Especially, an important task is comprehensive assessment, which is the basis for making a decision about changes in curricula.

The first step is information search of employers' requirements. The information can be found on vacancy sites, professional and social nets, analytic and infographic sites. In this paper vacancy descriptions are considered. Different algorithms can be used for searching. It is suggested to apply topical search, as it was realizes in [2].

The second step is pattern creation for employers' requirements data retrieval. It presumes knowing of competencies connected with certain major and needed level of education. Such information is accessible at university sites and in curricula. In this paper the position of project manager is used as an example.

The third step is data retrieval and pattern fill in. For automation of this process it is necessary to develop the domain ontology of requirements and competencies.

The fourth step is tracing competency requirements. In order to perform this, it is necessary to choose a scale for competencies estimation and describe the trace rules. For example, one requirement of employers can summarize several competencies. Vacancy descriptions do not contain requirements according to the whole set of competencies. But text of vacancy description should be converted into competency values.

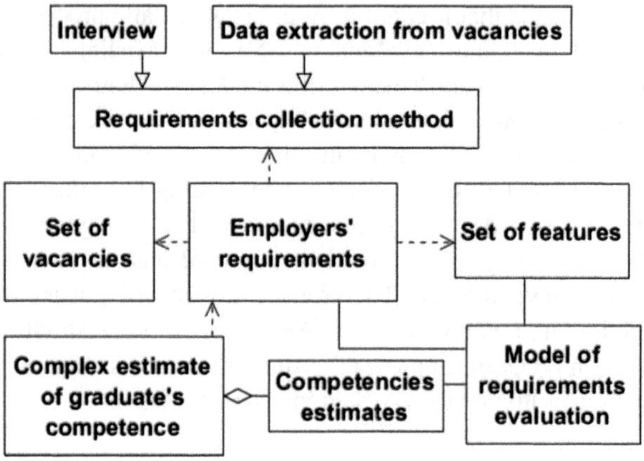

Fig. 1. Conceptual model of requirements evaluation.

The fifth step is defining each competency by using the trace table. The most important for employers groups of competencies can be defined in such a way. Identification and estimation of employers' requirements is the basement for changes and adaptation of curriculum.

Taking into consideration the experience in development of web-based quality monitoring IS [2, 3] it is suggested comparator identification method as a basis for this approach [12, 13]. The most promising today is the usage of models and methods of information technology, based on the results obtained in solving the problems of artificial intelligence (AI). The existence of finite predicates algebra discovers an opportunity for transition from the algorithmic description of information processes to their description as equations which set the relationship between variables. All necessary notions of predicates algebra, which is a proven mathematical tool for recording the relationships, were introduced in [12, 13].

As a rule employers' requirements are represented in the verbal form. They are based on the qualitative estimates of the desired competencies. So, requirements have to be formalized and structured in order to be used for learning process management. In this paper the ontology of employers' requirements which can be used for their automatic collection from the web is suggested.

Actually, for studying process management purposes it is necessary to compare employers' requirements with the level of qualification of university graduates. Therefore, it is reasonable to express the requirements through the graduates competence formed during the studying process. In fact, any vacancy implies the definition of the minimal required level of different competencies that an applicant should possess.

It is supposed that all competencies are estimated through indicators. All indicators of competencies are divided into two groups: soft and professional. Soft competencies may include the following indicators: ability to find and use necessary information, ability to learn, planning skills, ability to use equipment, language skills, etc. Professional competencies include: ability to use innovations, ability to analyze the results of work, orientation on professional level improvement.

Formally the ontology of employers' requirements can be presented in the following way:

$$O\_Requirements = \langle T,\ R,\ C \rangle \tag{1}$$

where

$T = \{requirement,\ competency,\ indicator,\ scale\}$;
$R = \{is\_part\_of,\ is\_expressed\_through,\ is\_measured\}$;
$C_1 = \exists c \exists i(Indicator(i) \wedge is\_part\_of(i, c) \wedge Competency(c))$;
$C_2 = \exists r \exists c(Requirement(r) \wedge is\_expressed\_through(r, c) \wedge Competence(c))$;
$C_3 = \exists i \exists s(Indicator(i) \wedge is\_measured(i, s) \wedge Scale(s))$.

Such ontology can be used for requirements collection, analysis and estimation. Vacancy description is the source of employers' requirements.

Each vacancy, which is analyzed in order to develop the curriculum of some major, is represented through the set of competencies. Let $R = \{R_1,\ R_2,\ \ldots, R_k\}$ be a set of considered vacancies that reflect employers' requirements related to some major. To describe these vacancies we define p discrete groups of features $Q_1,\ Q_2,\ \ldots,\ Q_p$ that reflect employers' requirements. Each group of features $Q_s = \{q_s^l |\ l = \overline{1, h_s}\},\quad s = \overline{1,p}$ determines the content of competency which is necessary for the job, for example, years of experience, level of foreign language skills. The peculiarity of vacancies is that there can exist r instances of each vacancy R_j. In the description of each instance R_j only one value of a feature from each group $Q_1,\ Q_2,\ \ldots,\ Q_p$ can be met.

Formally the problem can be formulated in the following way. It is necessary to find the competence indicators which support employers' requirements. This problem can be considered as a problem of multicriteria classification of vacancies' description R_j based of the values of features provided that Q_s is considered as a scale of s-th feature.

The model of requirements collection based on the method of comparator identification is suggested [12, 13]. It allows matching the data from a vacancy description and the template. It based on the relation between the keywords and the placement of these words on the webpage. This method models the extraction process as a human intelligent activity, since a human looking through a vacancy can easily determine whether it is corresponds to the template or not.

A template of the particular vacancy describes the features and their scales (Fig. 2).

To build a comparator model E is defined as a set of elements, W is a set of keywords. A binary relation "requirements and keywords extracted from vacancy" $R_{VACANCY} \subseteq E \times W$ for which

$$R_{VACANCY} = \{(e_1, w_1), \ldots, (e_s, w_j)\},$$
$$\forall (e_i, w_j) \in R_{VACANCY}:\ F(e_i) = \begin{cases} c_1,\ if\ (w_{i1} \wedge w_{i2} \wedge \ldots) \vee (w_{j1} \wedge w_{j2} \wedge \ldots) \vee \ldots \\ \ldots \\ c_m,\ if\ (w_{im} \wedge w_{im} \wedge \ldots) \vee (w_{jm} \wedge w_{jm} \wedge \ldots) \vee \ldots \end{cases}$$

Let $C = \{c_1, \ldots, c_m\}$ be a set of templates, $E_p = \{e_j \in E | c = F(e_j), c \in C\}$.

Let $R_{PATTERN} \subseteq E \times C$ be a binary relation "requirements contain template's keywords", $R_{PATTERN} = \{(e_i, c_j) | e_i \in E_p, c_j \in C\}$. The comparison of vacancy and a pattern is made on the basis of predicate $P_{pattern} = \begin{cases} 1, & if\ (\exists e_1 \exists e_2 \exists e_3 (E(e_1, e_V) \wedge E(e_1, e_R) \wedge E(e_1, e_I))) \equiv 1, \\ 0, & otherwise. \end{cases}$

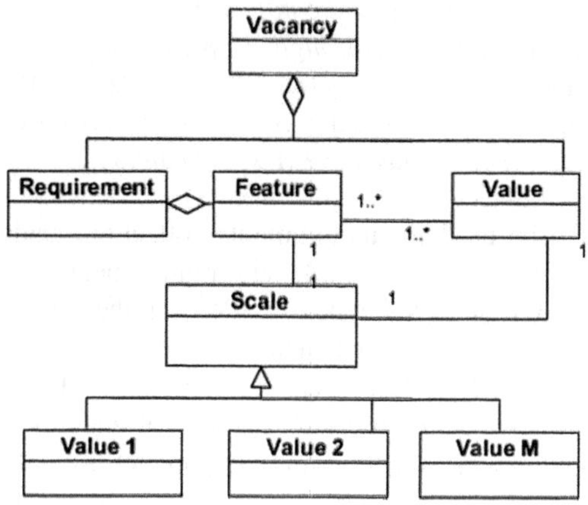

Fig. 2. The model of the template of a vacancy.

The proposed model can be used for construction of vacancies patterns in some major. In such way data about employers' requirements are collected and further analyzed their compatibility with graduates' competence.

5 Experiments and Results

Competency modelling is a popular method used to examine job-related information and employee skills [14, 15]. Competency evaluation presumes that competencies have to fit the required vacancy characteristic [16]. We consider the examples of vacancy requirements for project manager position. At [17] all the project manager competencies and their definitions were determined. There are 29 competencies. For the experiment we enumerated them from K1 to K29.

The pipeline of the experiment is: to choose a site of job vacancies, run parser to collect data from vacancies, extract and classify data on requirements for vacancies, compare competency to requirements, determine the level for each competency. There are a lot of sites, where vacancy descriptions for a project manager could be found. One of the most popular in Ukraine is www.work.ua (Fig. 3) [18].

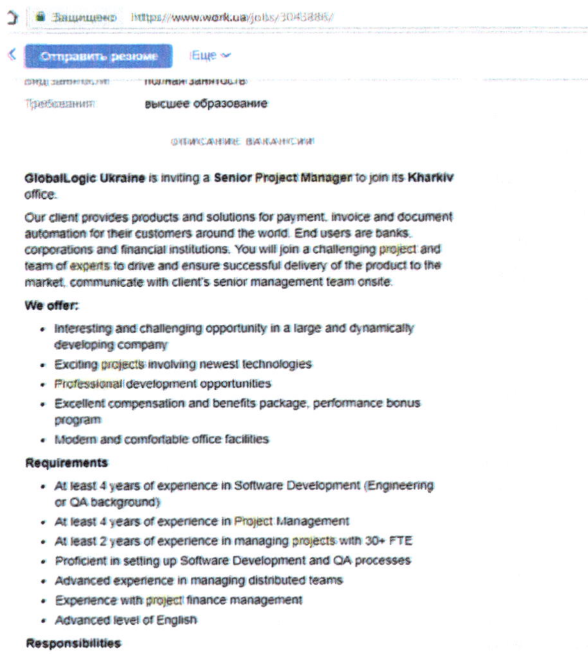

Fig. 3. Example of vacancy description.

The next step is to get the content of a specific HTML page, which contains the necessary information for its further retrieval and processing. Then the processing and cleaning of the previously received content are performed. This step involves removing all HTML-tags from the content obtained in the previous step. To parse the HTML page, the Jsoup library is used. Received data from HTML-page is stored in an external JSON file (Fig. 4).

So, the vacancy description of IT project manager can be as follows [18]:

"Qualifications:

- Experience in Software Development (Engineering or QA background);
- Experience in Project Management;
- Experience in managing projects with 30 + FTE;
- Proficient in setting up Software Development and QA processes;
- Advanced experience in managing distributed teams;
- Experience with project finance management;
- Advanced level of English.

From vacancy description the relations between required qualifications and competencies could be defined. For example, Advanced level of English correspond to Language proficiency (K15), Advanced experience in managing distributed teams at the same time correspond to Production efficiency (K10) and Stress management (K12), etc.

To evaluate in which extend the applicant should master each competency the scale can be used. In order to simplify the case in the example all competencies are evaluated in four-dimension category: high mastering(a1), middle(a2), low (a3), not given (a4). In order to estimate value (level of mastering) of required competency the thesaurus for Project Manager Vacancy was developed. For example, in vacancy description there such word concerning the level of foreign language mastering: Upper-Intermediate, Advanced, Fluent or Excellent it means the competency of Language proficiency should be on the high level of mastering.

In our example 20 IT project management vacancies were researched and value of competencies were distinguished. Based on the received values of competencies, groups of competencies were detached. Competencies with high value belong to a strongly required group (a1 $\geq$ 80%). Such competencies are required by almost all employers in a high level of mastering. Group of desired competencies (a1 and a2 $\geq$ 70%) is present at most vacancy descriptions. Group of additional competencies (a3 and a4 $\geq$ 50%) gives some advantages to an applicant. Thus, it was defined that the strongly required competencies are Self-confidence, Flexibility, Responsibility, Stress management, Conceptual thinking, Language proficiency, Communications, Conflict management, Management, Leadership, Collaboration.

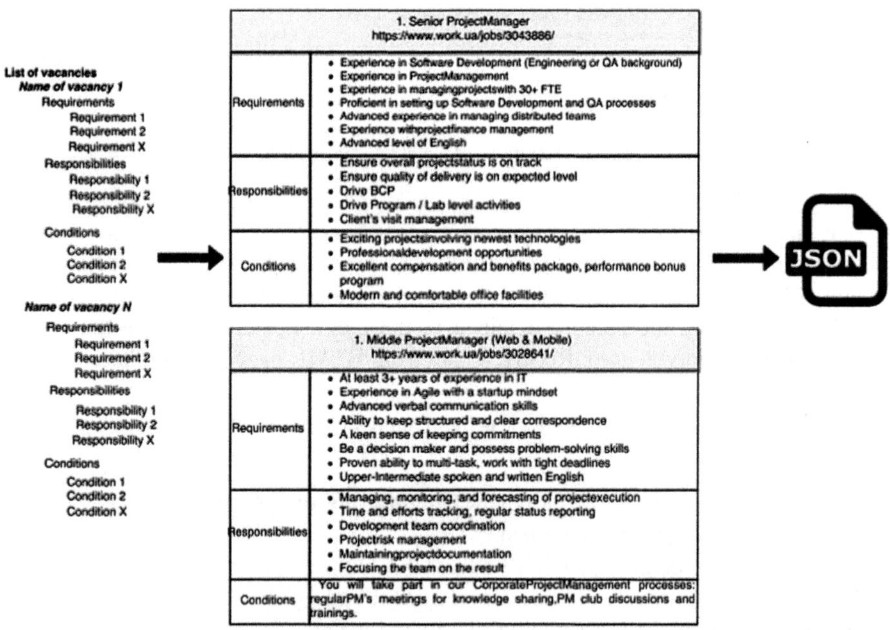

Fig. 4. Structure of a JSON file for storing vacancy data.

Having groups of competencies for IT project management position it is possible to define how to correct curriculum, which competencies should be formed at university teaching process and which competencies students can develop by self-learning.

Thus, it is possible schematically describe the technology of data collection from vacancies description that are located on job search sites (Fig. 5). The description of the entities of the data flow diagram is as follows:

1. Parsing of vacancies is the process of collecting web pages with advertisements for job search.
2. Extracting vacancies' description is the process of extracting data about vacancies (requirements of basic knowledge, experience, etc.)
3. Classification of received data is the process of classifying extracted data based on ontologies.
4. Data comparison is matching data on requirements vacancies with competences and learning outcomes; adjustment training programs in accordance with the received data.

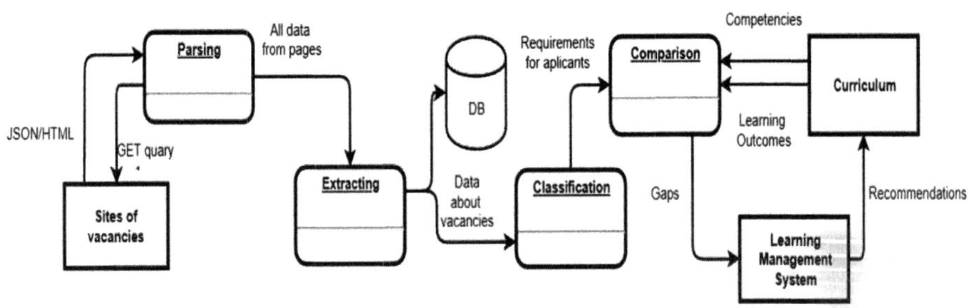

Fig. 5. Technology of data collection on vacancies.

6 Discussion

In this paper, the necessity of communication between higher education and business is proved. To be more competitive on the labor market graduates of universities should get knowledge and skills requested by their future employers. So, university curricula should be responsive in some extent to the requirements of employers.

In works [19, 20] the importance of creating an information model of interaction between employers and universities was considered, which should be based on an active mutual exchange of information on the status and prospects of labor market development, requirements for the fundamental and professional components of training specialists. The authors [21] consider the modeling of informational processes for the employment of graduates on the basis of an analysis of the requirements of employers. As a result, it is possible to obtain an assessment of the compliance of competencies, and thus the quality of graduate training, with the real requirements of the labor market. However, it is not always possible to get a feedback from graduates, and the information system does not implement a mechanism for automatically updating the basic requirements of employers.

Nowadays the retrieval system is one of the most effective tools for identifying the information for decision-making. The approach based on factual information systematization is proposed [22]. It is explored how can natural language processing methods help to check contradictions and mismatches in facts automatically. The reference model of the analytical system is proposed. It consists of such basic components as Document Search component, Fact retrieval component, Fact Analysis component,

Visualization component, and Control component. Such reference model could be used for vacancy descriptions processing and data retrieval.

In order to organize a proper connection between companies and universities a methodology of employers' requirements collection and evaluation is suggested. To get information about the requirements, we propose to analyze the open vacancies of companies. The model of requirements collection based on the method of comparator identification allows sorting all vacancies and form the patterns of vacancies with the values of their features for various specialties.

The model of requirements evaluation is based on the ontology of competencies, since it is supposed that requirements are in fact expressed by employers in terms of competencies formed during the educational process.

7 Conclusions

Thus the drawbacks related to the suggested methodology are the following. First of all, it is necessary to determine the indicators of the graduates' competencies. It is needed to define the scales and methods of assessment. It is obvious that experts must be involved in this process. However, the question about the appropriate method of experts' judgment is still open.

Another problem is creation of thesaurus of requirements, vacancies and features for different specialties in different languages. Description of vacancies in IT sphere is unified enough worldwide, i.e. companies use the same terms in English which became a standard. But it is not necessarily true for all other areas of industry. That's why thesaurus of equivalent terms and their translations have to be created.

Finally, an open issue is invariance of the developed models for different specialties. Since the features of vacancies of different specialties can be different, it needs to adapt the templates of vacancies and change the patterns in the requirements collection model. So, it is needed to make domain ontologies. Future work will be directed on the implementation of the suggested approach in the domain of IT specialists.

References

1. Cherednichenko, O., et al.: Monitoring and Evaluation Problems in Higher Education-Comprehensive Assessment Framework Development, pp. 455–460. CSEDU (2013)
2. Cherednichenko, O., et al.: Models of research activity measurement: web-based monitoring implementation. In: EuroSymposium on Systems Analysis and Design, pp. 75–87. Springer, Cham (2014). http://dx.doi.org/10.1007/978-3-319-11373-9_7
3. Cherednichenko, O., Yanholenko, O.: Information technology of web-monitoring and measurement of outcomes in higher education establishment. In: EuroSymposium on Systems Analysis and Design, pp. 103–116. Springer (2015). http://dx.doi.org/10.1007/978-3-319-24366-5_8
4. Matrosova, E., Naumov, D., Tikhomirova, A.: Decision support system in the sphere of vacancy selection for graduates of technical universities. Procedia Comput. Sci. **145**, 337–341 (2018). https://doi.org/10.1016/j.procs.2018.11.081

5. Shatovska, T., Repka, V., Kamenieva, I.: Intelligent recruitment services system. In: International United Information Systems Conference, pp. 411–419. Springer, Heidelberg (2009). http://dx.doi.org/10.1007/978-3-642-01112-2_42
6. Abdul-Hamid, H.: What matters most for education management information systems: a framework paper (2014)
7. Martins, J., Branco, F., Gonçalves, R., Au-Yong-Oliveira, M., Oliveira, T., Naranjo-Zolotiv, M., Cruz-Jesus, F.: Assessing the success behind the use of education management information systems in higher education. Telematics Inform. (2018). https://doi.org/10.1016/j.tele.2018.10.001
8. White, B., Larusson, J.A.: Strategic directives for learning management system planning. Research Bulletin. EDUCAUSE Center for Applied Research, Boulder, CO (2010). http://neteducause.edu/ir/library/pdf/ERB1019.pdf
9. Pinho, C., Franco, M., Mendes, L.: Web portals as tools to support information management in higher education institutions: a systematic literature review. Int. J. Inf. Manag. **41**, 80–92 (2018). https://doi.org/10.1016/j.ijinfomgt.2018.04.002
10. Al-Emran, M., Mezhuyev, V., Kamaludin, A., Shaalan, K.: The impact of knowledge management processes on information systems: a systematic review. Int. J. Inf. Manag. **43**, 173–187 (2018). https://doi.org/10.1016/j.ijinfomgt.2018.08.001
11. Cheung, K.S.: Implementation of electronic workflow systems in higher education institutions: issues and challenges. In: Advanced Research on Electronic Commerce, Web Application, and Communication, pp. 306–311. Springer, Heidelberg (2011). http://dx.doi.org/10.1007/978-3-642-20370-1_50
12. Bondarenko, M.F., Shabanov-Kushnarenko, U.P.: Theory of Intelligence: A Handbook. SMIT Company, Kharkiv (2006)
13. Bondarenko, M.F., et al.: About brain-like computers. Radioelectronics Inform.–Kharkov: KHNURE **2**, 89–105 (2004)
14. Campion, M.A., et al.: Doing competencies well: best practices in competency modeling. Pers. Psychol. **64**(1), 225–262 (2011). https://doi.org/10.1111/j.1744-6570.2010.01207.x
15. Embo, M., Helsloot, K., Michels, N., Valcke, M.: A Delphi study to validate competency-based criteria to assess undergraduate midwifery students' competencies in the maternity ward. Midwifery **53**, 1–8 (2017). https://doi.org/10.1016/j.midw.2017.07.005
16. Hertzum, M., Simonsen, J.: Configuring information systems and work practices for each other: what competences are needed locally? Int. J. Hum. Comput. Stud. **122**, 242–255 (2019). https://doi.org/10.1016/j.ijhcs.2018.10.006
17. Liikamaa, K.: Developing a project manager's competencies: a collective view of the most important competencies. Proc. Manuf. **3**, 681–687 (2015). https://doi.org/10.1016/j.promfg.2015.07.305
18. Job Search Web Site. www.work.ua
19. Grossman, A.M., Johnson, L.R.: How employers perceive online accounting education: evidence from Kentucky. J. Account. Educ. **40**, 19–31 (2017). https://doi.org/10.1016/j.jaccedu.2017.06.002
20. Singh, P., Thambusamy, R.X., Ramly, M.A.: Fit or unfit? Perspectives of employers and university instructors of graduates' generic skills. Proc. – Soc. Behav. Sci. **123**, 315–324 (2014). https://doi.org/10.1016/j.sbspro.2014.01.1429
21. Druzhinina, E.N.: The system of interaction between the university and the employer to facilitate the adaptation of graduates of a pedagogical university to the labor market. Pedagogical Education in Russia, №. 6 (2015)
22. Sharonova, N., Doroshenko, A., Cherednichenko, O.: Issues of fact-based information analysis. In: CEUR Workshop Proceedings (2018). ceur-ws.org/Vol-2136/10000011.pdf

Cyber Security and Safety

Provision of Cybersecurity in Ukraine: Issues of Legal Responsibility

T. G. Katkova(✉), A. M. Stiebieliev(✉), S. E. Chmykhun(✉), and M. Zh. Mkrtchan(✉)

National Aerospace University "Kharkiv Aviation Institute", Kharkiv, Ukraine
tatyana.katkova07@gmail.com,
Stebelev.advokat@gmail.com,
svitlana.chmykhun@gmail.com,
marinamkrtcan264@gmail.com

Abstract. The article deals with the issues of legal responsibility in the field of cybersecurity. The Law of Ukraine "On the Main Principles of Cybersecurity Provision in Ukraine" of October 5, 2017, was analyzed and it was stated that art.12 presupposes responsibility for any breaches of legislation in the field of national security, electronic communications and information security in case the cyberspace is the place and/or the means of committing any legal offence, is the blanket article and refers to the articles of the Code of Ukraine on Administrative Offence, the Criminal Code of Ukraine, the Civil Code of Ukraine. The authors of the article offer characteristics of the kinds of legal responsibility: administrative, criminal and civil, for violations in the field of electronic communication and information security in the cyberspace, emphasizing that the issues of the robots' legal responsibility are not regulated by the general Ukrainian law.

Keywords: Cybersecurity · Criminal liability · Administrative liability · Torts liability

1 Introduction

Along with the development of the information infrastructure available via the Internet beyond any territorial borders and the progress of digital activity in the network and penetration of technologies in all spheres of social and government action, the role of cybersecurity grows, namely protection of the expressive interests of all people and citizens as well as the society and the state while using the cyberspace, which provides for the sustainable development of the information society and the digital communicative environment as well as timely detection and elimination of real and potential threats to the national security of Ukraine in the cyberspace.

The Law of Ukraine "On the National Security of Ukraine" of 21 June 2018 [1] determines the fundamental principles of the national security, with cybersecurity to be distinguished among them. Cybersecurity provision as the crucial task of the national security is indissolubly related to the complex of legal arrangements, among which legal responsibility holds a special place as it covers a wide range of social relations in the virtual environment.

M. Nechyporuk et al. (Eds.): *Integrated Computer Technologies in Mechanical Engineering*, AISC 1113, pp. 243–254, 2020.
https://doi.org/10.1007/978-3-030-37618-5_22

Moreover, the present-day challenge is the issue of the legal regulation of robots' legal capacity and legal responsibility for the infliction of harm by robots. This issue is under scrutiny all over the world. The work by Ryan Calo [2] who conducted a fundamental analysis of the litigation practice on the use of robots in the USA, is worth particular attention.

The issue of robots' responsibility was paid particular attention in the UN and the European Union. Thus, the World Commission on the Ethics of Scientific Knowledge and Technology (COMEST) published "Preliminary Draft Report of COMEST on Robotics Ethics" in 2016, which considers the ethical issues connected with the application of autonomous robots and their interaction with people. The Report states that, robots' autonomy is likely to develop to such an extent that there will be a need to install a system of ethical norms by means of programming with the purpose-built ethical codes for unsafe behaviour prevention (e.g. causing harm to human beings or the environment) [3].

The European Parliament adopted a resolution of 16 February 2017 with recommendations to the Commission on Civil Law Rules on Robotics [4]. The resolution proposes to lay primary liability on the producer and force producers and owners of robots to have insurance. However, it also presupposes division of liability between the producer and the owner with regard to the situation. The document introduces the status of "electronic personality" for the most sophisticated autonomous robots responsible for any harm caused when robots make independent decisions in the interaction with the third party.

Regardless of the international recommendations, the domestic law of Ukraine does not regulate the issue of robots' liability at all – robots are not considered legal subjects in Ukraine and, thus, cannot have liability.

The aim of the present research is the definition of problematic issues of the existing legal responsibility mechanism for any legal offence in the cyberspace as well as perspectives for regulation of robots' responsibility for any damage caused. The legal responsibility mechanism for legal offences in the cyberspace must reflect the present state of development of the digital society, be effective, and demonstrate the needs of the national security and, on the other hand, protection of the human rights and freedoms in the cyberspace. The components of the legal responsibility mechanism are criminal, administrative and torts liability, which arrange various sanctions depending on the regulatory environment (Fig. 1).

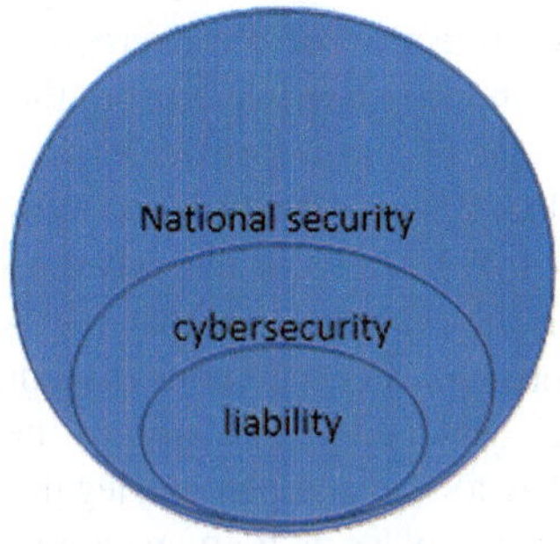

Fig. 1. The role of legal responsibility for legal offence in the cyberspace in the national security mechanism

2 Liability

Article 12 of the Law of Ukraine "On the Main Principles of Cybersecurity Provision in Ukraine" of October 5, 2017 states that individuals guilty of any violations of legislation in the field of national security, electronic communications and information security in case the cyberspace is the place and/or the means of committing any legal offence punishable under the civil, administrative and criminal legislation, incur responsibility by act of law. Therefore, article 12 is the blanket article and refers, in particular, to the Code of Ukraine on Administrative Offence, the Criminal Code of Ukraine, the Civil Code of Ukraine.

2.1 Administrative Liability

It is worth noting that the Law on Administrative Liability in Ukraine fails to develop along with the rapid development of the virtual environment. It was adopted on 23 December 1980, in the time of the USSR, and, despite all amendments and attachments it does not correspond to the current level of development of information technologies and requires fundamental improvement.

The Law of Ukrainc "On thc Main Principlcs of Cybersecurity Provision in Ukraine" defines cybercrime (as a socially dangerous guilty act in the cyberspace) but it does not define cyber infraction. Thus, on the basis of analogy (with the definition of infraction in the Code of Ukraine on Administrative Offence), we may consider a cyber infraction to be a wrongful guilty act in the cyberspace, which trespasses against the public order, property, citizens' rights and freedoms as well as the established control order and which creates administrative liability determined by the law. Article 1 of the Law of Ukraine "On the Main Principles of Cybersecurity Provision in Ukraine" defines the cyberspace as the environment (virtual space), which enables communication and/or realization of social relations, and formed as a result of functioning of concurrent (combined) communication systems and provision of electronic communication with the use of the Internet and/or other global data transfer systems. Having analyzed the provisions of the Code of Ukraine on Administrative Offence [6], the following sets of elements of administrative offence in case the cyberspace is the place and/or means of their committing:

- article 14. Violation of conditions and rules, which regulate activity in the sphere of telecommunications and radiofrequency resources exploitation in Ukraine provided by licenses and permits;
- article 146. Violations of rules of radio electronic facilities and radiating circuits implementation and exploitation as well as the radiofrequency resources exploitation in Ukraine;
- article 148-1. Violations of Regulations of Telecommunication Services Rendering and Receiving;
- article 148-2. Violations of the order and terms of communication services rendering within public networks;

- article 148-3. Exploitation of communication means with the aim adverse to the national interests, with the aim of violating public order and infringement of citizens' honour and dignity;
- article 148-4. Exploitation of technical facilities and equipment employed in public-service telecommunication networks without compliance certification;
- article 148-5. Violation of regulations on public-service telecommunication networks interworking;
- article 163-14. Violation of the procedure for e-cash operations;
- article 164-9. Unauthorized distribution of audiovisual works, audio and video records, software and databases;
- article 164-17. Violation of terms and regulations, which determine the procedure of cessation of copyright and/or allied rights infringements in the Internet;
- article 164-18. Deliberate provision of unreliable information in claims for cessation of copyright and/or allied rights infringements in the Internet;
- article 188-39. Infringements of regulations in the sphere of private data protection.

It is worth distinguishing, in particular, the set of elements of administrative offence connected with the subjects that directly implement cybersecurity provision measures within their competence:

- article 188-7. Failure to comply with the legal requirements of the National Commission for the State Regulation of Communications and Informatization;
- article 188-31. Failure to comply with the legal requirements of public officials in the bodies of the State Service for Special Communication and Information Protection of Ukraine.

The analysis of a series of the abovementioned articles enables to conclude that the legislative body has not provided for the norms that arrange administrative liability for:

Failure to comply with the legal requirements for organization and provision of technical and/or cryptographic data protection.

Failure to comply with the legal requirements of public officials of the Security Service of Ukraine in the sphere of cybersecurity (as the subject which directly implements cybersecurity provision measures within its competence).

The Law of Ukraine "On the Main Principles of Cybersecurity Provision in Ukraine" states in section 4 of article 6 that: "Liability for cybersecurity provision for communication and manufacturing systems of the objects of critical infrastructure, for technological information protection in compliance with the legal requirements, for prompt reporting to Computer Emergency Response Team of Ukraine (CERT-UA) on cybersecurity incidents as well as for the arrangement of independent information security audit at such institutions is laid on the owners and/or executives of enterprises, institutions and organizations referred to as the objects of critical infrastructure". In our opinion, this provision of the law must be reflected in in the articles of the Code of Ukraine on Administrative Offence, namely, determination of the set of elements of the offence for non-reporting or late reporting to Computer Emergency Response Team of Ukraine (CERT-UA) about the cybersecurity incidents by the owners and/or executives of enterprises, institutions and organizations referred to as the objects of critical infrastructure.

Another debating issue of administrative liability in the field of cybersecurity is formation of the mechanism for bringing authorized officers to responsibility for private information disclosure.

Case study. The Deputy Principal for Academic Work and Discipline at one of the Ukrainian schools asked class masters to complete the text document stored in the section "School Documentation" on the school site, which contained information about disabled war veterans and combat veterans (Kiev's anti-terrorist operation). These measures were explained by the Deputy Principal for Academic Work and Discipline to be necessary for academic work and discipline training. The Deputy Principal was brought to administrative responsibility for the administrative offence and imposition of a fine of 5100 UAH (around 207 USD). This amount of fine is most often imposed by the court for data privacy violation in Ukraine. This is a small sum in comparison with the amount of fines arranged by General Data Protection Regulation in the European Union. GDPR arranges for the fine of 20,000,000 euros. In out opinion, the Ukrainian legal regulations in the field of data privacy protection must be harmonized with the European law and the amount of fines in Ukraine must be increased.

To sum up the issue of administrative liability, it can be stated that the legal and regulatory framework for administrative liability in the field of cybersecurity does not fully cover the suite of rights of individuals and citizens as well as the society and the state long with the national interests in the cyberspace with administrative rules.

To our mind, the future Administrative Code of Ukraine must include a special separate section, which would provide for the penalty for wrongful acts in the cyberspace as well as cover the issue of administrative liability of the robot manufacturers and owners with regard to the EU recommendations concerning robots.

2.2 Criminal Liability

Criminal liability for legal offence in the spheres of electronic communication and information security in case the cyberspace is the place and/or the means of committing any legal offence is created by the independent Chapter XVI of the Criminal Code of Ukraine "Crimes in the Sphere of Electronic Computing Machines (Computers), Systems and Computer Networks Application". The given chapter of the Criminal Code of Ukraine includes 6 articles [7] namely:

- article 361. Unauthorized tampering with electronic computing machines, automated systems, computer networks or telecommunications networks;
- article 361-1. Creation of malware and technical devices with the aim of their use, distribution or sale as well as their distribution or sale;
- article 361-2. Unauthorized distribution of restricted data stored in electronic computing machines (computers), automated systems, computer networks or carriers of such information;
- article 362. Unauthorized operations with information processed in electronic computing machines (computers), automated systems, computer networks or stored on carriers of such information by authorized individuals;

- article 363. Misuse of electronic computing machines (computers), automated systems, computer networks or telecommunications networks or violation of stored data protection order or rules;
- article 363-1. Resistance to the operation of electronic computing machines (computers), automated systems, computer networks or telecommunications networks by means of mass communication.

According to the statistics of the Uniform Report of Criminal Offence in the state for January-August 2019 the following number of crimes were reported: 981 under article 361; 138 under article 361-1; 31 under article 361-2; 428 under article 362; 4 under article 363; 2 under article 363-1 [8].

Content analysis conducted in the field under study enabled to define threats, which enterprises, organizations and private entrepreneurs faced in 2018-2019. They were primarily connected with information systems infection with viruses, hacking computer mailboxes, company sites attacks and internet-fraud. The traditional types of cybercrime include: (1) theft of computer data and technical devices; (2) damage of devices and all information stored on them; (3) unauthorized access to information resources; (4) stealing passwords and money from banks or credit cards; (5) data espionage; (6) electoral fraud [9, p. 114].

Case study. Kyivskiy district court of Kharkiv in 2017 found three individuals guilty of criminal offence – crimes provided for in section 2 of article 361-1 and section 2 of article 361 of the Criminal Code of Ukraine. The criminal plot was aimed at the distribution of malicious software for unauthorized tampering with computer systems and computer networks by means of selling through the Internet. The abovementioned individuals posted and advertisement in the Internet for «Overflow Bot» malware aimed at the organization of DOS-attacks at the information resources, sites, forums, online applications etc. for the purpose of their blocking by means of sending a large number of email queries (packets) to communication channels; provided technical specifications and functional details for «Overflow Bot» by sending a web link on a file hosting service; arranged demonstrations of trial DDOS-attacks at sites; after the money transfer to the electronic wallet in «Web Money» payment system, using an account in Jabber-client, they sent a message with the key attributes (login and password) of the web-admin access panel for the given software [10].

As scientists in the field of criminal law aptly note, despite the availability of current laws and regulations, the domestic prospective legislation partially meets present-day needs in the sphere of state cybersecurity infrastructure formation. Thus, criminal law, while determining four basic principles of criminal statute operation in space, in compliance with which bringing to criminal liability depends on locus delicti (the territory), the citizenship of the liable person, type of criminal offence and the possibility of harm inflicted to the interests of Ukraine, sets substantial limits to the possibilities of the criminal code of Ukraine in force [10]. Article 203-2 of the Criminal Code of Ukraine creates responsibility for the organization of gambling business. The public danger of this offence is undoubted, which requires law enforcements bodies to step up measures at struggle with the abovementioned phenomenon. At the same time, the given article is effective only against real gambling venues on the territory of Ukraine. Today, the vast majority of betting terminals and other kinds of gambling business have been transferred to the Internet. With account taken of the fact that such betting terminal is created by foreign residents and software resides on servers physically located outside the territory of Ukraine, there are no grounds for the application of this law to the organizers of such an illegal gambling business regardless of the fact that a lot of Ukrainians actively participate in such events. The given problems are also connected with the issue of liability for unauthorized access to computer information stored on servers located outside Ukraine and distribution of prohibited articles via the Internet ("soft" drugs, non-public information retrieval means etc.).

Acts connected with criminal information distribution in the Internet are also considered problematic for classification. In compliance with article 26 of the Criminal Code of Ukraine, criminal complicity refers to joint participation of several subjects of crime in committing a deliberate crime. The given definition as a mandatory characteristic presupposes general criminal intent of the accomplices. In the meantime, today, it is relatively easy to find in the Internet detailed instructions for making explosive devices, deleting crime traces or criminal offence organization. Absence of general criminal intent, i.e. accomplices' awareness of their participation and aims of every accomplice, excludes the liability of the authors of the publication for criminal complicity in the offence committed by the person who used these instructions.

Besides the abovementioned issues of the mechanism of criminal liability for computer crimes, at present, determination of criminal liability for the individuals who will be liable for the damage resulting from action or omission of a certain algorithm (computer program, artificial intelligence) is highly relevant. These individuals may be the developer or the user, or both. However, if a program is capable of searching and correcting errors inside it, i.e. of self-improvement (i.e. it is not the initial program released by the developer or is not fully controlled by the user) what should be done?

The authors of the article share the idea of Radutny [12], a Ukrainian scientist, who forecasts emergence of a chapter in the Criminal Code of Ukraine with the number of XIV-2 and a working title being "Criminal Measures Against Electronic Bodies" by analogy with the existing chapter Chapter XIV-1 "Criminal Measures Against Corporate Bodies".

2.3 Civil Liability

The issue of torts liability for criminal offence in the cyberspace is another issue of cybersecurity provision in Ukraine being of equal importance. Intellectual property protection procedure takes the he integral place in the system of cybersecurity in Ukraine.

It is worth noting that only the objects of intellectual property law, which can be presented in the digital format can reside in the cyberspace: items subject to copyright and related rights, means of identification, research and development findings. The objects of intellectual property law can be used in the Internet as (a) website content, (b) domain names characterizing clause and (c) metatags [13, c. 317].

In the Civil Code of Ukraine, relations in the sphere of intellectual property are established in a separate structural unit (book 4 "Intellectual Property Law"), which testifies their importance for the private law [14, c. 41].

Violations of intellectual property rights for the objects of intellectual property in the cyberspace may be classified by the objects, the rights for which are most frequently violated. The first group must include violations of copyright and related rights. These are the rights for musical compositions (lyrics and music), related rights of performers of these musical compositions, audio and video records, data compilations (databases), audiovisual works, copyright for works of visual and phonographic art as well as the copyright of the authors of various literary and artistic works – articles in journals and newspapers etc. Besides, the particular place among violations of the rights for the objects of intellectual property is taken by software, which supports websites functioning in the Internet. A copyright violation refers to unauthorized use of an artistic work. It is worth noting that the use of an artistic work in the cyberspace is its upload, in particular, in the Internet and, thus, this upload is only possible upon authorization of the copyright holder (except for its use and reproduction for non-commercial purposes – for private purposes namely studying etc.).

The second group of violations includes violations of rights for an industrial model. An industrial model may be protected as the physical configuration of a certain manufactured item (pattern or colour, or their combination) placed on a website in the Internet, as well as the graphic interface of the site itself, which facilitates the interaction of website users with complex software programs.

The third group of violations includes violations of rights for trademarks (marks for goods and services) and trade (firm) names including their use in domain names, website content (e.g. banner advertisements) and metatags [13, c. 329].

In order to distinguish crimes in the sphere of domain names, the term of "cybersquatting" was created, which is characterized by the following features: an individual's activity in the Internet aimed at registration of a certain name, as a rule, a domain name, as well as an individual's acts of registering such name are unlawful, i.e. done without any right for that domain name. In addition, these acts must be deliberate, wrongful and aimed, in particular, at making a profit from the possible resale of the wrongfully registered domain name or attracting attention to a certain information resource identified by the corresponding domain name, or inflicting harm to the competitor's reputation by means of publishing unsavory information on the corresponding web resource.

The most widespread procedure for intellectual property rights protection in domain name disputes is judicial protection implemented in civilian and economic judicial procedure.

Article 432 of the Civil Code of Ukraine arranges the right of an individual to turn to court for protection of their intellectual property right in compliance with article 16 of the Civil Code of Ukraine. Article 16 of the Civil Code of Ukraine creates the following remedies: (1) recognition of rights; (2) unvalidation of the dealership; (3) barring the action interfering with the right; (4) legal restitution; (5) compulsory enforcement of liabilities specifically; (6) change of legal relations; (7) discontinuation of legal relations; (8) compensation of losses and other methods of property damage compensation; (9) redress for the non-pecuniary damage; (10) unvalidation of decisions, actions or failure to act of an executive government agency or an agency of local self-government and their public officials. The court may protect the civil right or interest by any other means established by the agreement or the court in cases established by the law [15].

Case study. in 2010, "MEDIA HRUPA "RIA" Ltd filed a lawsuit against PARTY_1 with the participation of a third party without independent demand on the side of the defendant – "Television and Radio Broadcasting Company "Somyy Kontinent" Ltd for their obligation to commit action and recognition of the right for domain name registration. The lawsuit was motivated by the fact that "MEDIA HRUPA "RIA" Ltd is the owner of the trademark "RIA" proved by the trademark certificate №76602, the disposition of the State Department of Intellectual Property of 07.09.2007 and the extract from the public register of trademarks for goods and services of Ukraine of 16.10.2007. However, the defendant uses the domain name "ria.ua", which, in compliance with the "Ukrainian Latinic Alphabet" established by the State Statistic Service of Ukraine on 18.10.1995, is "PIA". The given grounds were the reason for filing the lawsuit. The court ordered to sustain the claim. To cease the right of PARTY_1 for the domain name "ria.ua" by obliging them to apply to the administrator of the domain of Vinnytsia Chamber Of Commerce and Industry and "Hostmaster" Ltd. for sub-delegation of the domain name in behalf of "MEDIA HRUPA "RIA" Ltd. To recognize the right for the registration (delegation, sub-delegation) of the domain name "ria.ua" [16].

What are the current problems in the sphere of intellectual property protection in the cyberspace? Firstly, this is the problem of entrenching the offence. Independent entrenchment will not be taken by the court as evidence and, correspondingly, entrenchment must be conducted by an appointing authority or an export on condition that the evidence is not turned down. In addition, the procedure must be paid for in the amount, which comes within the issue of "acceptable price". Secondly, the problem is the amount of compensation for the individual whose rights were violated. The law arranges liability in the form of a compensation in the amount of 10 to 50000 minimum national living wages instead of damage reparation or collecting profits. In case the subject whose rights were violated could not ask for well-grounded damages or lost profit, as a rule, the latter claims a financial compensation. In 90% of cases for compensation, the court orders a compensation in the amount of 10 minimum national living wages, i.e. the minimum sum regardless of the number of objects of offence [17].

Apart from the issue of intellectual property rights protection in the cyberspace, another problem widely discussed in the world is the problem of recognizing robots as the subjects of law, which can have rights and liabilities, and bear responsibility. Today, the Civil Code of Ukraine sees the robot as the object of law in respect of which deals are made.

Supporters of recognizing robots before the law provide a historic analogy of corporate body formation. The professional association actively discusses the creation of the first decentralized autonomous organizations (DAO) namely an electronic corporate body. The reason of this attention is obvious: this is one of the first companies to be manages by self-running smart contracts, without any traditional management bodies such as the CEO or the Board of Directors. In a traditional company, its activity and interaction with shareholders are regulated by the law, the statute, the corporate agreement and other documents. In a decentralized organization, corporate rules are written in the form of self-running smart contracts. For example, DAO is established via a program-contract, which regulates incoming cryptocurrency used for company financing and functioning as well as converting the incoming cryptocurrency into tokens (the analogy of shares in a traditional company). Smart contracts also regulate autonomous organization management. The program code determines the decision-making process concerning the company agreements and profits distribution. The code stablishes the quorum for voting, terms of its holding etc. The rules of changing the program code, which manages DAO may be the same as the rules concerning the statute and internal documents in a traditional company. Minor changes (e.g. minor errors correction) are made without shareholder vote. Agreement of the majority of token holders will be needed for major amendments. The given rules will be integrated in the program.

Therefore, with the account taken of the abovementioned, civil law must be brought in line with the present-day realia of public relations in the cyberspace and development of models for DAO introduction in the Civil Code of Ukraine is currently highly relevant (Table 1).

Table 1. Kinds of liability and issues of their legal regulation

Kinds of liability	Regulatory regime	Problem
Administrative liability	The Code of Ukraine on Administrative Offence	1. Absence of a chapter in the Code of Ukraine on Administrative Offence, which would arrange articles for wrongful acts in the cyberspace and which would cover the issues of administrative liability of robots' producers and owners with the account taken of the European Union recommendations on robots 2. Low sanctions for violations of private data protection procedure in the cyberspace
Criminal liability	The Criminal Code of Ukraine	1. Limited capacity of the criminal law due to the principles of effectuating criminal law in space 2. Presence of provisions of general criminal intent of accomplices in the criminal law 3. Absence of legal regulation of an electronic body's responsibility in the Criminal Code of Ukraine
Civil liability	The Civil Code of Ukraine	1. Problem of registering violations of intellectual property rights in the cyberspace 2. Indetermination of electronic corporate bodies (DAO) in the civil law

3 Conclusions

Legal responsibility for legal offence in the spheres of telecommunications and data protection in case the cyberspace is the place for committing this offence has the integral part in cybersecurity provision in Ukraine. Therefore, formation of an effective mechanism for legal regulation of liability for offence in the cyberspace is important. The Code of Ukraine on Administrative Offence, the Criminal Code of Ukraine and the Civil Code of Ukraine, which arrange liability for offence in the cyberspace, must be updated with regard to the modern trends of the development of a digital state and public relations in the cyberspace as well as the established practice in applying the law, which already formed in the criminal, administrative and civil proceedings. Moreover, the abovementioned legislative instruments must cover the issues of legal regulation of robots/electronic bodies/electronic corporate bodies' responsibility.

The authors consider that the legal cybersecurity provision mechanism can be improved by a series of legislative developments and attachments namely:

1. Update of the Code of Ukraine on Administrative Offence by creating a separate chapter on liability for wrongful acts in the cyberspace, which would also cover the issue of administrative liability of robots' producers and owners with the account taken of the EU recommendations on robots.
2. Harsher administrative liability for private data protection – increase of the amount of fines for private data disclosure in the cyberspace with the account taken of the European experience with GDPR.
3. Developments to the provisions of the Criminal Code of Ukraine on criminal law operation principles in Ukraine.
4. Developments to the Criminal Code of Ukraine on the general criminal intent of accomplices.
5. Addition of the chapter "Criminal Measures Against Electronic Bodies" to the Criminal Code of Ukraine.
6. Attachment of provisions for independent registration of intellectual property rights violations in the cyberspace to the civil law of Ukraine.
7. Attachment of provisions on electronic corporate bodies – decentralized autonomous organizations to the civil law.

References

1. Pro natsional'nu bezpeku Ukrayiny. Zakon Ukrayiny vid 21 chervnya 2018 (On the National Security of Ukraine. The Law of Ukraine of 21 June 2018). https://zakon.rada.gov.ua/laws/show/2469-19
2. Calo, R.: Robots in American Law. Legal Studies Research Paper No. 2016-04, 44 p. (2016)
3. Preliminary Draft Report of COMEST on Robotic Ethics. unesdoc.unesco.org/images/0024/002455/245532E.pdf
4. European Parliament Resolution of 16 February 2017 with Recommendations to the Commission on Civil Law Rules on Robotics. http://www.europarl.europa.eu/doceo/document/TA-8-2017-0051_EN.html?redirect#title1

5. Pro Osnovni Zasady Zabezpechennya Kiberbezpeky Ukrayiny. Zakon Ukrayiny vid 5 zhovtnya 2017 r. (On the Main Principles of Cybersecurity Provision in Ukraine. The Law of Ukraine of 5 October 2017). https://zakon.rada.gov.ua/laws/show/2163-19
6. Kodeks Ukrayiny Pro Administratyvni Pravoporushennya vid 07.12.1984 r. (Code of Ukraine on Administrative Offence of 07.12.1984). https://zakon.rada.gov.ua/laws/show/80731-10
7. Kryminal'nyy Kodeks Ukrayiny vid 05.04.2001 r. (The Criminal Code of Ukraine of 5.04.2001). https://zakon.rada.gov.ua/laws/show/2341-14
8. Yedynyy Zvit pro Kryminal'ni Pravoporushennya po Derzhavi za Serpen' 2019 roku. (Uniform Report of Criminal Offence in the State for August 2019). https://www.gp.gov.ua/ua/stst2011.html?dir_id=113897&libid=100820&c=edit&_c=fo
9. Nashynets'-Naumova, A.Yu.: Protydiya Komp"yuterniy Zlochynnosti v Kibernetychnomu Prostori (Fighting Computer Crime in the Cybernetic Space). Visnyk Kryminolohichnoyi Asotsiatsiyi Ukrayiny **1**(20), 108–117 (2019)
10. Vyrok Kyyivs'koho Rayonnoho Sudu m. Kharkova po Spravi № 640/953/17 vid 21.03.2017 r. (Verdict of Kyivskyi District Court of Kharkiv on Case № 640/953/17 of 21.03.2017). http://reyestr.court.gov.ua/Review/65496457
11. Pavlykovs'kyy, V.I., Selevko, V.B.: Kiberzlochynnist' ta Obmezheni Mozhlyvosti Tradytsiynoho Kryminal'noho ta Kryminal'no-Protsesual'noho Prava (Cybercrimes and Limited Possibilities of the Traditional Criminal Law and Criminal Procedure). Visnyk kryminolohichnoyi asotsiatsiyi Ukrayiny **1**(20), 128–133 (2019)
12. Radutnyy, O.E.: Artificial Intelligence (Shtuchnyy Intelekt) ta Inshi Zahrozy (Kryminal'no-Pravovyy Vymir) (Artificial Intelligence and Other Threats to (Criminal Law Dimension). http://aphd.ua/publication-354/
13. Benedysyuk, I.M.: Posibnyk dlya Suddiv z Intelektual'noyi Vlasnosti (Guidelines for Judges on Intellectual Property), 424 p. K. I. S., Kyyiv (2018). ta in
14. Kodynets', A.O.: Pravovi Zasady Reformuvannya Zakonodavstva u Sferi Intelektual'noyi Vlasnosti (Legal Framework for Reforming Legislation in the Sphere of Intellectual Property). In: Furashev, V.M., Petryayev, S.Yu. (eds.) Kiberbezpeka ta Intelektual'na Vlasnist': Problemy Pravovoho Zabezpechennya: Materialy Mizhnarodnoyi Naukovo-Praktychnoyi Konferentsiyi. 21 kvitnya 2017 r., m. Kyyiv, v 2-kh chastynakh.. Chastyna persha. Uporyad, 146 p. Natsional'nyy Tekhnichnyy Universytet Ukrayiny «Kyyivs'kyy Politekhnichnyy Instytut imeni Ihorya Sikors'koho» Vyd-vo «Politekhnika», Kyyiv (2017)
15. Tsyvil'nyy Kodeks Ukrayiny vid 16.01.2003 r. (Civil Code of Ukraine of 16.01.2003). https://zakon.rada.gov.ua/laws/show/435-15
16. Zaochne Rishennya Lenins'koho Rayonnoho Sudu m. Vinnytsi vid 26.04.2010 Roku po Spravi № 2-3374/10 (Judgement in Default of Lenins'kyi District court on Case № 2-3374/10). http://reyestr.court.gov.ua/Review/9910421
17. Katrashov, N.: Problemy Zakhystu Avtors'koho Prava v Merezhi Internet. (Problems of Copyright Protection in the Internet). https://blog.liga.net/user/nkartashov/article/26575

Cyber-Physical System and IoE

Prototyping and Rapid Development of IoT Systems in Context of Edge Computing

Plakhteyev Anatoly(✉), Heorhii Zemlianko(✉), and Vyacheslav Kharchenko(✉)

National Aerospace University "Kharkiv Aviation Institute", Kharkiv, Ukraine
{A.Plakhteev, G.Zemlynko, V.Kharchenko}@csn.khai.edu

Abstract. Currently, Cloud, IoT, Dew, Edge and touch networks of "things" are allocated as part of the program system. Data at the touch network level is associated with physical objects. At the Edge level, data is separated from objects to form a Digital Twin representation of their state, properties, and behavior with minimal latency. Copies of Digital Twin "things" are stored and used at different levels, but must reflect their current state. Interaction with "things" in IoT, data transport between sensor networks through their boundary nodes is considered. For Edge computing research it is proposed to create physical and virtual models of fragments of touch networks, their boundary nodes, processing of data and organization of access of devices as part of "things" to remote IoT resources. The paper shows how to build a virtual device model in a Proteus environment and assess adequacy versus a physical model. Implemented a piece of wired sensor network with TCP/IP protocol based on open platforms.

Keywords: IoT · Edge computing · Digital Twin · Sensor network · Boundary node

1 Introduction

1.1 Motivation

Centralizing resources and synchronizing processes in cloud-based IoT runs counter to the distributed nature of data, leads to excessive traffic and unacceptable delays in data collection and management processes for a number of applications. Approaching data sources and consumers in Fog and Dew technologies partially addresses the problem of reducing data access latency, but data analysis computing resources are still removed from sources abroad of virtual and physical worlds.

Edge computing separates the physical sensors and executive devices of Things from the virtual space, forming a digital representation (Digital Twin) of their state, properties, and behavior. Digital Twins, thanks to Dew Fog and Cloud analytics, are refined, synchronized, and matched to summarize and highlight knowledge about the properties and behavior of "things" classes. Edge computing and Digital Twin are currently among the IoT's key technologies. For IoT, the Edge level is mobile. It may be some device having access to IoT services, or a sensor network boundary unit of a

M. Nechyporuk et al. (Eds.): *Integrated Computer Technologies in Mechanical Engineering*, AISC 1113, pp. 257–267, 2020.
https://doi.org/10.1007/978-3-030-37618-5_23

mechanism, a machine (on board computer), a production site (industrial computer), a house, etc., connected to the IoT.

Data-related sensors/activators can be represented in virtual space as part of a touch network with access through a boundary node or multiple nodes. Touch network edge nodes can simultaneously be part of multiple networks with different interfaces and exchange protocols, providing data transport between networks and with IoT. At the Edge level, there are no established standards for presenting and sharing data, building platforms for processing and storing data. Many studies and experiments with prototype IoT devices are required to choose rational solutions in different applications.

1.2 State of the Art

Sensor networks can be wired and wireless. The most popular networks are Ethernet, RS485 and similar, CAN, LIN, etc. Wireless networks can be proprietary in the ISM bands 315, 433, 868 MHz, as well as 2.4 GHz networks based on the IEEE802.15.4 standard (ZigBee, 6LoWPAN, etc.), IEEE802.15.1 (Bluetooth v.2 .. 5), IEEE802.11 (WiFi). Low-speed sensor networks in industry and home automation are combined using high-speed interfaces - Ethernet, WiFi. In fact, a heterogeneous network is formed, where it is necessary to provide access to various nodes in each network and their interconnectivity. Boundary nodes are bridges (gateways) between segments of sensor networks [1].

Web technologies implement a variety of IoT services for storage, data processing and remote access (Amazon Web Services IoT Platform, Microsoft Azure IoT Hub, Google Cloud IoT, IBM Watson IoT Platform, CISCO IoT Cloud Connect, ThingSpeak, etc.). This simplifies the development of applications for IoT. But these technologies have a lot of disadvantages [2]. There are restrictions on the intensity of the data flow for storage in the Cloud stores and a significant delay in access to these data. For a growing number of IoT sites the permanent access to the Internet is required. Also the increasing of bandwidth is needed. It prevents the use of Cloud technologies in real-time management systems and critical appointments. Excessive traffic arising in the process of access to remote resources causes increasing energy costs and the cost of access to information. The solution is to approach resources to their consumers.

For rapid development and prototyping of IoT nodes, the platforms included in ecosystems are designed; ARM Mbed [3] data processing facilities Arduino for data collection and exchange [4, 5]; WiFi Espressif microcontrollers Reyax for access to Bluetooth and WiFi networks, as well as various Starter kits [6, 7].

The objectives of the paper are to analyze IoT technology and positioning of its components in context of edge computing, and to discuss stages and examples of prototyping and rapid development of IoT systems. The paper is structuring as follows. Section 2 describes interaction of end devices with IoT, Edge computing for association sensor networks, also types and structure of IoT devices. Section 3 describes stage of prototyping, Sect. 4 describe case study. In Sect. 5 we summarize.

2 IoT Devices and Edge Computing

2.1 Interaction of End Devices with IoT

In the general scheme (Fig. 1) of the interconnection of nodes of sensor networks (SN), individual devices (Dev), sensors (S), actuators (A) are showed. At the Edge, Dew, Fog, Cloud (Cloudlet) levels, the state of S and A is represented by digital copies (Digital Twin). Thus, the IoT fragment can be represented as a hierarchical structure from a variety of disparate sensor networks SN = {SN1, SN2, SNi, …}. Each network consists of a set of nodes: SNi = {Ni, 1, … , Ni, j, …} connected by the network communication interface Ci from the set C = {C1, C2, …}, by the protocol Pi from set P = {P1, P2, …}. The Ni, j node serves a set of sensors (S) and actuators (A).

To implement the functions of the node a microcontroller platform MCUi, j is used that has a network interface Ci, as well as analog and digital interfaces, that form data streams from sensors S and for controlling actuators A. Networks are designed to collect data from sensors primary processing, accumulation, presentation in some form (indication, sound, video, etc.), control and management of executive devices. The network can be embedded in some object (robot, tool, machine, etc.), and the state of set S and A determines the state of the object. Let the current state of the node of one network be determined by the state of the sensor S, and the node of the other network by the state of the actuator A. In so doing, the state S is displayed on the state A. The simplest example is that the state of the switch determines by the state of the lamp, which can be realized by their direct connection. Alternatively, the state S is determined by the MCU that is associated with the MCU that controls A.

The logical link of the sensor to the executive device can be implemented: interconnectivity by sending a message S′ that displays the state S to another network as a package A′ representing the state A. In the case of different interfaces and protocols of associated sensor networks, a series of transformations S′ → A′ is required. In the

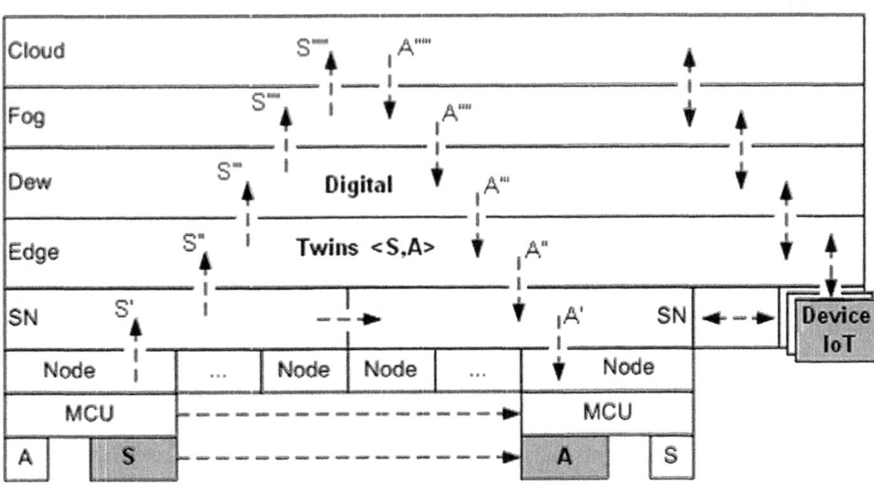

Fig. 1. Methods of an interaction of network nodes

absence of direct network connection, Edge computing is used, operating with a digital representation of $<S'', A''>$, that is called Digital Twin [9]. Accordingly, Digital Twin at the levels Dew, Fog, Cloud, will be used in various forms, but displaying the current state of $<S, A>$. Synchronization $<S', A'>$, $<S'', A''>$, … , $<S'''', A''''>$ requires certain computational costs and the expenditure of traffic, and hence - time costs. Generally, there are time intervals of the desynchronization in the meaning of various Digital Twin, which can affect the operation of systems sensitive to such uncertainties. Here, it should be entrusted Edge computing with critical to communication delays and direct interaction of sensor networks.

2.2 Edge Computing for Association of Sensor Networks

The IoT feature is the importance of the level boundary interaction of TCP/IP – oriented components and services with sensor networks (SN) and individual devices – stationary, mobile, moveable (EDGE computing). This level of interaction is difficult for formalizing because of variety of devices types, interfaces, network protocols, numerous vulnerabilities and strong requirements to power of IoT devices. Acquiring necessary skills for building the boundary level of IoT is the pressing challenge of training specialists in networking technologies.

Let's consider independent sensor networks SN_1 (head node 1, internal nodes 11, 12, 13 and boundary nodes BN_1, BN_{12}, BN_{13}), SN_2 (head node 2, internal nodes 21, 22, 23 and boundary nodes BN_2, BN_{12}, BN_{23}), SN_3 (head node 3, internal nodes 31, 32, 33 and boundary nodes BN_3, BN_{13}, BN_{23}). In Fig. 2, double lines show inter-network data flows. Let the data for node 21 come from the access point through the boundary node BN1. The chain is constructed: Data1 $\rightarrow$ BN1 $\rightarrow$ BN12 $\rightarrow$ 21. Data is delivered from node 31 via the chain: $31 \rightarrow 3 \rightarrow BN_{23} \rightarrow 2 \rightarrow BN_{12} \rightarrow 1 \rightarrow BN_1 \rightarrow$ Data1. A shorter way, in the presence of BN13 is: $31 \rightarrow 3 \rightarrow BN_{13} \rightarrow 1 \rightarrow BN_1 \rightarrow$ Data1.

Networks SN_1, SN_2, SN_3 can be used for data exchange between the head and inner nodes and external networks via BN_1, BN_2, BN_3. The following conditions must be met: support for packet switching in selected networks; sufficient length of network messages for organizing data transfer over SN_1–SN_3 network protocols. Data transfer rate in networks ensures an acceptable delay in transmission of data packets; network traffic must have sufficient redundancy to accommodate additional traffic; head nodes 1, 2 and 3 allow the extension of the basic set of functions; energy costs for implementation of additional services should not extend beyond established limits.

A number of sources [6, 8–10] describe the results of the interaction of wired and wireless networks. The principal possibility of reliable transportation of CAN-packets through the IP network is shown. However, many sensor networks have too limited capabilities of interfaces and protocols to implement additional functions. This may require profound changes to the services of elements 1, 2 and 3, will affect a number of existing protocols and will lead to the emergence of new protocols that support prospective platforms of sensory networks.

Thus, the problem of constructing a common information field from independent heterogeneous networks is solved as follows. Network interfaces are assumed by the boundary nodes (BN). There are BN_1, BN_2, BN_3 for external access to SN_1, SN_2, SN_3

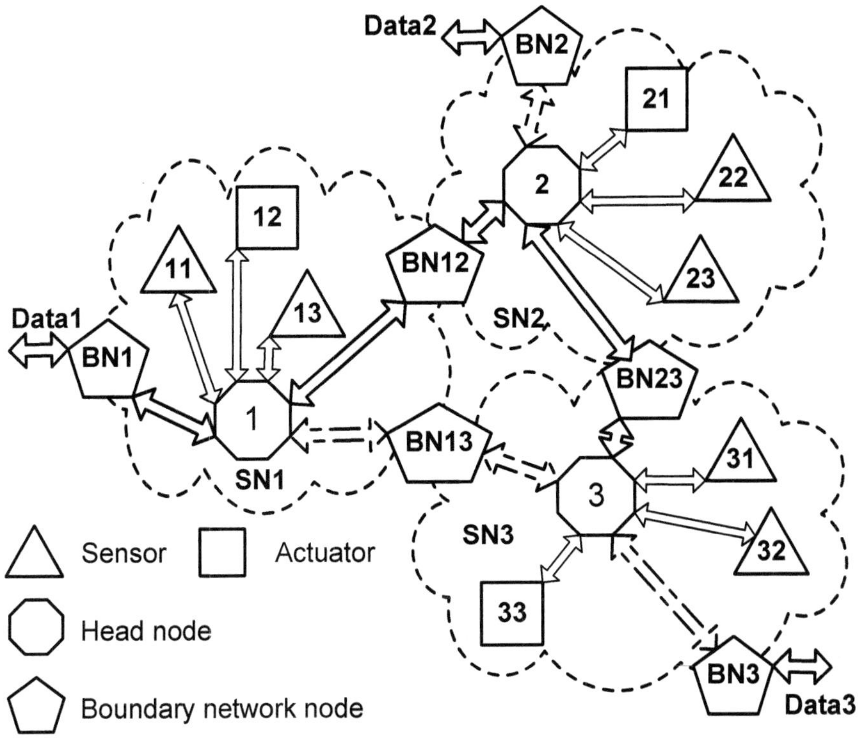

Fig. 2. Combining of sensor networks with internetwork data transport

and there are BN_{12}, BN_{13}, BN_{23} for the interaction of networks. Boundary nodes that perform the function of gateways have the possibility of a simultaneous presence in at least two networks between which interaction is organized.

Each network controller through a network interface is related to the nodes of its network - sensors and drives and it performs a set of basic functions (services). Expansion of the set of functions gives access to the network controller from the side of the boundary node, which provides transport of data from one network to another. Head nodes (network controllers) form requests (commands) to sensors and drives receive response messages in accordance with the internal logic of the network functioning. To provide communication with the global network of all nodes, without exception, that generates and receive data, it is advisable to use one entry point for a cluster of nodes within one or more sensor networks [13, 14].

2.3 Types and Structure of IoT Devices

Devices on which IoT conditionally is under construction share on: simple attached device, intelligent device, border gateway. The simple attached device generates data, performs instant operations and carries out data transmission. As a rule, contains the microcontroller with limited resources, built in by software, does not demand big costs of the equipment, provides basic functions of connection, basic tools of safety. These are the most mass devices to which specific, often contradictory requirements are imposed.

The intelligent device contains the microprocessor or SoC, the operating system and considerable hardware resources (Ready IoT). Provides data analysis on peripheral sections, support connectivity across multiple networks, makes decisions and carries out local calculations. Provides the maximum level of safety, controllability, interaction and compatibility, reliable work of solutions, support of cloud computing, the user interface and reduces data transmission cost. The border gateway is the intelligent device for computing Edge with the high level of safety, minimizes the problems connected with interaction of the physical and virtual world and scaling of the IoT systems. IoT projects increasingly rely on existing out-of-the-box solutions. Benefits are the following [16]: quicker Time To Market, access to crucial skills, secure by design, optimized to work with wider ecosystem, scale with ease, enable a more end-to-end offering, etc.

The choice for independent development usually is accepted for simple and parts of intelligent devices. The structure of these devices is presented in a general view on Fig. 3. Kernel of devices (CPU) can be ready or independently projectable modules on the basis of 8-32 bit microcontrollers [13–15]. Samples of sensors, indication and data entry, communication means in modular or submodular execution for prototyping or creation of end devices are widely presented at the market.

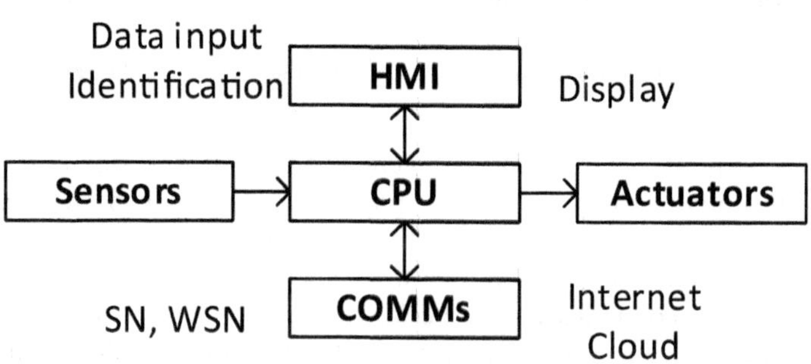

Fig. 3. Block scheme of device IoT

3 Stages of Prototyping and Rapid Development of IoT Systems

There are following six stages in every IoT based system development life cycle model [16, 17]: requirement gathering and analysis, design, implementation or coding, testing, deployment, maintenance. Development of IoT of applications is the iterative process allowing eliminating errors and mismatching to requirements at different stages. Errors of initial stages of development are most difficult eliminated. Design of the IoT components of systems includes:

1. Providing functional requirements:

- modeling for decision-making (Matlab, Simulink);
- distribution of functions between the IoT components of systems, use of support from mobile devices (smartphones, tablets and so forth);
- rational distribution of functions between equipment rooms and software (minimization of hardware expenses);
- use of the previous developments.

2. Depreciation of components:

- rational choice of element base;
- use of open platforms.
- reduction of weight and dimensional parameters:
- rational configuration;
- choice of cases of elements;
- replacement of bulky elements (power supply, indication, management).

3. Decrease in terms of development of components:

- use of the previous developments, resources of ecosystems;
- rational choice of development tools, compilers, simulators.

4. Use of the previous developments.
5. Decrease in energy consumption (collecting energy for a power supply).
6. Reliability augmentation (resistance to failures, power failures and so forth).
7. Reduction of expenses on service.
8. Work in severe conditions of the environment.
9. Adaptation to new requirements.
10. Standardization of interfaces for Sensors, Actuators, network and between network interactions.
11. Interaction with services Edge, Dew, Fog, Cloud.
12. Complex use of different platforms.

At different development stages focus of fast prototyping is transferred to different components. Existence of lightweight IoT middleware for rapid application development is important [18, 19].

4 Prototyping of IoT Systems. Case Study

Physical prototypes of simple devices can have virtual analogs, for example, in the environment of Proteus (Fig. 4). Virtual devices cannot reflect fully behavior of physical prototypes, but considerably accelerate intermediate prototyping. Availability of components of an ecosystem of Arduino, Breadboards to fast assembly of prototypes cause their wide circulation, especially in education [14–16].

All largest vendors of microprocessors, microcontrollers, SoC, communication means are guided by IoT and offer both end-to-end solutions, and means of fast prototyping. Elements of compatibility with shields of an ecosystem Arduino are often entered and in the same format own payments are offered. Semiconductor provides configurable, end-to-end, rapid prototyping platforms for the Internet of Things [17]. These platforms enable development of energy efficient solutions for smart homes/buildings, smart cities, industrial IoT (Predictive Maintenance, Asset Monitoring, etc.) and personal IoT (Wearables, activity monitors, etc.).

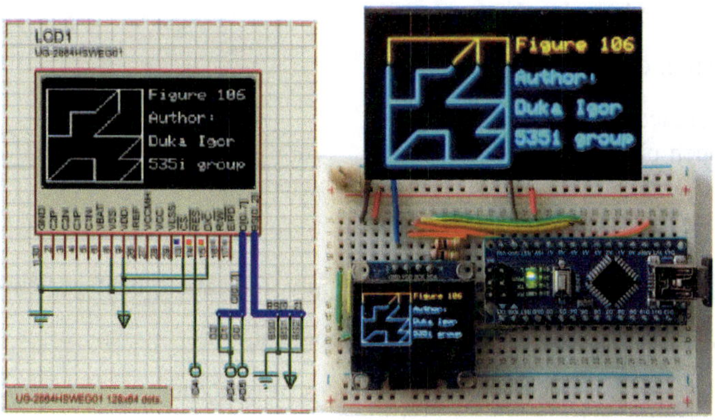

Fig. 4. Means of prototyping of the simple IoT device

The IDK baseboard can be connected with different shields depending on the required IoT application. The IDK baseboard allows the user to create many types of IoT nodes and/or gateways depending on which shields are used with the baseboard. Programing/configuring the IDK requires the ON Semiconductor IDE software.

Based on the company's highly sophisticated NCS36510 system-on-chip (SoC) with a 32-bit ARM® Cortex® M3 processor core, it has all the necessary hardware resources for constructing highly effective, differentiated IoT systems, along with a comprehensive software framework to attend to interfacing with the cloud (Fig. 5).

By attaching different shields to the IDK baseboard, a wealth of connectivity (WiFi, Sigfox, Ethernet, ZigBee and Thread protocols, etc.), sensor (motion, ambient light, proximity, heart rate, etc.) and actuator (with stepper and brushless motor driving, plus the ability to drive LED strings) options can be added to the system. This means that compromises do not have to be made, and the most suitable technology can be chosen.

Offering a wide range of choices including configurable hardware, multiple cloud connectivity, easy-to-use development software, and application examples, these platforms reduce time-to-market and allow rapid deployment of IoT-enabled products.

Designed for expert makers, entrepreneurs, and industrial IoT companies, the Intel Edison module provides easier prototyping with a fully open source hardware and software development environment. It supports WiFi and BLE 4.0 connectivity. This kit contains eleven, selected Grove sensors and actuators. It can be used to track indoor environment as well as to create smart-home applications [19].

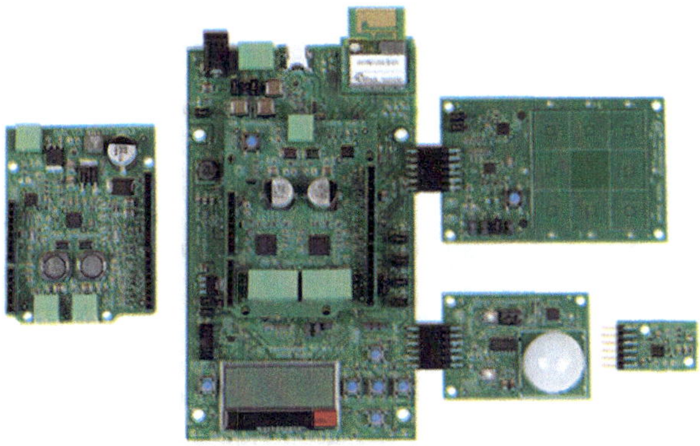

Fig. 5. EVBUM2497/D IoT prototyping platforms

At production of single copies or the small IoT series of solutions these platforms are final option. The option of rapid prototyping of a network fragment using wired and wireless access that realized with use a Wi-Fi router shown in Fig. 6. There are four Ethernet ports for connecting the end devices (Ethernet MCU) and the local server, the WiFi access point for connecting the IoT wireless devices:

Fig. 6. Rapid prototyping of the IoT network fragment

WiFi MCU (ESP 8266, Espressif ESP 32, etc.), SoM Raspberry Pi, laptops, smartphones, tablet computers. As the access point WiFi mobile devices GSM, DSL modems and WiFi MCU can be used. This creates a variety of tasks for building various Edge-level network configuration for building and analyzing IoT fragments, mastering promising IoT platforms.

Simple Ethernet MCUs are built using Ethernet - SPI converters Wiznet w5100, w5500, Microchip ENC28J60 and microcontroller platforms [26]. Converters implement TCP/IP protocol in hardware. Microcontrollers can be connected to sensors and actuators, perform the functions of the boundary nodes of sensor networks, and

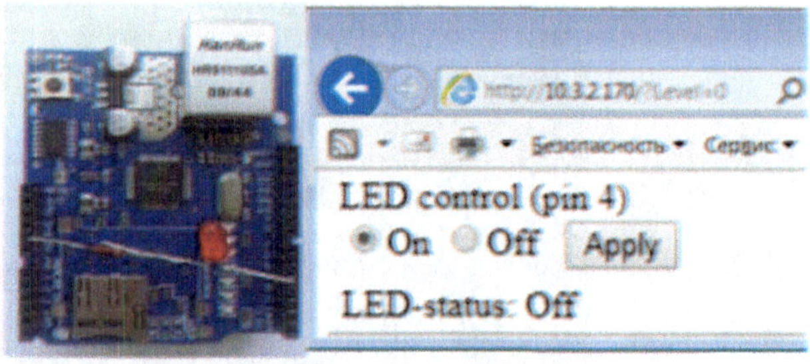

Fig. 7. Example of use of shield w5100 for remote control via the Internet.

exchange information among themselves using built-in interfaces. Using multiple routers and connected devices allows the local server to simulate interaction at the boundary level of higher IoT levels (Fig. 7).

The prototype is the base for the development and research of industrial automation systems, a smart home that is based on technologies of the Internet of Things.

5 Conclusions

Objects of the physical world can be connected by one or several touch networks in about tens of sensors and actuation mechanisms and programmable computing modules (the robot, the car, the house, the machine, etc.). The state and behavior of object - "thing" is defined by data flows in networks, and on Edge, Dew, Fog and Cloud levels of global network are formed copies of his digital double (Digital twin).

Digital doubles have to reflect adequately a condition of physical objects, and impact on doubles – to cause the corresponding reaction physical objects and change of conditions of all copies of Digital twin for representation to users. Development and prototyping of such IoT components of systems and their deployment are very difficult. On the other hand, sensor networks and elements of network interconnection consist of rather simple devices with available development tools and prototyping. Rapid development assumes availability of functionally full range of elements cuts of fast assembly of devices rapid developments of programs of their debugging. Classical approach is – based on development of the device with the program languages of the low level, but with visually way programming is considered. The example of modern approach on the basis of the open platform allows to implement quickly devices of monitoring and remote control with access to Web services.

The IoT platform marketplace remains fragmented; McKinsey counts more than 100 providers, including broad-based and industry- and device-specific providers. By paying close attention to the characteristics of vendor solutions such as use of industry standards and protocols, integration, flexibility, security, privacy and partner ecosystem, hoteliers can start an IoT journey with specific, problem-solving use cases, while leaving the door open to add, and manage a much more diverse set of solutions. The following are 5 steps to establishing an effective IoT environment: set your vision, assess the current state, consider open versus closed networks, ensure the right ecosystem, take a stringent approach to security.

The following steps of research can be dedicated to embedding of developed decisions in "Cloud-to-Things continuum" [12] and assessment of variants according with different criteria.

References

1. Vermesan, O., Friess, P.: Internet of Things: Converging Technologies for Smart Environments and Integrated Ecosystems. River Publishers, Aalborg (2013)
2. Kulkarni, S., Kulkarni, S.: Communication models in internet of things: a survey. IJSTE – Int. J. Sci. Technol. Eng. **3**(11), 3 (2017)
3. Kienzler, R.: Digital Twins and the Internet of Things (2019). https://developer.ibm.com/articles/digital-twins-and-the-internet-of-things/. Accessed 25 June 2019
4. Botta, A., Donato, W., Persico, V., Pescap, A.: Integration of cloud computing and internet of things: a survey. J. Future Gener. Comput. Syst. 1–54 (2015)
5. Sahni, Y., Cao, J., Zhang, S.: Edge mesh: a new paradigm to enable distributed intelligence in internet of things. IEEE Access **5**, 16441–16458 (2017)
6. Higuchi, T., Yamaguchi, H., Higashino, T.: Mobile devices as an infrastructure: a survey of opportunistic sensing technology. J. Inf. Process. **23**(2), 94–104 (2015)
7. Maharrey, B.K., Lim, A.S., Gao, S.: Interconnection between IP networks and wireless sensor networks. Int. J. Distrib. Sensor Netw., 4 December 2012. http://journals.sagepub.com/doi/full/. https://doi.org/10.1155/2012/567687. Accessed 25 June 2019
8. Schwartz, M.: Internet of Things with Arduino Cookbook. Packt Publishing, Birmingham (2016)
9. Waher, P.: IoT: Building Arduino-Based Projects (+Code). Apress, New York (2016)
10. Xiao, P.: Designing Embedded Systems and the Internet of Things (IoT) with the ARM Mbed. Wiley, Hoboken (2018)
11. Pramudianto, F.: Rapid Application Development in the Internet of Things: A Model-Based Approach. https://publications.rwth-aachen.de/record/464316/files/464316.pdf. Accessed 25 June 2019
12. Padraig, S., Lueth, K.: Guide to IoT solution development (2016). https://iot-analytics.com/wp/wp-content/uploads/2016/09/White-paper-Guide-to-IoT-Solution-Development-September-2016-vf.pdf. Accessed 25 June 2019
13. Guan, G., Dong, W., Gao, Y., Fu, K., Cheng, Z.: TinyLink: A Holistic System for Rapid Development of IoT Applications. https://ieeexplore.ieee.org/document/8116508. Accessed 25 June 2019
14. Karvinen, K., Karvinen, T.: IoT rapid prototyping laboratory setup. Int. J. Eng. Educ. **34**(1), 263–272 (2018)
15. Configurable Rapid Prototyping Platform for The Internet of Things. https://www.rs-online.com/designspark/iotidk-kit. Accessed 25 June 2019
16. EVBUM2497/D: IoT Development Kit (IDK). Quick Start Guide. https://www.mouser.com/pdfdocs/ONSemi_IDK_QuickStart.pdf. Accessed 25 June 2019
17. Intel Edison and Grove IoT Starter Kit Powered by AWS. http://wiki.seeedstudio.com/Grove_IoT_Starter_Kits_Powered_by_AWS/. Accessed 25 June 2019
18. Plakhteyev, A., Frolov, V., Perepelitsyn, A.: Edge computing for IoT: an educational case study. In: DESSERT 2018, p. 26. http://dessert.ieee.org.ua/wp-content/uploads/2018/05/DESSERT2018program-final.pdf. Accessed 20 Oct 2019
19. Boyarchuk, A., Illiashenko, O., Kharchenko, V., Maevsky, D., Phillips, Ch., Plakhteyev, A., Vystorobska, L.: Internet of things for industry and human applications: ALIOT based vertically integrated education. In: Dependable IoT for Human and Industry: Modeling, Architecting, Implementation, pp. 535–559. River Publishers (2018)

Information Modeling

Mathematical Modelling of Residual Lifetime of Pumping Units of Electric Power Stations

Andrii Kelin[1], Oleksiy Larin[2], Raisa Naryzhna[1], Oleksandr Trubayev[2], Oleksii Vodka[2(✉)], and Mariya Shapovalova[2]

[1] Engineering and Technical Center “KORO”, Kharkiv, Ukraine
[2] Department of Dynamics and Strength of Machines, National Technical University “Kharkiv Polytechnic Institute”, Kharkiv, Ukraine
oleksii.vodka@gmail.com

Abstract. With long-term operation of pumping and other equipment, there is often a need to extend the life and assess the residual life. Modern life-time assessment methods are based on three-dimensional modelling and the finite element method. Within the framework of these methods, geometric models are constructed that take into account the thinning of the walls of the pumps as a result of operation. By constructing geometric models, finite element models are constructed. These models are made taking into account the conditions of loading of the structure, tightening of bolted connections, technological and temperature loads. The constructed models allow to assess the strength of the pump casing and the main bolted connections during normal operation and gyro testing. The paper assesses the cyclic strength of these structural elements. The results of the work confirmed a sufficient residual life-time for safety pump operation in the next 15 years.

Keywords: Life-time prediction · Long-term operation · Finite element model · Pumps

1 Introduction

The problem of the exhaustion of the design life-time of the equipment of nuclear and thermal power plants is relevant for Ukraine. The limited financial resources do not allow building new power units. A rational solution in this case is to conduct a set of measures to assess the residual life-time and extend the design life-time [1, 2].

For pumping units of thermal power plants and nuclear power plants during operation, thinning of the walls of the housing and the cover is typical. The cause of thinning is corrosion and erosion wear. As a result of wall thinning, the redistribution of mechanical stresses in the walls and pump cover occurs. The change in stresses has a negative effect on the static and cyclic strength of the pump unit. Thus, to assess the residual life-time, it is necessary to calculate a stress state. Stress data should be used to evaluate static and cyclic strength.

M. Nechyporuk et al. (Eds.): *Integrated Computer Technologies in Mechanical Engineering*, AISC 1113, pp. 271–288, 2020.
https://doi.org/10.1007/978-3-030-37618-5_24

The depletion of the life-time is significantly affected by the operation of the pump in various modes. Pumps operate in the following modes. There are normal operating conditions (NOC), hydrotest mode (HM).

To assess the life-time various methods are using [3–14]. These methods based on different ideas. Some works use the technique of assessing the reliability and probability of failure [3, 4, 7, 9, 10]. Others based on deterministic models [5, 6, 8]. All these works use the principle of determining the stress state, on the basis of which the life-time is estimated.

2 Objectives

The objectives of the paper are to estimate the residual static strength and residual life-time of the X45/90a centrifugal pump. This pump operates in the NOC and HM. For the assessment of wall thickness due operation the results of ultrasonic test is used.

The investigation of a residual life-time has made on example of the centrifugal pump of type X45/90a. This pump was made in 1985 by the Kataysky pumping plant. Design service life is 30 years. Technical characteristics and parameters of the pump are given in Table 1.

Table 1. Technical characteristics and parameters of the X45/90a pump

Characteristics	Values
Weight	218 kg
Working environment	Boron Water
Supply of working fluid	45 m^3/h
Pressure of pumped liquid on an input	–
Pressure of pumped liquid on an input	from −40 to +90 °C
Temperature of pumped liquid on an input	0,9 MPa
Pressure of pumped liquid at the exit	0,675 MPa
Temperature of hydro tests	10 °C

The planned life-time extension of the X45/90a pump is 15 years (until 2032). For details of pump case is selected walls thickness for assessment of residual strength. According to the thickness measuring of the pump the minimum sizes of walls are determined (the procedure of measurements is carried out in 2017). It is accepted that wall thinning evenly in all body parts by 3.7% per 100 working hours (the average operating time of this pump makes 135 h a year). These data correspond to the average speed of erosion-corrosive wear of pumps of similar types. The accepted values of uniform wall thickness of case shaped parts for the X45/90a pump taking into account the planned term of extension of 15 years are given in Table 2. Values have selected as rounded to the next smaller value.

Table 2. The accepted values of thickness of walls case of the x45/90a pump

Case shaped part	Actual wall thickness, mm	The expected operating time taking into account the planned extension term, h	The expected thinning of walls by the end of the planned extension term, %	Wall thickness for assessment of residual strength, mm
Case wall	14.2	2025	7.5	13.1
Cover	16.4	2025	7.5	15.1
Inlet pipe	15.8	2025	7.5	14.6
Outlet pipe	20.5	2025	7.5	18.9

3 Development of the Pump Model

For carrying out pump strength calculations the technology of three-dimensional finite-element modeling has been used. ANSYS software for engineering calculations is used.

The three-dimensional geometrical model for carrying out calculations is given in Fig. 1 in a general view and in Fig. 2 in a section. Walls thicknesses of case parts are set according to data of Table 3.

Fig. 1. Three-dimensional geometric model of the pump

Fig. 2. Three-dimensional geometric model of the pump by section of the vertical plane

For calculation of stress the finite element (FE) mesh has applied on geometrical model. A linear finite element of hexagonal and tetrahedral shape has been used. Cover and the entering branch pipes are constructed with regular mesh of hexagonal FE, and the pump casing is meshed by free tetrahedral mesh. After splitting into a mesh of finite elements all details of the pump have no degenerate FE. The basic structural elements have at least two elements on thickness and the ratio of the parties is sustained. The mesh used for calculations is presented in Fig. 3.

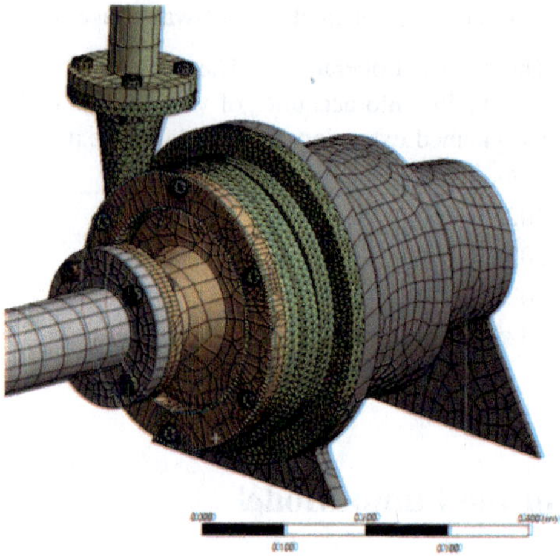

Fig. 3. The finite element model mesh (general view).

Table 3. Mechanical characteristics of the material 12Cr18H9Ti

Temperature, T, °C	Elastic modulus, *E*, GPa	Poisson ratio, ν	Density, ρ, kg/m^3	Yield strength, σ_T ($R_{0.2}$), MPa	Ultimate tensile strength, σ_B (R_m), MPa	Allowable stresses[a], [σ], MPa	Coef. of thermal conductivity, λ, W/(m °C)	Coef. of linear expansion, α, mkK^{-1}
20	205	0.3	7900	196	441	131	15	16.4
50	202	0.3	7900	196	422	131	–	16.4
100	200	0.3	7900	196	412	131	16	16.6
150	195	0.3	7900	177	402	118	–	16.8
200	190	0.3	7900	167	383	111	18	17

Table 4. Mechanical characteristics of steel 1020

Temperature, T, °C	Elastic modulus, *E*, GPa	Poisson ratio, ν	Density, ρ, kg/m^3	Yield strength, σ_T ($R_{0.2}$), MPa	Ultimate tensile strength, σ_B (R_m), MPa	Allowable stresses[b], [σ], MPa	Coef. of thermal conductivity, λ, W/(m °C)	Coef. of linear expansion, α, mkK^{-1}
25	200	0.3	7859	216	402	144	52	11.5
50	197	0.3	–	206	392	137	–	11.5
100	195	0.3	7834	206	392	137	50.6	11.9
150	192	0.3	–	206	392	137	–	12.2
200	185	0.3	7803	196	373	131	48.6	12.5
250	180	0.3	–	196	373	131	–	12.8
300	175	0.3	7700	177	363	118	46.2	13.1

The following materials as steel grade 12Cr18H9Ti (for the case) and steel 1020 (for bolted connections) are selected for calculations. Their physical and mechanical characteristics of materials are respectively presented in Tables 3 and 4. Rated allowable stress for the elements of the equipment and pipelines loaded with internal pressure is accepted minimum of the following values calculated by (1). Where R_m is the tensile strength; and $R_{0.2}$ is the yield strength.

$$[\sigma] = \min\{R_m/2.6;\ R_{0.2}/1.5\} \tag{1}$$

For bolted and stud connections pressure calculated by (2).

$$[\sigma]_w = R_{0,2}/2 \tag{2}$$

4 Mathematical Models

To calculate the stresses, displacements and temperature fields finite element method is used. In the framework of this method, the volume of the investigated part is meshed into finite elements. For each finite element, its resolving equations are written. The solution of the equations is stitched in the mesh nodes. This leads to the solution of a system of linear equations of order equal to the number of degrees of freedom of all finite elements.

The work consistently solves the problem of stationary heat conduction (3) and the problem of static stress distribution (4).

$$[M]\{\dot{\mathrm{T}}\} + [\mathrm{K}]\{\mathrm{T}\} = \{\mathrm{f}_T\} \tag{3}$$

$$[\mathrm{K}]\{\mathrm{q}\} = \{\mathrm{f}_N\} + \{\mathrm{f}_{Temp}(\{\mathrm{T}\})\} \tag{4}$$

where $[M]$ – global thermal capacitance matrix;

$[K]$ – global stiffness matrix;
$\{T\}$ – nodal temperature vector;
$\{q\}$ – nodal displacement vector;
$\{\mathrm{f}_T\}$ – temperature loading vector;
$\{\mathrm{f}_T\}$ – mechanical loading vector;
$\{\mathrm{f}_{Temp}(\{\mathrm{T}\})\}$ – mechanical loading due thermal expansion vector.

5 Analysis of Static Strength

For assessment of residual strength of the pump taking into account the possible extension of service life, calculations on static and cyclic strength at two options of loading are carried out:

1. to the normal operation conditions (NOC);
2. to hydrotest mode (HM).

For calculation on static strength under normal conditions and during hydro tests the group of strength characteristics should be used. They are applied under different conditions of loading and/or different actual or conditional types of the stress-strain state arising in construction.

5.1 The Analysis of Static Strength of the Case at NOC

The following calculation parameters are set:

- rigid sealing on the supporting surfaces of pump racks;
- rigid fastening in the axial direction for the pipe supplying the working fluid and its outlet pipe. In the plane perpendicular to the axis of these pipes, elastic supports with a stiffness of 0,1 MN/m have placed;
- volume force is gravity;
- the bolt tightening torque according to Table 5;
- on internal surfaces convective heat exchange "water – steel" is set (coefficient of heat emission, 500 W/ m^2 °C);
- on outer surfaces convective heat exchange "steel – air" is set (coefficient of heat emission, 5 W/ m^2 °C);
- ambient temperature is accepted by 25 °C;
- external pressure is specified as normal.
- internal pressure to equally calculated pressure in each step and pump outlet pressures.

Table 5. Standard forces and torques of tightening

Standard size	Preload force, kN	Tightening torque, N·m
M12	9.0	41
M16	17.0	102

At the first stage the problem of the prestressed state caused only by tightening of studs has solved. In Fig. 4 distribution of strain intensity (equivalent strain by Mises's criterion) which is formed in the studs under this loading mode.

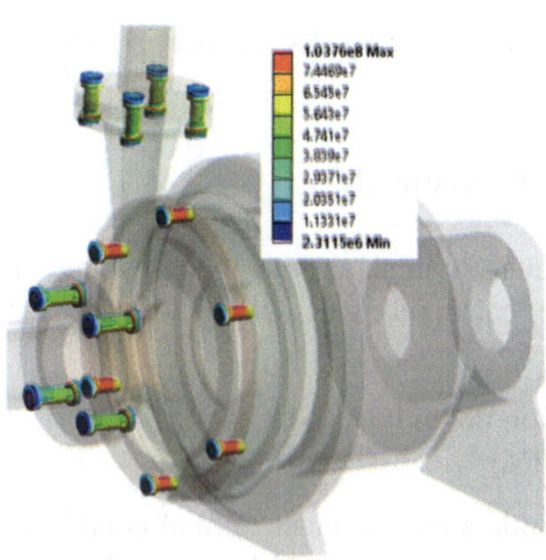

Fig. 4. von Mises stress in the bolts and studs after its tightening, Pa.

The analysis of the pre-stressed state of the studs showed the presence of compressive stresses in the studs equal 103,7 MPa, which is less than the maximum allowable value of 137 MPa. The high level of stress has formed in the pump casing (Fig. 5), at the same time the received stress (98,5 MPa) is also less then limit one (131 MPa).

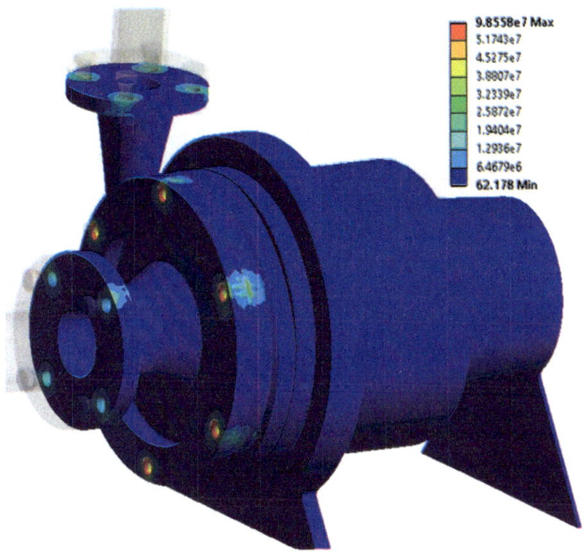

Fig. 5. von Mises stress in the pump casing and covers after bolts and studs tightening, Pa.

In Fig. 6 shows the temperature distribution across the pump casing.

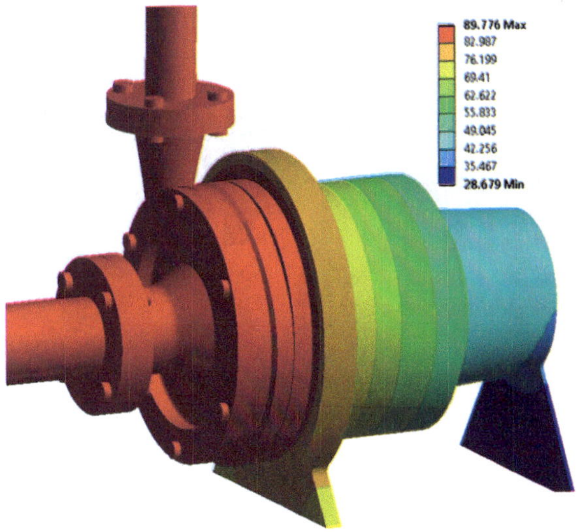

Fig. 6. Temperature distribution over the pump casing at NOC, degrees of Celsius.

The pressure distribution scheme is shown in Fig. 7.

Fig. 7. Loading the pump with internal pressure at the normal operation conditions.

The results of calculations with this loading scheme are shown in Figs. 8 and 9. Maximum stresses formed in the elements of the body.

According to the results of the calculation (Figs. 8 and 9), the strength condition is satisfied (the maximum stress value is (σ) = 98.9 MPa, which is less than the limit value [σ] = 131 MPa.

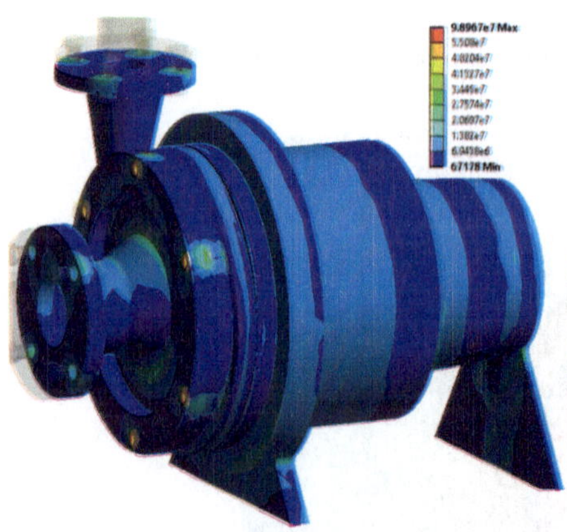

Fig. 8. von Mises stress in the pump case under normal conditions operation, a general view, Pa.

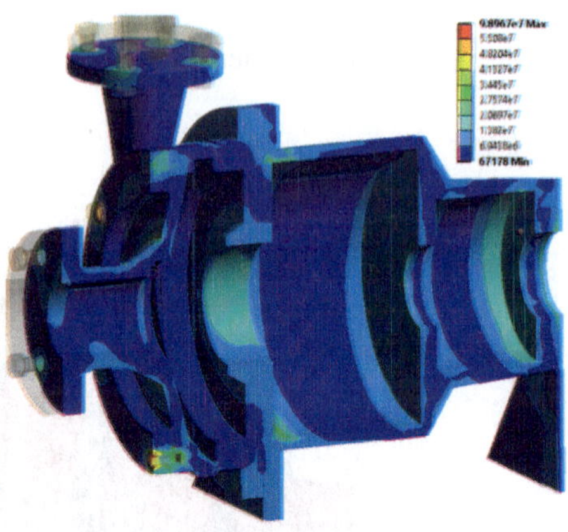

Fig. 9. von Mises stress in the pump case under NOC, a sectional view, Pa.

5.2 Analysis of the Strength of Bolts

Figure 10 shows the von Mises equivalent stress distribution in bolts and studs under NOC.

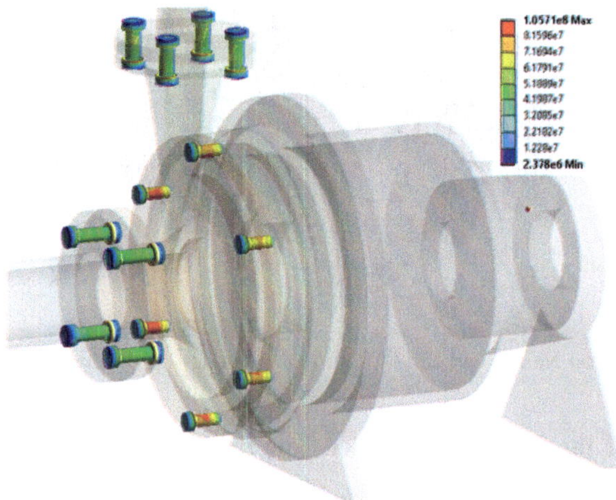

Fig. 10. von Mises stress in bolts and studs at NOC, Pa.

The maximum stress in the bolted joint is 105.7 MPa, which is less than the limit value [σ] = 137.0 MPa. Thus, the strength condition is satisfied.

5.3 Static Strength Analysis with Hydrotests

To check the system integrity, it is tested by increased pressure. It is also calling hydrotesting. To assess the strength in pump hydrotesting, such boundary conditions are used that corresponded to normal operating conditions, with the exception of internal pressure. Internal pressure is set constant throughout all sections of the pump.

The results of calculations of the stress-strain state under such a loading scheme are shown in Fig. 11.

Maximum stresses are formed in the back of the pump. According to the results of the calculation (Fig. 11), the strength condition is executed (the maximum stress value is σ = 100.9 MPa, which is less than the limiting value of 177 MPa).

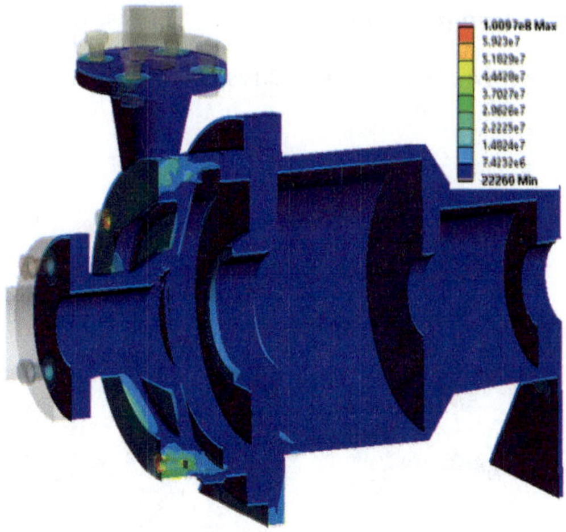

Fig. 11. von Mises stress in the pump case under during hydrotesting, Pa.

5.4 Analysis of the Strength of Bolted Joints with Hydrotests

The stress distribution in bolted joints during hydrotesting is shown in Fig. 12.

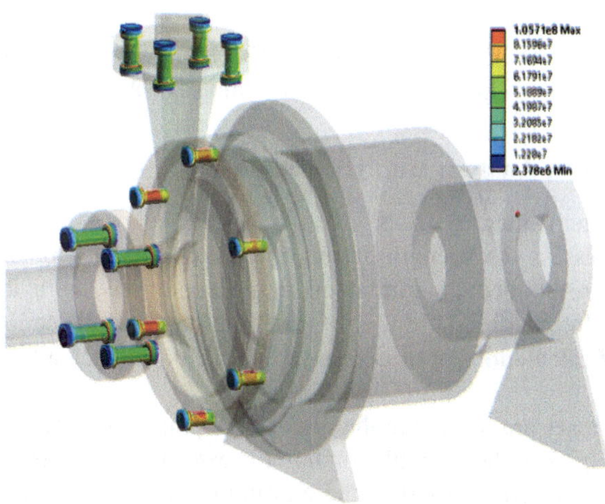

Fig. 12. von Mises stress in bolts and studs during hydrotesting, Pa.

The maximum stress in the bolted connection is 105.7 MPa, which is less than the limit value 144 MPa. Thus, the strength condition is satisfied.

6 Analysis of Cyclic Strength

When calculating on cyclic durability the following operation modes presented in Table 6 are considered.

Table 6. Pump operation modes

Mode	Approbation of the pump, with inclusion in work on operating modes	Hydrotesting
	6 launches per year	*1 test per 8 years*
Quantity of cycles in the mode for 30 years	6·30 = 180	4
Quantity of cycles in the mode for the planned extension term in 15 years	6·15·1.15 = 104	2
Total	284	6

For calculation the allowed tension amplitude [σ_{aF}] at the set allowed number of cycles [N_0] $\leq 10^{12}$ is defined on cyclic strength at least from two received values according to (3) and (4):

$$[\sigma_{aF}] = \frac{E \cdot e_c}{n_\sigma (4 \cdot [N_0])^m} + \frac{R_c}{n_\sigma \left[(4 \cdot [N_0])^{m_e} + \frac{1+r}{1-r} \right]} \quad (5)$$

$$[\sigma_{aF}] = \frac{E \cdot e_c}{(4 \cdot n_N \cdot [N_0])^m} + \frac{R_c}{(4 \cdot n_N \cdot [N_0])^{m_e} + \frac{1+r}{1-r}} \quad (6)$$

And at [N0] $\leq 10^6$ it is allowed to use the minimum value received from dependences for finding of stress amplitude [σ_{aF}] according to (5) and (6):

$$[\sigma_{aF}] = \frac{E \cdot e_c}{n_\sigma (4 \cdot [N_0])^m} + \frac{R_{-1}}{n_\sigma \left[1 + \frac{R_{-1}}{R_m} \cdot \frac{1+r}{1-r} \right]} \quad (7)$$

$$[\sigma_{aF}] = \frac{E \cdot e_c}{(4 \cdot n_N \cdot [N_0])^m} + \frac{R_{-1}}{1 + \frac{R_{-1}}{R_m} \cdot \frac{1+r}{1-r}} \quad (8)$$

Where n_σ is safety factor of strength on tension;

n_N is safety factor of strength on number of cycles;
m and m_e (10) are also characteristics of material;
r is a coefficient of asymmetry of cycle of tension;
R_c is a characteristic of strength (7);
e_c is a characteristic of plasticity (8);

R_{-1} is limit of endurance at symmetric cycle (9);
Z_c is relative throat of lateral section of sample at static destruction (Z_c = Z at Z < 50% and Z_c = 50% at Z > 50%).

$$R_c = R_m\left(1 + 1.4 \cdot 10^{-2}Z\right) \tag{9}$$

$$e_c = 1.15\,\lg\frac{100}{100 - Z_C} \tag{10}$$

$$R_{-1} = 0.4 \cdot R_m \tag{11}$$

$$m_e = 0.132\,\lg\left[\frac{R_m}{R_{-1}}\left(1 + 1.4 \cdot 10^{-2}Z\right)\right] \tag{12}$$

Use of (7–10) and information about physical characteristics of material (Tables 3 and 4) allows defining design characteristics of fatigue parameters for the studied pump (Table 7).

Table 7. Characteristics of fatigue at temperature 100 °C

Steel	$R_{m,}$ MPa	$R_{p0.2,}$ MPa	e_c	m	m_e	n_σ	n_N	Z, %
12CR18H9TI	461	189	0.002	0.5	0.0528	2.0	10	40
1020	392	206	0.0019	0.5	0.0528	1.5	5	38

Settlement curves of fatigue are defined for the cyclic deformation set like asymmetry. The coefficient of asymmetry is defined by (13):

$$r = \frac{(\sigma)_{\max} - 2(\sigma_a)}{(\sigma)_{\max}} \tag{13}$$

Where is $(\sigma)_{max}$ is the maximum specified stress in the studied cycle; (σ_a) is amplitudes of the specified stress in this cycle.

Settlement probes by (5–8), allow, having chosen minimum of couples (5)–(6) and (7)–(8) values to construct curves Weller (Fig. 13). The analysis of the given curves on graphics allows choosing as curve fatigue minimum of given (means to $[N_0] < 10^7$) curve calculated by (5), and $10^7 < [N_0]$ by (7).

The strength condition in the presence of various cyclic loads is checked by (14)

$$\sum_{i=1}^{k}\frac{N_i}{[N_0]_i} = a \le [a_N] \tag{14}$$

Where N_i is a number of cycles for i type during operation;

k is total number of types of cycles,
$[N_0]_i$ is the allowed number of cycles for i type,
a is the accumulated fatigue damage, which extreme value $[a_N]$ = 1.

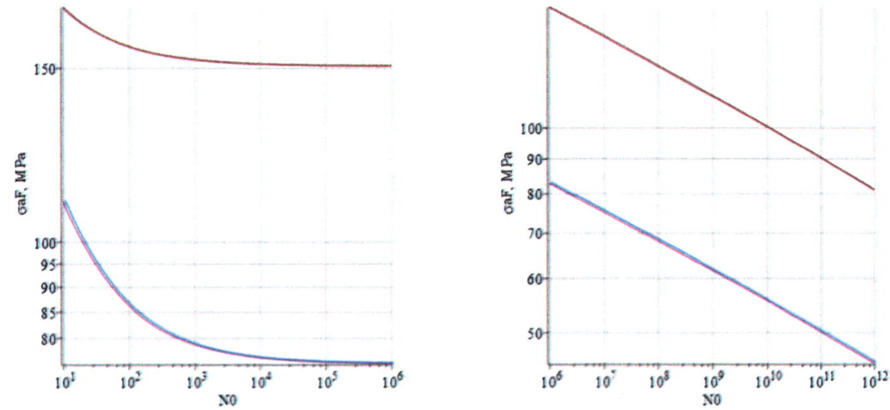

a) $N_0 = 10^1..10^6$ b) $N_0 = 10^6..10^{12}$

Fig. 13. Settlement curves of fatigue (Weller's Curves) for steel 12Cr18H9Ti at temperature of 100 °C.

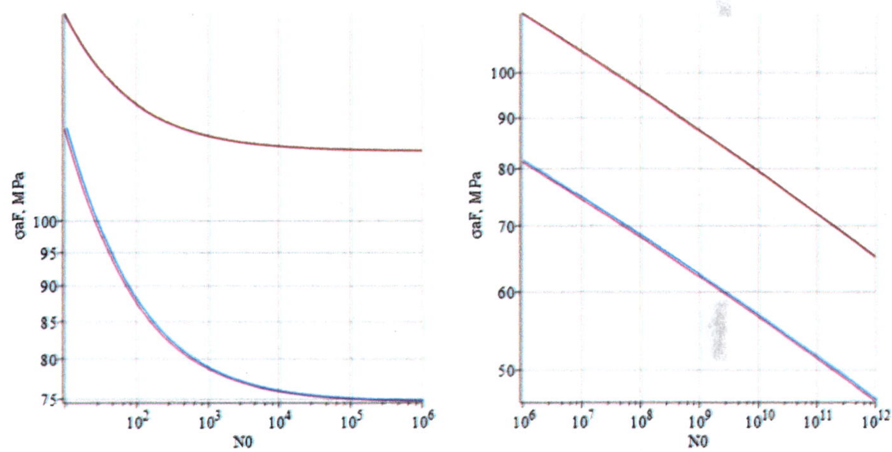

a) $N_0 = 10^1..10^6$ b) $N_0 = 10^6..10^{12}$

Fig. 14. Settlement curves of fatigue (Weller's Curves) for steel 1020 at temperature of 100 °C.

Amplitude of tension of cycle is defined as (15):

$$[\sigma_{aF}] = (\sigma^{\max} - \sigma^{\min})/2 \tag{15}$$

For the case under normal operation conditions σ_{aF} = 20.7 MPa, the number of cycles exceeds 10^{12} therefore have accept $N_0 = 10^{12}$. On the other hand for situation of hydrotests σ_{aF} = 32.7 MPa, the number of cycles exceeds 10^{12} therefore have accept $N_0 = 10^{12}$. Thus, for the considered pump casing structural elements taking into account the planned extension term at quantity of cycles of inclusions (284 cycles) and accounting of hydrotests (6 cycles), value close to zero (a ≈ 0) that is less then permissible value $[a_N] = 1$. The condition of strength is satisfied (Fig. 14).

For the bolted joints admissible $\sigma_{aF} = 3.5$ MPa, the number of cycles exceeds 10^{12} therefore have accept $N_0 = 10^{12}$. On the other hand for situation of hydrotests $\sigma_{aF} = 3.8$ MPa, the number of cycles exceeds 10^{12} therefore have accept $N_0 = 10^{12}$. Thus, for the considered pump casing structural elements taking into account the planned extension term at quantity of cycles of inclusions (284 cycles) and accounting of hydrotests (6 cycles), value close to zero (a ≈ 0) that is less then permissible value $[a_N] = 1$. The condition of strength is satisfied.

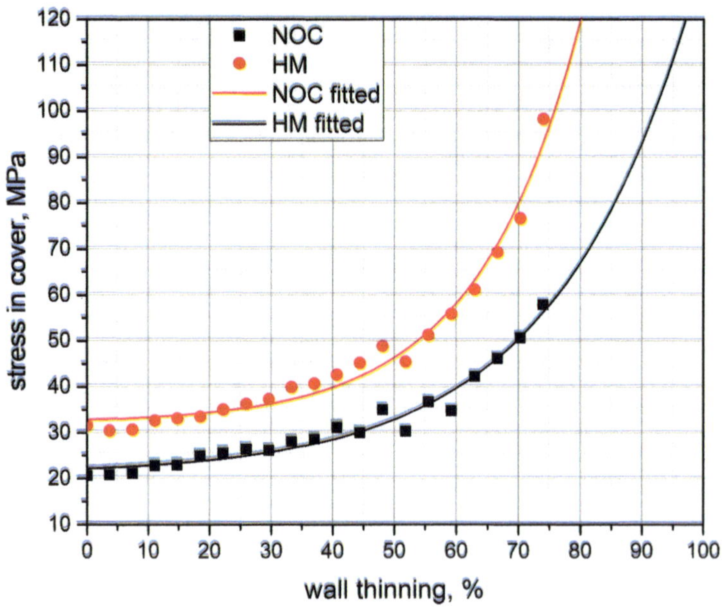

Fig. 15. Dependence of stresses on wall thinning

7 Lifetime Estimation

One of the high-loaded elements of the pump is the cover. It is also subject to great wear during operation. To determine the effect of wear of the cover on its stressed state, a series of calculations have been performed. These calculations have been performed for the two operation modes NOC and HM. To determine the dependence, the cover wall thickness has been varied, after which the maximum stresses in it has been determined. The result of the calculations is shown in Fig. 15. To summarize the obtained data, the data was approximated by a function of the form (16). The obtained parameters of this function are presented in Table 8.

$$\sigma_{aF} = a_1 \exp(x/a_2) + a_3 \quad (16)$$

To determine the number of cycles to failure, dependence (16) has been substituted into Eqs. (5)–(13). As a result of the calculations, the dependence presented in Fig. 16 has been obtained. As can be seen from the figure, with an increase in wear, the number of cycles to failure decreases. The NOC operation mode has a large life-time, because the lower pressure is applied to the cover. For HM mode, higher pressures lead to smaller life-time.

Table 8. Parameter of approximation

	a_1, MPa	a_2	a_3, MPa
NOC	1.2707	22.264	20.917
HM	0.6912	16.504	31.642

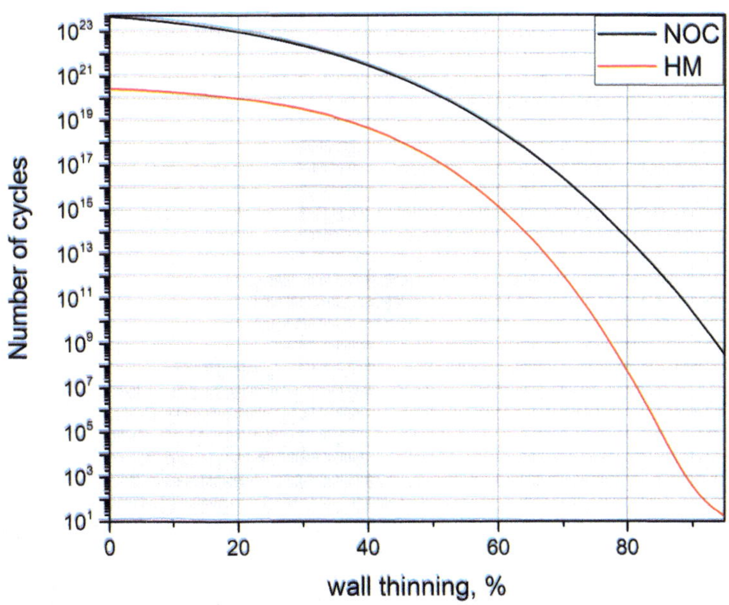

Fig. 16. Dependence of number of loading cycles on wall thinning

8 Seismic Strength Analysis

An important element in ensuring the safety standards of nuclear and thermal power plants is the determination of the seismic stability of the structure. For this purpose, the stresses that occur during typical earthquakes are estimated. Figure 17 shows typical

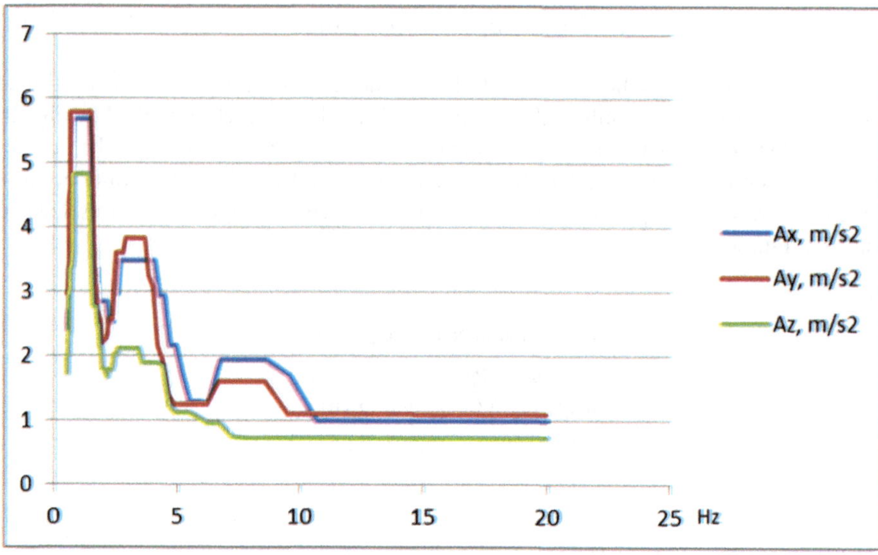

Fig. 17. Typical spectrum of room vibration in accelerations by directions, m/s^2

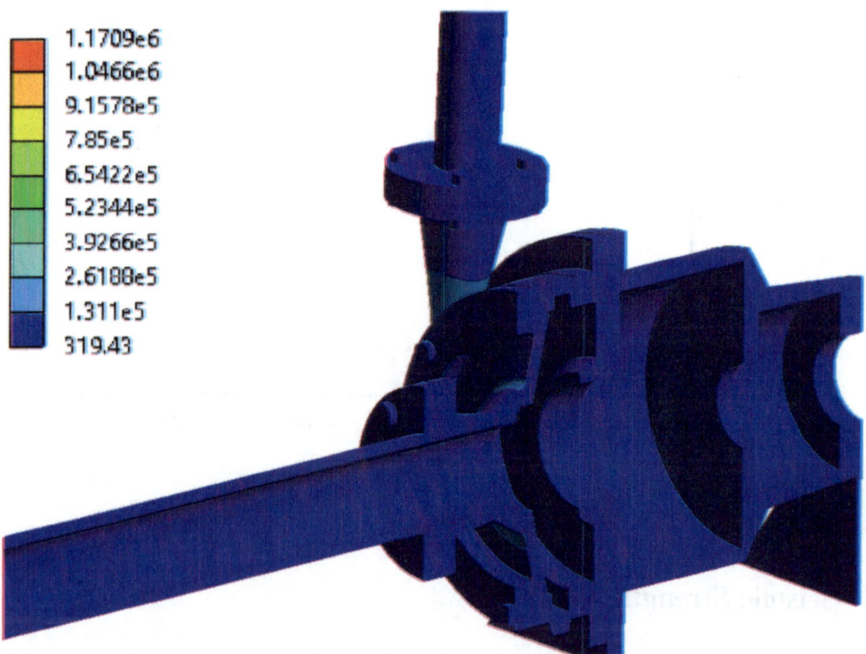

Fig. 18. von Mises stress in pump caused by seismic loading, Pa

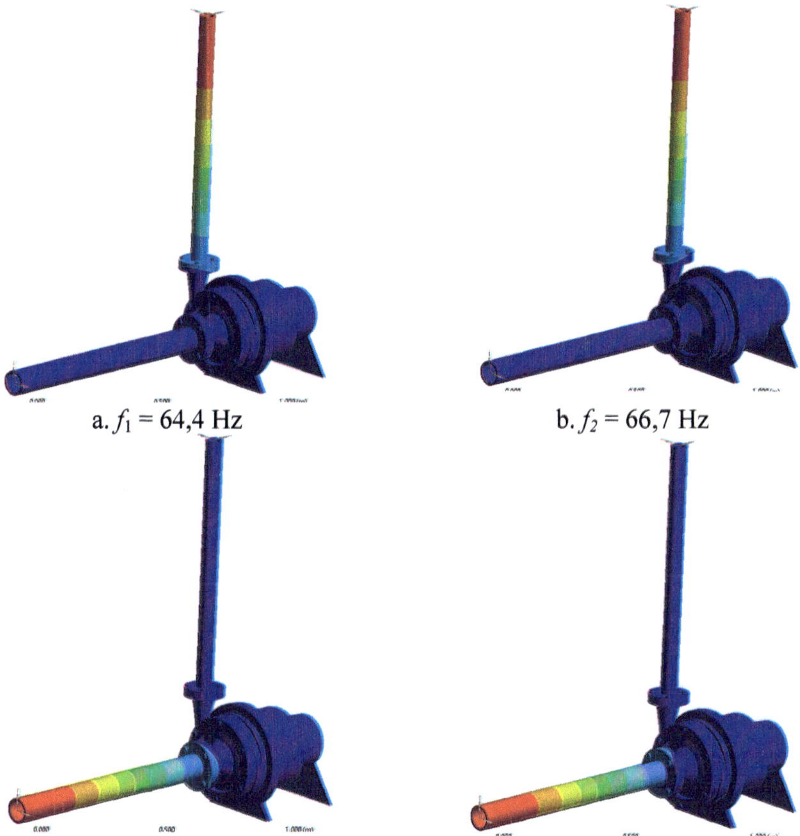

a. f_1 = 64,4 Hz

b. f_2 = 66,7 Hz

c. f_3 = 100 Hz

d. f_4 = 102 Hz

Fig. 19. Eigenfrequencies and forms of pump

acceleration spectra for equipment room. The first four eigen forms and frequencies of the structure are shown in Fig. 19.

Figure 19 shows that the first eigenfrequencies are higher than the frequencies of the earthquake spectrum (Fig. 17). This means that during earthquake the resonance phenomenon will not be observed. Therefore, the expected stresses will be low.

From Fig. 18 it can be seen that the stress due to earthquakes throughout the structure is less than 1.1 MPa. Therefore, their contribution to the total stress state is less than 1% and they do not affect the strength of the structure.

9 Conclusions

The paper discusses steps to determine the residual life-time of a centrifugal pump. Distributions of equivalent stresses in the pump construction elements under normal operating conditions and hydrotesting mode are obtained. The amplitudes of stress cycles has been calculates, on the basis of which the life-time has been determined, as well as the possibility of extending the design life-time. According to the results of calculations, the design life-time of pumps with a corresponding thinning of the walls can be extended for 15 years.

The paper also estimates the number of cycles for rupture at various levels of wear on the pump cover. The dependences of stresses on wall thickness are obtained.

The results of seismic strength calculations showed that the stresses resulting from earthquakes contribute about 1% to the total stress state and may not be taken into account in further modelling.

References

1. Verhuelsdonk, B.: Increasing the operational lifetime of rotary lobe pumps. World Pumps **2005**, 42–44 (2005)
2. Kahlman, L.: Extending the life of pumps and fans. World Pumps **2016**, 32–33 (2016)
3. Patel, S.M., Allaire, P.E., Wood, H.G., Throckmorton, A.L., Tribble, C.G., Olsen, D.B.: Methods of failure and reliability assessment for mechanical heart pumps. Artif. Organs **29**, 15–25 (2005)
4. Cheng, Q., Wang, S., Yan, C.: Robust optimal design of chilled water systems in buildings with quantified uncertainty and reliability for minimized life-cycle cost. Energy Build. **126**, 159–169 (2016)
5. Jacobs, J.A., Mathews, M.J., Kleingeld, M.: Failure prediction of mine de-watering pumps. J. Fail. Anal. Prev. **18**, 927–938 (2018)
6. Larin, O., Kelin, A., Naryzhna, R., Potopalska, K., Trubayev, O.: Analysis of the pump strength to extend its lifetime. Nucl. Radiat. Saf. **3**, 30–35 (2018)
7. Pourgol-Mohammad, M., Makarachi, P., Soleimani, M., Ahmadi, A.: Reliability enhancement of centrifugal pumps by multi-objective genetic algorithm optimization. Int. J. COMADEM **20**, 23–30 (2017)
8. Yin, F., Nie, S., Ji, H., Huang, Y.: Non-probabilistic reliability analysis and design optimization for valve-port plate pair of seawater hydraulic pump for underwater apparatus. Ocean Eng. **163**, 337–347 (2018)
9. Saldanha, P.L.C., de Simone, E.A., Frutuoso e Melo, P.F.: An application of non-homogeneous Poisson point processes to the reliability analysis of service water pumps. Nucl. Eng. Des. **210**, 125–133 (2001)
10. Rathore, A., Patidar, N.P.: Reliability assessment using probabilistic modelling of pumped storage hydro plant with PV-Wind based standalone microgrid. Int. J. Electr. Power Energy Syst. **106**, 17–32 (2019)
11. Getman, A.: Resource of operation of vessels and pipelines of nuclear power plants. Energoatomizdat, Moscow (2000). (in Russian)
12. Vodka, A.A., Kelin, A.A., Naryzhnaya, R.N., Trubaev, A.I.: Features of cable calculations in an open strip for strength during their seismic qualifications. Theor. Appl. Mech. (32), 134–140 (2017). (in Russian)
13. Larin, O.O., Vodka, O.O., Trubayev, O.I.: The fatigue life-time propagation of the connection elements of long-term operated hydro turbines considering material degradation. PNRPU Mech. Bull. **1**, 167–193 (2014)
14. Larin, O., Vodka, O.: A probability approach to the estimation of the process of accumulation of the high-cycle fatigue damage considering the natural aging of a material. Int. J. Damage Mech. **24**, 294–310 (2015)

Modelling of Bird Strike on an Aircraft Glazing

Natalia Smetankina[1(✉)], Igor Kravchenko[2], Vyacheslav Merculov[2], Dmitry Ivchenko[2], and Alyona Malykhina[1]

[1] A. Podgorny Institute of Mechanical Engineering Problems of the National Academy of Sciences of Ukraine, 2/10 Pozharsky Street, Kharkiv 61046, Ukraine
nsmetankina@ukr.net

[2] SE Ivchenko-Progress, 2 Ivanova Street, Zaporozhye 69068, Ukraine

Abstract. Collisions between birds and aircrafts during the taking-off, cruising and landing phases have become an increasingly serious and catastrophic issues for aircrafts safety. Aviation standards in force require that the aircraft construction would allow the crew to conclude the flight safely after collision with 1.81 kg bird. A method for analysis of the stress-strained state of laminated airplane glazing at different operational factors is presented. The method includes a technique for strength analysis of the laminated airplane glazing at the bird impact, and a technique for analysis of superfluous pressure. The model of laminated glazing is based on the refined theory of the first-order accounting transverse shear strains, thickness reduction and normal element rotation inertia each layer. The mathematical model of pressure impulse authentically reproducing bird impact is based on the experimental researches. Theoretical results are in good agreement with experimental data that allows recommending the method for working out new airplane glazing elements.

Keywords: Bird strike · Laminated windshield · Strength

1 Introduction

Bird strike is one of the most important concerns about safety in the aviation industry. According to the statistical data from the Federal Aviation Administration, the number of bird strike accidents annually has increased by six times from 1795 cases to 10,856 cases in year 1990 and 2013 respectively, with total accident number of 138,257 cases with 14 years. Such intensive bird strike accidents have caused huge fatalities, namely, at least 103 aircrafts and 262 lives were lost in civil aviation field from year 1912–2008 where annual property loss was increased from 614 million to 1.28 billion US dollars [1–4].

It is estimated that nearly 40% of the major bird strike incidents for civil aircrafts can be attributed to engine ingestion, 33% to collision with wings, 16% with windshields, 7% with fuselages. For military aircrafts, there are different statistics: engines – 55%, fuselages – 11%, windshields – 10%, wings – 14% [2–4].

M. Nechyporuk et al. (Eds.): *Integrated Computer Technologies in Mechanical Engineering*, AISC 1113, pp. 289–297, 2020.
https://doi.org/10.1007/978-3-030-37618-5_25

All modern aircraft structures are designed with account of likely collision with birds [4–6]. Aviation standards in force require that the aircraft construction would allow the crew to conclude the flight safely after collision with 1.81 kg bird [7]. Also, the reliable protection from pressurization, namely a static loading, which arises through the pressure difference outside and inside the aircraft cockpit, is of great importance for ensuring the normal flight.

2 Literature Review

Currently, certification of new aircraft parts is usually done empirically [1–5]. This is an expensive process as several tests may be required to evaluate the effectiveness of the windshield. The idea of thus is to replace the expensive empirical verifications by computer simulations. If simulation of bird strike is accurately able to predict the behaviour of bird strike on windshields, then the windshield design can be optimized before an actual test is carried out. This will lower the costs and expedite the design and certification processes.

It is stated that the behavior of birds impacting a rigid target is described by using a hydrodynamic model. At high-speed impacts the bird behaves like a fluid with insignificant viscosity [8].

With the advent of highly effective FEM-based software packages, further consideration was given to the problem of joint deformation of the bird and target. In doing so, the focus was on the accuracy of describing the process of bird damage. Lagrangian, Arbitrary Lagrangian Eulerian (ALE) and Smooth Particle Hydrodynamics (SPH) formulations have found wide application [9–16].

Windshield is an important element of an aircraft and some basic features are depended on its quality. The important quality characteristics of windshield are visibility through the canopy, structure rigidity and reliability, bird impact resistance. The most widely used materials for an aircraft windshield are glass and polymers.

Yang et al. [14], elaborated an experimental and FEM of windshield subjected to high-speed bird impact. Authors of works [15] and [16] focused on the analysis of an effective numerical method to simulate bird impact aircraft windshield events, using the SPH approach and the explicit finite element program PAM-CRASH. Salehi et al. [6] investigated the effect of the bird strike on different aircraft windows both numerically and experimentally.

The aim of the present study is to devise the method of calculating the stress-strained state parameters for laminated aircraft windshields at bird impact and operating static load on the basis of an immersion method.

3 Research Methodology

3.1 Mathematical Model of a Bird Impulse

A fluid dynamic model of a bird impulse is proposed. The target is a laminated glass constant thickness subjected to impulse loads simulating impact action. The glass has a complex form in plan and it is considered in the Cartesian system of coordinates related to its outer surface subjected to a bird strike (Fig. 1). A bird with mass M collides with the glass with velocity V. The bird's trajectory of motion is at angle α to the glass.

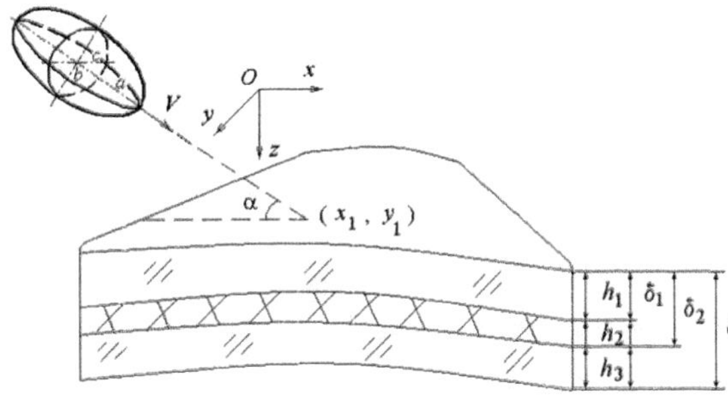

Fig. 1. Scheme of a bird collision with a laminated glass.

The components of the vector of the external load $\mathbf{P} = \{p_j(x,y,t)\}$ $(j = \overline{3I+3})$ acting on the glass during a strike are represented as

$$p_1 = p_2 = p_{3+i} = p_{3+I+i} = p_{3+2I+i} = 0,\ i = \overline{1,I};\ p_3 = \frac{1}{2}[1 + \mathrm{sign}(\tau_b - t)]F(t), \tag{1}$$

where t is time; I is the number of layers; τ_b is time of the bird-and-glass interaction; $\tau_b = \frac{2\sqrt{a^2+k^2b^2}}{V}$; $k = ctg\alpha$; $F(t)$ is a function of the contact pressure over the load area;

$$F(t) = P_0\left[1 - \frac{(x-x_1)^2}{u_b^2} - \frac{(y-y_1)^2}{v_b^2}\right]^{1/2};$$

u_b and v_b are lengths of semi-axes of the elliptical load area; x_1 and y_1 are coordinates indicating the point where the trajectory of motion of the bird's centre of mass intersects the glass;

$$u_b = \frac{ab}{a^2+k^2b^2}\sqrt{(1+k^2)\left(2Vt\sqrt{a^2+k^2b^2} - V^2t^2\right)},\ v_b = c\sqrt{\frac{2Vt\sqrt{a^2+k^2b^2} - V^2t^2}{a^2+k^2b^2}}.$$

The impact force of the bird and glass relates to load intensity as follows:

$$P_b(t) = \iint\limits_{\Omega(t)} p_3 d\Omega = \frac{2}{3}P_0\pi u_b v_b. \tag{2}$$

where $P_b(t)$ is contact force; $\Omega(t)$ is the area of bird and glass contact.

According to the fluid dynamic theory suggested, the first approximation of the contact interaction force $P_b(t)$ is assumed represented by the value obtained from the fluid dynamic theory. It takes the form

$$P_b(t) = \rho_b V^2 \sin^2 \alpha \, \pi \, u_b v_b, \tag{3}$$

where ρ_b is bird tissue density; $\rho_b = \frac{3M}{4\pi abc}$.

3.2 Mathematical Model of a Laminated Glass

We examine the laminated glass as an open-ended laminated cylindrical shell with R radius. It comprises isotropic layers with constant thickness h_i (Fig. 1). The behavior of a laminated shell is described by the first-order theory accounting for transverse shear strain, thickness reduction and normal element rotation inertia in each layer

$$u_k^i = u_k + \sum_{j=1}^{i-1} h_j u_{3+I(k-1)+j} + (z - \delta_{i-1}) u_{3+I(k-1)+i}, k = 1,\, 2,\, 3, \quad i = \overline{1,\, I},$$

where $\delta_i = \sum_{j=1}^{i} h_j$, $\delta_{i-1} \le z \le \delta_i$; $u_k = u_k(x,\, y,\, t)$ $(k = 1,\, 2,\, 3)$ are displacements of coordinate surface points to coordinate axes; $u_{3+I(k-1)+i} = u_{3+I(k-1)+i}(x, y, t)$ $(k = 1,\, 2)$ are angles of rotation of the normal element in the ith layer about the coordinate axes; $u_{3+2I+i} = u_{3+2I+i}(x,\, y, t)$ is normal element reduction within the ith layer; t is time.

The equations of motion of a laminated shell affected by impact load, well as the respective boundary conditions on boundary Γ are derived by Hamilton's variational principle

$$\mathbf{\Omega}^{\rho} \mathbf{U}_{,tt} - \mathbf{\Lambda} \mathbf{U} = \mathbf{P}, (x,\, y) \in \Omega,\, \mathbf{U} = \mathbf{U}_{,t} = 0,\, t = 0, \tag{4}$$

$$\mathbf{B}^{\Gamma} \mathbf{U} = \mathbf{P}^{\Gamma}, \quad (x,\, y) \in \Gamma, \tag{5}$$

where $\mathbf{\Omega}^{\rho}$ and $\mathbf{\Lambda}$ are symmetric matrices; $\mathbf{U} = \{u_j(x,y,t)\}$; $\mathbf{P} = \{p_j(x,y,t)\}$; $j = \overline{1, 3I+3}$.

The problem of investigating non-stationary vibrations of a laminated shell subjected to a bird impact is reduced to integrating a system of motion Eq. (4) with account of boundary conditions (5).

The analytical solution of the problem is obtained by the immersion method [17]. According to this method, a non-closed cylindrical laminated shell is immersed into an auxiliary enveloping cylindrical shell with the same composition of layers. It is loaded within domain Ω similar to that for the primary shell. An auxiliary shell is one whose contour shape and boundary conditions yield a simple analytical solution. In this case, the auxiliary shell is a simply supported non-closed cylindrical laminated with rectangular plan-view shape, allowing to find the problem solution as trigonometric series.

To satisfy actual boundary conditions, additional distributed compensating loads $\mathbf{Q}^{comp} = \left\{ q_j^{comp}(x, y, t) \right\}$ $(j = \overline{1,\ 3I + 3})$, the intensity of which are to be found, are applied to the auxiliary shell over the boundary Γ. The compensating loads appear in the motion equations as curvilinear distributions. Based on the condition of satisfying boundary conditions on the boundary Γ (5), we form a system of integral equations for determining the intensities of compensating loads

$$\mathbf{B}^{\Gamma}\, \mathbf{U}[\mathbf{Q}^{comp}(x,\ y,\ t) = \mathbf{P}^{\Gamma},\ (x,\ y) \in \Gamma. \tag{6}$$

Displacements and loads (1) are expanded in the auxiliary shell domain in trigonometric series for functions satisfying simply supported conditions. The compensating loads are expanded into a series along the boundary Γ

$$q_j^{comp}(s, t) = \sum_{\alpha=1,2} \sum_{\mu=0}^{\infty} q_{j\alpha\mu}(t)\, b_{\alpha\mu}(s),\ j = \overline{1,\ 3I + 3}, \tag{7}$$

where

$$b_{1\mu} = \sin\left[\mu\gamma(s)\right],\ b_{2\mu} = \cos\left[\mu\gamma(s)\right],\ \gamma(s) = 2\pi \int_0^s d\tilde{s} \Big/ \oint_{\Gamma} d\tilde{s},\ 0 \le \gamma(s) \le 2\pi,\ \mu = \overline{0,\ \mu^*}$$

Hence, the system of integral Eq. (6) is transformed to a system of algebraic equations with respect to the expansion coefficients of the compensating loads (7). The system of motion Eq. (4) is integrated by a method of expanding the solution into Taylor's series [17, 18].

4 Results

Numerical results, demonstrating the theoretical and experimental approach, were obtained for the windshield of an AN-178 aircraft under the bird strike. Experimental studies were carried out with dynamic wide-band strain gauging [18, 19].

The windshield is treated as an seven-layers elastic supported glass with radius $R = 1.34$ m and dimensions $s_1 = 0.695$ m, $s_3 = 0.54$ m, $s_5 = 0.61$ m, $s_7 = 0.545$ m, $r_1 = 0,05$ m, $r_2 = 0.03$ m, $r_3 = 0.04$ m, $r_4 = 0.045$ m (Fig. 2). Layers of the windshield are made of silicate glass (layers 1, 3, 5 and 7) and polymer material (layers 2, 4, 6). Data for the glass layers are as follows: $E_i = 6.12 \times 10^4$ MPa, $v_i = 0.22$ and $\rho_i = 2.5 \times 10^3$ kg/m^3 for $i = 1, 3, 5, 7$; $E_i = 2.8 \times 10^2$ MPa, $v_i = 0.38$ and $\rho_i = 1.2 \times 10^3$ kg/m^3 for $i = 2, 4, 6$; $h_1 = h_6 = h_7 = 5$ mm, $h_2 = 4$ mm, $h_3 = 12$ mm, $h_4 = 2$ mm and $h_5 = 6$ mm. Here E_i is Young's modulus for the material, v_i is Poisson's ratio, and ρ_i is density of the i th layer material. The strike was made in the middle of the external windshield surface parallel to the aircraft fuselage axis. During the experiment, bird bodies were launched against the laminated glass (Fig. 2). Strains were measured. A rosette of strain gauges (Fig. 3) was affixed with an adhesive to the glass back surface (the point C on Fig. 2).

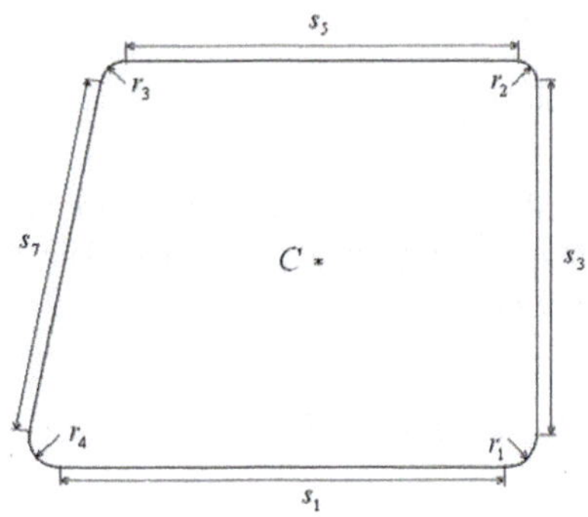

Fig. 2. Scheme of a laminated glass.

Fig. 3. A rosette of strain gauges.

In Fig. 4 the pneumatic gun for launching birds before testing is shown. Figure 5 shows the glass after testing.

Fig. 4. Pneumatic gun.

Fig. 5. Laminated windshield after testing.

Figure 6 shows the theoretical and experimental results for strain ε_1^5 vs. time during collision with a bird having a mass of 1.81 kg at the collision velocity of 157 m/s and an impact angle of $\alpha = 40°$ in the point C.

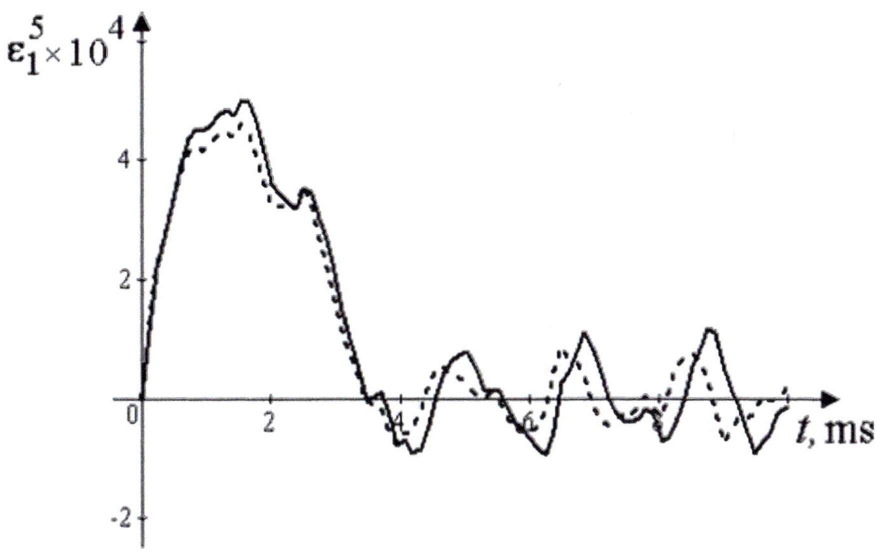

Fig. 6. Laminated windshield strain vs. time

The solid line is the theoretical curve, and the dashed line denotes the experimental data. A good match of theoretical and experimental results confirms the feasibility and effectiveness of the method for to evaluating the stress-strained stresses of aircraft glasses.

In accordance with international requirements the cockpit windows should stand a maximum operating excessive pressure (pressurization) of $P_{op} = 0.0618$ MPa and the design one of $P_d = 0.247$ MPa. Thus, instead of dynamic loading (1), we consider the uniform distributed static pressure. Maximum normal tensile stresses under operating and design pressures are 8.6 MPa and 34.3 MPa, respectively. Stresses in the windshield induced by cockpit pressurization did not exceed feasible values. Thus, the windshield meets the operating requirements.

5 Conclusions

A method of evaluating the stress-strained state of laminated aircraft windshield is devised that is based on the refined windshield model accounting for the effect of different operating factors. The method includes the procedure of strength calculations for laminated aircraft cockpit windows on the bird strike and cockpit pressurization. A model of the load impulse arising from the collision of laminated windshields with a bird is constructed. The problem solution has been obtained in the form of trigonometric series.

The stress-strained state of the laminated windshield of modern aircrafts was investigated at real operating loads. It was established that the stresses did not exceed feasible values. Comparison of calculation results and experimental data demonstrates their good agreement.

The advanced approach and calculation results can reduce costs and time for calculations, pre-design and full-scale tests of laminated aircraft windshields.

References

1. Dolbeer, R.A.: Birds and aircraft – fighting for airspace in ever more crowded skies. Hum.-Wildl. Confl. **3**(2), 165–166 (2009)
2. Heimbs, S.: Computational methods for bird strike simulations: a review. Comput. Struct. **89** (23–24), 2093–2112 (2011)
3. Hedayati, R., Sadighi, M.: Bird strike: An Experimental, Theoretical and Numerical Investigation. Elsevier Science, Cambridge (2015)
4. El-Sayed, A.F.: Bird Strike in Aviation: Statistics, Analysis and Management. Wiley, New Jersey (2019)
5. Georgiadis, S., Gunnion, A.J., Thomson, R.S., Cartwright, B.K.: Birdstrike simulation for certification of the Boeing 787 composite moveable trailing edge. Compos. Struct. **86**(1–3), 258–268 (2008)
6. Salehi, H., Ziaei-Rad, S., Vaziri-Zanjani, M.-A.: Bird impact effects on different types of aircraft bubble windows using numerical and experimental methods. Int. J. Crashworthiness **15**(1), 93–106 (2010)

7. Standard Airworthiness Certification Regulations. Part 25. Airworthiness Standards: Transport Category Airplanes. U.S. Federal Aviation Administration. https://www.faa.gov/aircraft/air_cert/airworthiness_certification/std_awcert/std_awcert_regs/regs/. Accessed 15 Nov 2018
8. Wilbeck, J.S., Rand, J.L.: The development of a substitute bird model. ASME J. Eng. Power **103**(4), 725–730 (1981)
9. Abrate, S.: Soft impacts on aerospace structures. Prog. Aerosp. Sci. **81**(2), 1–17 (2016)
10. Riccio, A., Cristiano, R., Saputo, S.: A brief introduction to the bird strike numerical simulation. Am. J. Eng. Appl. Sci. **9**(4), 946–950 (2016)
11. Dar, U.A., Zhang, W., Xu, Y.: FE analysis of dynamic response of aircraft windshield against bird impact. Int. J. Aerosp. Eng. **2013**(4), 1–12 (2013)
12. Guida, M., Marulo, F., Meo, M., Grimaldi, A., Olivares, G.: SPH–Lagrangian study of bird impact on leading edge wing. Compos. Struct. **93**(3), 1060–1071 (2011)
13. Hedayati, R., Ziaei-Rad, S.: A new bird model and the effect of bird geometry in impacts from various orientations. Aerosp. Sci. Technol. **28**(1), 9–20 (2013)
14. Yang, J., Cai, X., Wu, C.: Experimental and FEM study of windshield subjected to high speed bird impact. Acta Mech. Sin. **19**(6), 543–550 (2003)
15. Liu, J., Li, Y., Xu, F.: The numerical simulation of a bird-impact on an aircraft windshield by using the SPH method. Adv. Mater. Res. **33**(3), 851–856 (2008)
16. Li, Z.-Q., Han, Q., Yang, J.-L., Yao, X.: SPH–based numerical simulation of aircraft windshield under bird impact. J. South China Univ. Technol. (Nat. Sci.) **37**(12), 146–151 (2009)
17. Smetankina, N.V.: Non-stationary deformation, thermal elasticity and optimisation of laminated plates and cylindrical shells. Miskdruk Publishers, Kharkiv (2011)
18. Rodichev, Y.M., Smetankina, N.V., Shupikov, O.M., Ugrimov, S.V.: Stress-strain assessment for laminated aircraft cockpit windows at static and dynamic load. Strength Mater. **50**(6), 868–873 (2018)
19. Smetankina, N., Ugrimov, S., Kravchenko, I., Ivchenko, D.: Simulating the process of a bird striking a rigid target. In: Ivanov, V., Trojanowska, J., Machado, J., Liaposhchenko, O., Zajac, J., Pavlenko, I., Edl, M., Perakovic, D. (eds.) Proceeding of the 2nd International Conference on Design, Simulation and Manufacturing: The Innovation Exchange, DSMIE-2019, pp. 711–721. Springer, Cham (2019)

Information Technology in the Design and Manufacture of Engines

The Use of Information Technology for the Design of a Prototype Engine with Rotor in Magnetic Bearings

G. Yu. Martynenko, O. M. Marusenko(✉), Yu. M. Ulyanov, and L. V. Rozova

National Technical University "Kharkiv Polytechnic Institute", NTU «KhPI», Kharkiv, Ukraine
alexeymars.am@gmail.com

Abstract. The questions of planning and calculation of prototype magnetic rotor suspension of small engines at making of elaborate machine-building designs in particular pumps, turbo-expanders or expanders of compressor units of small and medium sizes. The preliminary estimation of dynamic descriptions of the rotor systems requires development of adequate mathematical models with taking into account numerous parameters, including design features, such as supports, stiffness and damping characteristics. A laboratory installation of a complete magnetic suspension of a horizontal rotor according to a combined magnetic-electromagnetic type is proposed and implemented. It uses two passive magnetic bearings created using two annular permanent magnets with axial magnetization as two radial bearings. The axial support is an active magnetic bearing (AMB) with stators in the form of armored cores. The functionality of this AMB uses a discrete method and an analog control system. Finite-element modeling was carried out and the dynamic characteristics of the structure under study were calculated. An experimental installation of the rotor propotype was developed and manufactured, with the help of which experimental testing of the developed model was carried out and the reliability of the calculated results was verified. Comparison of calculated and experimental results shows the effectiveness of the application of information technology in the design of complex engineering structures.

Keywords: Rotary systems · Engine · Electromagnetic suspension · Laboratory installation · Finite element model

1 Introduction

The development of modern mechanical engineering cannot be imagined without the use of information technology in the design and production of complex machine-building structures, in a particular to model and study their dynamic behavior. Rotary systems are used in many areas of modern industry [1]. At their work it is necessary to take into account features of a design of a rotor and its behavior under the influence of different loads, ways of support.

M. Nechyporuk et al. (Eds.): *Integrated Computer Technologies in Mechanical Engineering*, AISC 1113, pp. 301–309, 2020.
https://doi.org/10.1007/978-3-030-37618-5_26

The rotor systems are the main unit of various machines. used in power engineering, transport engineering, machine tool, aircraftbuilding and shipbuilding, chemical and gas industry, metallurgy [1]. They are used to create compressors, centrifuges, engines, turbine units, turboexpander, turbopump gas-turbine engines, etc.

The reliability and durability of the entire structure is generally determined by the rotor-support system. Creation and improvement of non-contact suspensions has recently been one of the trends in the development of modern technology. There is a tendency to use magnetic bearings or electromagnetic suspension systems (EMS) for rotors as supports. When using magnetic bearings, there is no mechanical contact, the need for lubrication, and, conversely, there is a high speed factor, low friction, the ability to work in vacuum and in aggressive environments, as well as the ability to minimize dimensions of the structure.

Therefore, at present, AMBs are widely used in many regions of technology [1, 2], for example, in machine tools (milling machines and precision machining machines for small parts, electrospindles); high-speed engineering (turbomolecular pumps, turbogenerators, compressors); ultracentrifuges for the needs of the nuclear industry; inertial navigation (non-contact suspension of the rotor in gyroscopes, gyrocompasses and gyrodines); flywheel energy storage devices; medical equipment (as bearing units of an artificial heart pump). Of particular importance is the possibility of using rotor systems with electromagnetic bearings (EMB) in transport (railway, ground-based on a magnetic cushion), where EMB is used to suspend the rotors of turbochargers (turbochargers) of diesel engines [3].

2 Overview of Existing Solutions

Rotor systems are a rotating shaft with hinged elements fixed on it (dumis, wheels, thrust discs, etc.). Reliability and safety of operation of machines with rotor systems depends on their dynamic characteristics, which are largely determined by the mass and inertial characteristics of the rotor, the damping properties used in the manufacture of materials, the type and location of bearings, seals and dampers, design and working conditions, etc.

The main part of rotor system failures is caused by intense oscillations of rotors. Even with short-term excesses of the permissible limits of vibration levels, a loss of system performance occurs, rotors and attachments, as well as bearings and seals are destroyed.

A preliminary assessment of the dynamics of rotor systems requires the development of adequate mathematical models taking into account numerous parameters, many of which are very difficult to reliably evaluate. For example, the coefficients of stiffness, resistance and circulating forces of bearings, the coefficients of the forces of internal friction and the structure of its mathematical model, the interaction between the nodes of the machine and the environment.

An increase in the rotor speed leads to an increase in the oscillation amplitudes of the shafts, bearings, and the housing as a whole. Therefore, studies of the dynamic characteristics of rotor support systems, taking into account the influence of the damping properties of individual structural elements on them, the ability to predict the level of vibrations and increase the reliability of the structure as a whole are relevant tasks of modern engineering.

It was noted that a decrease in the level of vibrations in rotor systems is achieved by reducing the rigidity of the rotor support system and creating damping supports [1, 7]. As bearings of rotor systems, various types of bearings are used, which helps to avoid increasing the level of vibration and loss of stability during operation. The characteristic features of the three main types of bearings: rolling, sliding and magnetic, are described in [4–6].

Particular attention is currently being paid to non-contact supports: magnetic and electrical suspensions [6]. It was noted that at present, preference is given to active magnetic bearings (AMB) [7]. They represent a mechatronic system in which a stable state of the rotor is achieved by the action of magnetic forces of attraction. These forces act on the rotor from the side of the electromagnets, the electric current in which is set by the automatic control system on the signals of the sensors of the electronic control unit. Such control units can use analog and digital signal processing. Active magnetic bearings have good dynamic characteristics, reliability and high efficiency. In such supports, the constituent parts are sensor systems, a control unit, actuators of various types, with the help of which an impact on the control object is performed [8]. Mechatronic bearings include electromagnetic bearings (EMB) - a bearing assembly built on the basis of AMB, together with a control system. EMBs have various designs of the power electromagnetic part, differing in the way of creating the magnetic flux, power supply circuits of the field windings, the configuration of magnetic systems and the general layout of the nodes. These differences significantly affect the output characteristics of the devices.

3 Statement of the Problem

The object of the study is a prototype rotor on a combined passive-active magnetic suspension. Such rotors are used in diesel engines of vehicles, turbochargers, ultracentrifuges and other rotary machines. When calculating such structures, it is necessary to take into account a large number of factors, including design features, the type of support, which have a great influence on the dynamic characteristics.

The peculiarity of the rotor is the combined passive-active magnetic suspension, which is most effective in light high-speed rotary machines. Permanent magnetic bearings and active magnetic bearings have both advantages and disadvantages when used as supports for rotors. Therefore, when designing and creating rotary machines, combined magnetic suspensions are often used, using bearings of both types in different design variations.

The purpose of the study is to design a model of the rotor prototype, to determine the vibration characteristics of the rotor using numerical simulation, to build a laboratory installation and manufacture of a prototype rotor, to experimentally determine the dynamic characteristics of the rotor, to compare the experimental and calculated results to confirm the correctness of the developed model.

4 The Specifics of the Application of Information Technology for the Design of Rotor Systems

Among the most common packages today are CAD/CAM/CAE multi-purpose software packages. The mathematical basis of such packages is the finite element method, the most common and sufficiently versatile method for solving the problems of deformed solid mechanics, which is quite conveniently adapted to the description of geometry and boundary conditions of structures. CAD/CAM/CAE compatibility with leading vendor systems and a user-friendly interface have led to the widespread use of ANSYS, SolidWorks, Autodesk Inventor, COMPASS, CATIA, Altium Designer, OrCAD for scientific calculations and research.

Calculation software systems provide modeling of structures, creation of three-dimensional models that reproduce the properties of real objects, and their calculation.

Multipurpose functions of software systems are ensured by the fact that it coexists with many families of individual specialized programs that have many common functions, but the mathematical support of which is designed to solve individual classes of problems.

As a result of the simulation, the project was completed and a geometric model of the individual units of the laboratory installation was created, as well as the layout and assembly of the assembly into the product was completed (Fig. 1).

Fig. 1. Complete geometric model of the laboratory installation.

In Fig. 2 presents a scheme of combined magnetic and electromagnetic rotor suspension, where 1 is a rotor made of non-magnetic material, 2 are moving rings of radial magnetic bearings on permanent ring magnets located on the rotor, 3 - fixed rings of radial magnetic bearings), 4 - a disk of axial AMP made of ferromagnetic material located on the rotor, 5 and 6 - stators of axial AMP, 7 - windings of axial AMP, 8 - disk of the measuring system (BC), located on the rotor, 9 - sensor of the aircraft.

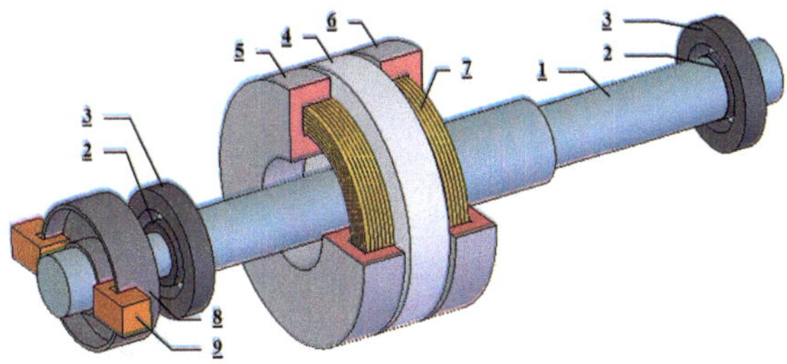

Fig. 2. Scheme of combined magnetic rotor suspension.

AMB ensures the stability of the rotor of the laboratory unit in the axial direction.

The rotor was accelerated by an electric motor with a change in the angular speed of rotation in the range from 0 to 3000 rot/min.

Annular permanent magnets and stator axial AMB with racks are fixed on the surfaces of the coupling of imposed restrictions on all degrees of freedom corresponding to linear displacements.

The forces of magnetic interaction between the parts of the magnetic bearings of the model are resilient elements with constant values of rigidity. The choice of stiffness values for each type of magnetic bearing is made on the basis of an analysis of the strength and stiffness characteristics.

As a result of the simulation, a finite element model of the rotor was constructed on a combined passive-active magnetic suspension (Fig. 3). The physical model consists of a solid-state rotor model with all attachments - a shaft with an optical sensor, an axial disk drive, a magnet rotor. When constructing a finite element model does not take into account the control and comparison devices, sensors, amplifiers.

The action of magnetic fields is modeled by the action of elastic elements.

The finite element model of the rotor is shown in Fig. 3.

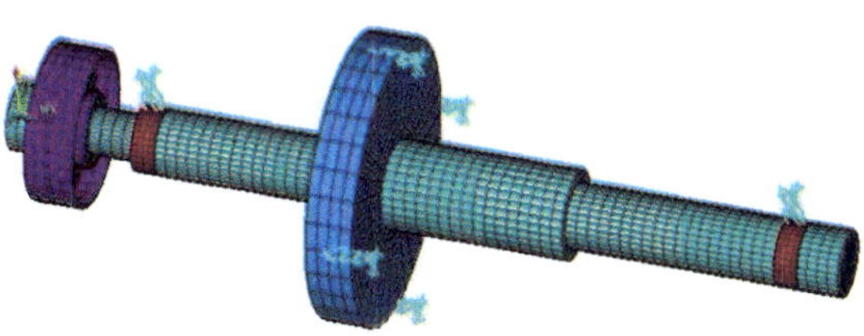

Fig. 3. The finite element model of the rotor.

The calculation of dynamic characteristics (stress-strain state, natural oscillation frequencies and critical rotational speeds) was per-formed with using the constructed model, to further compare the obtained data with the results of experimental studies of the developed prototype at the laboratory facility.

The experimental laboratory unit is shown in Fig. 4.

Fig. 4. Experimental laboratory installation.

5 Experimental Data

The combined passive-active magnetic rotor suspension of the experimental installation consists of two radial magnetic bearings on permanent ring magnets (PMB 1, 2) and one axial active magnetic bearing (AMB 3) of double action with two windings L1, 2 (Fig. 5).

PMB rings have axial magnetization and provide self-centering in the radial direction by repulsive forces. The stabilization of the position of the rotor in the axial direction is carried out by the forces of magnetic attraction, and its stability is provided by a control device (CD) with a negative feedback that implements the adopted control algorithm, that is, the algorithm of change of control voltages U1, 2. They are applied to the windings of the AMB 3 depending on the position of the rotor, which is determined by optical sensors. The centers of mass of the moving and fixed magnets are coincide at the central position of the rotor, which is connected by a fixed right Cartesian coordinate system OXYZ, and the disk is located in the middle between the stators.

Figure 5 presents a scheme of the model rotor in the radial PMBs and the axial AMB. The dimensions of the annular permanent magnets used in the laboratory installation are D_1 = 58, D_2 = 40, D_3 = 29, D_4 = 15, H = 10.5 mm with a rotor length of 0.3 m and a mass of 2.5 kg. The rings are made of the NdFeB alloy with a residual induction Br = 1.07 T and a coercive force Hc = 808000 A/m.

Figure 5 also shows the power characteristics of the PMB within the gaps, which are obtained by calculating the magnetic forces at different positions of the rotor magnet using the finite element method [10, 11]. The other installation parameters have the following values: $l_1 = 0.106$ m, $l_2 = 0.176$ m, $J_1 = 0.0107$ kg m^2, $J_3 = 0.0034$ kg m^2, $\delta_r = 5.5 \cdot 10^{-3}$ m, $\delta_a = 3 \cdot 10^{-3}$ m, $e = 6 \cdot 10^{-5}$ m, $\gamma = 0.003$ rad and $Q_{Rqj} = b_{qj} \partial q_j / \partial t$, $b_{qj} = 2.325$ kg/s, control voltages $u_{c1,2}$, given according to the accepted control law: $u_{c2,1} = (u_{\max} - 2u_{\min}) z_3^2 / (2\delta_a^2) \pm \pm u_{\max} z_3 / (2\delta_a) + u_{\min}$, at $r_{c1} = r_{c2} = 5\ \Omega$, $U_0 = 24$ V, $k_i = 2/5$, $u_{\max} = k_i u_{\min}$.

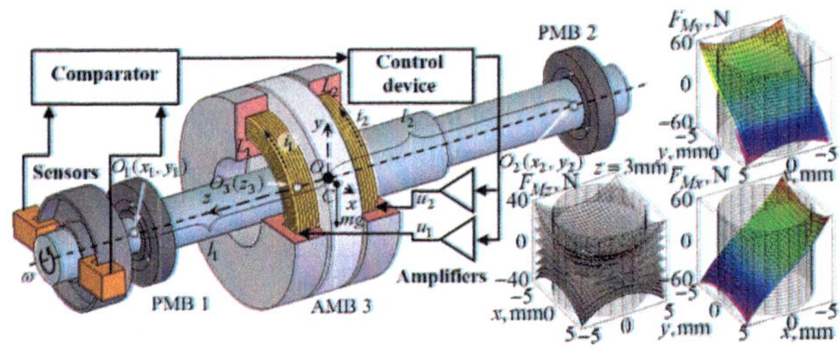

Fig. 5. Calculation scheme of the model rotor in the radial PMBs and the axial AMB and power characteristics of the PMB within the gaps.

Figure 6 shows the results of experiments in the form of amplitude-frequency characteristics of the rotor.

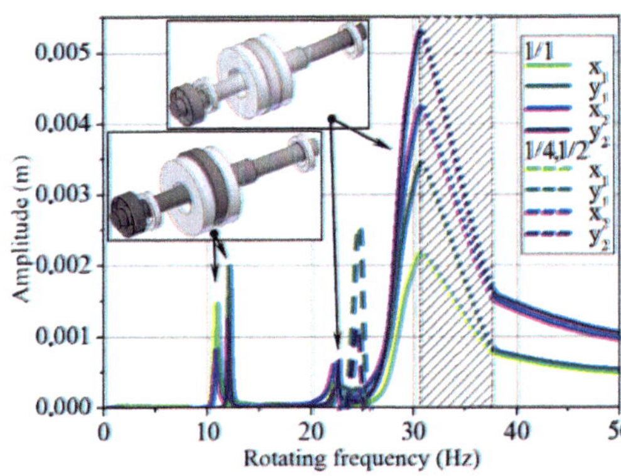

a)

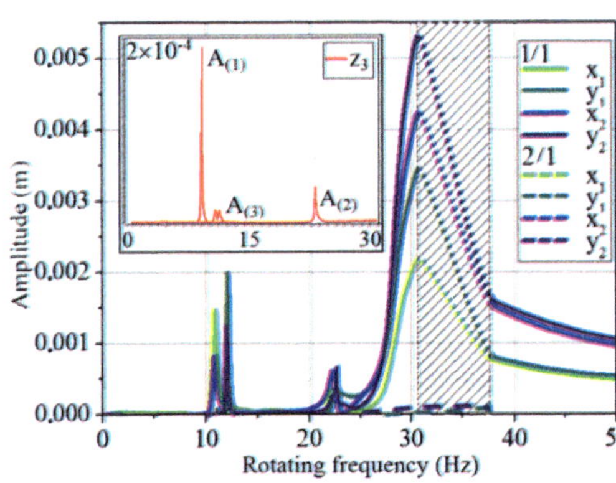

b)

Fig. 6. Experimental amplitude-frequency characteristic of the rotor (of a fundamental 1/1 harmonics), as well as a dependence of amplitudes of subharmonics 1/4, 1/2 (a) and a superharmonics 2/1 (b) on a rotational speed.

All five degrees of freedom of the rotor in the MBs were controlled – horizontal and vertical displacements of the centers of the radial supporting regions (the left one x_1, y_1 and the right one x_2, y_2) and an axial displacement (z_3). An acceleration of the rotor was carried out up to 3000 rpm (50 Hz). The region of maximum amplitudes with a breakdown of oscillations in the range of ~1850–2280 rpm (~31–38 Hz) is marked by the dashed lines in Fig. 6. In addition, an analysis of the results made it possible to detect various nonlinear phenomena in the system. Sub- (Fig. 6a) and superharmonic (Fig. 6b) oscillations, multiple sub- and super-resonances, oscillation breaking with a transition from one stable mode to another stable one are observed along with harmonic oscillations with the frequency of an excitation (rotation). A relationship between radial and axial oscillations was also found (Fig. 6b).

The results of the finite element model calculation, the experimental installation data and the calculation precision are presented in Tables 1 and 2.

Table 1. Natural frequencies of oscillation of the rotor.

	Results of calculations	Experimental data	Precision, %
Natural frequencies of oscillation of the rotor, (Hz)	8,8599	8,87	0,11
	10,848	10,5	7,0
	12,183	12,0	1,5
	23,897	22,5	5,8
	31,854	33,0	3,6

Table 2. Critical rotor speed.

	Results of calculations	Experimental data	Precision %
Critical rotor speeds, (2π rad/s)	9,22	9,2	0,22
	11,289	10,75	4,77
	12,182	11,9	2,32
	23,74	22,6	4,8
	31,596	30,9	2,2

6 Conclusion

Based on the use of information technology for design, a laboratory installation of a complete magnetic suspension of a horizontal rotor according to a combined magnetic-electromagnetic type was proposed and physically implemented. It uses two passive magnetic bearings created using two annular permanent magnets with axial magnetization as two radial bearings. The axial support is an active magnetic bearing with stators in the form of armored cores. The functionality of this AMB uses a discrete method and an analog control system. This laboratory installation is a prototype of the magnetic suspension of small engine rotors. It can be used for any other rotor systems and machines, for example, as well as pumps, turbo-expanders or expander compressor units of small and medium sizes.

The creation of a laboratory installation was made possible thanks to a preliminary detailed analytical and numerical analysis of linear and nonlinear rotor dynamics [5, 6]. These project studies allowed us to choose the optimal (rational) parameters and characteristics of individual nodes – PMB, AMB, system and control algorithm for a specific horizontal rotor.

References

1. Schweitzer, G., Maslen, E.H.: Magnetic Bearings: Theory, Design, and Application to Rotating Machinery, pp. 1–24. Springer, Berlin (2009)
2. Kimman, M.H., Langen, H.H., Munnig Schmidt, R.H.: A miniature milling spindle with active magnetic bearings. Mechatronics **20**(2), 224–235 (2010)
3. Sotelo, G.G., et al.: Experimental and theoretical levitation forces in a superconducting bearing for a RealScale Maglev system. IEEE Trans. Appl. Supercond. **21**(5), 3532–3540 (2011)
4. Provenza, A.J., Montague, G.T., Jansen, M.J., Palazzolo, A.B., Jansen, R.H.: High temperature characterization of a radial magnetic bearing for turbomachinery. In: Proceedings of the ASME/IGTI Turbo Expo, Atlanta, GA, 16–19 June (2003)
5. Falkowski, K., Henzel, M.: High efficiency radial passive magnetic bearing. Solid State Phenom. **164**, 360–365 (2010)
6. Realisation and test of a passive magnetic bearing. https://documents.epfl.ch/users/p/pa/parafita/www. Accessed 21 Sept 2019
7. Bleuler, H.: Magnetic levitation: a challenge for control design in mechatronics. Toshiba Chair Intell. Mechatron. **44**(12), 578–583 (2011)
8. Wu, A., Cai, Z.: Model-based control of active tilting-pad bearings. IEEE/ASME Trans. Mechatron. **12**, 689–695 (2008)
9. Detoni, J.G., Impinna, F., Amati, N., Tonoli, A., Piccolo, M.P., Genta, G.: Stability of a 4 degree of freedom rotor on electrodynamic passive magnetic bearings. In: 14th International Symposium on Magnetic Bearings, Linz, Austria, 11–14 August (2014)
10. Martynenko, G.: Accounting for an interconnection of electrical, magnetic and mechanical processes in modeling the dynamics of turbomachines rotors in passive and controlled active magnetic bearings. In: 2018 IEEE 3rd International Conference on Intelligent Energy and Power Systems (IEPS 2018), pp. 326–331. IEEE, Kharkiv, Ukraine (2018). https://doi.org/10.1109/ieps.2018.8559518
11. Martynenko, G.: Resonance mode detuning in rotor systems employing active and passive magnetic bearings with controlled stiffness. Int. J. Automot. Mech. Eng. **13**(2), 3293–3308 (2016). https://doi.org/10.15282/ijame.13.2.2016.2.0274
12. Martynenko, G.: The interrelated modelling method of the nonlinear dynamics of rigid rotors in passive and active magnetic bearings. East.-Eur. J. Enterp. Technol. Appl. Phys. **2/5**(80), 4–13 (2016). https://doi.org/10.15587/1729-4061.2016.65440

Thermal-Stress State of the Piston During Transient Diesel Operation, Synthesis of the Piston Profile

Nguyen Van Duong(✉), O. Bilohub(✉), and Ye. Martseniuk(✉)

N. E., Zhukovsky National Aerospace University "KhAI", Kharkiv, Ukraine
dongcomaybay@gmail.com, av.belogub@gmail.com,
ev.martsen@gmail.com

Abstract. This paper deals with simulating the thermal and stress states of the piston and cylinder of a two-stroke diesel engine type D100 for a locomotive using two different methods. The methods are based on the use of various models of piston-cylinder arrangement: one of them uses assembly of the piston-cylinder arrangement and includes all components, but the other uses two models separately: the cylinder and the piston and rings assembly. It was shown that for both methods, the piston temperature fields differ slightly both in zone of the combustion chamber and on the inner surface, and the maximum difference was 8 K in the zone of the first piston ring, but the difference in the cylinder temperature fields is significant, especially in the middle section where the difference reaches 35 K. Transient thermal-stress states of the piston were determined for three programs of engine starting from the cold state (with various initial temperatures equal to 20 °C; 0 °C; −20 °C) and heating it up to its maximum operating mode, and for one program with cooling from maximum mode to idling, every time with stepped loading. Warming up of oil and anti-freeze were taken into account. It is shown that temperatures and stresses reach their peaks and then drop in some zones of the piston. The stress reaches a maximum level of 380 MPa at the center of the piston surface from the combustion chamber side when heated with an initial temperature −20 °C, which is two times more than the stress at steady-state maximum operating mode (192 MPa). The profile of the piston side surface was synthesized taking into account the heating rate. The piston design with different materials of the piston head – VCh60-2, 25H2G2FL, 12DH1MFL – was proposed for D49 and D80 engines. The mean temperature of the piston hottest zone drops by about 40 K in proposed design. The stress also decreased.

Keywords: Diesel · Piston · Cylinder · Transient mode · Heat transfer coefficient · Temperature · Stress

1 Introduction

The resource requirements for the pistons of modern internal combustion engines are constantly increasing. Two factors most strongly affect the durability of the piston are the load and the wear. In the previous works [1, 2] the authors of this study proposed an

M. Nechyporuk et al. (Eds.): *Integrated Computer Technologies in Mechanical Engineering*, AISC 1113, pp. 310–324, 2020.
https://doi.org/10.1007/978-3-030-37618-5_27

alternative piston design for the D100 engine and the technique of calculating the boundary conditions for modeling the stress state of the piston. This work continues the research started. The study analyzed the thermal and stress state of the piston-cylinder arrangement (PCA) parts during accelerating and unloading engine. According to the results of studies [3], the stress in the piston under transient conditions can reaches a value of 2…3 times more in comparison with the steady-state maximum mode. This has negative affect on piston strength and lifetime. The issues of working capacity and lifetime of the piston are solved in different ways, for example, by controlling the fuel supply, but first of all, by choosing (optimizing) the design and choosing the material. Important design parameters include the profile of the side surface. A correctly created profile should ensure the absence of contact between very heated piston head and sleeve at all engine operating modes [4–6] as well as the optimal contact surface of the piston-sleeve friction pair along the skirt [7–9]. On the other hand, it is known that in ICE the temperature of the combustion chamber (CC) surface affects the indicator parameters, the temperature of the working fluid and the heat transfer coefficient from the working fluid to the walls of the CC. At different temperatures of the CC surface, the equivalent boundary conditions (BC) will be different.

2 Research Goals

1. To develop the algorithm for calculating the thermal boundary conditions on the surfaces of the piston and cylinder during transient diesel operation.
2. To determine temperature, strain and stress in piston of a D100 type two-stroke engine in transient modes.
3. To optimize the profile of the piston side surface, taking into account its operation in pair with the cylinder liner.
4. To propose an alternative piston design for D49 and D80 diesel engines operated at Ukrzaliznytsia.

3 Analysis of the Transient Thermal and Stress States of the Piston. Optimization of the Piston Side Surface for the Engine Type D100

3.1 Statement of the Problem

The four transients were considered, three of which – from cold engine with initial component temperatures of −20 °C, 0 °C and +20 °C to maximum operating mode (hereinafter – variants 1, 2, 3), and variant 4 – from maximum mode to idle. The type of loading is stepped. Figure 1 shows the piston geometric parameters of diesel engine type D100 at 20 °C, that determine the initial profile. Here is indicated: R_p – the piston radius, mm; h_p – the piston height, beginning from CC, mm; θ – the angular coordinate with beginning ($\theta = 0°$) from wrist pin axis, deg.

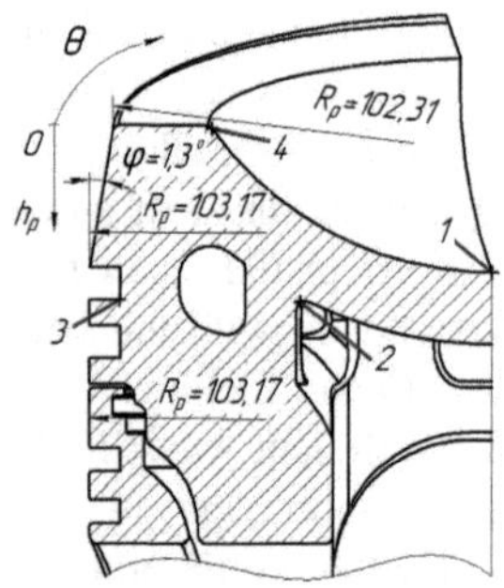

Fig. 1. The geometric parameters of the piston

Based on the researches carried-out for piston of a diesel type D100, the design of alternative piston for diesel engines of the D49 and D80 families was proposed and investigated.

3.2 Materials

All parts of piston-cylinder arrangement of D100 type engine are made of inoculated or high-strength cast iron with spherical graphite that has hardness HB = 207…255 [10]. The exception is the piston skirt, which is made of aluminum alloy AK4. For modeling the piston head, cylinder, rings and wrist pin, the VCh60-2 cast iron was chosen [11, 12], physical properties of which are presented in Table 1.

Table 1. The physical properties of VCh60-2 cast iron at different temperatures

Temperature T, °C	Thermal conductivity λ, W/(m K)	Specific heat C_p, J/(kg K)	Thermal expansion coefficient α, 1/K·10^{-6}	Elasticity modulus E, MPa
20	54.5	390	10	180000
370	46.2	505	11.5	163650
537	42.4	523	12.25	152410
650	40.3	568	12.45	142500

The oil used for the 10D100 engine is M14V2. Its thermophysical properties – density ρ, specific heat C_p, thermal conductivity λ, kinematic viscosity ν – can be calculated using the following formulas:

$$\rho = \exp(6.818 - 0.000718 \cdot T), \text{kg/m}^3; \tag{1}$$

$$C_p = \exp(7.475 + 0.002272 \cdot T), \text{ J/(kg K)}; \tag{2}$$

$$\lambda = 0.1427 \cdot \exp(-0.0009971 \cdot T), \text{ W/(m K)}; \tag{3}$$

$$\upsilon = 2.29042 \cdot T^{-2.60745},\ \mathrm{m^2/s} \tag{4}$$

As a coolant, the antifreeze «−65 °C» was chosen, specific heat of which is determined by formula (5)

$$C_p = 7.6106 \cdot T + 769,\ \mathrm{J/(kg\,K)}. \tag{5}$$

3.3 Algorithm for Calculating Thermal BC for Transient Engine Operating Modes

The problem of determining the heat exchange boundary conditions on the surfaces of diesel combustion chamber for transient operating modes is associated with following:

- significant influence of temperature on the properties of both the structural elements that limit volume of the combustion chamber and the cooling liquids (oil, antifreeze);
- time-varying heat transfer boundaries;
- time-varying parameters of gas in combustion chamber.

To correctly determine the thermal loads in the specified operating conditions, it is necessary to ensure the thermal balance of the system "gas – combustion chamber parts – coolants". Since the task is unsteady, the heat balance must be achieved at each calculated time, given the variable amount of heat generated and removed from the system.

Heat Transferring to Oil and Cooling Liquid. The change in the temperature of oil and antifreeze during the *i*-th time interval depends on the amount of heat ΔQ_i transferred during this interval and can be determined by the formula:

$$\Delta T_i = \Delta Q_i / (mC_i), \tag{6}$$

where, m is the mass of oil or antifreeze;

C_i is the specific heat of oil or antifreeze at *i*-th time interval.
Then the temperature at the next solving time interval will be:

$$T_{i+1} = T_i + \Delta T_i, \tag{7}$$

where, Ti is the temperature at the current i-th interval.

Heat Transferring on the Surfaces of Combustion Chamber. Due to the piston movement the inner surface of the cylinder changes as well as the state of the working fluid, therefore the following technique [13] was applied for calculating the BC for a multi-zone surface. The Fig. 2 shows the principal scheme of heat exchange process that occurs at *i*-th time moment on multi-zone boundary surface *A* with time-varied BC – the local heat transfer coefficients α_{Aji} and local bulk temperatures T_{Aji}, and on single-zone surface *B* with constant BC α_B, T_B.

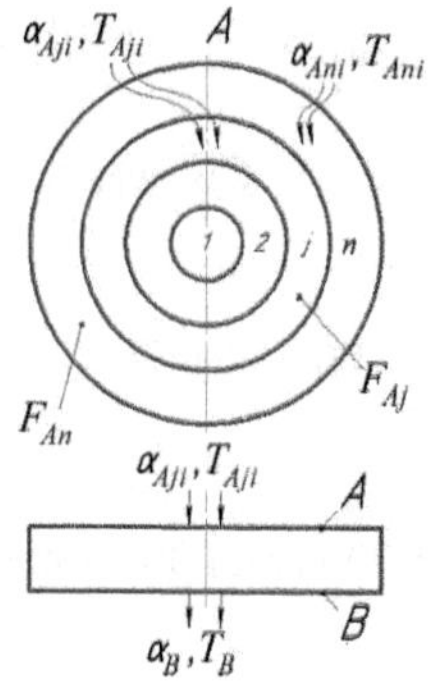

Fig. 2. The scheme of solving the BC

The surface A with total area F_A is divided on n number of zones A_j with areas F_{Aj}. The areas F_A and F_{Aj} are not constant and changed with time. The total amount of heat Q_A transferred through surface A equal to

$$Q_A = \sum_{j=1}^{n}\sum_{i=1}^{m} \alpha_{Aji} \cdot F_{Aji} \cdot \left(T_{Aji} - \overline{T}_{Aj}\right) \cdot \Delta t = \overline{\alpha}_A \cdot F_A \cdot \left(\overline{T}_{A.e} - \overline{T}_A\right) \cdot \tau, \qquad (8)$$

and the local amount of heat Q_{Aj}, transferred through j-th zone of surface A, equal to

$$Q_{Aj} = \sum_{i=1}^{m} \alpha_{Aji} \cdot F_{Aji} \cdot \left(T_{Aji} - \overline{T}_{Aj}\right) \cdot \Delta t = \overline{\alpha}_{Aj} \cdot F_{Aj} \cdot \left(\overline{T}_{Aj.e} - \overline{T}_{Aj}\right) \cdot \tau, \qquad (9)$$

where, m is a number of time intervals i; τ is the total time of described process;

F_{Aji} is the current area of the j-th zone on the surface A at the time i;
$\overline{T}_A$, $\overline{T}_{Aj}$ is the mean temperature of the surface A and j-th zone of surface A during the time τ;
$\overline{\alpha}_A$ and $\overline{\alpha}_{Aj}$ is the mean value of the heat transfer coefficient at surface A and j-th zone of surface A during the time τ;
$\overline{T}_{A.e}$ and $\overline{T}_{Aj.e}$ is the equivalent bulk temperature of the ambient acts on surface A and j-th zone of surface A during time τ;
Δt is the time step.

Then the mean equivalent BC for surface A and its zones can be determined by next equations:

$$\overline{\alpha}_A = \frac{Q_A}{\sum_{j=1}^{m}\sum_{i=1}^{n} F_{Aji} \cdot \left(T_{Aji} - \overline{T}_{Aj}\right) \cdot dt}; \qquad (10)$$

$$\overline{\alpha}_{Aj} = \frac{Q_{Aj}}{\sum_{i=1}^{n} F_{Aji} \cdot \left(T_{Aji} - \overline{T}_{Aj}\right) \cdot dt}; \qquad (11)$$

$$\overline{T}_{A \cdot e} = \frac{Q_A}{\overline{\alpha_A} \cdot F_A \cdot \tau} + \overline{T}_A; \tag{12}$$

$$\overline{T}_{Aj.e} = \frac{Q_{Aj}}{\overline{\alpha}_{Aj} \cdot F_{Aj} \cdot \tau} + \overline{T}_{Aj}. \tag{13}$$

The Algorithm. The special algorithm was developed to calculate the boundary conditions for heat exchange surfaces the schematic diagram of which is presented at Fig. 3.

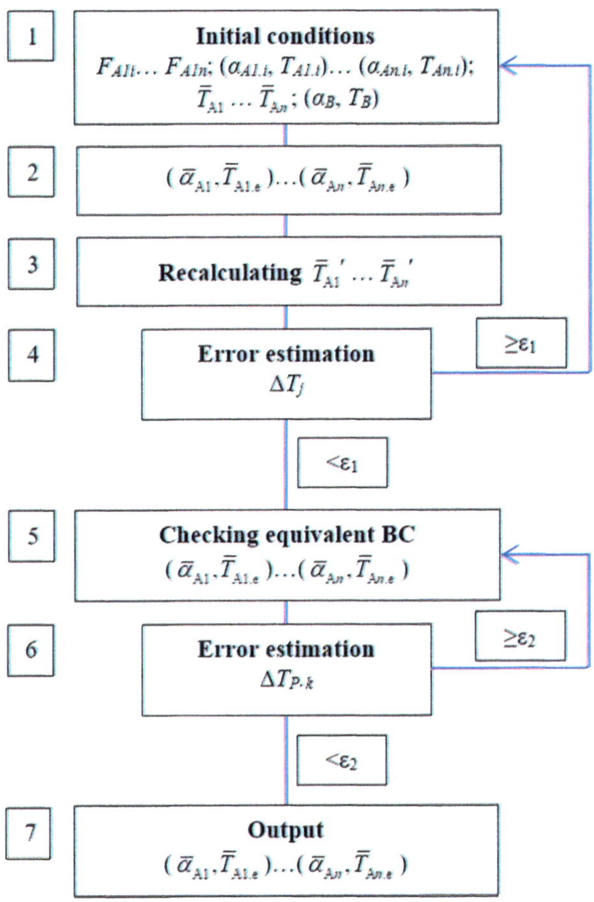

Fig. 3. Algorithm of determining an equivalent BC for A surface

The initiating of the algorithm is begins from defying areas F_{A1i}… F_{A1n} of all *j*-th zones (from 1 to *n*) and the corresponding time-varied boundary conditions ($\alpha_{A1.i}$, $T_{A1.i}$) … ($\alpha_{An.i}$, $T_{An.i}$) at the current time moment *i*, also the mean surface temperatures

$\overline{T}_{A1} \ldots \overline{T}_{An}$ for all j-th zones and BC (α_B, T_B) for surface B – block 1. The block 2 calculates the equivalent BC, $(\overline{\alpha}_{A1}, \overline{T}_{A1.e}) \ldots (\overline{\alpha}_{An}, \overline{T}_{An.e})$, using (9), (11), (13). The block 3 is associated with CAE module that provides thermal analysis of structure according to equivalent BC received from block 2, and then recalculates the surfaces mean temperatures $\overline{T}'_{A1} \ldots \overline{T}'_{An}$. The next is block 4 that compares the levels of the recalculated temperature $\overline{T}'_{Aj}$ with the initially defined $\overline{T}_{Aj}$ at each zone. If the case of temperature differences $\Delta T_j = \overline{T}_{Aj} - \overline{T}'_{Aj}$ exceeding the accuracy level, the algorithm redefines the initial data in block 1 and repeats. The blocks 5 and 6 are for checking the correctness of equivalent BC. The procedure is carried out in to steps. The equivalent BC $(\overline{\alpha}_{Aj}, \overline{T}_{Aj\cdot e})$ are used in steady-state thermal analysis of PCA to calculate the initial condition for further transient thermal analysis with time-varied BC ($\alpha_{Aj.i}$, $T_{Aj.i}$) corresponded to steady-state operating mode of the engine. The error estimated as temperature differences in local points $P.1 \ldots P.k$ of the piston surface at start (initial state) and end (time τ) of the transient analysis – $\Delta T_{P.k} = T_{P.k\ start} - T_{P.k\ end}$. The block 7 gives the final equivalent boundary conditions that ensure needed accuracy of thermal state for PCA components.

For the piston internal surface B, cooled by oil, the thermal loads are also variable over time. However, modeling the processes of heat transferring through this surface in CAE module, as for a region divided by a n number of zones with variable heat exchange boundaries, is extremely difficult due to the complex configuration of the piston crown. Therefore, it is recommended to complete the determination of equivalent BC on surface B after block 4.

This algorithm can be also used when the time-varied BC with variable heat exchange boundaries are defined on both the surface of combustion chamber and the surface of piston crown. In this case, the determining of BCs must be carried out in parallel for both surfaces with matching the results.

3.4 Simulating the Steady Operating Mode of Diesel Engine

This task is solved by high level CAE software.

Two models are investigated in the work: the first one is the assembly that consists of all parts of PCA (Fig. 4); the second one is presented by two separate sub-models – piston and rings assembly and the cylinder (Fig. 5). For the first model, the calculated surfaces are the piston surface from side of CC and the inner surface of the cylinder. For the second model, the calculated surfaces are the piston surface from side of CC (one surface), the rings surfaces from the side of cylinder (four surfaces) and the inner surface of the cylinder divided on 10 zones.

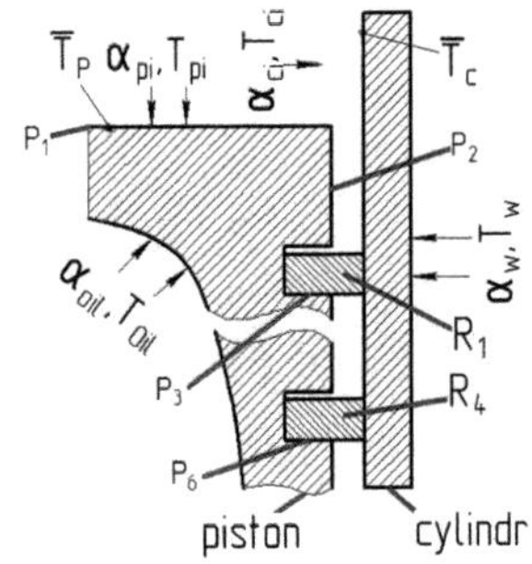

Fig. 4. The model 1

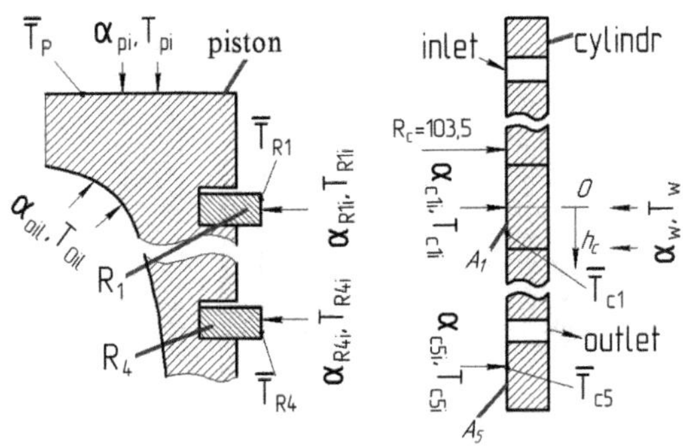

Fig. 5. The model 2

The time-varied BC in combustion chamber on the piston surfaces (α_{pi}, T_{pi}), on the rings (α_{R1i}, T_{R1i}) … (α_{R4i}, T_{R4i}) and on the cylinder surface (α_{c1i}, T_{c1i}) … (α_{c5i}, T_{c5i}) are determined in result of preliminary workflow calculation [2].

According to [14] and based on workflow calculation [2] the mean emission of heat to the cooling system at maximum mode is defined as Q_w = 1000 kW, and to the oil system – Q_{oil} = 420 kW. The mass of the antifreeze m_w = 1250 kg, and the oil mass m_{oil} = 1250 kg.

The heat transfer coefficient from the cylinder wall to the antifreeze is 18000 W/(m^2 K) [15]. The heat transfer coefficient between the oil and the inner surface of the piston head is calculated as described in [13].

The thermal analysis results for the PCA components are presented in Table 2 and on the Figs. 6 and 7. From Table 2 and Fig. 6 it is seen, that the temperature fields on the piston head and the mean temperatures of the piston surface $\overline{T}_p$ from CC side differ slightly. The maximum difference is 8 K in the grove for 1-st ring (point P_3, Fig. 4) – 414 K in model 1 versus 422 K in model 2. The difference in temperatures of the piston skirt is less – only 4 K at point P_5 and 3 K at point P_6. However, the temperature field of the cylinder differ significantly (Fig. 7). In middle zone this difference reaches 35 K.

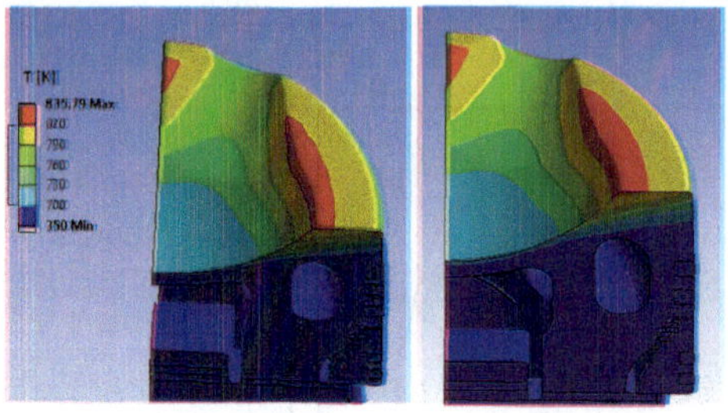

a b

Fig. 6. The piston temperature field: (a) model 1; (b) model 2

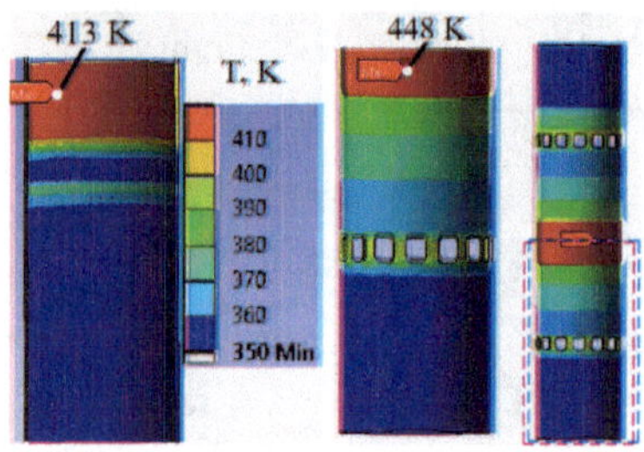

a b

Fig. 7. The cylinder temperature field: (a) model 1; (b) model 2

The other difference between the models is next. Use of the model 2 it is necessary to determine and redefine the BCs for (n + m + 1) surfaces while, as for model 1 – only one surface. So, volume of work with model 2 is more. Summarizing the noted above, it is recommended to use model 1 for thermal stress analysis of the piston.

Table 2. The piston temperature in different points (Figs. 4 and 5)

The temperature of piston, K		
Point or zone number	Model 1	Model 2
$\overline{T}_p$	775	776
P_1	710	712
P_2	668	670
P_3	414	422
P_4	376	382
P_5	366	370
P_6	364	367

The amount of heat transferred at maximum mode to inner surface of the cylinder from the ring – is 634 J/cycle, from combustion products – is 6260 J/cycle. In view of small heat transferring through the rings (about 10% according to Table 2) and usability for analysis of heating rate the model 2 is chosen for the cylinder.

3.5 Analysis of the Transient Modes

Using the method described above, the equivalent time-dependent BC for the piston were obtained for heating variants 1…3, noted in (Sect. 3.1). The Table 3 shows results for variant 1, where: Q_p is the amount of heat transferred to the piston per 1 cycle (1 revolution of the crankshaft); $\overline{\alpha}_p$, $\overline{T}_{pe}$ are the equivalent heat transfer coefficient and the mean temperature.

Table 3. Varying the equivalent boundary conditions at the piston surface during transient mode

Mean temperature of the piston and amount of heat transferred to the piston				
τ, *sec*	$\overline{T}_p$, *K*	$\overline{Q}_p$, *J*	$\overline{\alpha}_p$, W/(m² K)	$\overline{T}_{\text{э}p}$, *K*
0	253	2700	1850	870
80	775	925	1890	980
150	815	771	1892	987
500	775	925	1890	980
1000	775	925	1890	980

Figures 8 and 9 show the graphs of temperature and stress varying in point P.1 (see Fig. 1) during heating up. The analysis is carried-out taking into account the acting pressure of 10 MPa and lateral force of 22800 N. Figures 10 and 11 show the loading variants 1, 2, 3, 4 according to (Sect. 3.1).

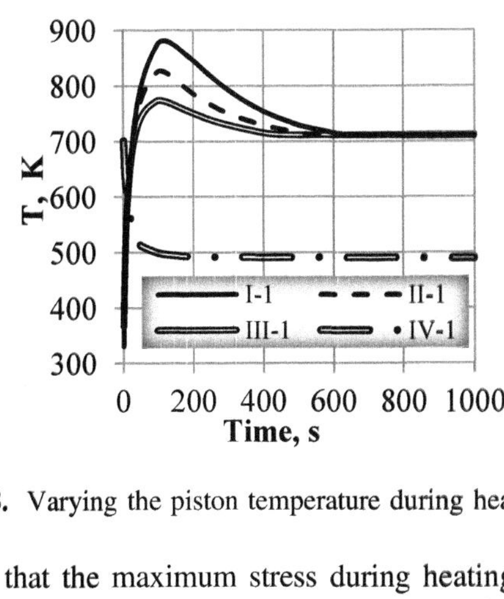

Fig. 8. Varying the piston temperature during heating up

The Fig. 9 shows that the maximum stress during heating up at two times more than in steady state mode. In variant I the maximum stress in point P.1–380 MPa (at

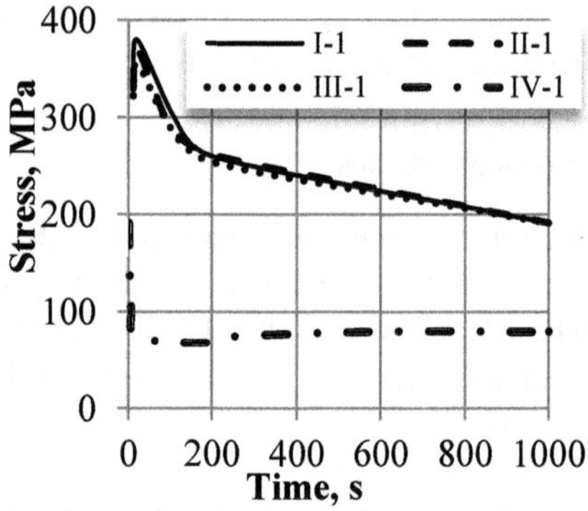

Fig. 9. Varying the stress in the piston during heating up

τ = 20 s) when in steady-state it is no more 192 MPa. Variant IV does not generates the high level of stress.

The additional analysis was carried out to examine the effect of transition time from initial conditions (cold engine) to maximum operating mode on stress owerswing. For this, the three additional variants *V*, *VI*, *VII*, were simulated with transition time from mode to mode equal to 50, 100 and 200 s respectively. The maximum stress in piston head drops (Fig. 10) from 380 MPa in variant *I* to 300 MPa in variant VII.

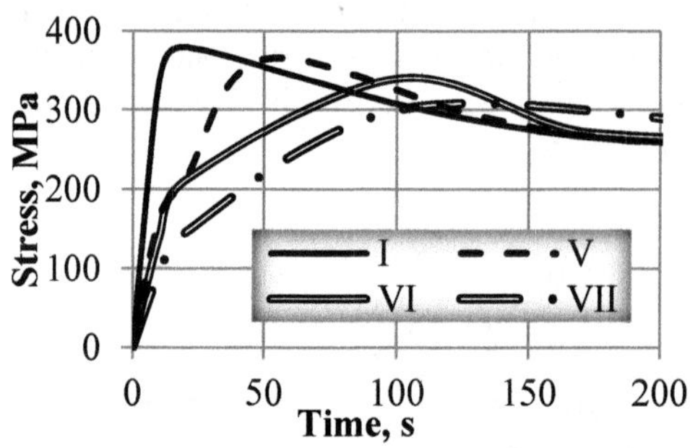

Fig. 10. Stress varying in piston head at different transition time

3.6 Synthesis of the Piston Profile

The modeling of thermal, stress and strain states under the heating by variants 1…3 allowed us to synthesize the profile of the piston outer surface and the cylinder inner surface. In each variant the piston-to-cylinder clearance is exanimated. Figure 11 shows the profiles of the piston and cylinder liner. It is shown that at τ = 200 s the points are appeared on the piston surface at distance h_p = 34 mm. These points penetrate into cylinder at TDC position. It is necessary to decrease diameter in this cross-section and change in the surface inclination angle φ of the piston head. After corresponding changes of the piston geometric parameters the dangerous zone is not

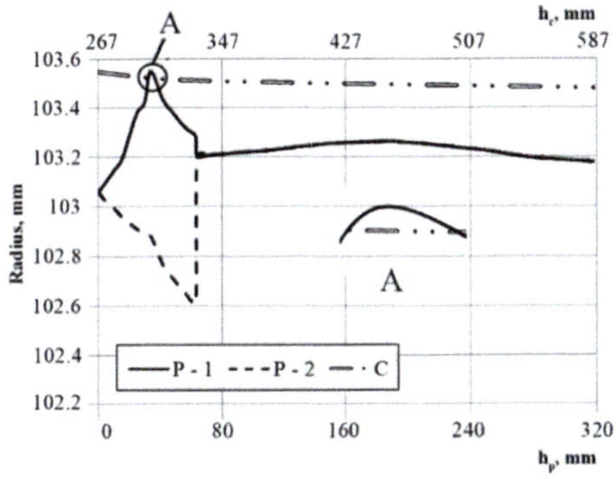

Fig. 11. Varying the piston and cylinder liner profiles in cross-section normal to the wrist pin axis. Heating time – 200 s

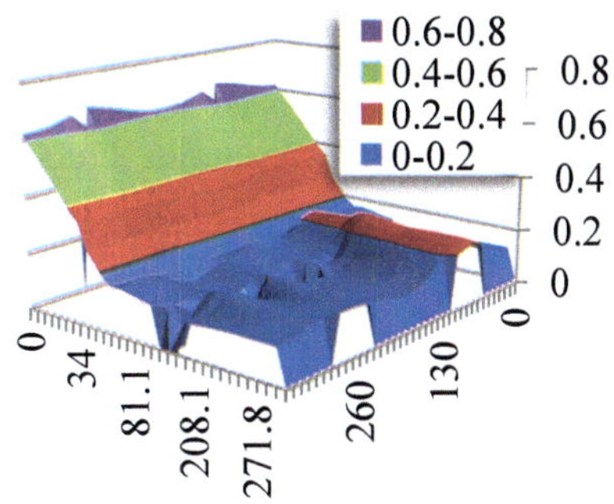

Fig. 12. Varying the piston profile at maximum operating mode

appeared and the minimum piston-to-cylinder clearance no less 0.13 mm at angle $\theta = 90°$ and operating time $\tau = 1000$ s. Figure 12 shows the deformation of piston (the profile) at maximum operating mode.

4 Development of the Piston Design for Engine Type D49 and D80

The Fig. 13a, b shows the design of the original and proposed piston respectively. In proposed design, the outer surfaces are kept without changes, but inner zone was modified and circulated oil cooling was replaced by jet.

In line with [14], the piston head of prototype is manufactured from steel 20Kh3MVF and the piston skirt – from aluminum alloy AK4. The different materials for piston head were examined in proposed design. These materials are VCh60-2, 25Kh2G2FL, 12DKh1MFL. The thermal-stress state of the piston was determined by [13], including workflow calculation, choosing the α-equation [2], determining the thermal BC. Table 4 presents the BC in zones shown on Fig. 13.

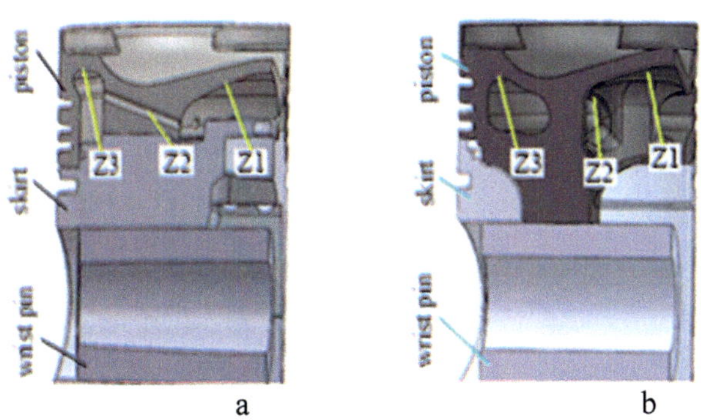

Fig. 13. The D80s pistons design: a – original; b – proposed

The pistons stress states are presented on Fig. 14 and in Table 5. It is shown that proposed design is 2 kg lighter than the original, if materials 25Kh2G2FL, 12DKh1MFL are used, and 4 kg lighter, if material VCh60-2 is used. The proposed design from cast iron is colder, the mean temperature is about 40 K less (675…685 K versus 715 K). As a result, the stress in proposed design from material VCh60-2 was dropped about in 3 times (156 MPa versus 475 MPa).

Table 4. The boundary conditions in chosen zones

Zone	T_{oil}, K	Original	Proposed
		α_{oil}, W/(m^2 K)	α_{oil} α_{oil}, W/(m^2 K)
Z1	350	530	1000
Z2	350	540	550
Z3	350	330	550

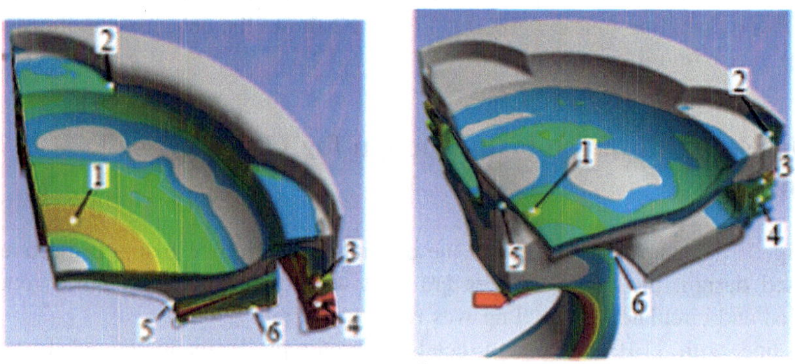

a b

Fig. 14. Stress fields: a – original; b – proposed

Table 5. Comparing the thermal-stress states of the original and proposed piston design

Point num.	Original design		Proposed design			
	20Kh3MVF		VCh60-2		25Ch2G2FL	
	T, K	σ, MPa	T, K	σ, MPa	T, K	σ, MPa
1	710	142	650	112	660	134
2	740	110	705	85	710	100
3	430	475	470	156	460	190
4	400	305	450	192	440	232
5	475	257	530	126	530	117
6	460	685	400	220	400	245

5 Conclusions

The paper deals with choosing the model for calculating equivalent BC at diesel transient operating mode for piston-cylinder arrangement.

The analysis results show that both the temperature and the stress of the piston reach their maximum level and then decreased when engine heated up. In variant *I* in central point of piston the temperature reaches level of 885 K (at 110 s) and the stress is 380 MPa (at 20 s), and then these levels decrease up to 710 K and 192 MPa.

The transition time has significant effect on maximum level of stress: 380 MPa at stepped loading, while only 300 MPa when transition time is 200 s.

The piston design was modified to avoid penetration of piston into cylinder in high temperature zones at all operating modes.

The proposed piston design for D80 engine is colder than the original.

Acknowledgment. The authors are grateful to Professors Andrei Marchenko and Vladimir Pylev (National Technical University «Kharkiv Polytechnic Institute») for supporting this work and helping to provide factual data on heat transferring for D80 diesel engines.

References

1. Bilohub, O.V., Nguyen, V.D., Linkov, O.Yu., Kravchenko, S.A.: Development of the "lightweight" piston design for D100 diesel engines. Intern. Combust. Engines **1**, 50–55 (2016)
2. Van Duong, N., Bilohub, O.V.: Calculation of the heat transfer process in a D-100 diesel engine using the well-known & α-formulas. Intern. Combust. Engines **1**, 14–20 (2018)
3. Mordvintseva, I.A.: CAE-modeling of transient thermal loads and life-time strength of a high-speed diesel pistons. Ph.D. diss. Kharkov, p. 167 (2018)
4. Pearson, S.R., Shipway, P.H., Abere, J.O., Hewitt, R.A.A.: The effect of temperature on wear and friction of a high strength steel in fretting. Wear **303**, 1–2, 622–631 (2013)
5. Rodriguez, J., Martin, A., Llorca, J.: Modeling the effect of temperature on the wear resistance of metals reinforced with ceramic particles. Acta Mater. **48**, 993–1003 (2000)
6. Liu, R., Wu, X.J., Kapoor, S., Yao, M.X., Collier, R.: Effects of temperature on the hardness and wear resistance of high-tungsten stellite alloys. Metall. Mater. Trans. A **46**, 587–599 (2015)
7. Moskalenko, I.N., Dotsenko, V.N., Bilohub, O.V.: Overview of methods for profiling pistons skirts of ICE. Intern. Combust. Engines **2**, 75–81 (2013)
8. Rozhdestvensky, Yu.V.: Computer Simulation of the Dynamics of "Piston-Cylinder" Tribological Conjugation: Training Manual, p. 50. Publishing House of SUSU, Chelyabinsk (2009)
9. Kwangsoo, K., Paras, S., Takiguchi, M., Aoki, S.: A study of friction and lubrication behavior for gasoline piston skirt profile concepts. SAE World Congress & Exhibition (2009)
10. Avrunin, A.G.: Locomotive Diesel Engines 2D100 and 10D100, p. 320. Transportation Public, Moscow (1970)
11. Girshovich, N.G.: Handbook of Iron Casting, p. 758. Engineering Publication, Leningrad, Leningrad (1978)

12. Bukhmirov, V.V., Rakutina, D.V., Sonyshkova, Yu.S.: Reference Materials for Solving Problems on the Course "Heat and Mass Transfer", p. 102. GOU VPO "V.I. Lenin Ivanovo State Energy University", Ivanovo (2009)
13. Nguyen, V.D., Bilohub, O.V.: A method for predicting temperature and stress fields of ICEs piston. Systems and means of transport. Problems of operation and diagnostics. Kherson, KhSMA, 9–27 (2019)
14. Dorofeev, V.M.: Locomotive Diesel Engines of D49 Family. Design, Maintenance, Repair, p. 380. FSBEI Educational and Methodological Center for Education in Railway Transport, Moscow (2016)
15. Rosenblit, G.B.: Heat Transfer in Diesel Engines, p. 216. Mechanical Engineering Publication, Moscow (1977)

Analysis of the Error in the Gas Temperature and the Thermocouple Time Constant Measuring Through Gas Turbine Engine Tests

Sergiy V. Yepifanov(✉) and Qijie Li

Kharkiv Aviation Institute, National Aerospace University, 17, Chkalov Street, Kharkiv 61070, Ukraine
yepif_khai@ukr.net, alexleekhai@gmail.com

Abstract. The purpose of this paper is to analyze the error in the gas temperature measured by thermocouples. Correlations were established analytically between the dynamic measurement error and the parameters of a thermocouple temperature measuring system. The accuracy of the thermocouple time constant estimation was identified as a crucial factor of the dynamic error. Further, an analysis of the error in an experimentally obtained estimate of the thermocouple time constant was carried out. It revealed that the thermocouple time constant, which well represents its dynamic characteristics, can be accurately estimated with rationally planned experiments. This study provides an *in situ* analysis that considers the thermocouple as a part of the engine instead of a stand-alone element. The analytical solution demonstrates the correlation behind the errors and dynamic parameters, and pictures a guideline for optimizing the gas temperature measuring system of gas turbine engines.

Keywords: Turbine engine · Gas temperature · Thermocouple · Dynamic model · Parameter identification

1 Introduction

Gas temperature is one of the thermodynamics parameters of great importance to the gas turbine engine. This temperature is used for estimating thrust and specific fuel consumption (SFC), indicating the thermal state of components and giving information about the intensity of thermal stress also the materials' performance under static and cyclic loading. As a result of its importance, monitoring, and controlling temperature within an optimal range is one main task in nowadays engine control systems [1, 2].

However, as a widely accepted engine temperature sensor, thermocouple itself provides a raw signal that is severely attenuated and lagged due to its poor response time and heat transfer with surroundings [3]. For many years its thermodynamics characteristics have been extensively studied. The National Aeronautics and Space Administration (NASA) conducted numerous experiments to develop thermocouples with high accuracy. Goldstein and Scherrer designed and calibrated a total temperature probe for the supersonic flow in 1949 [4]. Their design dramatically improves the recovery factor of thermocouples to 0.992 ± 0.002, whose value indicating its

M. Nechyporuk et al. (Eds.): *Integrated Computer Technologies in Mechanical Engineering*, AISC 1113, pp. 325–336, 2020.
https://doi.org/10.1007/978-3-030-37618-5_28

capability in recovering the dynamic temperature to total temperature. G. Glawe contributed systematic work on the recovery factor, correction coefficients of radiation error and times constant of several shielded and unshielded thermocouples [5], and also the influence of thermocouple geometric features [6]. These technical notes helped to formulate an empirical correlation between dimensionless number (such as Nusselt number (*Nu*) and Reynold number (*Re*)) under Mach number ranging 0.1–0.9 and Reynolds number ranging 250–30,000 [7]. Petit et al. [8] used the theoretical model of Yule et al. [9] to study the transient performance of thermocouples and concluded that the conduction error caused by heat transfer through support is insignificant if thermocouples own long wires. Simbirskyi et al. [10] carried out methods for experimentally determining the dynamic characteristics of thermocouples in turbojet engines at transient mode. More recent works mainly focused on studying these problems with well-developed numerical method. Through the application of conjugated heat transfer, Zou et al. [11] separately studied the difference source of steady-state error under Mach number ranging from 0.1 to 0.7 and proposed three physical models for velocity error, conduction error and radiation error. Braun et al. [12] numerically emulated and compensated the dynamic response of several similar bare thermocouples of different wire diameters under a step-change in static temperature.

However, although the steady-state error and dynamic response to a step temperature change have been many times either experimentally or numerically examined, as often emphasized, the to-be-applied thermocouple probes should be related with the engineering systems where probes are applied. With an incompatible coupling, the dynamic error alone can lead to unacceptable measuring error. It is why, from a view of aircraft engine control, the thermocouple should be optimized regarding the engine dynamics. Yepifanov et al. [13] proposed a method for identifying the dynamic characteristics of thermocouples by correcting its time constant based on engine testing data. This study presented that the thermocouple time constant can be accurately calibrated as a part of the engine, and it provided a possibility for further investigation on the correlation behind these dynamic parameters.

This paper introduced a mathematical model for analyzing parameters of the temperature measurement channel and their effects on dynamic temperature measurement error. Solutions to this model correlated the dynamic temperature measurement error with the parameters of the engine model and also the temperature measurement channel. A thermocouple was used as an inertial temperature sensor to extend the universality of this solution. In results, this analysis can be applied to any specific channel for satisfying accuracy requirements towards the temperature measurement in gas turbine engines.

2 Mathematical Model

2.1 Formulation of the Investigated Problem

Temperature control plays a key role in the engine automatic control system. At the designing stage of these control systems, this task is often solved by building separated closed-loop control, whose function is to maintain the gas temperature at a specific level. This temperature level is defined by both temperature limiter and set-point controller, which is why it may vary with engine operation modes and surrounding conditions. Thus a high-accuracy temperature control is a vital issue towards a high-precision control on the thermal cycle, strength, and life-time of engines.

The main focus of this paper is the error in the measured gas temperature, which is the fundamental factor affecting the temperature measuring accuracy. The accuracy of gas temperature control relies on not only the accuracy of the temperature measurement channel but also the parameters of the engine model. In this study, the temperature sensor is a thermocouple, and it is well known that during measurement thermocouple signal T_{TC} differs from actual gas temperature T_g by an error. The static component of this error that occurs when measuring constant temperature is due to the following reasons: designing features of the stagnation tube, the temperature difference between gas flow and electrodes, thermoelectric effect in electrodes contact zone, heat exchange between electrodes and casing and others. As this error has been carefully investigated, thus correcting this error in measuring scheme is possible [13]. In further study, the steady-state error was adapted as zero. Therefore, this study aims to analyze the dynamic error $\Delta T_{dyn} = T_g - T_{TC}$, which appears during transient operation modes due to the thermal inertia. In this case, the control signal is an increment in the throttle angle, which leads to a temperature increment in gas flow.

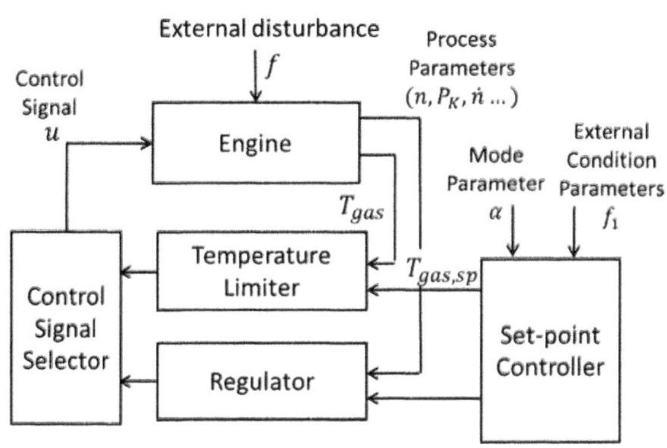

Fig. 1. Structure of the engine control system with a temperature limiter

Increasing engine power output by increasing the throttle angle is done by resetting the mode parameters α, and as a result, the fuel consumption increases. However, due to the rotor inertia, it accelerates relatively slower than fuel combustion and consequently decreases the excess air ratio in the combustion chambers, with which accompanying an increment in T_g. To limit this temperature, a regulator with feedback is always necessary. However, thermocouples used in engines are often with considerable inertia, which is why if thermocouple signal were applied directly as regulating signal, it would lead to overheating in hot section, causing additional life depletion to its components. Thus the temperature measurement channel should include itself with correcting elements to avoid this scenario, whose characteristics are adjusted according to thermocouple dynamic performance. In this way, a processed signal of the gas temperature $\widehat{T}_{gas}$ is formed from the measurement channel and sent to the temperature regulators. The structure of forming this temperature signal is presented in Fig. 2. This open-loop system will be the object for further analysis.

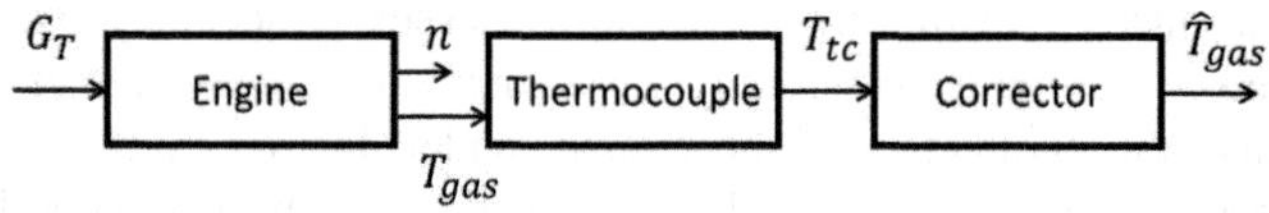

Fig. 2. Structural schema of the temperature measurement channel

The value of thermocouple time constant used in the corrector setting may differ from its actual value. Such a difference leads to a dynamic error between the actual and estimated temperatures. Therefore, the second chapter is devoted to identifying influencing factors of the dynamic error in the corrected temperature. A correlation between the corrector's parameters and the dynamic temperature error is established. Meanwhile, the time constant of thermocouples varies together with the gas flow it measures. As the incoming flow parameters (such as temperature, pressure, velocity) cover a wide range of various operation modes and external conditions, it requires a series of particular experiments to determine the temporal time constant [13]. For a rational arrangement of these experiments, it is necessary to correlate all mentioned parameters with the experimentally obtained time constant estimate. This part became the third chapter of this paper.

2.2 Research Method and Model

In this study, the analytical method was used to solve the previously described problem. On the one hand, to apply the analytical approach, the object under investigation has to be simplified and thus limited its applicability; but on the other hand, such a mathematical model and its solution are universal, and it can provide criteria for analyzing systems with any given dynamic parameters.

A single-shaft turbojet engine was taken as the testing base, whose input parameter is fuel consumption G_{fuel}, and output parameters are rotor speed n and gas temperature T_g. Thermocouple was described as a first-order inertial element with only time constant as the parameter. Though different correctors are applied in practice, whose classification can be found in [2], in this paper, for a dynamic error analysis, it is sufficient to consider the corrector of the simplest structure.

In reality, the actual transient process inside this engine is governed by complex fuel consumption adjustments, because such control is carried out through control programs of individual parameters and interaction between regulators, as in Fig. 1. In this study, an open-loop system was considered: Its control signal was a step change to fuel consumption, whereas, in a real system, fuel consumption often changes less intensively, and the deviation of the estimated temperature will be less significant. In other words, by choosing the step change to fuel consumption, the largest error in temperature measurement occurred.

To obtain an analytical description for dynamic processes, a mathematical model consisted of differential equations has to be used. However, because of the nonlinearity in those thermo-aerodynamics processes, the mathematical model of the engine is nonlinear. The same situation suits for thermocouple: because its dynamic characteristics vary with the measured flow, the thermocouple is also a nonlinear object. Therefore, the system presented in Fig. 2 started from a local steady-state. Then these equations could be reorganized with a deviation Δx from its base x_0 in place of the absolute parameter x. This base x_0 here corresponded with a specific installed mode. Thus, the model was linearized.

Thus, the mathematical model of the engine was expressed as

$$\tau \frac{dn}{dt} + \Delta n = K \cdot \Delta G_{fuel}; \tag{2.1}$$

$$\Delta T_{gas} = c \cdot \Delta n + d \cdot \Delta G_{fuel}, \tag{2.2}$$

where τ is the engine time constant; K is the gain coefficient of the rotor speed while feeding fuel; c, d are gas temperature gain coefficients.

Thermocouple was described using an inertial element model

$$\varepsilon \frac{dT_{TC}}{dt} + \Delta T_{TC} = \Delta T_{gas}, \tag{2.3}$$

where ε is the thermocouple time constant.

The model of correcting element was described as [2]

$$\nu \frac{d\hat{T}_{gas}}{dt} + \Delta \hat{T}_{gas} = e \frac{dT_{TC}}{dt} + \Delta T_{TC}, \tag{2.4}$$

where e is the estimate of thermocouple time constant, which is used to adjust the corrector; ν is the time constant of the corrector.

It is necessary to note that, the time constant of the corrector is introduced to stabilize the system; therefore, ν is far smaller than e, indicating $\nu \ll e$.

3 Temperature Dynamic Error Analysis

In this section, it was assumed that a step-change in fuel consumption ΔG_0 was applied to this system, and there was no initial variation in the system at moment $t = 0$, which set all considered parameters to zero as an initial condition. Then our solution to Eqs. (2.1) and (2.2) are:

$$\Delta n(t) = K \cdot \Delta G_0 (1 - e^{-\frac{t}{\tau}}); \tag{3.1}$$

$$\Delta T_{gas}(t) = \Delta G_0 [cK(1 - e^{-\frac{t}{\tau}}) + d]. \tag{3.2}$$

These expressions can be rewritten in a dimensionless form as

$$\bar{T}_{gas}(t) = A(1 - e^{-\bar{t}}) + 1, \tag{3.3}$$

where $\bar{T}_{gas} = \Delta T_{gas}/(d \cdot \Delta G_0)$ is dimensionless temperature change; $\bar{t} = t/\tau$ is the dimensionless time; $A = c \cdot K/d$ is an engine coefficient.

In this way, those parameters representing the dynamic characteristics of the engine were integrated as single parameter A, and it showed the engine's influence on the correlation between dimensionless time and dimensionless temperature.

Further, by substituting temperature (3.3) into (2.3), we obtained

$$\varepsilon \frac{dT_{TC}}{dt} + \Delta T_{TC} = \Delta G_0[cK(1 - e^{-\frac{t}{\tau}}) + d]. \tag{3.4}$$

Thus, the solution to the differential equations was

$$\Delta T_{TC}(t) = \Delta G_0[d + cK - (d - cK\frac{\varepsilon}{\tau - \varepsilon})e^{-\frac{t}{\varepsilon}} - cK\frac{\tau}{\tau - \varepsilon}e^{-\frac{t}{\tau}}]. \tag{3.5}$$

Rewritten in a dimensionless form it became

$$\bar{T}_{TC} = 1 + A - (1 - A\frac{\bar{\varepsilon}}{1 - \bar{\varepsilon}})e^{-\frac{\bar{t}}{\bar{\varepsilon}}} - A\frac{1}{1 - \bar{\varepsilon}}e^{-\bar{t}}, \tag{3.6}$$

where $\bar{T}_{TC} = \Delta T_{TC}/d \cdot \Delta G_0$ is the dimensionless temperature signal from the thermocouple; $\bar{\varepsilon} = \varepsilon/\tau$ is the dimensionless time constant of the thermocouple.

The derivative of the temporal thermocouple signal is

$$\frac{dT_{TC}(t)}{dt} = \Delta G_0[(\frac{d}{\varepsilon} - \frac{cK}{\tau - \varepsilon})e^{-\frac{t}{\varepsilon}} + \frac{cK}{\tau - \varepsilon}e^{-\frac{t}{\tau}}]. \tag{3.7}$$

By substituting this derivative term into (2.4), the equation of corrector dynamic under step fuel consumption change could be written as

$$\nu\frac{d\hat{T}_{gas}}{dt} + \Delta\hat{T}_{gas} = \Delta G_0\{d + cK + [d(\frac{e}{\varepsilon} - 1) - cK\frac{e - \varepsilon}{\tau - \varepsilon}]e^{-\frac{t}{\varepsilon}} - cK\frac{\tau - e}{\tau - \varepsilon}e^{-\frac{t}{\tau}}\}. \tag{3.8}$$

Therefore, its solution within the given initial conditions is obtained as

$$\Delta\hat{T}_{\text{gas}} = \Delta G_0 \begin{pmatrix} d + cK - \{d + cK + [d(\frac{e}{\varepsilon} - 1) - cK\frac{e-\varepsilon}{\tau-\varepsilon}]\frac{\varepsilon}{\varepsilon-\nu} - cK\frac{\tau-e}{\tau-\varepsilon}\frac{\tau}{\tau-\nu}\}e^{-\frac{t}{\nu}} \\ + [d(\frac{e}{\varepsilon} - 1) - cK\frac{e-\varepsilon}{\tau-\varepsilon}]\frac{\varepsilon}{\varepsilon-\nu}e^{-\frac{t}{\varepsilon}} - cK\frac{\tau-e}{\tau-\varepsilon}\frac{\tau}{\tau-\nu}e^{-\frac{t}{\tau}} \end{pmatrix} \tag{3.9}$$

In a dimensionless form:

$$\begin{aligned} \bar{T}_{gas} = 1 + A - A\tfrac{1-\bar{e}}{1-\bar{\varepsilon}}\tfrac{1}{1-\bar{\nu}}e^{-\bar{t}} + [(\tfrac{\bar{e}}{\bar{\varepsilon}} - 1) - A\tfrac{\bar{e}-\bar{\varepsilon}}{1-\bar{\varepsilon}}]\tfrac{\bar{\varepsilon}}{\bar{\varepsilon}-\bar{\nu}}e^{-\frac{\bar{t}}{\bar{\varepsilon}}} \\ -\{1 + A + [(\tfrac{\bar{e}}{\bar{\varepsilon}} - 1) - A\tfrac{\bar{e}-\bar{\varepsilon}}{1-\bar{\varepsilon}}]\tfrac{1}{1-\frac{\bar{\nu}}{\bar{\varepsilon}}} - A\tfrac{1-\bar{e}}{1-\bar{\varepsilon}}\tfrac{1}{1-\bar{\nu}}\}e^{-\frac{\bar{t}}{\bar{\nu}}} \end{aligned}, \tag{3.10}$$

where $\bar{T}_{gas} = \Delta T_{gas}/(d \cdot \Delta G_0)$ is the dimensionless temperature from the corrector; $\bar{e} = e/\tau$, $\bar{\nu} = \nu/\tau$ are the dimensionless corrector coefficients.

An example is presented in Fig. 3: In this example, dynamic parameters were assigned as $\tau = 4$ s, $K = 10$ rpm/(kg/h), $c = -0.2$ K/rpm, $d = 3$ K/(kg/h), $\varepsilon = 0.3$ s, $e = 0.4$ s, $v = 0.001$ s, $\Delta G_0 = 100$ kg/h, whose dimensionless values correspond with $A = -0.667$, $\bar{\varepsilon} = 0.075$, $\bar{e} = 0.1$.

Figure 3 shows that the error ΔT is a temporal parameter, and has global and local maximum; the global one locates at the initial moment, and the local one after intersection:

$$\Delta T = G_0\left\{\left[d\left(\frac{e}{\varepsilon} - 1\right) - cK\frac{e-\varepsilon}{\tau-\varepsilon}\right]e^{-\frac{t}{\varepsilon}} + cK\frac{e-\varepsilon}{\tau-\varepsilon}e^{-\frac{t}{\tau}}\right\}. \tag{3.11}$$

Maximal error at the initial moment:

$$\Delta\bar{T}(0) = \bar{e}/\bar{\varepsilon} - 1. \tag{3.12}$$

By solving $\Delta\dot{T} = 0$, the second maximal error moment t_2 is

$$t_2 = \left(\frac{1}{\varepsilon} - \frac{1}{\tau}\right)^{-1}\ln\left[\left(1 - \frac{\tau-\varepsilon}{A\varepsilon}\right)\frac{\tau}{\varepsilon}\right]. \tag{3.13}$$

By substituting (3.13) into (3.11), the dynamic error at the moment t_2 is

$$\Delta T_{t=t_2} = G_0\left\{\begin{array}{l}\left[d\left(\frac{e}{\varepsilon} - 1\right) - cK\frac{e-\varepsilon}{\tau-\varepsilon}\right]\left[\left(1 - \frac{\tau-\varepsilon}{A\varepsilon}\right)\frac{\tau}{\varepsilon}\right]^{-\frac{1}{1-\frac{\varepsilon}{\tau}}} + \\ + cK\frac{e-\varepsilon}{\tau-\varepsilon}\left[\left(1 - \frac{\tau-\varepsilon}{A\varepsilon}\right)\frac{\tau}{\varepsilon}\right]^{-\frac{1}{\frac{\tau}{\varepsilon}-1}}\end{array}\right\}. \tag{3.14}$$

Applying obtained solutions to the previously presented example, we obtained: $t_1 = 1$ s; $t_2 = 1.81$ s; $\Delta T_{t=t_2} = 6.57$ K. The analytical results correspond to the numerical results solved by Simulink.

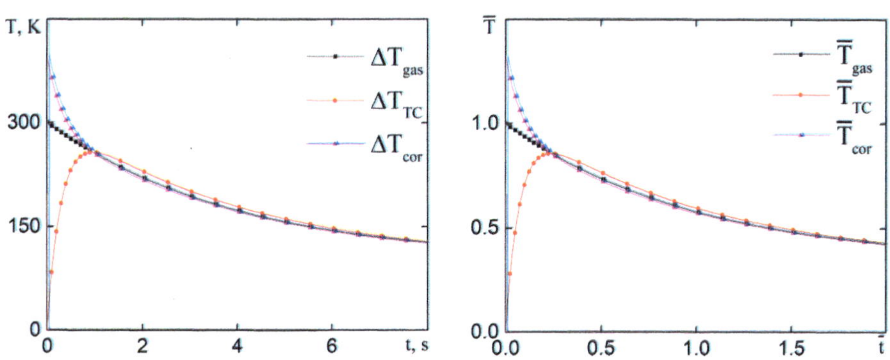

(a). absolute value (b). dimensionless value

Fig. 3. Temperature responses of the temperature measurement channel elements towards a step-change in fuel consumption.

In practice, the main focus is given to the initial error, because at this moment, control actions are imposed intensively with maximal errors. Analysis explicitly stated that this error is directly proportional to the error in the thermocouple time constant estimate.

4 Thermocouple Time Constant Estimation Analysis

Thermocouple is a widely used temperature measuring tool. Its dynamic characteristics are often described in the attached technical documentation, in which the time constant variation regarding the flow parameters is present. However, these correlations are often in a simplified type, which strongly limits its usage in different engine operation modes. The temperature limitation is a problem of reality not only maximal mode but also take-off mode due to the low flow velocity. That is why simply applying the formulas in documentation to these operation modes would lead to significant error.

Previous analysis indicated that the quality of temperature correction strongly relies on the accuracy of the thermocouple time constant under relevant operation mode. Thus series of experiments should be conducted to calibrate the thermocouple as an integrated element of the engine [13]. So an experiment was investigated in this chapter to analyze the error in the estimation of the time constant from testing data: the engine dynamics are described by (2.1) and (2.2) with known parameters. The time constant ε of the thermocouple will be determined basing on measured temperature results, which is obtained under a process of step increment ΔG_0 in fuel consumption. Temperature is measured discretely in moments $t_1, \ldots, t_j, \ldots, t_N$; the measuring results are the thermocouple signals T_{TC}, which denoted as $y_1, \ldots, y_j, \ldots, y_N$ (Fig. 4).

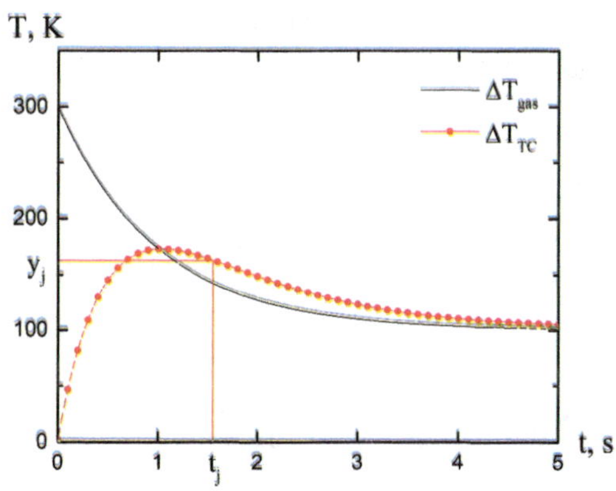

Fig. 4. Thermocouple signal measuring process

In a general view, the task of determining the thermocouple parameter can be formulated as a task of solving N-equation system respect to this parameter:

$$y(\varepsilon, t_j) = y_j, \text{ j} = 1, \ldots, \text{N}. \tag{4.1}$$

The previous analysis showed that the signal from thermocouple could be formulated as (3.5). This function would be used as a fitting model, and its only parameter was the thermocouple time constant ε. The task of solving the N-equation system respect one parameter ε is mathematically overdetermined. We defined $\Phi(\varepsilon)$ as the sum of the squares of the residuals. Thus the least-squares method (LSM) has to be applied:

$$\Phi(\varepsilon) = \sum_{j=1}^{N} (y_j - y(\varepsilon, t))^2 \rightarrow \min. \tag{4.2}$$

The variable $\hat{\varepsilon}$ was the parameter under LSM searching in (4.2). As (4.1) described a nonlinear dependence respect to ε, the solving process became nonlinear and had to be solved iteratively: $\hat{\varepsilon}^{s+1} = \hat{\varepsilon}^s + \Delta\hat{\varepsilon}^s$. The $\Delta\hat{\varepsilon}^s$ is the correction to the previous estimation.

The solution to the linearized problem was approximated with a first-order Taylor polynomial expansion:

$$\frac{\partial y}{\partial \varepsilon}(\hat{\varepsilon}^s, t_j)\Delta\varepsilon^s = y_j - y(\hat{\varepsilon}^s, t_j), \text{ j} = 1, \ldots, \text{N}: \tag{4.3}$$

$$\Phi(\varepsilon) = \sum_{j=1}^{N} \left(y_j - y(\varepsilon^s, t_j) - \frac{\partial y}{\partial \varepsilon}(\varepsilon^s, t_j)\Delta\varepsilon^s \right)^2 \rightarrow \min. \tag{4.4}$$

Therefore, the LSM solution could be obtained as

$$\Delta\hat{\varepsilon}^s = \frac{\sum_{j=1}^{N} \left[\frac{\partial y}{\partial \varepsilon}(\varepsilon^s, t_j) \cdot (y_j - y(\varepsilon^s, t_j)) \right]}{\sum_{j=1}^{N} \left[\frac{\partial y}{\partial \varepsilon}(\varepsilon^s, t_j) \right]^2}. \tag{4.5}$$

And the variance of estimation σ_ε^2 could be defined with the help of linearized equations

$$\sigma_\varepsilon^2 = \frac{\sigma_y^2}{\sum_{j=1}^{N} \left[\frac{\partial y}{\partial \varepsilon}(\hat{\varepsilon}, t_j) \right]^2} = \frac{\sigma_y^2 \cdot \Delta t}{\sum_{j=1}^{N} \left[\frac{\partial y}{\partial \varepsilon}(\hat{\varepsilon}, t_j) \right]^2 \cdot \Delta t} \approx \frac{\sigma_y^2 \cdot \Delta t}{\int_0^{t_N} \left[\frac{\partial y}{\partial \varepsilon}(\hat{\varepsilon}, t_j) \right]^2 \cdot dt}, \tag{4.6}$$

where σ_y^2 is the variance of the thermocouple signal; Δt is the time interval between two measurements.

For a relatively long sampling period ($t_s > 3\tau$), the exponential terms approach to zero, which indicated that the error in the estimated time constant is not affected by further measurement; in other words, after this period of time, there is no new information about our desired parameter. Therefore, taking the derivative of (3.5) and substituting into (4.7), limiting $t_N \to \infty$, the expression for the estimation variance is

$$\sigma_\varepsilon^2 = \frac{\sigma_T^2 \cdot \Delta t \cdot \varepsilon^4}{(\Delta G_0)^2 \cdot d^2 \cdot \tau^2} \left\{ \begin{array}{c} \dfrac{A^2\varepsilon^4}{2(\tau+\varepsilon)(\tau-\varepsilon)^2} + \dfrac{1}{\tau^2}\left[1 - \dfrac{A\varepsilon}{(\tau-\varepsilon)}\right]^2 \dfrac{\varepsilon^3}{4} \\ + \dfrac{2}{\tau}\dfrac{A\varepsilon}{(\tau-\varepsilon)^2}\left[\dfrac{A\varepsilon}{(\tau-\varepsilon)} - 1\right]\left[\dfrac{\varepsilon^2}{4} - (\dfrac{\tau\varepsilon}{\tau+\varepsilon})^2\right] \end{array} \right\}^{-1},$$

$$\sigma_\varepsilon^2 = \sigma_T^2 \cdot \Delta t \cdot (\Delta G_0)^{-2} \cdot d^{-2} \cdot \tau^{-2} \cdot f(\varepsilon). \tag{4.7}$$

Figure 5 shows the variance σ_ε^2 of the experimental estimate of the thermocouple time constant respect to parameters ε and τ, while $A = -0.6667$, $\sigma_y^2 = 20\ \text{K}^2$, $\Delta t = 0.2$ s. In Fig. 5(a), the engine time constant τ was set as 4 s; In Fig. 5(b), the thermocouple time constant ε was set as 0.3 s. Notice that variance σ_ε^2 is plotted on a logarithmic scale. Especially, a case considering the time constants of the engine and the thermocouple are of the same values was studied. The results are denoted with blue triangles in Fig. 5.

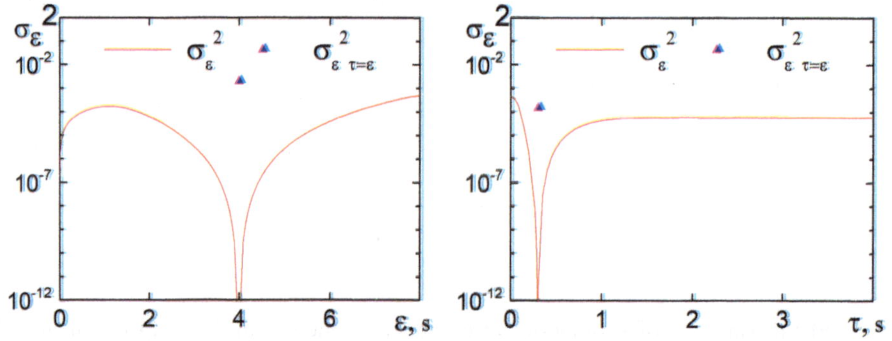

(a). Variance σ_ε^2 respect to ε (b). Variance σ_ε^2 respect to τ

Fig. 5. Influence of the engine and thermocouple dynamic parameters on the error in the thermocouple time constant estimation.

The setting used here is consistent with the previous example in the measurement error section. Despite the low sampling frequency (5 Hz), the variance of this thermocouple time constant estimate is still limited below $6 \times 10^{-6}\ \text{K}^2$, whose relative error merely reached 2.5%. This result proved the sufficiency of our model.

5 Conclusion

In this paper, an analysis of the temperature measuring error is presented for thermocouples integrated with a turbojet engine. The result indicates that the dynamic temperature error appearing in engine transient modes can be well confined if the corrector in the temperature measurement channel is carefully justified. Further study on the determination process for thermocouple time constants presented a small divergence in the time constant estimation, which indicates a high accuracy in the estimate of thermocouple time constants.

From the results, it might be noticed that by increasing sampling frequency and the magnitude of step-change to fuel consumption decrease the estimation variance dramatically, but over increasing sampling frequency leads to enhanced interference between neighboring measurement and over increasing the change in fuel consumption invalidates the linearization assumption by intensifying the nonlinearity in all thermodynamics process. Moreover, a sudden increment in the gas temperature could push the compressor into an unstable working condition. To our surprise, the special case which considered the engine and the thermocouple share the same inertia value leads to a discontinuous point in that location. The mechanism behind this phenomenon is still under consideration.

Besides, some mathematical approximations were included in the solving process, so the results may deviate from reality at a certain level. It is necessary to verify this result with further numerical simulations and experiments.

References

1. Jaw, L., Mattingly, J.: Aircraft Engine Controls: Design, System Analysis, and Health Monitoring. American Institute of Aeronautics and Astronautics Inc, Reston (2009)
2. Sinyakov, A.N., Shaimardanov, F.A.: Automatic Control Systems of Flying Vehicles and Their Power Units. Russian Federation, Moscow (1991)
3. Von Moll, A., Behbahani, A.R., Fralick, G.C., Wrbanek, J.D., Hunter, G.W.: A review of exhaust gas temperature sensing techniques for modern turbine engine controls. In: 50th AIAA/ASME/SAE/ASEE Joint Propulsion Conference. American Institute of Aeronautics and Astronautics, Inc., Cleveland, OH, USA (2014)
4. Goldstein, D.L., Scherrer, R.: Design and calibration of a total-temperature probe for use at supersonic speeds. NACA, Technical Note 1885. Washington, D.C., USA (1949)
5. Glawe, G.E., Simmons, F.S., Stickney, T.M.: Radiation and recovery corrections and time constants of several chromel-alumel thermocouple probes in high-temperature, high-velocity gas streams. NACA, Technical note. 3766. Washington, D.C., USA (1956)
6. Glawe, G.E., Glawe, G.N., Holanda, R., Krause, L.E.: Recovery and radiation corrections and time constants of several sizes of shielded and unshielded thermocouple probes for measuring gas temperature. NACA, Technical paper 1099. Washington, D.C., USA (1978)
7. Scadron, M.D., Warshasky, I.: Experimental determination of time constants and nusselt numbers for bare- wire thermocouples in high-velocity air streams and analytic approximation of conduction and radiation errors. NACA, Technical note 2599. Washington, D.C., USA (1952)

8. Petit, C., Gajan, P., Lecordier, J.C., Paranthoen, P.: Frequency response of fine wire thermocouple. J. Phys. **15**(7), 760–770 (1982)
9. Yule, A., Taylor, D.S., Chigier, N.A.: Thermocouple signal processing and on-line digital compensation. J. Energy **2**(4), 223–230 (1978)
10. Simbirsky, D.F., Skripka, A.I.: Determination of the dynamic characteristics of the sensitive elements of the gas flow temperature sensors for the automatic control system of turbojets. National Aerospace University, Kharkiv, Ukraine (1993)
11. Zou, Z., Yang, W., Zhang, W., Wang, X., Zhao, J.: Numerical modeling of steady state errors for shielded thermocouples based on conjugate heat transfer analysis. Int. J. Heat Mass Transfer **119**, 624–639 (2018)
12. Braun, J., Lu, S., Paniagua, G.: Development of high frequency virtual thermocouples. In: ASME Turbo Expo 2017: Turbomachinery Technical Conference and Exposition. ASME, Charlotte (2017)
13. Yepifanov, S.V., Shpylovyi, A.A., Suchovyi, S.I.J.: An indentification of thermocouple dynamic characteristics based on engine testing data. Aerosp. Tech. Tech. **66**(9), 166–171 (2009)

Enhancing the Efficiency of Marine Diesel Engine by Deep Waste Heat Recovery on the Base of Its Simulation Along the Route Line

Roman Radchenko[1(✉)], Victoria Kornienko[2], Maxim Pyrysunko[2], Mykola Bogdanov[3], and Andrii Andreev[2]

[1] Admiral Makarov National University of Shipbuilding, 9 Heroes of Ukraine Avenue, Mykolayiv 54025, Ukraine
nirad50@gmail.com
[2] Kherson Branch of Admiral Makarov National University of Shipbuilding, 44 Ushakova Avenue, Kherson 73000, Ukraine
[3] National University «Odessa Marine Academy», Odessa, Ukraine

Abstract. The efficiency of integrated cooling air at the intake of Turbocharger and Scavenge air at the inlet of working cylinders of the main diesel engine of dry-cargo ship by transforming the waste heat into a cold by an Refrigerant Ejector Chiller as the most simple in design and reliable in operation and by complex in design but more efficient Absorption Lithium-Bromide Chiller was analyzed. A ship power plant of cogeneration type using the relatively low-grade heat of water of a heat supply system with a temperature of about 90 °C, that significantly complicates the problem of its conversion into cold were considered. A new approach is proposed to improve the efficiency of integrated cooling Intake Air of turbocharger and Scavenge Air at the inlet of the working cylinders of the ship main engine of a transport ship, which consists in comparing the required cooling capacity and the corresponding heat needs during the trade route with the available heat of exhaust gases and scavenge air of the cogeneration power plant, determining the deficit and excess cooling capacity of heat utilizing cooling machines of various types, that allows to identify and realize the reserves of improving the efficiency of cooling intake air of the turbocharger and the scavenge air of the main diesel engine through the joint use of chillers of various types. The minimum value of wall temperature and minimum value of exhaust gas temperature at the exit from exhaust gas boiler is determined, at which the permissible speed of low-temperature corrosion at a level of 0.25 mm/year is ensured.

Keywords: Marine diesel engine · Exhaust gas boiler · Water-fuel emulsion

1 Introduction

The operation conditions of ship low-speed diesel engines are characterized by a significant change in the ambient temperature during the voyage and corresponding temperatures of air at the inlet of turbocharger (TC) and the scavenge air after the TC. At high temperatures of outboard seawater the scavenge-air coolers (SAC) are not able

M. Nechyporuk et al. (Eds.): *Integrated Computer Technologies in Mechanical Engineering*, AISC 1113, pp. 337–350, 2020.
https://doi.org/10.1007/978-3-030-37618-5_29

to maintain a scavenge air temperature at the low temperature level sufficient for damping elevated air temperatures at the inlet of TC and to provide the engine operation at a stable air temperature at the intake to the working cylinders, which would ensure high fuel efficiency of the engine.

According to the manufacturers of ship low-speed engines "MAN" and "Wartsila" [1–3], an increase in air temperature at the inlet of TC by 10 °C causes an increase in specific fuel consumption b_e by 0.2…0.7%, and similarly, an increase in the scavenge air temperature after SAC leads to an increase in specific fuel consumption b_e by approximately 0.5% and a corresponding decrease in the efficiency η_e of engines which poses the acute problem of complex cooling of the engine cyclic air: at the inlet of TC and scavenge air at the inlet of the engine cylinders.

At the same time, with an increase in the temperature of ambient air and outboard water for cooling scavenge air, the heat, removed from the scavenge air and exhaust gases, increases. So it is quite expedient to utilize it with a waste heat recovery chiller to reduce the temperature of engine intake air and scavenge air at the inlet of engine cylinders. The use of low-boiling fluids (LBF) in the waste heat recovery chiller provides deep cooling of engine cyclic air.

Additional energy saving by utilizing the exhaust heat in internal combustion engines (ICE) of the ship power plants (SPP) allows to save fuel. This, accordingly, leads to reduction of harmful substances emissions into the atmosphere. That is why it is important to ensure a maximum utilization of the heat losses of main engine (ME). Due to significant decreasing of exhaust gas temperature of modern ICE and sharp increasing the low-temperature corrosion (LTC) rate (up to 1.5 mm/year at wall temperatures below 130 °C with the combustion of sulfur fuels with water content W^r = 2%), decreasing of exhaust gas boilers (EGB) steam capacity and depth of gas heat utilization has taken place.

The level of combustion fuel heat utilization can be increased by deep cooling of combustion products below the dew point temperature, due to the maximum utilization of not only the gas sensible heat, but also the latent heat of sulfuric acid vapor, which are contained in them. The condensation of sulfuric acid vapor limits the exhaust gas temperature at the wall temperatures below a dew point temperature of sulfuric acid vapor H_2SO_4.

Therefore, decreasing of the LTC intensity at the surface temperature below a dew point temperature of sulfuric acid vapor H_2SO_4 is practically the only possibility of reducing the EGB exhaust gas temperature and improving ecology and economy of SPP.

2 Literature Review

A lot of investigations are devoted to development of resource and energy saving and emissions reduction technologies [4–7]. Various ways of cooling the cycle air of engines by conversion of waste heat with refrigerant Ejector Chiller (ECh) have been investigated as the most simple in design and reliable in operation so as more complex Absorption Lithium-Bromide Chiller (ACh) [8].

The cooling air systems both for intake air of the TC and scavenge air of low-speed engines by the conversion of the waste heat of exhaust gases and scavenge air by waste heat recovery chillers were proposed and the efficiency of this direction in marine power engineering was shown [8].

However, the problem of complex cooling of cyclic air (air at the intake of TC and scavenge air) of the low-speed engines of cogeneration type by using as a source of relatively low-grade water heat of the heat supply system (with a temperature of about 90 °C) is still unsolved. Because of the insufficiently high efficiency of conversion of the hot water heat compared with steam, the resulting cooling capacity may not be sufficient for complex cooling air at the engine intake air and scavenge air, that raises the problem of the rational distribution of heat loads between cooling circuits (subsystems) for chilling the engine intake and scavenge air and the need for the use of the waste heat recovery chiller of various types.

According to data of [9, 10] heat loss of exhaust gases represents a high proportion of the total heat loss of combustion engines, turbines and boilers in power stations and SPP. The Rankine cycle system can convert the exhaust gas heat to electricity and improve the unit efficiency [11]. Authors [12] introduced the technique of exhaust gas heat utilization in gas cogeneration unit based on absorption heat exchange. The heat exchangers used to heat condensation are called low-temperature economizers [13–16] or low-pressure economizers [17]. The condensed acid vapor also glues the ash in exhaust gas, and adheres on heating surface [18], which increases the resistance of exhaust gas flow and heat transfer [19], affecting the units' reliable and economical operation. The experience of using WFE in boilers and diesel engines indicates the undeniable advantages of this type of fuel: the effective specific fuel consumption decreases by about 8% [20, 21], in the exhaust gases the concentration of nitrogen oxides is significantly reduced in 1,4…3,1 times [22, 23], the concentration of CO - in 1,3…1,5 times [24], smoke - in 1,3…2,4 times [23]. In [25] the influence of parameters of WFE combustion process in existing low capacity boiler plants on the level of formation of nitrogen oxides, carbon monoxide and soot was studied. The analysis of literary sources showed, that there is no quantitative data of LTC intensity of codensing low-temperature heating surfaces (LTHS) of EGB while WFE combustion.

3 Research Methodology

The waste heat recovery refrigerant Ejector Chiller (ECh) is the most simple in design [8]. However, the efficiency of conversion of waste heat into cold in Ejector Chiller (ECh) is relatively low: the coefficient of ECh performance ζ_E = 0.2…0.3, which requires a significant amount of heat. A coefficient of performance $\zeta = Q_0/Q_h$ – is the ratio of generated cold Q_0 to the consumed heat Q_h, extracted from exhaust gases, scavenge air. At the same time, the efficiency of transformation of heat into cold in quite complex Absorption Lithium-Bromide Chiller (ACh) is two to three times higher: $\zeta_A \approx$ 0.7…0.8. It would be expedient to use the advantages of each of the chillers: deep cooling of the air to t = 7…10 °C in the ECh and high efficiency of the transformation of heat into cold in ACh (ζ = 0.7…0.8).

It is obvious that the combined use of ACh and ECh would be optimal. To do this, it is necessary to analyze the efficiency of the integrated cooling of the cyclic air of the low-speed diesel engines (intake air of the TC and the scavenge air) by transforming the waste heat of exhaust gases and scavenge air into the cold, comparing the necessary cooling capacity and the corresponding heat needs during an actual voyage with the amount of available waste heat of diesel engines (heat of exhaust gases and scavenge air), the results of which reveal a deficit or excess of the available waste heat for each of the methods for cooling the air (by using ECh or ACh), and then reserves for additional cooling of the air (intake air or scavenge air) due to the corresponding excess waste heat.

So, at the first stage of the methodology for determining the thermal capacity of waste heat recovery chiller the comparison of the necessary cooling capacities for complex cooling of cycle air of diesel engines (simultaneously intake air of the TC and the scavenge air) and the corresponding total needs of waste heat of diesel engines with the amount of available heat of exhaust gases and scavenge air or excess of the existing waste heat for each of the methods of air cooling (using ECh or ACh), and then additional reserves of cooling capacity for cooling intake air or scavenge air due to respond accordingly excess waste heat.

The efficiency of engine inlet and scavenge air cooling has been considered for the ship's low-speed diesel engine MAN B&W 6S60MC6.1-TI with economical power output N_e = 10 MW, specific fuel oil consumption (SFOC) b_e = 169.8 g/(kWh) and air consumption 23.9 kg/s under ISO conditions and engine load – NMCR = 90% [1, 3]. To analyze the parameters of the cooling air system, as well as the characteristics of the main engine with air cooling for changeable ambient conditions the "mandieselturbo" software package was used [3].

For the 6S60MC6.1-TI engine, according to the data of the MAN company (according to the calculations by using "mandieselturbo" software package), cooling inlet air for every 10 °C results in reduction of specific fuel consumption 1.1–1.2 g/(kWh) [3].

Characteristics of the work of the ECh was chosen taking into account the features of cooling air for ship diesel engines: refrigerant – R142b; boiling temperature in the evaporator-air cooler t_0 = 5 °C; condensing temperature in a water cooled condenser with seawater – t_c = 25–45 °C; boiling temperature in the generator – t_g = 90 °C.

Thus, for comparative analysis the following values of coefficient of performance were accepted: ζ = 0.7 for Absorption Lithium-Bromide Chiller (ACh) and ζ = 0.2 for Ejector Chiller (ECh) at boiling temperature of the refrigerant R142b in the evaporator t_0 = 5 °C and in the generator t_g = 90 °C.

The calculation of the characteristics of the engine with cyclic air cooling was carried out on the operating mode during the voyage of the dry-cargo ship on the route Odessa-Yokogama. The changes in ambient air temperature t_a, temperature of outboard water t_w, ambient air absolute humidity d_a and relative humidity φ_a during the trade route are presented in Fig. 1.

When climatic conditions change during the trade route, the heat load of the cooling system changes, that is, cooling capacity Q_0 is necessary for cooling the air, as well as the corresponding heat consumption Q_h, transformed into cold in waste heat recovery chiller: $Q_h = Q_0/\zeta$.

Characteristics of the ECh performance was chosen taking into account the features of the design and use of air pre-cooling for ship diesel engines: refrigerant – R142b; boiling temperature in the evaporator-air cooler $t_0 = 5$ °C; condensing temperature in a water cooled condenser with seawater – $t_c = 25–45$ °C; boiling temperature in the generator – $t_g = 90$ °C.

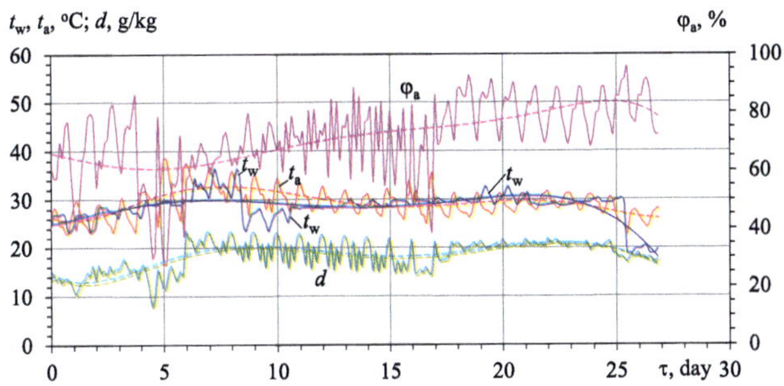

Fig. 1. The changes in ambient air temperature t_a, temperature of outboard water t_w, ambient air absolute humidity air d_a and relative humidity φ_a during the trade route Odessa-Yokogama

Thus, for comparative analysis the following values of coefficient of performance are accepted: $\zeta = 0.7$ for Absorption Lithium-Bromide Chiller (ACh) and $\zeta = 0.2$ for Ejector Chiller (ECh).

The investigations of corrosion were carried out at the experimental installation in the Admiral Makarov National University of Shipbuilding. The processing of measurement results for study of corrosion process kinetic was carried out after 2, 4, 8 and 12 h, based on which the approximation equations were obtained. The verification of approximation equations reliability was carried out on the results of experiment for 100 h.

4 Results

The cooling capacity $Q_{0.15}$, required for cooling the Intake Air of the engine, was calculated based on the decrease in its temperature from the current temperature in the engine room t_{a1} ($t_{a1} = t_a + 10$ °C) to $t_{a2} = 15$ °C by the value Δt_{a15}. The available cooling capacity was calculated as $Q_{0(0.7)}$ – during the transformation of the exhaust gas heat Q_h to cold in ACh ($Q_0 = Q_h \cdot \zeta$ at $\zeta = 0.7$) and as $Q_{0(0.2)}$ – during the transformation of exhaust gas heat Q_h in the ECh (at $\zeta = 0.2$). The heat required for cooling the air at the engine inlet to the temperature $t_{a2} = 15$ °C in ACh was calculated as $Q_{h.15(0.7)} = Q_{0.15}/\zeta$ (at $\zeta = 0.7$), and in ECh – as $Q_{h.15(0.2)} = Q_{0.15}/\zeta$ (at $\zeta = 0.2$) (Fig. 2).

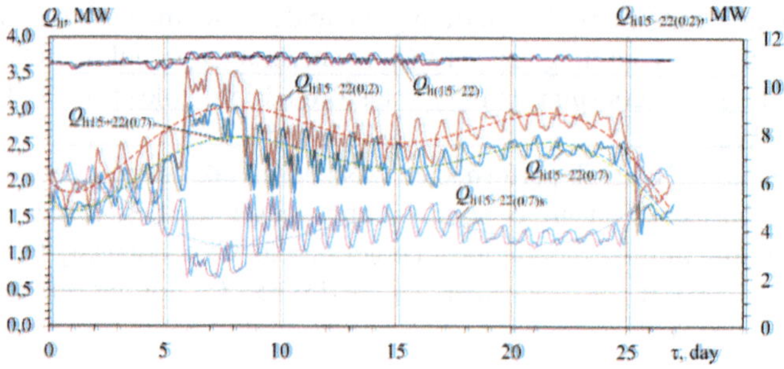

a

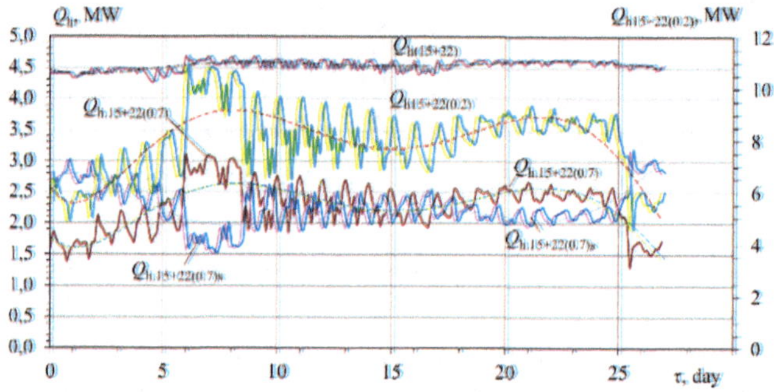

b

Fig. 2. The current values of available heat of exhaust gases and scavenge air $Q_{h(15+22)}$, heat required for cooling the intake air to t_{a2} = 15 °C and scavenge air in the SAC_{LT} to t_{a2} = 22 °C in the ECh $Q_{h15+22(0.2)}$ at $\zeta = 0.2$ (ECh at t_h = 90 °C and t_0 = 5), in the ACh $Q_{h15+22(0.7)}$ at $\zeta = 0.7$, surplus (excess) heat available for ACh $Q_{h.15+22(0.7)s}$ and heat deficit for ECh $Q_{h15+22(0.2)d}$ during the voyage Odessa-Yokohama at scavenge air temperatures at the outlet of the SAC_{HT} t_{h2}: a – t_{h2} = 140 °C; b – t_{h2} = 110 °C

As it is seen, the value of the available (existing) heat of exhaust gases $Q_{h.15}$ and scavenge air Q_{h22} are close to each other $Q_{h22} \approx Q_{h15} \approx 2.2\ldots2.6$ MW and their total amount $Q_{h(15+22)} \approx 4.5\ldots5.0$ MW is almost twice higher the heat $Q_{h15+22(0.7)}$, required for cooling the intake air to a temperature t_{a2} = 15 °C and scavenge air in the SAC_{LT} to a temperature t_{a2} = 22 °C in the ACh with a coefficient of performance $\zeta = 0.7$, as evidenced by a significant excess of the available heat $Q_{h15+22(0.7)s} = Q_{h(15+22)} - Q_{h15+22(0.7)} \approx 3{,}5\ldots4{,}5$ MW for ACh.

At the same time, the total amount of available heat of exhaust gases and scavenge air $Q_{h(15+22)}$ is almost half the heat $Q_{h15+22(0.2)} \approx 9$ MW, necessary for cooling the intake air to t_{a2} = 15 °C and Scavenge Air in a SAC_{HT} to t_{a2} = 22 °C in a ECh with a coefficient of performance $\zeta = 0.2$, as evidenced by a significant deficit of heat $Q_{h15+22(0.2)d} = Q_{h15+22(0.2)} - Q_{h(15+22)} \approx 3\ldots4$ MW for the ECh (Fig. 3).

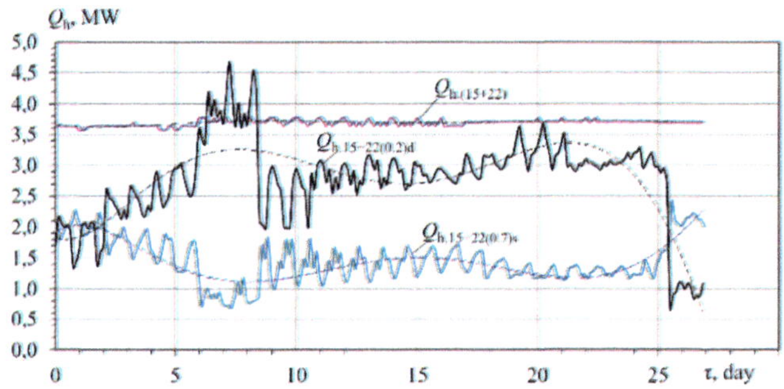

a

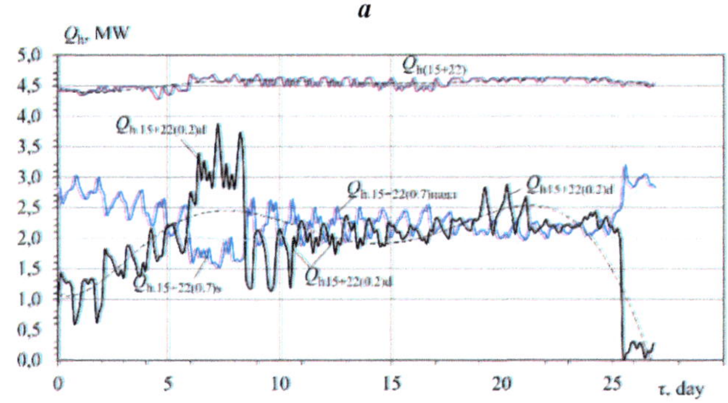

b

Fig. 3. The current values of the total amount of available heat of exhaust gases and scavenge air $Q_{h(15+22)}$, heat required for cooling the intake air to t_{a2} = 15 °C and scavenge air in the SAC_{LT} to t_{a2} = 22 °C in the ECh $Q_{h15+22(0,2)}$ at $\zeta = 0.2$ (ECh at t_h = 90 °C and t_0 = 5), surplus heat available for ACh $Q_{h.15+22(0.7)s}$ and heat deficit for ECh $Q_{h15+22(0.2)d}$ during the voyage Odessa-Yokohama at scavenge air temperatures at the outlet of the SAC_{HT} t_{h2}: a – t_{h2} = 140 °C; b – t_{h2} = 110 °C

The closeness of heat deficit values for ECh $Q_{h15+22(0.2)d}$ and excess available heat which is more then required for ACh $Q_{h.15+22(0.7)s}$ indicates the possibility of their mutual compensation through the joint use of ACh as an effective transformer of heat to cold (ζ = 0.7), however, at less depth of air cooling (up to t_{a2} = 15 °C), and less effective ECh (ζ = 0.2…0.3) for deeper cooling of the air (to the minimum potential t_{a2} = 10 °C and even lower).

The values of the available total cooling capacity, which can be obtained due to the total amount of heat of exhaust gases $Q_{h(15+22)}$ in a ACh $Q_{0.15+22(0.7)}$ (ζ = 0.7) and in ECh $Q_{0.15+22(0.2)}$ (ζ = 0.2) as well as the total cooling capacity $Q_{0.15+22}$, necessary for cooling intake air of the TC to t_{a2} = 15 °C and the scavenge air to t_{a2} = 22 °C with the total decrease in air temperature by the value of $\Delta t_{a15+22} = \Delta t_{a15} + \Delta t_{a22}$ according to the climatic conditions on the trade route (Fig. 4).

The values of ECh cooling capacity deficit $Q_{0.15+22(0.2)d} = Q_{0.15+22} - Q_{0.15+22(0.2)}$ and the ACh surplus (excess) cooling capacity $Q_{0.15+22(0.7)s} = Q_{0.15+22(0.7)} - Q_{0.15+22}$ during the voyage Odessa-Yokohama are presented in Fig. 5.

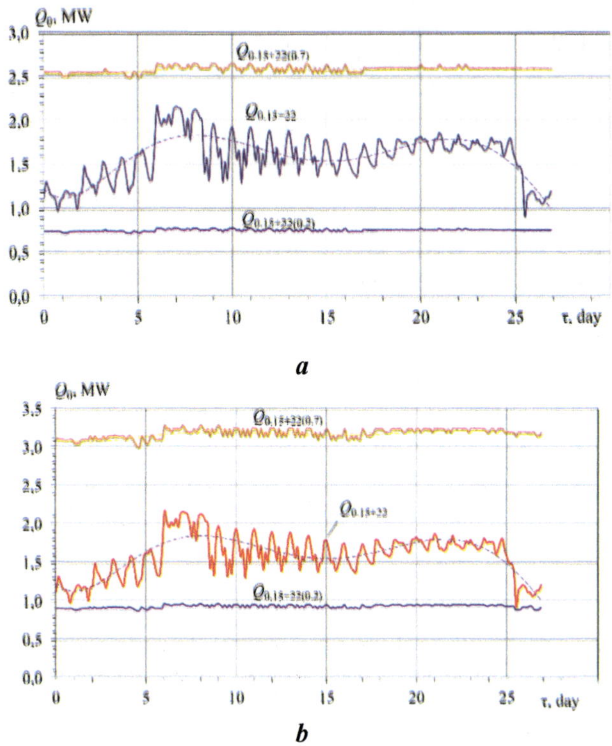

Fig. 4. The values of the available cooling capacity, which can be obtained by using the total amount of heat of exhaust gases $Q_{h(15+22)}$ in the ACh $Q_{0.15+22(0.7)}$ ($\zeta = 0.7$) and in the ECh $Q_{0.15+22(0.2)}$ ($\zeta = 0.2$), as well as the total cooling capacity $Q_{0.15+22}$, necessary for cooling the intake air to $t_{a2} = 15$ °C and scavenge air to $t_{a2} = 22$ °C at scavenge air temperatures at the outlet of the SAC_{HT} t_{h2}: a – $t_{h2} = 140$ °C; b – $t_{h2} = 110$ °C

It should be noted that correlation between the ECh cooling capacity deficit $Q_{0.15+22(0,2)d} = Q_{0.15+22} - Q_{0.15+22(0.2)}$ and the ACh surplus (excess) cooling capacity $Q_{0.15+22(0.7)s} = Q_{0.15+22(0.7)} - Q_{0.15+22}$ during the voyage (Fig. 5) can be changed by using the surplus cooling capacity of ACh $Q_{0.15+22(0.7)s}$ (compared with the total cooling capacity $Q_{0.15+22}$, necessary for cooling intake air to the temperature $t_{a2} = 15$ °C and scavenge air to $t_{a2} = 22$ °C) for completely eliminating the ECh cooling capacity deficit $Q_{0.15+22(0.2)d}$. It can be fulfilled through partial replacement of the ECh cooling capacity with the ACh cooling capacity by using the ECh only for cooling the engine intake air or scavenge air.

By rational redistribution of heat flows of waste heat (exhaust gases and scavenge air) between various types of waste heat recovery chillers (ECh and ACh), and the

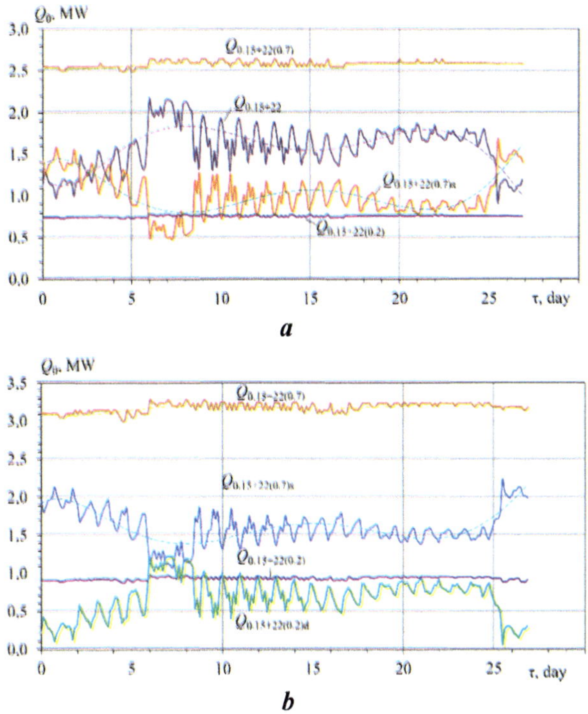

Fig. 5. The values of the available total cooling capacity, which can be obtained by using the heat of scavenge air and exhaust gases $Q_{h(15+22)}$ in the ACh $Q_{0.15+22(0.7)}$ ($\zeta = 0.7$) and in the ECh $Q_{0.15+22(0.2)}$ ($\zeta = 0.2$), cooling capacity deficit of the ECh $Q_{0.15+22(0.2)d} = Q_{0.15+22} - Q_{0.15+22(0.2)}$ and surplus cooling capacity of the ACh $Q_{0.15+22(0.7)s} = Q_{0.15+22(0.7)} - Q_{0.15+22}$ during the voyage Odessa-Yokohama: a – t_{h2} = 140 °C; b – t_{h2} = 110 °C

cooling capacity obtained by the chillers between the engine intake air and scavenge air cooling, it is possible to ensure simultaneous complex cooling of the cyclic air of the engine and obtain almost double effect in the form of reduced fuel consumption. In this case, the redistribution of heat flows during the actual conversion of waste heat into cold – the ratio of generated cooling capacity Q_0 and heat consumption for its production Q_h – is determined by the coefficient of performance $\zeta = Q_0/Q_h$, that is, by the type of waste heat recovery chiller.

The aggregate effect of complex cooling engine intake air to t_{a2} = 15 °C and scavenge air to t_{a2} = 22 °C with a summerised decrease in the cyclic air temperature by $\Delta t_{a15+22(0.2)}$ can be estimated by the current full reduction of specific Δb_{e15+22} and hourly B_{f15+22} fuel consumption and corresponding accumulated total reduction of fuel consumption $\sum B_{f.15+22}$ of the low-speed engine 6S60MC6.1-TI during the voyage for the climatic conditions on the trade route Odessa-Yokohama in Fig. 6.

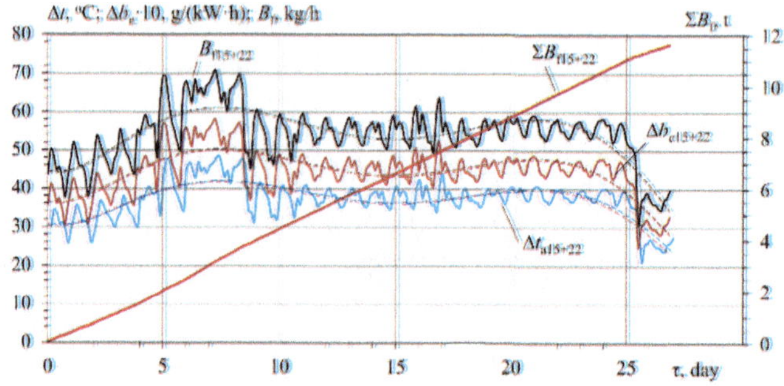

Fig. 6. The current values of the full reduction in specific $\Delta b_{e.15+22}$ and hourly $B_{f.15+22}$ fuel consumption due to simultaneously cooling the air at the intake of the TC to the temperature t_{a2} = 15 °C and scavenge air to t_{a2} = 22 °C with a summerised decrease in the cyclic air temperature by Δt_{a15+22} in ACh and ECh and the corresponding accumulated total reduction of fuel consumption $\sum B_{f.15+22}$ of the low-speed engine 6S60MC6.1-TI during the voyage for the climatic conditions on the trade route Odessa-Yokohama

As can be seen, the joint cooling of the intake air of the TC to the temperature t_{a2} = 15 °C and the scavenge air to t_{a2} = 15 °C ensures the reduction of the specific fuel consumption of the low-speed engine 6S60MC6.1-TI by the value $\Delta b_{e.15+22}$ = 4.0… 4.5 g/(kW·h) and hourly consumption $B_{f.15+22}$ = 50…60 kg/h and total fuel consumption $\sum B_{f.15+22} \approx 11$ t during the trade route Odessa-Yokohama.

Based on experimental studies the minimum value of wall temperature is defined (Fig. 7a), at which the permissible speed of LTC at a level of 0.25 mm/year and the minimum value of exhaust gas temperature at exit from EGB are ensured (Fig. 7b). Figure 7b show that the smallest values of corrosion intensity are observed for large values of water content in water-fuel emulsion 30%.

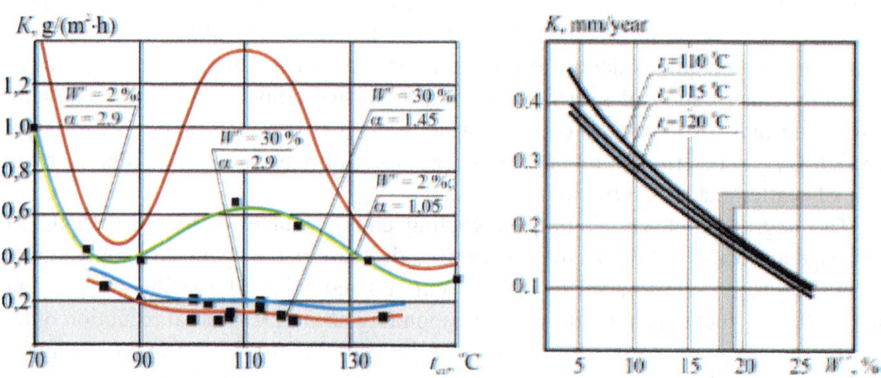

a b

Fig. 7. Correlation of corrosion rate K from wall temperature (a) and water content W^r in emulsion (b)

The wall temperature range of condensing low-temperature heating surface (LTHS) safe operation is determined, which opens opportunities for deep utilization.

When WFE is burnt, the rate of LTC and pollution decreases, which makes it possible to decrease exhaust gas temperature at the outlet from the EGB down to 80… 100 °C, to install condensing LTHS (evaporator with pressure p = 150 kPa and section of hot water) in the EGB (Fig. 8).

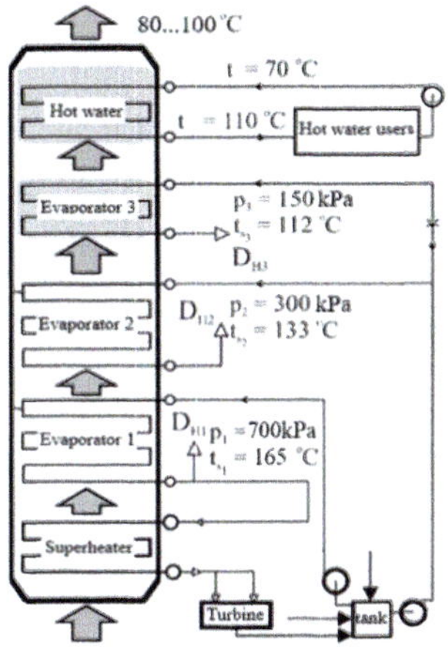

Fig. 8. Improved scheme of heat recovery circuit with condensation heating surface while WFE combustion

This scheme can be used in ships (bulk carriers, container ships, refrigerators) and autonomous cogeneration plants (industrial heat and power plants, municipal energy, factory boilers) for combined obtaining of thermal and electric energy.

Figure 9 shows the thermal diagrams of two types of EGB.

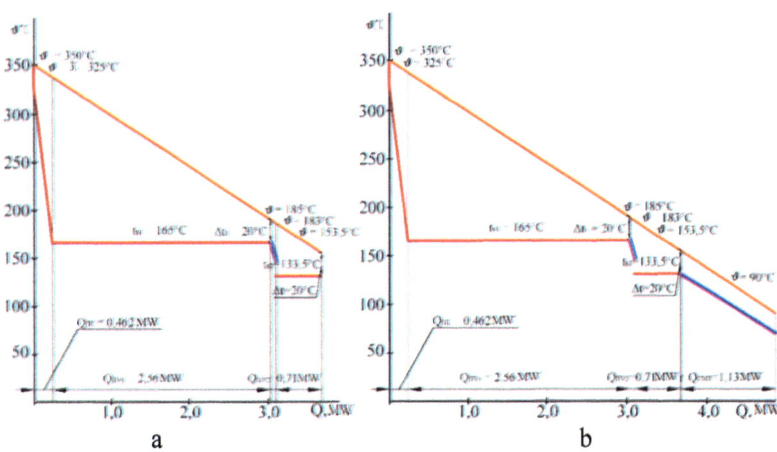

Fig. 9. Thermal diagrams of different schemes of EGB: (a) two-pressure EGB; (b) two-pressure EGB with hot water section (condensing surface)

The calculations showed that using of condensing heating surface in such schemes increases a depth of utilization by 23% (Fig. 9).

5 Conclusions

The efficiency of joint cooling the Intake Air of the turbocharger and Scavenge Air at the inlet of the working cylinders of the Ship Main Engine of the transport ship by the conversion of the waste heat by an Ejector Chiller and an Absorption Lithium-Bromide Chiller was analyzed. As a source of heat for chillers, water from the heat supply system, obtained from the heat of the Scavenge Air and engine exhaust gases, is used. The reserves to improve the cooling efficiency of the main diesel engine cyclic air through using the chillers of various types were reveiled. A new approach to improving the efficiency of cyclic air cooling of the Main Engine of a transport ship has been proposed. Studies show that the smallest values of corrosion intensity are observed for large values of water content in water-fuel emulsion 30%. The minimum value of wall temperature is determined, at which the permissible speed of low-temperature corrosion at a level of 0.25 mm/year is ensured and the minimum value of exhaust gas temperature at the exit from exhaust gas boiler. The wall temperature range of condensing LTHS that provides their safe operation is determined, which opens opportunities for deep exhaust gas heat utilization.

References

1. MAN Diesel & Turbo: MAN B&W Two-stroke Marine Engines. Emission Project Guide (2018). https://marine.man-es.com/applications/projectguides/2stroke/content/special_pg/7020-0145-09_uk.pdf. Accessed 9 Oct 2018
2. Wartsila: Wärtsilä Environmental Product Guide (2017). https://cdn.wartsila.com/docs/default-source/product-files/egc/product-guide-o-env-environmental-solutions.pdf. April 2017
3. MAN Diesel Turbo: CEAS Engine Calculations (2019). https://marine.man-es.com/two-stroke/ceas
4. Radchenko, R., Radchenko, A., Serbin,S., Kantor, S., Portnoi, B.: Gas turbine unite inlet air cooling by using an excessive refrigeration capacity of absorption-ejector chiller in booster air cooler. In: HTRSE-2018, E3S Web of Conferences, vol. 70, no. 03012, 6 p. (2018)
5. Radchenko A., Radchenko M., Konovalov A., Zubarev A.: Increasing electrical power output and fuel efficiency of gas engines in integrated energy system by absorption chiller scavenge air cooling on the base of monitoring data treatment. In: HTRSE-2018, E3S Web of Conferences, vol. 70, no. 03011, 6 p. (2018)
6. Radchenko, M., Radchenko, R., Ostapenko, O., Zubarev, A., Hrych, A.: Enhancing the utilization of gas engine module exhaust heat by two-stage chillers for combined electricity, heat and refrigeration. In: 5th International Conference on Systems and Informatics, ICSAI 2018, Jiangsu, Nanjing, China, pp. 240–244 (2019)
7. Radchenko, A., Bohdal, L., Zongming, Y., Portnoi, B., Tkachenko, V.: Rational designing of gas turbine inlet air cooling system. In: Tonkonogyi, V., et al. (eds.) Grabchenko's International Conference on Advanced Manufacturing Processes. InterPartner-2019. Lecture Notes in Mechanical Engineering, 10 p. Springer, Cham (2020)

8. Forduy, S., Radchenko, A., Kuczynski, W., Zubarev, A., Konovalov, D.: Enhancing the fuel efficiency of gas engines in integrated energy system by chilling cyclic air. In: Tonkonogyi, V., et al. (eds.) Grabchenko's International Conference on Advanced Manufacturing Processes. InterPartner-2019. Lecture Notes in Mechanical Engineering, 10 p. Springer, Cham (2020)
9. Luo, C., Luo, K., Wang, Y., Ma, Z., Gong, Y.: The effect analysis of thermal efficiency and optimal design for boiler system. Energy Proc. **105**, 3045–3050 (2017)
10. Shamsi, S.S.M., Negash, A.A., Cho, G.B., Kim, Y.M.: Waste heat and water recovery system optimization for flue gas in thermal power plants. Sustainability **11**(7), 1881 (2019)
11. Lecompte, S., Huisseune, H., van den Broek, M., Vanslambrouck, B., Paepe, M.: Review of organic Rankine cycle (ORC) architectures for waste heat recovery. Renew. Sustain. Energy Rev. **47**, 448–461 (2015)
12. Li, F., Duanmu, L., Fu, L., Zhao, X.: Research and application of flue gas waste heat recovery in cogeneration based on absorption heat-exchange. Proc. Eng. **146**, 594–603 (2016)
13. Levy, E., Bilirgen, H., Hazell, D., Goel, N., Jonas, G., Carney, B.: Performance and cost benefit analyses of condensing heat exchangers for cooling boiler flue gas. In: Proceedings 2012 Clearwater Clean Coal Conference, Clearwater, Florida (2012)
14. Baldi, S., Quang, T.L., Holub, O., Endel, P.: Real-time monitoring energy efficiency and performance degradation of condensing boilers. Energy Convers. Manag. **136**, 329–339 (2017)
15. Jeong, K., Kessen, M.J., Bilirgen, H., Levy, E.K.: Analytical modeling of water condensation in condensing heat exchanger. Int. J. Heat Mass Transf. **53**, 2361–2368 (2010)
16. Antonescu, N., Stănescu, P.-D.: Parametrical functioning study for a small boiler condensing unit. In: Bode, F., et al. (eds.) Sustainable Solutions for Energy and Environment Conference, EENVIRO 2016. Energy Procedia, vol. 112, pp. 563–570. Bucharest (2017
17. Fan, C., Pei, D., Wei, H.: A novel cascade energy utilization to improve efficiency of double reheat cycle. Energy Convers. Manag. **171**, 1388–1396 (2018)
18. Radchenko, M., Radchenko, R., Kornienko, V., Pyrysunko, M.: Semi-empirical correlations of pollution processes on the condensation surfaces of exhaust gas boilers with water-fuel emulsion combustion. In: Ivanov, V., et al. (eds.) Advances in Design, Simulation and Manufacturing II. DSMIE 2019. Lecture Notes in Mechanical Engineering, pp. 853–862. Springer, Cham (2020)
19. Kornienko, V., Radchenko, R., Stachel, A., Pyrysunko, M.: Correlations for pollution on condensing surfaces of exhaust gas boilers with water-fuel emulsion combustion. In: Tonkonogyi, V., et al. (eds.) Grabchenko's International Conference on Advanced Manufacturing Processes. InterPartner-2019. Lecture Notes in Mechanical Engineering 10 p. Springer, Cham (2020)
20. Miao, Y.C., Yu, C.L., Wang, B.H., Chen, K.: The applied research of emulsified heavy fuel oil used for the marine diesel engine. Adv. Mater. Res. **779**, 469–476 (2013)
21. Sugeng, D.A., Ithnin, A.M., Amri, N.S.M.S., Ahmad, M.A., Yahya, W.J.: Water content determination of steam generated water-in-diesel emulsion. J. Adv. Res. Fluid Mech. Thermal Sci. **49**(1), 62–68 (2018)
22. Baskar, P., Senthil Kumar, A.: Experimental investigation on performance characteristics of a diesel engine using diesel-water emulsion with oxygen enriched air. Alexandria Eng. J. **56** (1), 137–146 (2017)
23. Patel, K.R., Dhiman, V.: Research study of water- diesel emulsion as alternative fuel in diesel engine – an overview. Int. J. Latest Eng. Res. Appl. **2**(9), 37–41 (2017)

24. Wojs, M.K., Orliński, P., Kamela, W., Kruczyński, P.: Research on the influence of ozone dissolved in the fuel-water emulsion on the parameters of the CI engine. In: IOP Conference Series: Materials Science and Engineering, vol. 148, pp. 1–8 (2016)
25. Gupta, R.K., Sankeerth, K.A., Sharma, T.K., Rao, G., Murthy, K.M.: Effects of water-diesel emulsion on the emission characteristics of single cylinder direct injection diesel engine - a review. Appl. Mech. Mater. **592**, 1526–1533 (2014)

Increasing the Operation Efficiency of Air Conditioning System for Integrated Power Plant on the Base of Its Monitoring

Eugeniy Trushliakov[1], Andrii Radchenko[1(✉)], Serhiy Forduy[2], Anatolii Zubarev[1], and Artem Hrych[1]

[1] Admiral Makarov National University of Shipbuilding, 9 Heroes of Ukraine Avenue, Mykolayiv 54025, Ukraine
nirad50@gmail.com
[2] PepsiCo, Inc., Kiev, Ukraine

Abstract. The efficiency of reciprocating gas engines of integrated energy systems (IES) for combined electricity, heat and refrigeration generation is strictly influenced by their cyclic air temperatures. To evaluate the effect of gas engine cyclic air deep cooling, compared with conventional its cooling, the data on dependence of fuel consumption and power output of gas engine JMS 420 GS-N.L on its inlet air temperature at varying ambient air temperatures at the entrance of the radiator for scavenge air cooling were received. The results of treatment of gas engine efficiency monitoring proved non-effective operation of conventional chilling all the ambient air, coming into the engine room, because of increased air temperature at the inlet of turbocharger (TC), caused by heat influx from surroundings in the engine room. A new method of gas engine inlet air two-stage cooling at increased ambient air temperatures and advanced cyclic air cooling system with absorption lithium-bromide chiller and refrigerant ejector chiller was proposed. With this chilled water from absorption lithium-bromide chiller is used as a coolant in the first high-temperature stage of engine inlet air cooler and boiling refrigerant of ejector chiller in the second low-temperature stage.

Keywords: Gas engine · Integrated energy system · Fuel consumption · Air cooling · Chiller

1 Introduction

Reciprocating gas engines (GE) found a widespread application in integrated energy systems (IES), or so called trigeneration systems [1–4], for combined electricity, heat and refrigeration generation [5–9]. For application in plants of combined energy supply the engine manufacturers developed gas engines (GE) as cogeneration engine modules equipped with heat exchangers for producing hot water or steam by using heat of exhaust gas, charged gas-air mixture (scavenge air), engine jacket and lubricant oil cooling water [10].

With rise in intake air temperature a thermodynamic efficiency of gas engines and, respectively, power plants essentially decreases: electric power output decreases and

M. Nechyporuk et al. (Eds.): *Integrated Computer Technologies in Mechanical Engineering*, AISC 1113, pp. 351–360, 2020.
https://doi.org/10.1007/978-3-030-37618-5_30

specific fuel consumption increases. A cold, generated by absorption chiller, that recovers the heat released from the engine, could be used for cooling engine intake air. Application of the received cold for cooling demands of the engine itself (in-cycle trigeneration) would provide not only improvement of its fuel efficiency and a higher ratio of electric power to byproduct heat, but also prolong the time of efficient operation of trigeneration plant, since cooling demands for technological needs have, as a rule, periodic character.

The most of well-known concepts of increasing the efficiency of trigeneration plant are limited to engine out-cycle use of a cold (for external consumers) [9–15], and owing to this do not provide realization of the additional reserves through conversing heat, released from the engine, in its working cycle. A realization of in-cycle trigeneration concept would broaden the applicability of trigeneration systems due to their applications even without enough heat and cooling demands.

A conventional method of chilling all the ambient air, coming into the engine room, from were it is sucked by engine turbocharger (TC), is non-effective because of heat influx from surroundings to the air stream sucked, that results in increased air temperature at the inlet of turbocharger and enlarged cooling capacity required for chilling all the ambient air coming into the engine room. The problem of engine cyclic air cooling arises especially actually for increasing ambient air temperature above 25…30 °C.

To solve this problem a new method of gas engine inlet air two-stage cooling to stabilize its inlet temperature at increased ambient air temperatures with absorption lithium-bromide chiller and refrigerant ejector chiller was proposed. With this chilled water from absorption lithium-bromide chiller is used as a coolant in the first high-temperature stage of engine inlet air cooler and boiling refrigerant of ejector chiller in the second low-temperature stage.

The problem of engine cyclic air cooling arises especially actually when ambient air temperature increases above 25…30 °C. If scavenge air temperature exceeds the maximum temperature restriction of 50 °C the engine efficiency drops sharp. For protection of engine from negative impact of high scavenge air temperature on the thermal conditions in the combustion cylinders the scavenge air temperature of about 40 °C is maintained automatically by reducing gas supply to engine and, accordingly, engine load.

The purpose of the research is to estimate the enhancement of fuel efficiency of gas engine in integrated energy system due to cooling engine inlet air at increased ambient temperatures with using the monitoring data of actual gas engine performance.

2 Literature Review

An enhancement of fuel efficiency of combustion engines is possible by cooling inlet air in vapour-compression and waste heat recovery chillers. The absorption lithium-bromide chillers (ACh) are the most widely used and provide cooling air to the temperature of around 15 °C with a high coefficient of performance (COP = 0.7–0.8) [16–23, 27–29]. Vapour-compression chillers consume electrical energy to drive compressors and provide cooling air practically to any low temperature [3, 31]. The most simple refrigerant ejector chillers, which convert the exhaust heat of engine into a cold, provide cooling air to the temperature of 5 °C but with low efficiency: COP = 0.2–0.3 [4, 26, 30].

The efficiency of cooling systems and refrigeration machines can be improved through intensification of heat transfer processes in heat exchangers and by application of advanced scheme decisions [4, 24, 30] and energy conserving technologies like booster ambient air precooling by using an excess of refrigeration capacity, accumulated at decreased cooling loads, to cover peak heat loads [17, 25].

Various methodological approaches has been proposed for designing engine inlet air cooling systems and first of all for the choice of their rational cooling capacities [2, 5–8, 11, 13, 16, 18, 19, 21, 25, 32–36]. Some of them are based on the annular fuel savings due to engine inlet air cooling and is conducted by using the summarized annular fuel savings of the engine dependence on a design cooling capacity of the to choose its rational value, that provides closed to maximum annual fuel savings of the engine or maximum rate of the annual fuel saving increment due to increasing a design cooling capacity [25, 36].

Many researches deal with improving the performance efficiency of trigeneration systems for combined electricity, heat and refrigeration generation for space conditioning and technological and other needs [5–8, 11, 20, 22, 27, 28].

The most of well-known concepts of increasing the efficiency of trigeneration plants are limited to application for external consumers of cold [10, 12, 22] and owing to this do not provide realization of the additional reserves through conversing the heat released from the engine in its working cycle. A realization of in-cycle trigeneration concept would broaden the applicability of trigeneration systems due to their applications even without stable heat and cooling needs.

3 Research Methodology

A fuel efficiency of gas engine of integrated energy system for combined electricity, heat and cooling generation was investigated to analyze the influence of chilling air at the inlet of engine on its efficiency. A cold, generated by absorption lithium-bromide chiller (ACh), that recovers the heat released from the engine, was used for cooling intake air of engine.

To evaluate the effect of gas engine inlet air deep cooling, compared with conventional its procession in the central air conditioner for conditioning ambient air coming into the engine room, data on the dependence of fuel consumption and power output of gas engine on its inlet air temperature at varying ambient air temperatures at the entrance of the radiator for scavenge air cooling water were received by treating the results of gas engine JMS 420 GS-N.L efficiency monitoring.

A method for treatment of the monitoring data on fuel consumption and power output of gas engine was developed [23]. The results of monitoring of gas engine fuel efficiency were presented in the form of data sets on dependence of fuel consumption $B_e = f(t_{in})$ and engine power output $P_e = f(t_{in})$ upon the air temperatures t_{in} at the inlet of the turbocharger for the various ambient air temperature t_{amb} at the entrance to the radiator of scavenge air cooling.

The goal of monitoring data sets $P_e = f(t_{in})$ and $B_e = f(t_{in})$ treatment was to calculate the magnitude of the change in power ΔP_e and fuel consumption ΔB_e, caused by the change in the engine inlet air temperature t_{in} by 1 °C, that is $\Delta P_e/\Delta t_{in}$ and $\Delta B_e/\Delta t_{in}$, to evaluate the effect of the application of developed air cooling system.

A treatment of monitoring data of gas engine JMS 420 GS-N.L in integrated energy system for combined electricity, heat and cooling has proved a non-effective operation of conventional chilling all the ambient air, coming into engine room, because of increased air temperature at the inlet of engine turbocharger (TC), caused by heat influx from the engine room.

The two-stage cooling system of gas engine inlet air by chilled water from the absorption lithium-bromide chiller (ACh) as a coolant in the first high-temperature stage AC_{HT} of engine inlet air cooler (AC) and boiling refrigerant of ejector chiller (ECh) in the second low-temperature stage AC_{LT} was developed. To prove its efficiency the current values of air temperature at the inlet of gas engine turbocharger t_{in}, air temperature at the exit of central air conditioner CAC t_{AC2}, air temperature drop in the air cooler of CAC $\Delta t_{AC} = t_{amb} - t_{AC2}$, air temperature increase in the engine room $\Delta t_{ER} = t_{in} - t_{AC2}$ were compared with their values for the high-temperature stage AC_{HT} t_{HT2} and low-temperature stage AC_{LT} t_{LT2}, reduction of air temperature in the high-temperature stage AC_{LT} Δt_{HT} and low-temperature stage AC_{LT} Δt_{LT} and full temperature reduction in the proposed two-stage air cooler $\Delta t_{AC} = t_{amb} - t_{LT2}$, and corresponding heat loads on the high-temperature stage AC_{HT} $Q_{0.HT}$ and low-temperature stage AC_{LT} $Q_{0.LT}$ of the air cooler.

4 Results

4.1 Investigating the Efficiency of Typical Gas Engine Inlet Air Cooling System

The efficiency of cooling air at the inlet of gas engine was investigated for IES of combined energy supply at the factory "Sandora"–"PepsiCo Ukraine" (Nikolaev, Ukraine). The integrated energy system is equipped with 2 cogeneration Jenbacher gas engines JMS 420 GS-N.LC (rated electric power P_{eISO} = 1400 kW, heat power Q_h = 1500 kW) and absorption lithium-bromide chiller.

The heath taken away from exhaust gas by waste heat recovery boiler (economizer), from charged gas-air mixture in a high-temperature stage of an intercooler (IC_{HT}), from engine jacket and lubricant oil cooling water in corresponding coolant radiators, is used by absorption chiller for producing a chilled water with temperature of around 11…12 °C. Chilled water is used for technological process cooling and by central conditioners for cooling engine room intake air, from where cooled air is sucked by engine turbochargers (Fig. 1).

A typical scheme of gas engine inlet air cooling system is presented in Fig. 2. The engine room intake air is cooled in the central conditioner by chilled water from absorption lithium-bromide chiller and the engine turbocharger sucks the air from engine room.

a b

Fig. 1. Cogeneration gas engine module JMS GE Jenbacher (a) and waste heat recovery central conditioner for cooling engine room intake air (b)

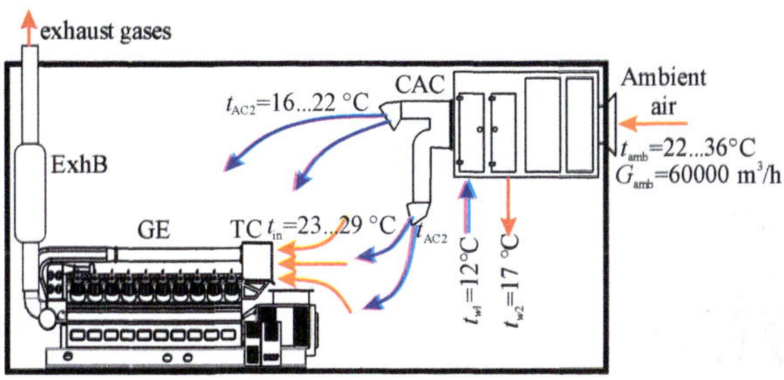

Fig. 2. A typical scheme of gas engine inlet air system with cooling the engine room intake air in the central conditioner by chilled water from the ACh and sucking the chilled air by the engine turbocharger from engine room: ExhB – exhaust heat boiler; GE – gas engine; CAC – central air conditioner; TC – turbocharger;

Because of heat influx from the engine room the temperature of engine intake air t_{in} is higher than its value at the outlet of central air conditioner (CAC) t_{AC2} (Fig. 3).

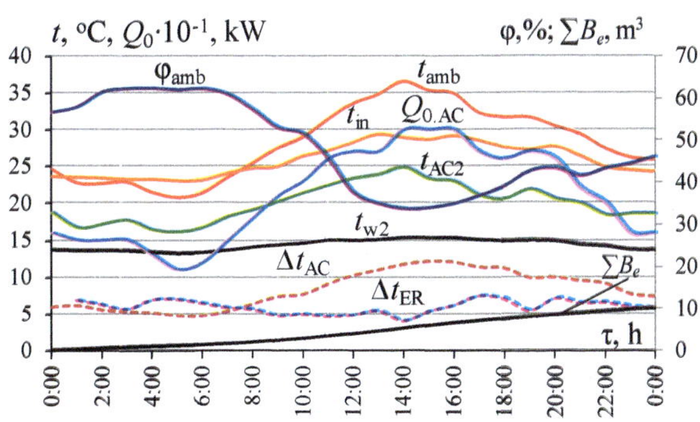

Fig. 3. Daily variation of temperature t_{amb} and relative humidity φ_{amb} of ambient air, temperature of air at the inlet of gas engine turbocharger t_{in}, air at the outlet of central air conditioner CAC t_{AC2}, cooling water at the outlet of air cooler of CAC t_{w2}, air temperature drop in the air cooler of CAC $\Delta t_{AC} = t_{amb} - t_{AC2}$, air temperature increase in the engine room $\Delta t_{ER} = t_{in} - t_{AC2}$, cooling capacity of air conditioner $Q_{0.AC}$ with air flow G_a = 60000 m³/h, $\Sigma\Delta B_e$ – full daily savings of natural gas due to cooling of air of ER, m³

4.2 Investigating the Efficiency of Improved Gas Engine Inlet Air Cooling System

The scheme of two-stage cooling system of gas engine inlet air by chilled water from the absorption lithium-bromide chiller (ACh) as a coolant in the first high-temperature stage AC_{HT} of engine inlet air cooler (AC) and boiling refrigerant of ejector chiller (ECh) in the second low-temperature stage AC_{LT} is presented in Fig. 4.

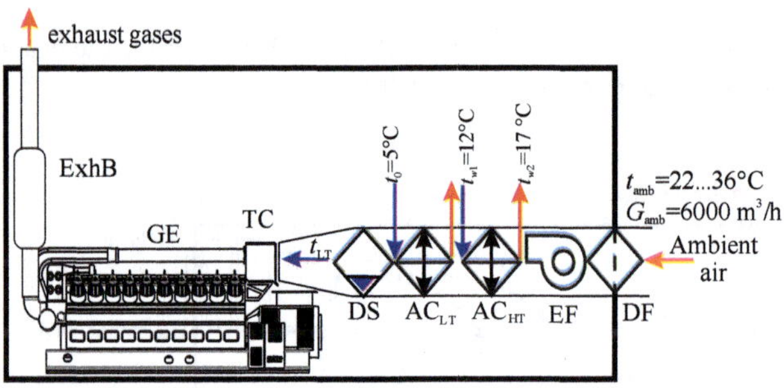

Fig. 4. The scheme of two-stage cooling system of gas engine inlet air by chilled water from the ACh in the first high-temperature stage AC_{HT} of engine inlet air cooler (AC) and boiling refrigerant of ejector chiller (ECh) in the second low-temperature stage AC_{LT}: ExhB – exhaust heat boiler; GE – gas engine; EF – electric fan; DF – dust filter; DS – droplet separator; TC – turbocharger; AC_{LT} and AC_{HT} – low- and high-temperature air coolers

Current values of air temperature at the exit from the high-temperature stage AC_{HT} t_{HT2} and low-temperature stage AC_{LT} t_{LT2} of two-stage air cooler, reduction of air temperature in the high-temperature stage AC_{HT} Δt_{HT} and low-temperature stage AC_{LT} Δt_{LT} and full temperature reduction in the air cooler $\Delta t_{AC} = t_{amb} - t_{LT2}$, heat load on the high-temperature stage AC_{HT} $Q_{0.HT}$ and low-temperature stage AC_{LT} $Q_{0.LT}$ and full heat load on the whole air cooler $Q_{0.AC}$, corresponding current reduction of specific fuel consumption due to cooling engine cyclic air in the high-temperature $\Delta b_{e.HT}$ and low-temperature $\Delta b_{e.LT}$ stages and in the whole air cooler Δb_e and the total daily reduction of fuel consumption ΣB_e are presented in Figs. 5 and 6.

A proposed two-stage cooling system of gas engine inlet air by chilled water from the absorption lithium-bromide chiller (ACh) in the first high-temperature stage AC_{HT} of engine inlet air cooler (AC) and boiling refrigerant of ejector chiller (ECh) in the second low-temperature stage AC_{LT} can provide decreasing engine inlet air temperature by about 20…25 °C compared with a typical scheme of gas engine inlet air system with cooling the engine room intake air in the central conditioner by chilled water from the ACh and sucking the chilled air by the engine turbocharger from engine room, that results in reduction of engine specific fuel consumption by about $\Delta b_e = (1\ldots2)$ g/(kW•h), i.e. about 2…3% decrease in specific fuel consumption at increased ambient air temperatures t_{amb} = 30…35 °C.

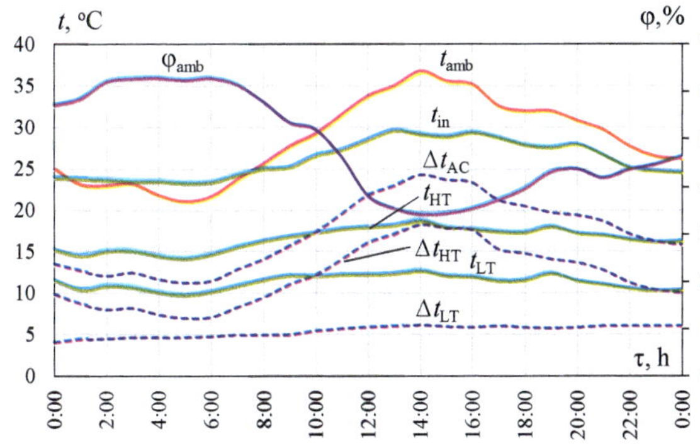

Fig. 5. Current values of temperature t_{amb} and relative humidity φ_{amb} of ambient air, temperature of air at the inlet of gas engine turbocharger t_{in} in basic version with central conditioner (Fig. 3), air temperature at the exit from the high-temperature stage AC_{HT} t_{HT} and low-temperature stage AC_{LT} t_{LT} of air cooler, reduction of air temperature in the high-temperature stage AC_{HT} Δt_{HT} and low-temperature stage AC_{LT} Δt_{LT} of air cooler and full temperature reduction in the air cooler $\Delta t_{AC} = t_{amb} - t_{LT}$ with air flow $G_a = 6000$ m^3/h

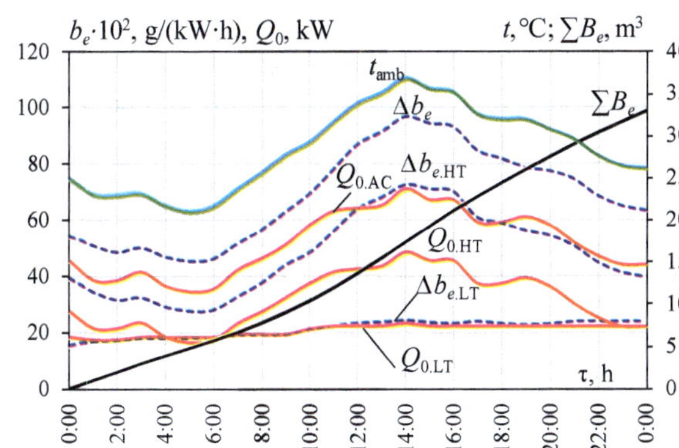

Fig. 6. Current values of ambient temperature t_{amb}, heat load on the high-temperature stage $Q_{0.HT}$ and low-temperature stage $Q_{0.LT}$ and full heat load on the air cooler $Q_{0.AC}$, current reduction of specific fuel consumption due to cooling air in the high-temperature $\Delta b_{e.HT}$ and low-temperature $\Delta b_{e.LT}$ stages and in the whole air cooler Δb_e and the total daily reduction of fuel consumption ΣB_e: air flow $G_a = 6000$ m^3/h

5 Conclusions

To evaluate the effect of gas engine inlet air deep cooling, compared with conventional its procession in the central air conditioner for conditioning ambient air coming into the engine room, data on the dependence of fuel consumption and electrical power output of gas engine on its inlet air temperature at varying ambient air temperatures at the entrance of the radiator for scavenge air cooling water were received by treating the results of gas engine JMS 420 GS-N.L efficiency monitoring in integrated energy system for combined electricity, heat and cooling generation in absorption lithium-bromide chiller.

The results of monitoring of gas engine fuel efficiency were presented in the form of data sets on dependence of fuel consumption $B_e = f(t_{in})$ and engine electrical power output $P_e = f(t_{in})$ on the air temperatures t_{in} at the inlet of the turbocharger for the various ambient air temperature t_{amb} at the entrance to the radiator of scavenge air cooling.

The monitoring data sets $P_e = f(t_{in})$ and $B_e = f(t_{in})$ were treated to calculate the magnitude of the change in electrical power ΔP_e and fuel consumption ΔB_e, caused by the change in the engine inlet air temperature t_{in} by 1 °C, that is $\Delta P_e/\Delta t_{in}$ and $\Delta B_e/\Delta t_{in}$, to evaluate the effect of the application of developed air cooling system.

A treatment of monitoring data on electrical power output and fuel consumption of gas engine JMS 420 GS-N.L in integrated energy system for combined electricity, heat and cooling generation in absorption lithium-bromide chiller has proved non-effective conventional method of chilling all the ambient air, coming into the engine room, from were it is sucked by engine turbocharger, is non-effective because of heat influx from surroundings to the air stream sucked, that results in increased air temperature at the inlet of turbocharger and enlarged cooling capacity required for chilling all the ambient air coming into the engine room.

A new method of gas engine inlet air two-stage cooling to stabilize its inlet temperature at increased ambient air temperatures with absorption lithium-bromide chiller and refrigerant ejector chiller was proposed. With this chilled water from absorption lithium-bromide chiller is used as a coolant in the first high-temperature stage of engine inlet air cooler and boiling refrigerant of ejector chiller in the second low-temperature stage.

An advanced gas engine inlet air cooling system with two-stage absorption-ejector chiller conversing waste heat of gas engine was proposed.

References

1. Angrisani, G., Rosato, A., Roselli, C., Sasso, M., Sibili, S.: Experimental results of a micro-trigeneration installation. Appl. Therm. Eng. **38**, 8 p. (2012)
2. Arteconi, A., Brandoni, C., Polonara, F.: Distributed generation and trigeneration: energy saving opportunities in Italian supermarket sector. Appl. Therm. Eng. **29**(8–9), 1735–1743 (2009)
3. Bassols, J., Kuckelkorn, B., Langrek, J., Schneider, R., Veelken, H.: Trigeneration in the food industry. Appl. Therm. Eng. **22**, 595–602 (2002)
4. Butrymowicz, D., Gagan, J., Śmierciew, K., Łukaszuk, M., Dudar, A., Pawluczuk, A., Łapiński, A., Kuryłowicz, A.: Investigations of prototype ejection refrigeration system driven by low grade heat. In: E3S Web of Conferences (HTRSE–2018), vol. 70, 7 p. (2018)
5. Canova, A., Cavallero, C., Freschi, F., Giaccone, L., Repetto, M., Tartaglia, M.: Optimal energy management. IEEE Ind. Appl. Mag. **15**, 62–65 (2009)
6. Cardona, E., Piacentino, A., Cardona, F.: Energy saving in airports by trigeneration. Part I: assessing economic and technical potential. Appl. Thermal Eng. **26**, 1427–1436 (2006)
7. Cardona, E., Piacentino, A.: A methodology for sizing a trigeneration plant in mediterranean areas. Appl. Thermal Eng. **23**, 15 p. (2003)

8. Carvalho, M., Serra, L.M., Lozano, M.A.: Geographic evaluation of trigeneration systems in the tertiary sector. Effect of climatic and electricity supply conditions. Energy **36**, 1931–1939 (2011)
9. Chua, K.J., Yang, W.M., Wong, T.Z., et al.: Integrating renewable energy technologies to support building trigeneration – a multi-criteria analysis. Renewable Energy **41**, 358–367 (2012)
10. Elsenbruch, T.: Jenbacher gas engines a variety of efficient applications. Bucureşti **28**, 73 p. (2010)
11. Freschi, F., Giaccone, L., Lazzeroni, P., Repetto, M.: Economic and environmental analysis of a trigeneration system for food-industry: a case study. Appl. Energy **107**, 157–172 (2013)
12. GE Jenbacher Company Overview: Economic utilization of Biomass and Municipal Waste for power generation. Some energy lasts for generations, June 13, 39 p. (2007)
13. Ghaebi, H., Karimkashi, Sh., Saidi, M.H.: Integration of an absorption chiller in a total CHP site for utilizing its cooling production potential based on R-curveconcept. Int. J. Refrig. **35**, 8 p. (2012)
14. Gluesenkamp, K., Yunho, H., Radermacher, R.: High efficiency micro trigeneration systems. Appl. Thermal Eng. **50**, 6 p. (2013)
15. IEA – International Energy Agency Report: Combined heat and power – Evaluating the benefits of grater global investment. International Energy Agency. http://www.iea.org/Papers/2008/chp_report.pdf
16. Kavvadias, K., Tosios, A., Maroulis, Z.: Design of a combined heating, cooling and power system: sizing, operation strategy selection and parametric analysis. Energy Convers. Manage. **51**, 833–845 (2009)
17. Lai, S.M., Hui, C.W.: Integration of trigeneration system and thermal storage under demand uncertainties. Appl. Energy **87**(9), 2868–2880 (2010)
18. Markis, T., Paravantis, J.A.: Energy conservation in small enterprises. Energy Build. **39**, 11 p. (2007)
19. Marques, R.P., Hacon, D., Tessarollo, A., Parise, J.A.R.: Thermodynamic analysis of tri-generation systems taking into account refrigeration, heating and electricity load demands. Energy Build. **42**, 2323–2330 (2010)
20. Mróz, T.M.: Thermodynamic and economic performance of the LiBr–H_2O single stage absorption water chiller. Appl. Thermal Eng. **26**, 7 p. (2006)
21. Ortiga, J., Bruno, J.C., Coronas, A.: Operational optimisation of a complex trigeneration system connected to a district heating and cooling network. Appl. Therm. Eng. **50**, 1536–1542 (2013)
22. Possidente, R., Roselli, C., Sasso, M., Sibilio, S.: Experimental analysis of micro-cogeneration units based on reciprocating internal combustion engine. Energy Build. **38**, 5 p. (2006)
23. Radchenko, A., Radchenko, M., Konovalov, A., Zubarev, A.: Increasing electrical power output and fuel efficiency of gas engines in integrated energy system by absorption chiller scavenge air cooling on the base of monitoring data treatment. In: E3S Web of Conferences (HTRSE–2018), vol. 70, 6 p. (2018)
24. Radchenko, N.: A concept of the design and operation of heat exchangers with change of phase. Arch. Thermodyn.: Pol. Acad. Sci. **25**(4), 3–19 (2004)
25. Radchenko, R., Radchenko, A., Serbin, S., Kantor, S., Portnoi, B.: Gas turbine unite inlet air cooling by using an excessive refrigeration capacity of absorption-ejector chiller in booster air cooler. In: E3S Web of Conferences (HTRSE–2018), vol. 70, 6 p. (2018)

26. Radchenko, M., Radchenko, R., Ostapenko, O., Zubarev, A., Hrych, A.: Enhancing the utilization of gas engine module exhaust heat by two-stage chillers for combined electricity, heat and refrigeration. In: The 5th International Conference on Systems and Informatics: ICSAI 2018, Jiangsu, Nanjing, China, pp. 227–231 (2018)
27. Rocha, M.S., Andreos, R., Simões-Moreira, J.R.: Performance tests of two small trigeneration pilot plants. Appl. Therm. Eng. **41**, 84–91 (2012)
28. Rodriguez-Aumente, P.A., Rodriguez-Hidalgo, M.C., Nogueira, J.I., Lecuona, A., Venegas, M.C.: District heating and cooling for business buildings in Madrid. Appl. Therm. Eng. **50**, 1496–1503 (2013)
29. Rouse, G., Czachorski, M., Bishop, P., Patel, J.: GTI integrated energy system for buildings. Modular system prototype. Gas Technology Institute, 495 p., January 2006
30. Smierciew, K., Gagan, J., Butrymowicz, D., Karwacki, J.: Experimental investigations of solar driven ejector air-conditioning system. Energy Build. **80**, 260–267 (2014)
31. Suamir, I.N., Tassou, S.A.: Performance evaluation of integrated trigeneration and CO_2 refrigeration systems. Appl. Therm. Eng. **50**, 1487–1495 (2013)
32. Sugiartha, N., Tassou, S.A., Chaer, I., et al.: Trigeneration in food retail: an energetic, economic and environmental evaluation for a supermarket application. Appl. Therm. Eng. **29** (13), 2624–2632 (2009)
33. Sugiartha, N., Tassou, S.A., Chaer, I., Marriott, D.: Trigeneration in food retail: an energetic, economic and environmental evaluation for a supermarket application. Appl. Therm. Eng. **29**, 2624–2632 (2009)
34. Tassou, A., Chaer, I., Sugiartha, N., Ge, Y.-T., Marriott, D.: Application of tri-generation systems to the food retail industry. Energy Convers. Manage. **48**, 2988–2995 (2007)
35. Tianhong, P., Dongliang, X., Zhengming, L., Shyan-Shu, S., Shi-Shang, J.: Efficiency improvement of cogeneration system using statistical model. Energy Convers. Manag. **68**, 7 p. (2013)
36. Trushliakov, E., Radchenko, M., Radchenko, A., Kantor, S., Zongming, Y.: Statistical approach to improve the efficiency of air conditioning system performance in changeable climatic conditions. In: The 5th International Conference on Systems and Informatics: ICSAI 2018, Jiangsu, Nanjing, China, pp. 1303–1307 (2018)

Monitoring the Fuel Efficiency of Gas Engine in Integrated Energy System

Andrii Radchenko[1(✉)], Dariusz Mikielewicz[2], Serhiy Forduy[3], Mykola Radchenko[1], and Anatolii Zubarev[1]

[1] Admiral Makarov National University of Shipbuilding, Heroes of Ukraine Avenue, 9, Mykolayiv, Ukraine
nirad50@gmail.com
[2] Gdansk University of Technology, Gdansk, Poland
[3] PepsiCo, Inc., Kiev, Ukraine

Abstract. Reciprocating gas engines are widely used in integrated energy systems (IES) for combined cooling, heat and electricity supply. The fuel efficiency of gas engine considerably depends on the air temperature at the inlet of turbocharger. To estimate the influence of inlet air on gas engine fuel efficiency the data on dependence of fuel consumption and power output of gas engine on its inlet air temperature were received for Jenbacher gas engine JMS 420 GS-N. LC of cogenerative type with using the heat of exhaust gas, scavenge air, engine jacket cooling water and lubricating oil for heating water to the temperature of about 90 °C. The hot water is used in absorption lithium-bromide chiller to produce a chilled water, which is spent for technological needs and feeding to the central air conditioner, that provides cooling ambient air incoming the engine room, from where it is sucked by the engine turbocharger. The data on dependence of fuel consumption and power output of gas engine on its inlet air temperature received by processing the monitoring data are very usefull for improving the efficiency of conventional cooling system with chilling the engine room incoming air followed by heat influx from surroundings, that leades to increased air temperature at the suction of turbocharger and enlarged cooling capacity spent for cooling.

Keywords: Monitoring · Combined cooling · Heat and electricity · Fuel consumption · Power output · Absorption chiller

1 Introduction

Reciprocating gas engines are widely used in integrated energy systems (IES) for combined cooling, heat and electricity supply [1–3]. They are manufactured as cogeneration modules with heat exchangers [4, 5] to produce hot water with temperature of about 90 °C by utilizing the heat of exhaust gas, scavenge air, engine jacket cooling water and lubricating oil. A hot water is used as a heat source for absorption lithium-bromide chiller (ACh) to produce a chilled water, which is spent for technological needs and feeding to the central air conditioner for cooling ambient air incoming the engine room, from where it is sucked by the turbocharger. The peculiarity

M. Nechyporuk et al. (Eds.): *Integrated Computer Technologies in Mechanical Engineering*, AISC 1113, pp. 361–370, 2020.
https://doi.org/10.1007/978-3-030-37618-5_31

of IES, providing chilled water for food processing, is increased its temperature of about 12 °C, that is higher than typical temperature 7 °C, that might be produced in ACh. In the case of using this chilled water as a coolant for cooling air at the inlet of turbocharger (TC) to inhance engine efficiency it is desirably to use reduced coolant temperature to provide lowered inlet air temperature as compared with its increased temperature, caused by heat influx from the engine room in conventional chilling the ambient air, coming into the engine room.

To evaluate the efficiency of gas engine cyclic air cooling the data on dependence of fuel consumption and power output of gas engine at varying its inlet air are needed. These data can be received by processing gas engine fuel efficiency monitoring.

2 Literature Review

A lot of energy saving technologies are developed to increase fuel efficiency of combustion engines: by utilization of exhaust heat [6–8] and cooling engine cyclic air in refrigerant electrically driven vapour-compression [9] and waste heat recovery chillers [10–14]. The most widespread ACh provide cooling air to the temperature of about 15 °C with a high coefficient of performance: COP = 0.7–0.8 [15–20]. The refrigerant vapour-compression chillers provide cooling air practically to any low temperature but they consume electricity to drive compressors [9]. The refrigerant ejector chillers (ECh) are the most simple in design and provide cooling air to the temperature of 5–10 °C but with low COP = 0.2–0.3 [21–23].

The efficiency of heat exchangers in cooling systems and refrigeration chillers can be improved by intensification of heat transfer [24–26] and application of advanced coolant (refrigerant) circuits [27–30].

Many researches are aimed on efficient operation of cooling systems in actual climatic conditions [31–33], energy simulation and load responsive control [34–36] and methods to determine rational loads [37–40].

To cover peak heat loads on engine cyclic air cooling systems some energy saving technologies with accumulating chilled water at decreased cooling loads are developed [41].

The most of well-known concepts of increasing the efficiency of IES are limited to application of refrigeration capacity generated by conversing the GE heat output for external consumers such as space conditioning and technological needs [4, 5] and do not reveal reserves for enhancing the efficiency of the combustion engines themselves due to cooling inlet air by using generated refrigeration capacity i.e. within engine cycles according to the concept of in cyclic trigeneration [10, 11].

To evaluate the efficiency of gas engine cyclic air cooling by using energy conversing technologies the data on dependence of fuel consumption and power output of gas engine on varying inlet air temperatures are needed. These data can be received by processing the results of gas engine fuel efficiency monitoring.

3 Research Methodology

The efficiency of cooling air at the inlet of gas engine was investigated for IES of combined energy supply at the factory "Sandora"–"PepsiCo Ukraine" (Nikolaev, Ukraine). The IES is equipped with two cogenerative Jenbacher gas engines JMS 420 GS-N.LC (each of rated electric power P_{eISO} = 1400 kW, heat power Q_h = 1500 kW), in which the heat of the exhaust gases, scavenge air-gas mixture, engine jacket cooling water and lubricating oil is used for heating water to the temperature of about 90 °C. The hot water is used in AR-D500L2 Century absorption Li-Br chiller to produce a chilled water of 12 °C, which is spent for technological needs and feeding to the central air conditioner that provides cooling ambient air incoming the engine room, from where it is sucked into the engine turbocharger.

A method for treatment of the monitoring data sets on fuel consumption and power output of gas engine was developed in [10] and modified in this research to take into account the influence of varying engine loading and air temperatures at its inlet.

To evaluate the influence of inlet air cooling on the fuel efficiency of gas engine the data on the variation of fuel consumption and power output of gas engine at varying engine inlet air temperatures and engine loading in relation to ISO parameters [42] were received by treating the results of gas engine JMS 420 GS-N.L efficiency monitoring.

The results of monitoring gas engine fuel efficiency are presented in the form of data sets on dependence of fuel consumption $B_f = f(t_{in})$ and engine power output $P_{el} = f(t_{in})$ and, as result, specific fuel consumption $b_e = B_f/P_{el}$ on the varying air temperature t_{in} at the inlet of turbocharger.

4 Results

The scheme of IES at the factory "Sandora"–"PepsiCo Ukraine" is presented in Fig. 1.

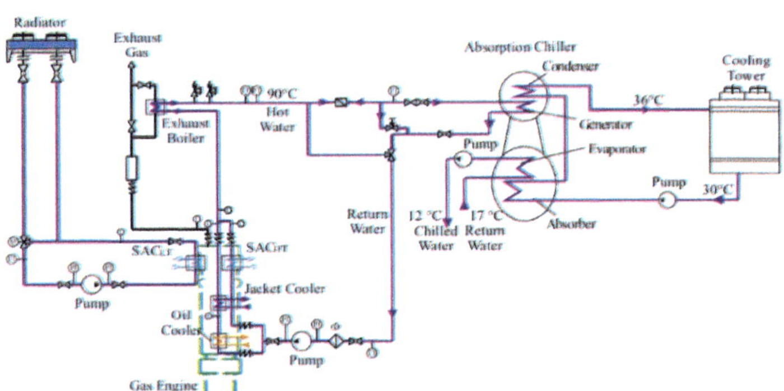

Fig. 1. The scheme of IES at the factory "Sandora"–"PepsiCo Ukraine": SAC_{LT} and SAC_{HT} – low- and high-temperature scavenge air coolers

The goal of monitoring data sets $P_{el} = f(t_{in})$ and $B_f = f(t_{in})$ treatment is to calculate the values of specific fuel consumption $b_e = B_f/P_{el}$ in dependence on the temperature t_{in} of air at the inlet of turbocharger: $b_e = f(t_{in})$.

The results of processing the monitoring data on electrical power output P_{el} and volume gas fuel consumption B_f are presented in Figs. 2 and 3.

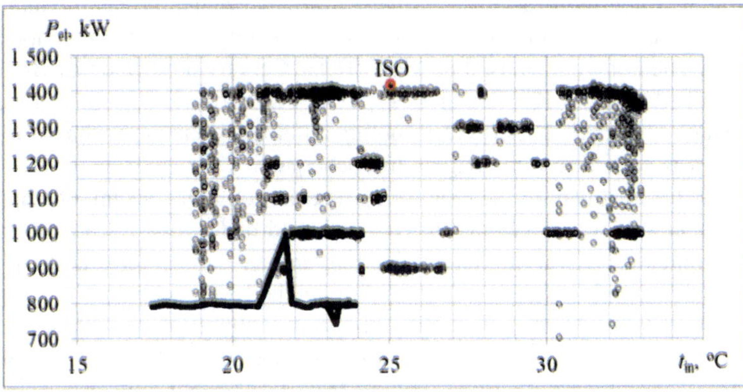

Fig. 2. Variation of electrical power output P_{el} versus turbocharger inlet air temperature t_{in} for Jenbacher gas engine JMS 420 GS-N.LC at various loading P_{el} = 800…1400 kW: point "ISO" corresponds to $P_{el.ISO}$ for "ISO" climatic parameters

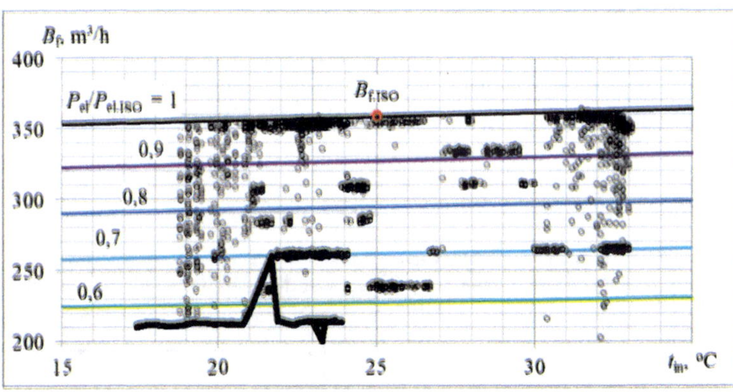

Fig. 3. B_f, t_{in}, $B_{f.ISO}$, $t_{in.ISO}$ Variations of volume gas fuel consumption B_e versus turbocharger inlet air temperature t_{in} for Jenbacher gas engine JMS 420 GS-N.LC at various loading $P_{el.ISO}/P_{el.ISO}$ = 0.6…1.0: point $B_{f.ISO}$ corresponds to "ISO" parameters

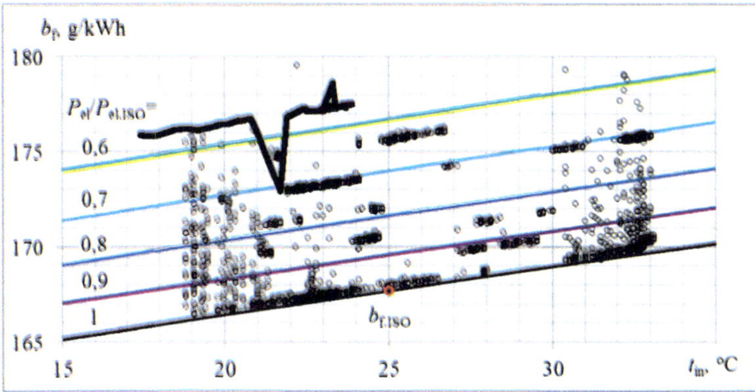

Fig. 4. Variation of mass specific fuel consumption b_f versus turbocharger inlet air temperature t_{in} for Jenbacher gas engine JMS 420 GS-N.LC at various loading $P_{el}/P_{el.ISO}$ = 0.6…1.0: point $b_{f.ISO}$ corresponds to "ISO" climatic parameters

As Fig. 4 shows, with arising engine inlet air temperatures t_{in} the mass specific fuel consumption b_f increases by about 0.25 g/(kW h) for 1 °C increment of engine inlet air temperature Δt = 1 °C.

With increasing engine inlet air temperatures t_{in} the electrical efficiency η_{el} decreases by about 0.006 of its absolute value or 1.36% in relative value for 10 °C increment of engine inlet air temperature Δt = 1 °C (Fig. 5).

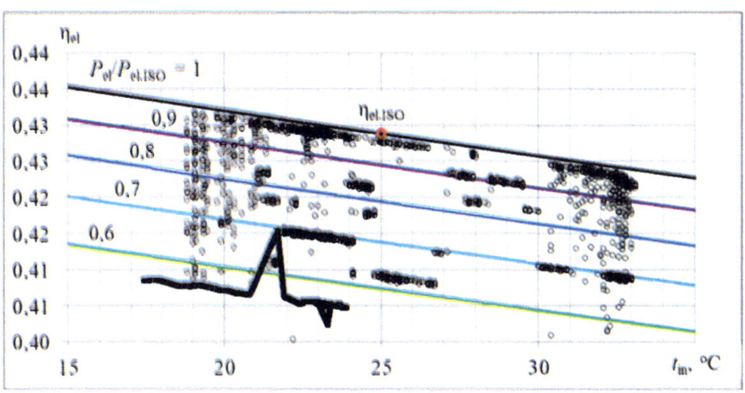

Fig. 5. Variations of engine electrical efficiency η_{el} versus turbocharger inlet air temperature t_{in} for Jenbacher gas engine JMS 420 GS-N.LC at various loading $P_{el}/P_{el.ISO}$ = 0.6…1.0: point $\eta_{el.ISO}$ corresponds to "ISO" climatic parameters

A deviation of calculated values of mass specific fuel consumption $b_{f.calc}$ from their monitoring data $b_{f.monit}$ is about 6…10%, that proves a satisfactory their agreement (Fig. 6).

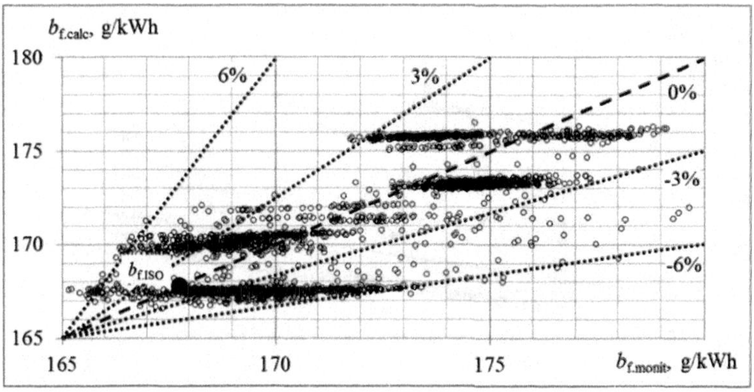

Fig. 6. The calculated values of mass specific fuel consumption $b_{f.calc}$ against their monitoring data $b_{f.monit}$: point $b_{f.ISO}$ corresponds to "ISO" climatic parameters

Issuing from decrease in specific fuel consumption b_f due to arising the engine efficiency η_{el} with lowering engine inlet air temperatures t_{in} (Figs. 5 and 6) a concept of addition subcooling engine inlet air down to lower temperatures t_{in} = 7…10 °C as compared with t_{in} = 20…25 °C and higher in the case of conventional inlet air cooling system in ABCh with chilled water temperature of 12 °C, used for technological needs.

This concept of subcooling engine inlet air, addition to its cooling by chilled water from ABCh with temperature of around 12 °C, used for technological needs, can be realized by two-stage cooling air with using refrigerant as a coolant in the second low temperature stage of air cooler and correspondingly two-stage waste heat conversion with refrigerant ejector chiller as low temperature stage [39, 40]. Thus, inlet air deep cooling allows to maintain gas engine at practically stable inlet air temperature at ambient air temperatures t_{amb} = 30…35 °C (in engine room above 35…40 °C caused by heat influx from surroundings) and provides inlet air temperature decrease of about 20…25 °C compared with conventional inlet air cooling system in ABCh with chilled water temperature of 12 °C. This results in reduction of engine specific fuel consumption by about 5 g/(kW h), i.e. about 3% decrease in fuel consumption at increased ambient air temperatures t_{amb} = 30…35 °C.

The enhancement of gas engine fuel efficiency due to inlet air deep cooling can be estimated by decreasing current specific fuel consumption b_e and its summarized daily values $\sum \Delta b_e$ due to engine inlet air cooling to the temperatures of 7, 10 and 15 °C compared with conventional cooling, that leads to increased air temperature t_{in} at the inlet of turbocharger and enlarged fuel consumption.

Some results of treating the monitoring data on cooling engine inlet air to the temperatures 7, 10 and 15 °C are presented in Fig. 7.

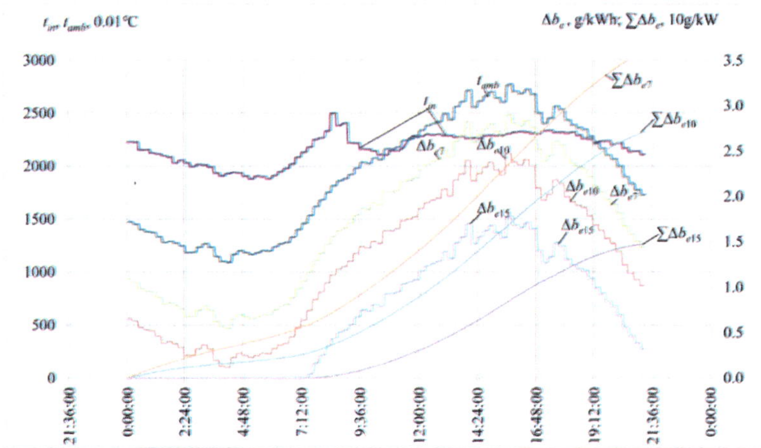

Fig. 7. Daily variation of decrease in current engine specific gas fuel consumption Δb_e and their summarized daily values $\sum \Delta b_e$ due to engine inlet air cooling to the temperatures of 7, 10 and 15 °C compared with traditional cooling all the engine room incoming air with increased engine inlet temperatures t_{in} with variation of ambient temperatures t_{amb} during time τ

As Fig. 7 shows, engine inlet air deep cooling provides decreasing current specific fuel consumption by the values Δb_e = 2.5–3.0 g/kWh at increased ambient air temperatures t_{amb}, that leads to their summarized daily values $\sum \Delta b_e$ more than 30 g/kW, related to 1 kW power of gas engine, or in absolute values for gas engine JMS 420 GS-N.L of 1400 kW power output it provides daily fuel saving above 40 kg.

5 Conclusions

To evaluate the efficiency of gas engine cyclic air cooling by using different energy conversing technologies the data on dependence of fuel consumption and power output of gas engine on varying inlet air temperatures are received by processing the results of gas engine fuel efficiency monitoring.

A method for treatment of the monitoring data sets on fuel consumption and power output of gas engine was modified to take into account the influence of varying engine loading and air temperatures at its inlet.

A treatment of monitoring data sets on fuel efficiency of gas engine JMS 420 GS-N.L in integrated energy system for combined electricity, heat and cooling generation has proved inefficient conventional cooling engine inlet air in ABCh with chilled water temperature of 12 °C, used for technological needs.

Issuing from decrease in specific fuel consumption b_f with lowering engine inlet air temperatures t_{in} a concept of addition subcooling engine inlet air down to lower temperatures t_{in} = 7…10 °C as compared with t_{in} = 20…25 °C and higher in the case of conventional inlet air cooling system in ABCh with chilled water temperature of 12 °C, used for technological needs. This concept is realized by two-stage cooling air with using refrigerant as a coolant in the second low temperature stage of air cooler at the inlet of engine turbocharger and correspondingly two-stage waste heat conversion with refrigerant ejector chiller as low temperature stage. Inlet air deep cooling allows to maintain gas engine at practically stable inlet air temperature at increased ambient air temperatures and provides inlet air temperature decrease of about 20…25 °C compared with conventional inlet air cooling system in ABCh with chilled water temperature of 12 °C. This results in reduction of engine specific fuel consumption by about 3% at increased ambient air temperatures t_{amb} = 30…35 °C.

References

1. Zellner, S., Burgtorf, J., Kraft-Schäfer, D.: Cogeneration & Trigeneration – How to Produce Energy Efficiently. A practical Guide for Experts in Emerging and Developing Economies, p. 144. Deutsche Gesellschaft für Internationale Zusammenarbeit (GIZ) GmbH (2016)
2. CIMAC Position Paper Gas Engine Aftertreatment Systems By CIMAC WG 17, Gas Engines (2017). https://www.cimac.com/cms/upload/Publication_Press/WG_Publications/CIMAC_WG17_2017_Aug_Position_Paper_Gas_Engine_Aftertreatment_Systems.pdf
3. Canova, A., Cavallero, C., Freschi, F., Giaccone, L., Repetto, M., Tartaglia, M.: Optimal energy management. IEEE Ind. Appl. Mag. **15**, 62–65 (2009)
4. Rouse, G., Czachorski, M., Bishop, P., Patel, J.: GTI Integrated Energy System for Buildings. Modular System Prototype. GTI Project report 15357/65118: Gas Technology Institute (GTI), p. 495 (2006)
5. Elsenbruch, T.: Jenbacher gas engines a variety of efficient applications. Bucureşti, p. 73 (2010)
6. Radchenko, M., Radchenko, R., Kornienko, V., Pyrysunko, M.: Semi-empirical correlations of pollution processes on the condensation surfaces of exhaust gas boilers with water-fuel emulsion combustion. In: Ivanov, V. et al. (eds.) Advances in Design, Simulation and Manufacturing II. DSMIE 2019. Lecture Notes in Mechanical Engineering, pp. 853–862. Springer, Cham (2020)
7. Kornienko, V., Radchenko, R., Stachel, A., Pyrysunko, M.: Correlations for pollution on condensing surfaces of exhaust gas boilers with water-fuel emulsion combustion. In: Tonkonogyi, V. et al. (eds.) Grabchenko's International Conference on Advanced Manufacturing Processes. InterPartner-2019. Lecture Notes in Mechanical Engineering. Springer, Cham, p. 10 (2020)
8. Radchenko, M., Radchenko, R., Ostapenko, O., Zubarev, A., Hrych, A.: Enhancing the utilization of gas engine module exhaust heat by two-stage chillers for combined electricity, heat and refrigeration. In: 5th International Conference on Systems and Informatics, ICSAI 2018, Jiangsu, Nanjing, China, pp. 240–244 (2019)
9. Suamir, I.N., Tassou, S.A.: Performance evaluation of integrated trigeneration and CO2 refrigeration systems. Appl. Thermal Eng. **50**, 1487–1495 (2013)
10. Radchenko A., Radchenko M., Konovalov A., Zubarev A.: Increasing electrical power output and fuel efficiency of gas engines in integrated energy system by absorption chiller scavenge air cooling on the base of monitoring data treatment. In: HTRSE-2018, E3S Web of Conferences, vol. 70, p. 03011 (2018). p. 6

11. Forduy, S., Radchenko, A., Kuczynski, W., Zubarev, A., Konovalov, D.: Enhancing the fuel efficiency of gas engines in integrated energy system by chilling cyclic air. In: Tonkonogyi, V. et al. (eds.) Grabchenko's International Conference on Advanced Manufacturing Processes. InterPartner-2019. Lecture Notes in Mechanical Engineering, p. 10. Springer, Cham (2020)
12. Konovalov, D., Kobalava, H.: Efficiency analysis of gas turbine plant cycles with water injection by the aerothermopressor. In: Ivanov, V. et al. (eds.) Advances in Design, Simulation and Manufacturing II. DSMIE 2019. Lecture Notes in Mechanical Engineering, pp. 581–591. Springer, Cham (2020)
13. Konovalov D., Trushliakov, E., Radchenko, M., Kobalava, G., Maksymov, V.: Research of the aerothermopressor cooling system of charge air of a marine internal combustion engine under variable climatic conditions of operation. In: Tonkonogyi, V. et al. (eds.) Grabchenko's International Conference on Advanced Manufacturing Processes. InterPartner-2019. Lecture Notes in Mechanical Engineering, p. 10. Springer, Cham(2020)
14. Carvalho, M., Serra, L.M., Lozano, M.A.: Geographic evaluation of trigeneration systems in the tertiary sector. Effect of climatic and electricity supply conditions. Energy **36**, 1931–1939 (2011)
15. Chua, K.J., Yang, W.M., Wong, T.Z., et al.: Integrating renewable energy technologies to support building trigeneration – a multi-criteria analysis. Renew. Energy **41**, 358–367 (2012)
16. Freschi, F., Giaccone, L., Lazzeroni, P., Repetto, M.: Economic and environmental analysis of a trigeneration system for food-industry: a case study. Appl. Energy **107**, 157–172 (2013)
17. Ghaebi, H., Karimkashi, Sh, Saidi, M.H.: Integration of an absorption chiller in a total CHP site for utilizing its cooling production potential based on R-curveconcept. Int. J. Refrig. **35**, 8 (2012)
18. Gluesenkamp, K., Yunho, H., Radermacher, R.: High efficiency micro trigeneration systems. Appl. Thermal Eng. **50**, 6 (2013)
19. Rocha, M.S., Andreos, R., Simões-Moreira, J.R.: Performance tests of two small trigeneration pilot plants. Appl. Thermal Eng. **41**, 84–91 (2012)
20. Rodriguez-Aumente, P.A., Rodriguez-Hidalgo, M.C., Nogueira, J.I., Lecuona, A., Vene-gas, M.C.: District heating and cooling for business buildings in Madrid. Appl. Thermal Eng. **50**, 1496–1503 (2013)
21. Butrymowicz, D., Gagan, J., Śmierciew, K., Łukaszuk, M., Dudar, A., Pawluczuk, A., Łapiński, A., Kuryłowicz, A.: Investigations of prototype ejection refrigeration system driven by low grade heat. In: E3S Web of Conferences, HTRSE–2018, vol. 70, p. 7 (2018)
22. Smierciew, K., Gagan, J., Butrymowicz, D., Karwacki, J.: Experimental investigations of solar driven ejector air-conditioning system. Energy Build. **80**, 260–267 (2014)
23. Elbel, S., Lawrence, N.: Review of recent developments in advanced ejector technology. Int. J. Refrig. **62**, 1–18 (2016)
24. Radchenko, N.: A concept of the design and operation of heat exchangers with change of phase. Arch. Thermodyn.: Pol. Acad. Sci. **4**(25), 3–19 (2004)
25. Trushliakov, E., Radchenko, M., Bohdal, T., Radchenko, R., Kantor, S.: An innovative air conditioning system for changeable heat loads. In: Tonkonogyi, V. et al. (eds.) Grabchenko's International Conference on Advanced Manufacturing Processes. InterPartner-2019. Lecture Notes in Mechanical Engineering, p. 10. Springer, Cham (2020)
26. Bohdal, T., Sikora, M., Widomska, K., Radchenko, A.M.: Investigation of flow structures during HFE-7100 refrigerant condensation. Arch. Thermodyn. Pol. Acad. Sci. **4**(36), 25–34 (2015)
27. Southard, L.E., Saab, R., Ali, M.I.H.: Variable-refrigerant-flow cooling-systems performance at different operation-pressures and types-of-refrigerants. Energy Proc. **119**, 426–432 (2017)

28. Enteria, N., Yamaguchi, H., Miyata, M., Sawachi, T., Kuwasawa, Y.: Performance evaluation of the variable refrigerant flow (VRF) air-conditioning system subjected to partial and unbalanced thermal loadings. J. Thermal Sci. Technol. **11**(1), 1–11 (2016)
29. Park, D.Y., Yun, G., Kim, K.S.: Experimental evaluation and simulation of a variable refrigerant-flow (VRF) air-conditioning system with outdoor air processing unit. Energy Build. **146**, 122–140 (2017)
30. Zhou, Y.P., et al.: Simulation and experimental validation of the variable-refrigerant-volume (VRV) air-conditioning system in energy plus. Energy Build. **40**, 1041–1047 (2008)
31. Liu, C., Zhao, T., Zhang, J.: Operational electricity consumption analyze of VRF air conditioning system and centralized air conditioning system based on building energy monitoring and management system. Proc. Eng. **121**, 1856–1863 (2015)
32. Zhou, Y.P., Wu, J.Y., Wang, R.Z.: Energy simulation in the variable refrigerant flow air-conditioning system under cooling conditions. Energy Build. **39**, 212–220 (2007)
33. Zhu, Y., Jin, X., Du, Z., Fang, X., Fan, B.: Control and energy simulation of variable refrigerant flow air conditioning system combined with outdoor air processing unit. Appl. Therm. Eng. **64**, 385–395 (2014)
34. Yun, G.Y., Lee, J.H., Kim, H.J.: Development and application of the load responsive control of the evaporating temperature in a VRF system for cooling energy savings. Energy Build. **116**, 638–645 (2016)
35. Lee, J.H., Yoon, H.J., Im, P., Song, Y.-H.: Verification of energy reduction effect through control optimization of supply air temperature in VRF-OAP system. Energies **11**(1), 49 (2018)
36. Yun, G.Y., Choi, J., Kim, J.T.: Energy performance of direct expansion air handling unit in office buildings. Energy Build. **77**, 425–431 (2014)
37. Sait, H.H.: Estimated thermal load and selecting of suitable air-conditioning systems for a three story educational building. Proc. Comput. Sci. **19**, 636–645 (2013)
38. Radchenko, A., Radchenko, M., Trushliakov, E., Kantor, S., Tkachenko, V.: Statistical method to define rational heat loads on railway air conditioning system for changeable climatic conditions. In: 5th International Conference on Systems and Informatics: ICSAI 2018, Jiangsu, Nanjing, China, pp. 1308–1312 (2018)
39. Trushliakov, E., Radchenko, M., Radchenko, A., Kantor, S., Zongming, Y.: Statistical approach to improve the efficiency of air conditioning system performance in changeable climatic conditions. In: 5th International Conference on Systems and Informatics: ICSAI 2018, Jiangsu, Nanjing, China, pp. 1303–1307 (2018)
40. Radchenko, A., Bohdal, L., Zongming, Y., Portnoi, B., Tkachenko, V.: Rational designing of gas turbine inlet air cooling system. In: Tonkonogyi, V. et al. (eds.) Grabchenko's International Conference on Advanced Manufacturing Processes. InterPartner-2019. Lecture Notes in Mechanical Engineering, p. 10. Springer, Cham (2020)
41. Radchenko, R., Radchenko, A., Serbin,S., Kantor, S., Portnoi, B.: Gas turbine unite inlet air cooling by using an excessive refrigeration capacity of absorption-ejector chiller in booster air cooler. In: HTRSE-2018, E3S Web of Conferences vol. 70, p. 03012 (2018). P. 6
42. ISO 3046-1:2002 Reciprocating internal combustion engines Performance Part 1: Declarations of power, fuel and lubricating oil consumptions, and test methods. Additional requirements for engines for general use (This standard was last reviewed and confirmed in 2014.) https://www.iso.org/standard/28330.html

Lifetime Prediction of Threaded Connections of Hydraulic Turbines Based on Stress State Monitoring System

Oleksandr Trubayev, Yurii Ulyanov, and Oleksii Vodka(✉)

Department of Dynamics and Strength of Machines,
National Technical University "Kharkiv Polytechnic Institute", Kharkiv, Ukraine
oleksii.vodka@gmail.com

Abstract. This work deals with development of a monitoring system for the stress-strain state of bolted connections of hydraulic turbines. The wireless strain gauge module is an autonomous hardware element, with its own embedded power supply. The transfer of information from the module to the data recording and processing equipment is performed by acoustic interlinking through the impeller body and the shaft of the hydrogenerator. Processing the results of measurements of the stress state of bolted joints is carried out using methods for assessing the residual life. The approaches proposed in this work allow constructing a system for monitoring the residual life of bolted connections of hydraulic turbines.

Keywords: Monitoring system · Acoustic signals · Autonomous tensometric modulus · Fatigue · Life-time · Threaded connections · Hydraulic turbine · Material degradation

1 Introduction

Reliability analysis for many types of engineering structures is performed in accordance with the parameters of static and dynamic loadings which occur under nominal operating conditions. The design is performed considering safety factors which provide high reliability for these operating modes. Nevertheless, fatigue failures often occur in the various elements of different machines. The practical and theoretical issues of the fatigue strength prediction for real mechanical systems are still relevant in spite of the detailed description of the history of occurrence and investigation of the causes of fatigue failures for different machines and mechanisms represented in numerous sources (for example, [1–4]). Among all the mechanisms experiencing fatigue strength problems, structural elements of hydraulic turbines represent a significant part. Depending on the specifics of construction of the rotor and the flow part, certain elements of a turbine are susceptible to fatigue.

A large number of fatigue failures, detected in various components of hydraulic turbines demonstrate that the models of reliability prediction of such elements are not accurate enough. It is caused by considerable complexity of the processes in hydraulic turbines during their considerable service life. In the Kaplan turbines the elements of

M. Nechyporuk et al. (Eds.): *Integrated Computer Technologies in Mechanical Engineering*,
AISC 1113, pp. 371–382, 2020.
https://doi.org/10.1007/978-3-030-37618-5_32

the blades rotating mechanism [5, 6], the blades [7], the drive elements [8] and the rotor as a whole [9] are most susceptible to fatigue.

An important problem is the fatigue strength of bolted joints, since their failure can lead to failure of the entire structure. There are many works on the problems of analysis, design and use of threaded connections [10] (this article provides links to more than 700 works published between 1990 and 2002), and also widely used recommendations on the design of bolted connections [11, 12], but the issues of analysis of the strength and reliability of bolt joint prediction remain relevant [13]. The calculation of the fatigue strength of the bolted connection of the rotor wheel has a number of parameters, such as the presence of an aggressive environment [14, 15]. Therefore, the assessment of the reliability of the bolted connections of the wheel rotor is an important engineering task.

Despite the large amount of work on the fatigue strength of hydraulic turbine elements, most approaches to the analysis of reliability and service life consider structural elements under dynamic loads with a fixed frequency, taking into account the random fatigue endurance limit [16]. However, this leads to the omission of a large class of problems in the theory of reliability, where the lifetime is determined by transient conditions. Naturally, the lifetime depends on the frequency of transients, which is determined by the operating conditions and, therefore, subject to change. Thus, the frequency of occurrence of transients is random. Thus, we can distinguish a number of cases where the amplitude of the load is a deterministic quantity, and the frequency of occurrence of a transient mode is a random process. In practice, operating of hydraulic turbines units in the CIS countries led to a number of failures of bolted connections and pin joints of the turbines, however their nominal (design) life-time has not got worked out [17].

According to JSC "Turboatom" over the past 30 years, the number of transient modes under which reliable operation of hydraulic units should be ensured has increased dramatically: from 600–1400 starts/stops per year (according to design standards that existed before 1991) to 1200–1460 (according to design standards adopted after 1996); the total number of transient regimes from 1200–1600 per year (before 1991) to 1100–7000 (after 1996); the number of transfers to the synchronous compensator mode and back to the generator mode is at least 10,200 per year according to the requirements established in the technical specifications starting from 2006. In [17], a method for predicting the lifetime of bolted connections of hydraulic turbine impellers has been developed. It has been shown that the start and stop modes of hydraulic units have the most damaging effect on bolted connections, and the number of such modes is distributed to individual hydraulic units randomly. In this regard, it is important to develop a monitoring system for the stress-strain state of threaded connections of hydraulic turbines, which allows to control their life-time. In this paper, the possibility of creating such a system is illustrated by the example of bolted joints of impellers of a turbine.

The idea of creating a bolt monitoring system is not new. The review article discusses the main methods for monitoring the technical condition of bolted joints [18]. The paper considers the acoustoelastic method, piezoelectric active sensing method and piezoelectric impedance method. There are a lot other papers based on these methods [19–22]. However, the proposed approaches do not take into account the features of the considered bolts. In a Kaplan turbine, bolted joints are located on a rotating shaft. Bolts are surrounded by water. Therefore, the proposed methods are poorly applicable in this situation.

2 Problem Statement

In this work, it is necessary to develop a hardware system for measuring the stress state parameters of the bolted connection of the shaft and the impeller in real time. The hardware system has to have a wireless channel for data transmission and autonomous power. It is also necessary to choose rational parameters of the transmission frequency and evaluate the possibility of receiving such a signal.

3 Development of the System for Hydraulic Turbine'S Threaded Connections the Stress-Strain State Monitoring

Based on the experience gained by the authors and the reviewed articles on monitoring the stress state of the bolted joints of the impeller and shaft, it is proposed to develop a system with wireless information transfer. It is proposed to use an acoustic communication channel to transmit information. This approach combines well with the designs of Kaplan turbines and does not require any significant design modification.

Chosen method of data transfer allows to solve the problem, but it carries with it a number of difficulties. Such difficulties include the reception of an acoustic signal from a rotating shaft in a non-contact manner. Thus it is possible to present the general scheme of such a system (Fig. 1).

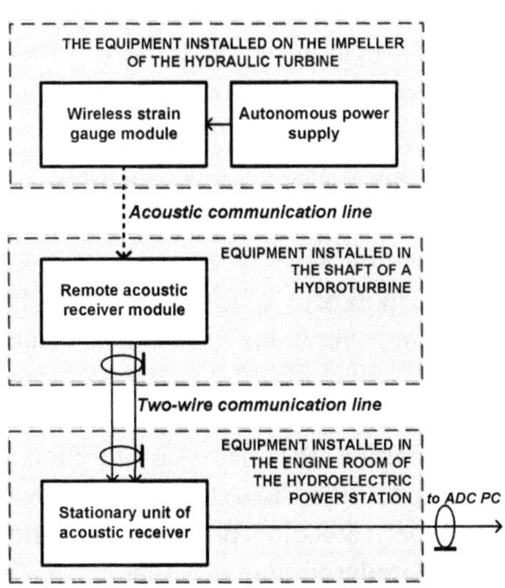

Fig. 1. Structure flowchart of the basic version of the measuring system

The wireless strain gauge (WSG) module is an autonomous hardware element, with its own self-contained power supply. The measuring system is designed as a multi-channel one, with the number of channels equal to the number of controlled bolted connection nodes. Every channel includes its own self-contained power supply block. The transfer of information from the module to the data recording and processing equipment is performed by acoustic interlinking through the impeller body and the shaft of the hydrogenerator. Detailed flowchart of WSG module is shown on Fig. 2.

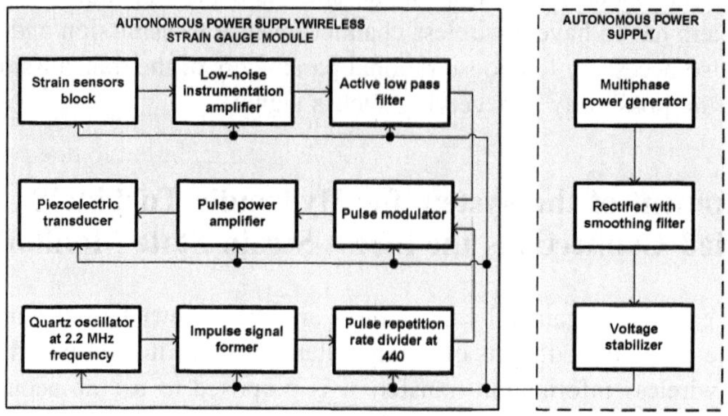

Fig. 2. Structure flowchart of the WSG module with self-contained power source.

Structure flowchart of the WSG module with self-contained power supply, includes:

- strain measuring bridge connected to instrumentation amplifier with low noise factor;
- active RC-filter with low-pass cutoff frequency below 50 Hz;
- pulse modulator with impulse power amplifier;
- the reference temporal stable signal forming unit.

Self-contained power supply consists of the multiphase power generator with permanent magnets, voltage rectifier with smoothing filter and the output voltage stabilizer.

A piezoelectric element glued directly to the test bolt is used as a transmitter. Using this piezoelectric element, sound vibrations are excited inside the rotor of a Kaplan turbine.

The second part of the monitoring system is a contactless acoustic signal receiving module. For its successful operation, it is necessary to choose a frequency for transmitting information. This can be done based on the following criteria. During the operation of the Kaplan turbine, cavitation processes are often observed. These processes have a spectral density similar to blue or violet noise. Different researchers give different spectra of cavitation noise [23–25]. However, most of them agree that a significant decrease in noise intensity is observed in the 10 kHz region.

On the other hand, in the lower part of the spectrum (up to 1 kHz) there are noises associated with the operation of the turbine itself, as well as frequencies that are multiples of the rotational speed.

At the same time it has to take into account characteristics of electroacoustic and acoustic-electric transducers. The practical verification of piezoelectric speaker Motorola KSN1165a [26] shows the maximum sensitivity also in 10 kHz region (Fig. 3).

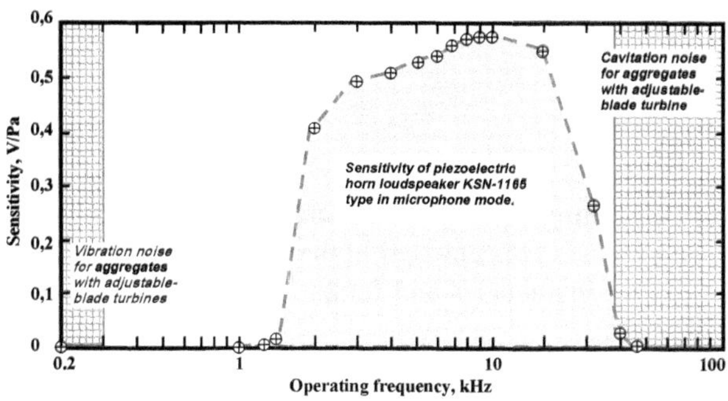

Fig. 3. Results of practical verification on the frequency selectivity and sensitivity the acoustic sensor KSN1165a

Figure 3 contains data on the possible noise of a typical Kaplan turbine. Sources of acoustic noise and vibration of structural elements are usually vibration of support structures, runout of a hydrogenerator shaft, mirror thrust bearing battle, state of steel structures of the generator, effects associated with the cavitation phenomenon. Based on the analysis of acoustic noise and taking into account the data of other researchers, the position of the frequency boundaries of typical noise was determined in the work. The low-frequency vibrational noise band extends from 0.4 Hz to almost 130 Hz, and the high-frequency band from 40 kHz to 900 kHz.

To verify the obtained data a finite element model of the turbine rotor has been built (Fig. 4). Mathematically the problem of calculating the eigen frequencies and form is formulated by the Eq. (1). To calculate harmonic response Eq. (2) has been used.

$$[M]\{\ddot{q}\} + [K]\{q\} = 0 \tag{1}$$

$$[M]\{\ddot{q}\} + [K]\{q\} = \{f\} \sin \omega t \tag{2}$$

where

[M] – global mass matrix;
[K] – global stiffness matrix;
{q} – nodal displacement vector;
{f} – nodal loading force vector;
ω – loading frequency;

Table 1. List of turbine rotor eigenfrequency

Number	Value
1	6.842
2	6.850
3	7.761
4	7.849
5	10.024
6	11.387
7	20.102
8	20.110
9	24.790

The results of calculating the eigen frequencies and forms confirm that the noise from the turbine operation is concentrated in the lower part of the spectrum (Table 1). The excitation of higher frequencies and forms are quite difficult.

Fig. 4. Finite element model of hydroturbine

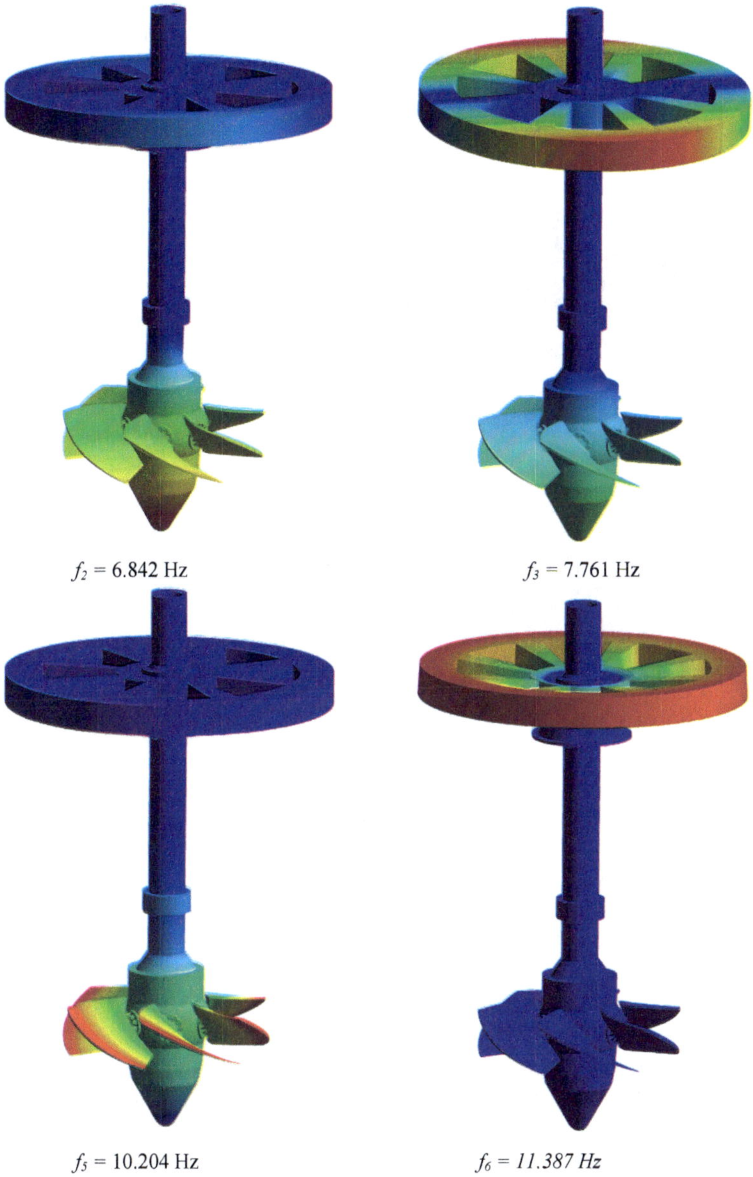

Fig. 5. Turbine rotor eigenforms according to Table 1

To confirm the possibility of receiving an acoustic signal from the turbine shaft, a calculation of the harmonic response has been made. For excitation a force of 1 N and a frequency of 10 kHz have been applied at the installation area of the piezoelectric exciter (Fig. 5).

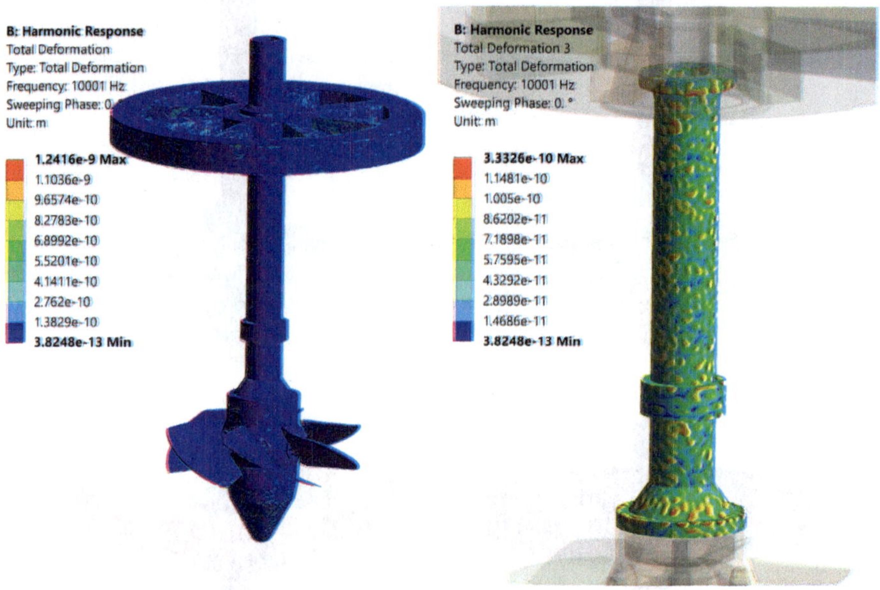

Fig. 6. Displacement of rotor on 10 kHz excitation

Figure 6 shows the results of numerical simulation. The color palette indicates the amplitude of displacement in meters on the surface of the turbine shaft. There is well visible spotted quasiregular structure of the intensity of acoustic oscillations on the surface of a generator shaft. The displacement distribution depends of excitation frequency. Anyway there is a lot of local maximum which can be used for acoustic signal recording.

The results of the numerical experiment allow to propose the method for recording acoustic signals. The essence of this method is illustrated in the diagram on Fig. 7. The acoustic signal is picking-up from the shaft by four acoustic directional sensors, similar by the principle of action to directed piezoelectric microphones. Such design and position of acoustic receivers makes it possible to ensure stable signal reception from a rotating shaft.

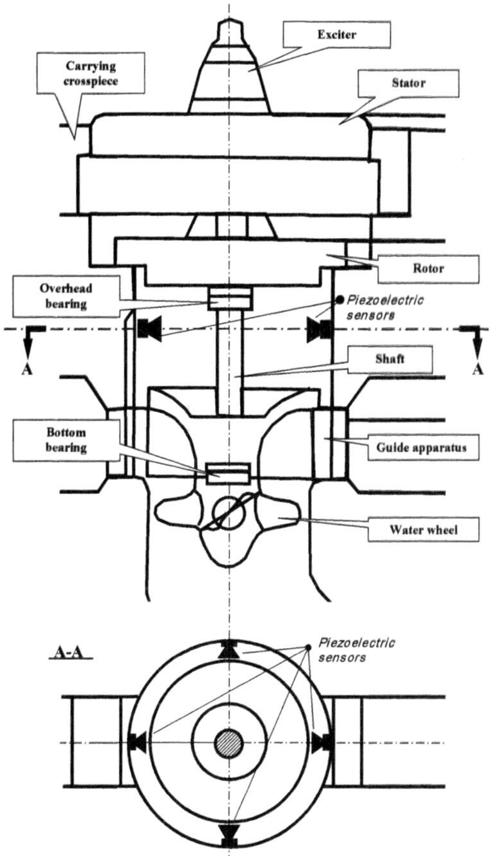

Fig. 7. The method of "taking" acoustic oscillations from the surface of the shaft generator

4 Lifetime Prediction of Threaded Connections of Hydraulic Turbines

The next stage of the work is to process the obtained data on the stress state of the bolted joint and assess the residual life-time. The life-time significantly depends on fatigue material properties. That the endurance limits of parts (σ_{-1p}) differ from the fatigue endurance of standard samples (σ_{-1}). This is due to the influence of various structural, technological and operational factors that can lead to both an increase and a decrease in fatigue endurance. The most influence is made by the following factors: absolute dimensions of the cross section (scale factor); surface finish; condition of the surface layer; operational factors (corrosion, temperature, loading frequency, etc.) [27–31]. The fatigue endurance of the bolted joint during operation is affected by almost all of the above factors: scale factor, surface condition, corrosive environment (fresh water), and fretting-corrosion in the thread. Thus, the fatigue endurance of the bolt with a symmetric cycle ($R = -1$) can be written in the form (3).

$$\sigma_{-1p} = \alpha\beta\gamma\varepsilon\sigma_{-1} \tag{3}$$

Bolted connections have a significant preliminary tightening. It leads to cycles with significant positive asymmetry coefficients are realized in bolts. To assess the influence of such cycles, in general, it is necessary to experimentally determine σ_{Rp} or use known models. In particular, the Gerber model is widely used for steels:

$$\sigma_{Rp} = \sigma_{-1p}\left[1 - \left(\frac{\sigma_m}{\sigma_u}\right)^2\right] \tag{4}$$

where σ_{Rp} – fatigue endurance of the part with cycle asymmetry factor R,

σ_m is the average stress of the cycle,

σ_u – ultimate strength of the material.

According to (3)–(4) the equation of the fatigue curve within the linear hypothesis of damage accumulation can be written in the next form,

$$N = N_0\left(\frac{\sigma_{RP}}{\sigma_a}\right), \quad \sigma_a > \sigma_{RP} \tag{5}$$

where σ_a – stresses amplitude (obtained from monitoring system),

N - the number of load cycles to failure at the σ_a stresses amplitude,

m – the Wöhler (S-N) curve parameter;

N_0 – the number of stress cycles with respect to the endurance limit.

Mathematical model Eq. (5) makes possible to estimate the residual life-time of bolted connection and, based on this information, plan repair work correctly.

5 Conclusions

In this work a specialized measuring system is proposed. This system uses the wireless acoustic transmission data about stress state in bolted connections. The practical feasibility of the proposed method has been confirmed.

The practical solution to the important technical problem associated with the taking of acoustic information from an excited shaft of the generator is found. It solved by the direction acoustic sensors modern horn piezoelectric tweeters.

References

1. Jones, D.R.H.: Failure Analysis Case Studies : A Sourcebook of Case Studies Selected from the Pages of Engineering Failure Analysis 1994–1996. Elsevier (1998)
2. Jones, D.R.H.: Failure Analysis Case Studies II : A Sourcebook of Case Studies Selected from the Pages of Engineering Failure Analysis 1997–1999. Elsevier, Pergamon (2001)
3. Jones, D.R.H.: Failure Analysis Case Studies III : A Sourcebook of Case Studies Selected from the Pages of Engineering Failure Analysis 2000–2002. Elsevier (2004)

4. Esaklul, K.A., ASM International: Handbook of Case Histories in Failure Analysis. ASM International, Cleveland (1992)
5. Luo, Y., Wang, Z., Zeng, J., Lin, J.: Fatigue of piston rod caused by unsteady, unbalanced, unsynchronized blade torques in a Kaplan turbine. Eng. Fail. Anal. **17**, 192–199 (2010). https://doi.org/10.1016/J.ENGFAILANAL.2009.06.003
6. Miclosina, C.-O., Campian, C., Frunzaverde, D., Cojocaru, V.: Fatigue Analysis of an Outer Bearing Bush of a Kaplan Turbine (2011)
7. Câmpian, V.C., Frunzaverde, D., Nedelcu, D., Marginean, G.: Failure analysis of a kaplan turbine runner blade. In: Proceedings of 24th IAHR Symposium on Hydraulic Machinery and Systems, pp. 1–10 (2008)
8. Diego, G., Serrano, M., Lancha, A.: Failure analysis of a multiplier from a Kaplan turbine. Eng. Fail. Anal. **7**, 27–34 (2000). https://doi.org/10.1016/S1350-6307(99)00006-0
9. Arsić, M., Bošnjak, S., Međo, B., Burzić, M., Vistać, B., Savić, Z.: Undefined: influence of loading regimes and operational environment on fatigue state of components of turbine and hydromechanical equipment at hydropower (2012) http://e2012.drustvo-termicara.com
10. Mackerle, J.: Finite element analysis of fastening and joining: a bibliography (1990–2002). Int. J. Press. Vessel Pip. **80**, 253–271 (2003). https://doi.org/10.1016/S0308-0161(03)00030-9
11. Kulak, G.L., Fisher, J.W., Struik, J.H.A., Fisher, J.W.: Guide to Design Criteria for Bolted and Riveted Joints. Wiley, Hoboken (1987)
12. Bickford, J.H.: Introduction to the Design and Behavior of Bolted Joints. CRC Press, Boca Raton (2008)
13. Casanova, F.: Failure analysis of the draft tube connecting bolts of a Francis-type hydroelectric power plant. Eng. Fail. Anal. **16**, 2202–2208 (2009). https://doi.org/10.1016/J.ENGFAILANAL.2009.03.003
14. Cetin, A., Härkegård, G.: Fatigue life prediction for large threaded components. Proc. Eng. **2**, 1225–1233 (2010). https://doi.org/10.1016/J.PROENG.2010.03.133
15. Zhao, L. Bin, Liu, F.R., Zhang, J.Y.: 3D numerical simulation and fatigue life prediction of high strength threaded bolt. Key Eng. Mater. **417–418**, 885–888 (2009). https://doi.org/10.4028/www.scientific.net/KEM.417-418.885
16. Zhang, L., Feng, F., Fan, X., Jiang, P.: Reliability analysis of francis turbine blade against fatigue failure under stochastic loading. In: 2012 International Conference on Quality, Reliability, Risk, Maintenance, and Safety Engineering, pp. 987–990. IEEE (2012). https://doi.org/10.1109/ICQR2MSE.2012.6246390
17. Larin, O.O., Vodka, O.O., Trubayev, O.I.: The fatigue life-time propagation of the connection elements of long-term operated hydro turbines considering material degradation. PNRPU Mech. Bull. **1**, 167–193 (2014)
18. Wang, T., Song, G., Liu, S., Li, Y., Xiao, H.: Review of bolted connection monitoring. Int. J. Distrib. Sens. Networks. **9**, 871213 (2013). https://doi.org/10.1155/2013/871213
19. Park, K.-T., Yu, Y.-J., Shin, H., Lee, J.-H., Lee, W.-S.: Monitoring system for bolt joints on steel structures, 24 March 2011. https://doi.org/10.1117/12.880301
20. Wang, T., Song, G., Wang, Z., Li, Y.: Proof-of-concept study of monitoring bolt connection status using a piezoelectric based active sensing method. Smart Mater. Struct. **22**, 087001 (2013). https://doi.org/10.1088/0964-1726/22/8/087001
21. Wang, B., Huo, L., Chen, D., Li, W., Song, G.: Impedance-based pre-stress monitoring of rock bolts using a piezoceramic-based smart washer—a feasibility study. Sensors **17**, 250 (2017). https://doi.org/10.3390/s17020250
22. Heyman, J.S.: A CW ultrasonic bolt-strain monitor. Exp. Mech. **17**, 183–187 (1977). https://doi.org/10.1007/BF02330995
23. Jablonská, J., Mahdal, M., Kozubková M.: Undefined: spectral analysis of pressure, noise and vibration velocity measurement in cavitation (2017). http://degruyter.com

24. Szmechta, M., Boczar, T., Frącz, P.: Frequency and time-frequency analysis of acoustic cavitation noise in insulating oils. Acta Phys. Pol. A **120**, 744–747 (2011). https://doi.org/10.12693/APhysPolA.120.744
25. Djakin, V.I., Zav'jalov, P.S., Bondarenko, A.V.: Measurement and spectral analysis of pressure pulsations in hydro turbines. Bull. NTU "KhPI." 45–50 (1973)
26. Model KSN 1165A - Bullet Tweeter - Piezo Source Store. http://piezosourcestore.bestgrouptechnologies.com/model-ksn-1165a-bullet-tweeter/. Accessed 15 Oct 2019
27. Schijve, J.: Fatigue of Structures and Materials. Springer, Heidelberg (2001)
28. Larin, O., Vodka, O.: A probability approach to the estimation of the process of accumulation of the high-cycle fatigue damage considering the natural aging of a material. Int. J. Damage Mech. **24**, 294–310 (2014). https://doi.org/10.1177/1056789514536067
29. Larin, O., Barkanov, E., Vodka, O.: Prediction of reliability of the corroded pipeline considering the randomness of corrosion damage and its stochastic growth. Eng. Fail. Anal. **66**, 60–71 (2016). https://doi.org/10.1016/j.engfailanal.2016.03.022
30. Larin, O., Kelin, A., Naryzhna, R., Potopalska, K., Trubayev, O.: Analysis of the pump strength to extend its lifetime. Nucl. Radiat. Saf. **3**(79), 30–35 (2018)
31. Vodka, O.: Computation tool for assessing the probability characteristics of the stress state of the pipeline part defected by pitting corrosion. Adv. Eng. Softw. **90**, 159–168 (2015). https://doi.org/10.1016/j.advengsoft.2015.08.012

Networks and Communication

The Problems of Control in Wireless Sensor and Mobile Ad-Hoc Networks

Oleksandr Lysenko[1], Valery Romaniuk[2], Olena Tachinina[3](✉), and Stanislav Valuiskyi[1]

[1] Department of Telecommunication, National Technical University of Ukraine «Igor Sikorsky Kyiv Polytechnic Institute», Kiev, Ukraine
lysenko.a.i.1952@gmail.com, samubf@gmail.com
[2] Military Institute of Telecommunications and Informatization, Kiev, Ukraine
romval2016@gmail.com
[3] Department of Automation and Energy Management, National Aviation University, Kiev, Ukraine
tachinina5@gmail.com

Abstract. One the main problem during development of self-organized networks such as wireless sensor (WSN) and mobile ad-hoc networks (MANET) is creation of effective control system, which can cover a lot of management tasks such as topology management, routing management, radio resource management, security management, quality of service management etc. For this purpose it proposed hierarchical model of intelligent control system for WSN and MANET described in Sect. 1. The novelty of the model lies in using the graph theory to make a formal description of the functional subsystems of the network (vertices of the graph) and their interaction processes (edges of the graph). Using of proposed model can accelerate and systemize the network design process considering their functioning environment and hierarchical structure of their control system. Using of intellectual agents technology and multiagent systems allows to minimize the service traffic and use network and node resources more efficiently. One particular management task intended to plan the optimal topology of network nodes is described in Sect. 2. There was formulated a mathematical statement of the problem of finding a rational WSN topology using unmanned aerial vehicles (UAVs), which consists of problem of sensor localization in monitoring zone and problem of UAV relays localization in the space for the organization of connected network with desired characteristics. Rational WSN topology, built according to these principles, will reduce traffic and reduce energy consumption at the nodes, which in turn will increase network lifetime and reduce the total cost of its maintenance.

Keywords: Networks · Wireless sensor · Mobile ad-hoc networks

M. Nechyporuk et al. (Eds.): *Integrated Computer Technologies in Mechanical Engineering*, AISC 1113, pp. 385–402, 2020.
https://doi.org/10.1007/978-3-030-37618-5_33

1 Section 1. The Hierarchical Model of Intelligent Control System for Wireless Sensor and Mobile Ad-Hoc Networks

1.1 Introduction

Dynamic nature of MANET and WSN with ground based and air based (e.g. UAVs) mobile nodes leads to necessity of hierarchical architecture of their control system (CS) (master nodes and slave nodes) [1, 2]. CS nodes make decisions based on gathering and processing the large volumes of service information about both node and entire network status; it is impossible to have full network status information in real time; therefore, CS should make decisions in uncertain conditions.

Network control process main requirement is that all management decisions for node and network resources must be carried out automatically by independent mobile nodes. Furthermore, during the management process every node's CS must consider not only its own goal function, but the goal function of all neighboring nodes [3], whose information is stored on the master node. In this scenario, MANET class radio network CS management decisions must be based on the intelligent ability to recognize and analyze different situations (on either node or network level).

Modern approach for intelligent node control system (ICS) design in view of the MANET functioning and mentioned above requirements is the use of the intelligent agents (IA) technology and multi-agent systems (MAS) [4]. Main feature of this technology is that an agent is considered as a hardware and software system that can make decisions in uncertain conditions. That is, IA and MAS can adapt to the changes in surrounding environment they interact with, even in the case when said changes are not defined in their behavior schemes.

There are many examples of IA and MAS used for gathering and processing information, as well as automatic management of different complex systems and processes [5]. But existing models of IA and MAS are designed using the intelligent methods that do not account for the MANET class network control features, and the lack of a method for designing corresponding models for ICS nodes delays the process of network development.

Therefore, the *purpose* of this article is to develop the hierarchical model for intelligent agents interaction for the MANET or sensor class radio network control system development.

1.2 Initial Data for the Model

According to the concept [2], network ICS is an aggregation of interacting node ICS that are deployed using the IA technology [6]. In this case, IA stands for a software product able to act to achieve a given goal and in addition to the main features (reactivity, proactivity and sociality) has [7]:

Mobility – IA can carry out its functions on another node on behalf of the initiator node;

Intelligence – the main feature of IA, that presume its ability to self-learn in the process of the mobile node operations so that it can find optimal behavior patterns for cases not foreseen at the design phase.

Every IA of a node ICS is designed for a specific type tasks (performs different functions, depending on OSI model levels) (Fig. 1), can interact with other IA for information exchange and make coordinated decisions, forming the executive layer of network ICS.

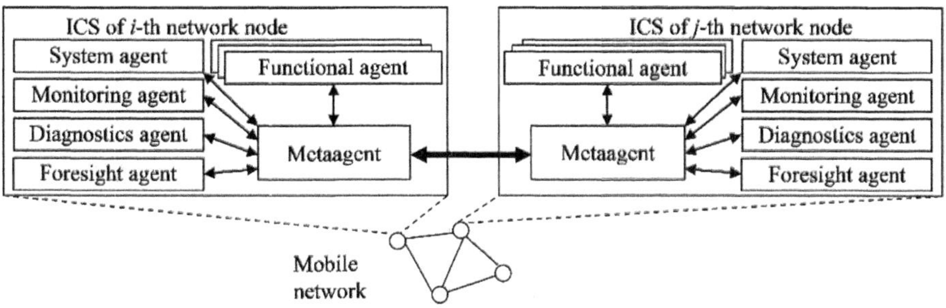

Fig. 1. Interaction of IA in the intelligent network control system

The coordination of IA operations on executive layer is managed by metaagents of node ICS. Multiple metaagents form the node layer of network ICS. In turn, coordination of metaagents' decisions of node ICS is managed by a master node. Any node of the network can be a master node, depending on its hardware or geographic location. As seen on Fig. 1, main management agents of the node ICS of the network can be distinguished as follows: functional agent, system agent, monitoring agent, diagnostics agent, foresight agent. Though, the quantity and composition of IA can vary drastically depending on the network node (mobile node, base station or sensor device) [8].

Metaagent takes care of coordination of IA operations to achieve common management goals using the management decision made by local agents and metaagents of neighboring nodes. Metaagent analyzes network information by communicating with neighboring nodes so that it is able to make a decision to provide a certain level of QoS.

System agent. Its main functions are: maintaining a database of neighboring node and network status (available resources), mobile node locations (topographic information); forming a knowledge base with rules of "behavior" of the given node under different circumstances; self-learning of the mobile node.

Monitoring agent – implements continuous monitoring of the network key performance indicators in real time; identifies different situation on the network, determines current and potential problems; gathers and analyzes service information (statistics).

Diagnostics agent – determines, localizes and analyzes nod malfunctions; runs tests of main functions of all of the mobile node's modules.

Foresight agent – uses the rules and algorithms of network performance analysis on all its layers to make a forecast of node and network status in the near future.

Functional agents – implements control methods for every layer of OSI model: topology management, routing management, data streams management, queue management, message priority and security, spectrum allocation, power allocation etc.

Most of the aforementioned IA are stationary, they are located on the node permanently. But for some functions (network zone monitoring, information route planning, etc.) system agent can generate a mobile IA (MIA). MIA is relocated to another network node, collects (and processes if necessary) the information of the given node and returns to the source node with a report (or, if necessary, is relocated to a new node). MIA life cycle is illustrated on Fig. 2.

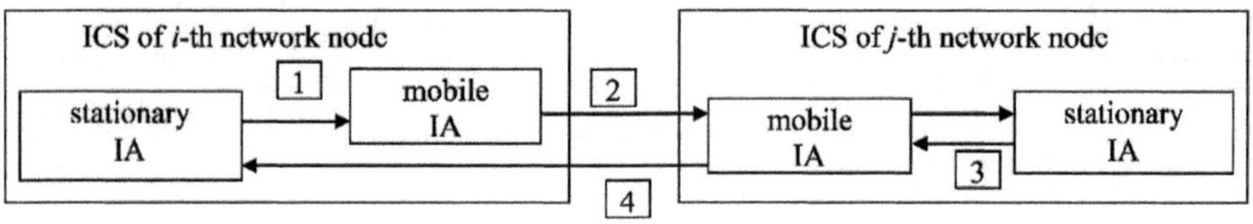

1 – node *i* generates a MIA, 2 – MIA relocation, 3 – data collection and processing, 4 – return to the node *i*

Fig. 2. MIA life cycle

Therefore, in view of the hierarchical concept of network ICS design [2] and aforementioned functional structure of node ICS with IA, formal description of MANET class radio network ICS can be presented as multiple IAs on different layers, that interact with each other by exchanging service information that is used to make management decisions. To achieve this we need to solve two problems: combine heterogeneous IA in the hierarchical network ICS and set up information exchange between IA in this structure. To solve these problems an hierarchical model of IA interaction is proposed, whose structure corresponds to the hierarchical network ICS design concept.

1.3 Hierarchical Model of IA Interaction

Formal description of the network ICS functional structure (with decentralized management) can be represented as a hierarchical IA structure with vertical relations between them. Given relations define the subordination of task that are resolved by IA at each layer [2]:

Zero (executive) layer – resolves management tasks according to the OSI model (routing, resource management, data streams management, security, etc.) by selecting the required values of node ICS subsystem parameters;
First (node) layer – consists of node ICS meta-agents that coordinate the zero layer IA by selecting optimal set of management actions and their implementation sequence on all node ICS subsystems;
Second (network) layer – consists of the master node that corrects the goal functions of first layer meta-agents in view of network status, as w hole, or its part.

Using graph theory, we can picture the given functional structure as shown on Fig. 3. Located at the root of the tree is a master node subsystem, at the vertices that are one edge away from the root are subsystems that represent Q meta-agents of node ICSs. Every mentioned subsystem of network ICS contains a control (identification) block and management block in turn, every first layer subsystem is connected to multiple functional subsystems of zero layer, that are located on two edges distance from the root. These subsystems represent IA interaction processes of every functional subsystem of node ICS [9]. This interaction consists of service information exchange and management decisions of each IA.

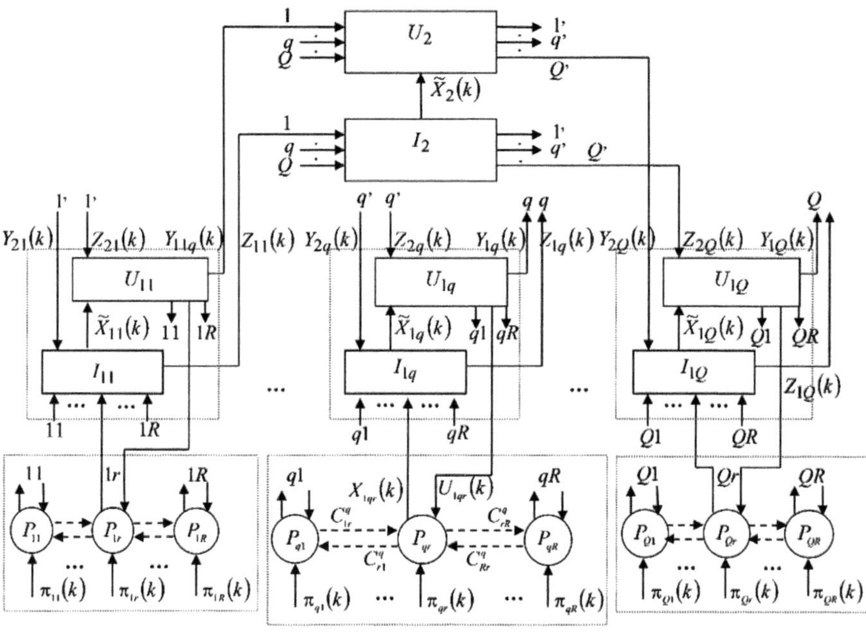

Fig. 3. Hierarchical model of IA organization of network ICS

For q-th management subsystem of the first layer $(I_{1q}, U_{1q}), q = \overline{1, Q}$ let us denote the following:

$X_{1qr}(k)$ - multiple state vectors of the qr-th IA, where the size of $X_{1qr}(k) = \left\{x^a_{1qr}(k)\right\}, a = \overline{1, a_{1qr}}$ is $a_{1q} \times 1$;
$\tilde{X}_{1q}(k)$ - multiple generalized estimated state vectors of q-th subsystem of the first layer (e.i. mobile node), where the size of $\tilde{X}_{1q}(k) = \left\{\tilde{x}^a_{1q}(k)\right\}, a = \overline{1, a_{1q}}$ is $a_{1q} \times 1$;
$U_{1qr}(k)$ - multiple management vectors of q-th subsystem of the first layer, that are directed to r-th IA of the zero layer, where the size of $U_{1qr}(k) = \left\{u^b_{1qr}(k)\right\}, b = \overline{1, b_{1qr}}$ is $b_{1qr} \times 1$;

$Y_{1q}(k)$ - multiple management vectors of q-th subsystem of the first layer, that are directed to the upper layer management subsystem (master node), where the size of $Y_{1q}(k) = \left\{ y_{1q}^{d}(k) \right\}, d = \overline{1, d_{1q}}$ is $d_{1q} \times 1$;
$Z_{1q}(k)$ - multiple estimated state vectors of q-th subsystem of the first layer, that are directed to upper layer management subsystem (master node), where the size of $Z_{1q}(k) = \left\{ z_{1q}^{d}(k) \right\}, d = \overline{1, d_{1q}}$ is $d_{1q} \times 1$;

For the second layer management subsystem (I_2, U_2) (master node), let us denote:

$\tilde{X}_2(k)$ - multiple generalized estimated state vectors of the first layer subsystems (metaagents of the node ICS), where the size of $\tilde{X}_2(k) = \{\tilde{x}_2^{l}(k)\}, l = \overline{1, l_r}$ is

$$l_r \times 1 = \left(\sum_{q=1}^{Q} a_{1q} \right) \times 1;$$

$Y_{2q}(k)$ - multiple management vectors of control variables, that are sent to lower layer management subsystems (metaagents of the node ICS), where the size of $Y_{2q}(k) = \left\{ y_{2q}^{d}(k) \right\}, d = \overline{1, d_{2q}}$ is $d_{2q} \times 1$;
$Z_{2q}(k)$ - multiple management vectors of variable estimated states, that are sent to the lower layer management subsystem (metaagents of the node ICS), where the size of $Z_{2q}(k) = \left\{ z_{2q}^{d}(k) \right\}, d = \overline{1, d_{2q}}$ is $d_{2q} \times 1$;

Finally, for qr-th subsystem of the zero layer $P_{qr}, q = \overline{1, Q},\ r = \overline{1, R}$ let us denote:

$C_{rp}^{q}(k)$ - multiple connections vectors (service information exchange between IA and their management decisions), where $C_{rp}^{q}(k) = \left\{ c_{rp}^{qmn}(k) \right\},\ m = \overline{1, m_r},\ n = \overline{1, n_r}$ between r-th and p-th subsystems $\left(r, p = \overline{1, Q}, p \neq r\right)$;
$\Pi_{qr}(k)$ - multiple external effects vectors, that are been measured by r-th IA of the q-th mobile node, where the size of $\Pi_{qr}(k) = \left\{ \pi_{qr}^{l}(k) \right\}, l = \overline{1, l_q}$ is $l_{qr} \times 1$

Wherein, multiple vectors of q-th IA states $X_q(k) = \bigcup_{r=1}^{R} X_{1r}(k)$ can be of different type depending on their state variables, that affect the channel quality and mobile node or network efficiency. Some of them are:

Network information load parameters vector

$$\Lambda(k) = \left\| \Lambda_1(k), \ldots, \Lambda_q(k), \ldots, \Lambda_Q(k) \right\|^T;$$

Information messages delays vector

$$\mathrm{H}(k) = \left\| \mathrm{H}_1(k), \ldots, \mathrm{H}_q(k), \ldots, \mathrm{H}_Q(k) \right\|^T;$$

Network radiofrequency environment parameters

$$\aleph(k) = \left\| \aleph_1(k), \ldots, \aleph_q(k), \ldots, \aleph_Q(k) \right\|^T;$$

Network spectrum resources vector

$$\Im(k) = \left\| \Im_1(k), \ldots, \Im_q(k), \ldots, \Im_Q(k) \right\|^T;$$

Network energy resources vector

$$\Re(k) = \left\| \Re_1(k), \ldots, \Re_q(k), \ldots, \Re_Q(k) \right\|^T;$$

Hardware resources vector (processor, battery capacity, RAM, etc.)

$$\mathrm{A}(k) = \left\| \mathrm{A}_1(k), \ldots, \mathrm{A}_q(k), \ldots, \mathrm{A}_Q(k) \right\|^T, \text{etc.}$$

As shown in the model (Fig. 3), any q-th management subsystem of the first layer $(I_{1q}, U_{1q}), q = \overline{1, Q}$ can be characterized by:

Mapping that describes the object being managed (metaagent of q-th mobile node)

$$\mathrm{O}_1^{(1)} : \tilde{X}_{1q} \times U_{1q} \times C_{qp} \times \Pi_q \rightarrow Y_{1q} \times Z_{1q}; \tag{1}$$

Mapping that describes the criteria used by q-th mobile node metaagent to determine the estimated state V_q and control influence W_q

$$\mathrm{O}_2^{(1)} : \tilde{X}_{1q} \times Z_{1q} \rightarrow V_q, \tag{2}$$

$$\mathrm{O}_3^{(1)} : U_{1q} \times Y_{1q} \rightarrow W_q; \tag{3}$$

Mapping that describes the generalized information Φ_q that arrives to upper layer subsystem (master node)

$$\mathrm{O}_4^{(1)} : Y_{1q} \times Z_{1q} \rightarrow \Phi_q; \tag{4}$$

Mappings that determine the constraints of input variables vectors Θ_q and control influence vectors Ψ_q, respectively

$$\mathrm{O}_5^{(1)} : X_{1qr} \rightarrow \Theta_q, \tag{5}$$

$$\mathrm{O}_6^{(1)} : U_{1q} \rightarrow \Psi_q. \tag{6}$$

Second layer subsystem (I_2, U_2) can be characterized by:

Mapping that describes the formation of the generalized estimated states vector of the mobile network

$$O_1^{(2)} : \Phi \to \tilde{X}_2, \tag{7}$$

where $\Phi = \bigcup_{q=1}^{Q} \Phi_q$;

Mapping that describes the criteria used by the (I_2, U_2) subsystem (master node) to determine the control influence destined for $(I_{1q}, U_{1q}), q = \overline{1, Q}$

$$O_2^{(2)} : U \times \tilde{X}_2 \to W, \tag{8}$$

where $U = \bigcup_{q=1}^{Q} Y_{2q}$;

Mappings that determine the constraints for generalized state and control vectors

$$O_3^{(2)} : \tilde{X}_2 \to \Theta, \tag{9}$$

$$O_4^{(2)} : U \to \Psi. \tag{10}$$

The functioning of the ICS of all mobile network elements (mobile or sensor nodes, mobile base station or network control center) [9] can be described by time intervals, as follows:

T_{1q} - time interval for performing management and control tasks (1–6) by metaagents of every node ICS ((I_{1q}, U_{1q}) subsystems);
$T_{1q}^{(2)}$ - time interval of generalized information transmission from metaagents (I_{1q}, U_{1q}) to the master node ((I_2, U_2) subsystem);
T_2 - time interval for performing the control and management tasks (7–10) by the master node;

The length of T_{1q} time interval is determined by the external influence vector $\Pi_{qr}(k)$ change rate, change of the control influences $Y_{2q}(k)$ and $Z_{2q}(k)$ from the master node (I_2, U_2), and the change of interconnection matrix $C_{rp}^{q}(k)$ structure. The length of $T_{1q}^{(2)}$ time interval is determined entirely by the methods and protocols of interaction between (I_{1q}, U_{1q}) and (I_2, U_2) subsystems, defined at appropriate levels of the OSI model.

Based on the information received by node ICS metaagents $(I_{1q}, U_{1q}), q = \overline{1, Q}$, the master node (I_2, U_2) checks the restraints (9), (10) and calculates the values of the indicator in (8) with control influences $U(k) = \{U_{1qr}(k)\}$, that are defined by subordinate node ICS on the previous time interval. If constraints are observed or a criterion has a deviation from the required value, a higher layer task is performed again, which defines the length of time interval T_2.

For a three layered network ICS (Fig. 3) the ratio between the aforementioned time intervals is as follows [9]:

$$T_2 \geq T_{1q}^{(2)},\ T_2 >> T_{1q},\ T_{1q}^{(2)} \geq T_{1q},\ \text{for}\ \forall q = \overline{1, Q}.$$

During those time intervals every element of the network ICS implements corresponding methods and algorithms of mobile network management, from mathematical methods and algorithms of link management (physical level of OSI model) to methods and algorithms of application level management (security management, power consumption management, QoS management, etc.).

1.4 Decision Making in the Network Intellectual Control System

In the general scenario, management decision making in the network ICS means providing a given quality of information exchange in MANET by determining the values of control variables of node ICS based on the analysis of current state of the radio network. But, as mentioned before, every node ICS is characterized by its own goal function, that is formed based on multiple factors:

Resources and hardware/software capabilities of the node, i.e. the totality of the devices for reaching the goal (RAM, processor performance, battery capacity, etc.);

Managed parameters: on physical level – transmitter power, modulation, transmission rate, etc., on channel level – access protocol, on network level – routing method, on transport level – transfer method, etc.;

Uncontrollable parameters: set exchange protocols, topology dynamics, network size, interference level, etc.;

Requirements for information exchange quality for different types of traffic (data, voice, video, graphics).

It leads to the inability to achieve global optimization of the entire mobile network in the case of decentralized management environment and with presence of contradiction between the optimal node ICS awareness and the timeliness of control influences. Thereby, it was proposed in [10] to decompose the main goal of mobile network management to multiple simpler goals. To achieve this, in the design phase of node ICS a goal structure (GS) is formed as a graph, where the vertices are goals, and edges are the influences of achieving a goal in a subgoal (Fig. 4).

In the previous research it was shown, that in an uncertain environment, where a mobile network functions, to describe a situation and make a management decision by the subsystems of node ICS it is advisable to use the methods of fuzzy logic [11]. Therefore, the goal structure (Fig. 4) can be mathematically interpreted as a list of fuzzy management goals of different levels $L_1, \ldots, L_k$ are connected by [10, 12]:

$$\begin{aligned} \mathrm{GS} = \{ & C_1, R_{2m_{(1)}}\{C_{21}, C_{22}, \ldots, C_{2m_{(2)}}\}, \\ & R_{3m_{(2)}}\{C_{31}, C_{32}, \ldots, C_{3m_{(3)}}\}, \ldots, \\ & R_{km_{(k)}}\{C_{k1}, C_{k2}, \ldots, C_{km_{(k)}}\}\} \end{aligned} \tag{11}$$

where C_1 – global goal of the network ICS, that is determined by the master node;

C_{il}, $i = \overline{1,k}$, $l = \overline{1,m_{(i)}}$ – l-th subgoal of i-th level of the goal structure, that is determined by the metaagent of the corresponding node ICS;

R_{ij}, $i = \overline{1,k}$, $j = \overline{1,m_{(i-1)}}$ – fuzzy relationship between the lax advantage of the objects on the i-th level over every object at the upper i − 1 level.

If R_{ij} describes the relationship only between the subgoals of neighboring levels, we should talk about a goal tree, otherwise the goal structure degenerates to a network.

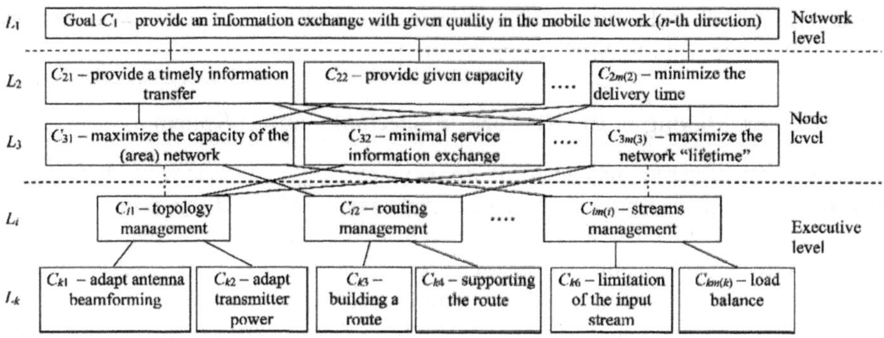

Fig. 4. Fragment of the goal structure of the network ICS

Let the goal system consist of k levels and every L_i level $i = \overline{1,k}$ consists of m_i objects (for first level $m_1 = 1$):

$$L_i = \left\{C_{i1}, C_{i2}, \ldots, C_{im_{(i)}}\right\}.$$

Goal structure (Fig. 4) can be described as a multiple of levels L_i:

$$\mathrm{GS} = \bigcup_{i=1}^{k} L_i = \bigcup_{i=1}^{k} \bigcup_{l=1}^{m_i} C_{il}.$$

As seen on Fig. 4, different elements of the goal structure are united under a global goal C_1, that can be described as providing the information exchange with given quality in the network. As mentioned before, a binary fuzzy relationship of a lax advantage R_{ij} is used to describe the relationship between global goal and lower level goals, that is given by a membership function $\mu_{R_{ij}}(C_{il}, C_{ir})$, $i = \overline{2,k}$, $j = \overline{1,m_{(i-1)}}$, $l, r = \overline{1,m_i}$.

It should be noted, that depending on the hierarchy layer (Fig. 3) there can be two types of relationship:

"goal - subgoal" relationship – appear between the elements of the network and node layers (between master node and subordinate nodes of a mobile network or its area) and create a goalforming part of the GS;

"subgoal – means to reach the goal" relationship – appear between elements of the node layer (metaagents of node ICS) and the elements of the executive layer (IA of corresponding functional subsystems) and create an implementing part of the GS.

And so, beginning with the second hierarchy layer (11), at every i-th layer there are as many fuzzy relationship of advantage R_{ij} as there are objects at i − 1 level of GS. In the general case, these relationships can be described as a matrix:

$$R_{ij} = \begin{vmatrix} 1 & \mu(C_{i1}, C_{i2}) & \ldots & \mu(C_{i1}, C_{im_{(i)}}) \\ \ldots & 1 & & \ldots \\ \mu(C_{im_{(i)}}, C_{il}) & \ldots & & 1 \end{vmatrix},$$

where $\mu_{R_{ij}}(C_{il}, C_{ir}) \in [0; 1]$, $i = \overline{2, k}$, $j = \overline{1, m_{(i-1)}}$, $l, r = \overline{1, m_i}$.

As a result, tasks of decision making of the network ICS are reduced to receiving of the priority vector of the lower layer elements in relationship to the global goal – the element of the first layer. To cope with this task in [12] it is proposed to use a weighting procedure of the hierarchy analysis method or fuzzy relationship convolution algorithm.

Section 1 Conclusion

Thereby, to respond to the features of the management in the MANET class mobile networks, the management system must have intellectual capabilities to recognize and analyze the situations in the radio network, and based on this, make management decisions to control the node and network resources. To design such management system, it is proposed to use the technology of intellectual agents and multiagent systems, that suggests that all subsystems of node ICS are implemented using multiple IA, that are defined by management functions depending on the level of the OSI network model.

To combine different IA in an intellectual network control system a hierarchical model of IA interaction was proposed in this article, whose essence lies in describing the network ICS as a hierarchical structure with vertical links, that indicate the subordination of management tasks.

The novelty of the model lies in using the graph theory to make a formal description of the functional subsystems of the network ICS (vertices of the graph) and their interaction processes (edges of the graph). Using the proposed model can accelerate and systemize the network design process considering their functioning environment and hierarchical structure of their ICS. Using the intellectual agents technology and multiagent systems allows to minimize the service traffic and use network and node resources more efficiently.

During future research a model for information resources organization of network ICS will be developed, to describe the circulation, processing and storage of the service information, that is used by the methods and protocols of corresponding subsystems for making management decisions in the mobile network.

2 Section 2. The Problem of Finding a Rational Topology of Wireless Sensor Networks Using UAVs

2.1 Introduction

In recent years WSN are widely used in various fields of human life, ranging from automation processes of data collection on various parameters of the observed objects (buildings, industrial equipment and processes) to the health sector, security and defense [13–15]. However, the use of WSN for monitoring environmental parameters over large geographical areas requires specific network architecture, consisting of the following levels. The first level is a network of low-power wireless sensors, deterministic or randomly located in the large geographical area. The second level is a network of gateways, which represent more powerful sensor units (by CPU performance and battery capacity of nodes) and are intended to gather monitoring information from sensor nodes and transfer it to the processing centers and ensure the connectivity of WSN topology. In this regard, depending on the application (difficult terrains or infected areas, fighting field, etc.), gateways can be executed in the form of mobile robots or UAVs, which can form the third level of WSN.

One of the main problems in the organization of such complex hierarchical WSN for remote collection of information is the task of finding the optimal network topology, which consists of two partial problems. The first is the search of optimal placement of sensors for complete coverage of monitoring area. Second is the search of the optimal placement of UAV repeaters to organize a connected network with specified data transmission characteristics (throughput, delay, etc.). This article will consider the mathematical statements of these problems and possible methods for their solution respectively in the first and second section of the article.

2.2 The Problem of Sensor Localization in the Monitoring Zone

Monitoring area of sensor (detection range) is usually represented as a circle of radius r centered at the location of the sensor. Then the problem can be formulated as follows: it is necessary to place a minimum number of circles that provide coverage of whole monitoring area of potentially dangerous object. Usually monitoring area has arbitrary form that can be easily approximated by multiply orthogonal polygons.

The problem of covering a polygon is easily reduced to a simpler problem of covering a rectangular area with obstacles (RO) (Fig. 5). The output polygon is denoted by **P**, a limited rectangular region – **A**. Addition **A\P** will be interpreted as a set of fictitious obstacles. Next, we will work with RO. Specified and fictitious obstacles are shaded in the figure.

On the plane, we introduce a coordinate system (OX, OY) such a way that the coordinate axes match with the bottom and left sides of rectangular area A. Output information of problem can be represented by the following set of data: $\{W, L, \mathbf{Z}, r\}$, where W and L – width and length of a rectangle that will covered; $\mathbf{Z}$ – obstacles that are given by set of rectangles $\mathbf{Z} = \{Z_1, Z_2, \ldots, Z_m\}$, where m – number of rectangle obstacles. The sides of rectangle from $\mathbf{Z}$ are parallel to the coordinate axes.

$Z_i = \{z_{xi}, z_{yi}, z_{li}, z_{wi}\}$ – rectangle, which simulates the obstacle, where $i = 1, \ldots, m$. (z_{xi}, z_{yi}) – coordinates of the lower left corner of the rectangle Z_i; (z_{li}, z_{wi}) – width and length of rectangle Z_i. Thus polygon **A\Z** is needed to be covered by minimum number of equal circles N of radius r.

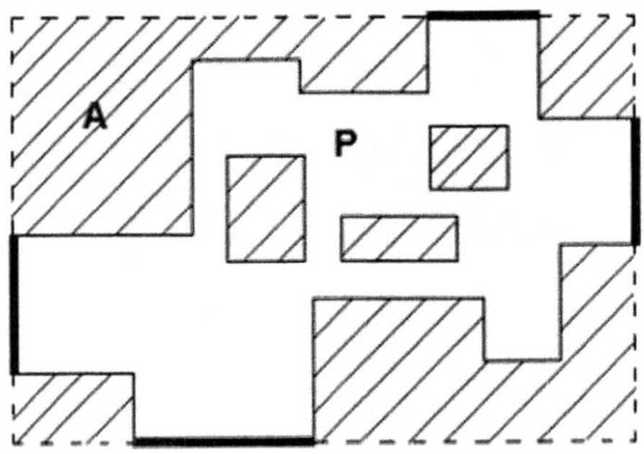

Fig. 5. Monitoring area as polygon.

Solution of the problem can be represented by a set of data: $C = \{N, X, Y\}$, where N – number of covering circles in solution; $X = \{x1, x2, \ldots, xN\}$, $Y = \{y1, y2, \ldots, yN\}$ – vectors of circle centers coordinates. Solution C is acceptable covering if the following conditions Ω are performed:

1. Circles are inside of rectangle **A**:

$$x_j \geq 0; y_j \geq 0; x_j \leq L; y_j \leq W \, \forall j = 1, \ldots, N.$$

2. Circle centers lie inside obstacles: at least one of the inequalities is performed:

$$(x_j - z_{xi})(x_j - z_{xi} - z_{li}) \geq 0 \text{ or } (y_j - z_{xi})(y_j - z_{xi} - z_{li}) \geq 0 \, \forall i = 1, \ldots, m, j = 1, \ldots, N.$$

3. Area **A\Z** is covered: if (p_x, p_y) – an arbitrary point from **A\Z**, then $\exists j : (p_x - x_j)^2 + (p_y - y_j)^2 \leq r^2$.

Candidate solution C is optimal if the number of N circles is minimum, i.e. $C_{opt} = \arg\min_{\mathbf{X},\mathbf{Y} \in \Omega} N$.

Example of coverage is shown in Fig. 6. Obstacles in the figure are shown by dark color.

The solution of such problems are engaged by such scholars as L.F. Toth, S.N. Astrakov, Ye.A. Muhacheva, V.Y. Kuznetsov et al. Summarizing their experience it is possible to allocate the following heuristic-based approaches: heuristic block coverage, heuristic hexagonal coverage, pseudo hexagonal heuristics coverage, evolutionary heuristics and others. The experimental results showed that all the proposed algorithms can be used in solving the problem of optimal placement of sensors in monitoring zone, the choice of algorithm is dictated by the specific requirements of the task. The resulting solution C_{opt} is initial data for UAV repeaters placement algorithms described in the next section.

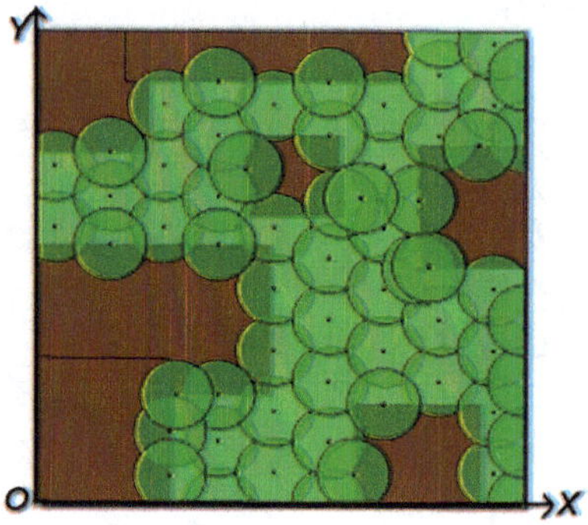

Fig. 6. Coverage of the polygon by circles.

2.3 The Problem of UAV Relays Localization in the Space

Let's consider an example of WSN architecture using UAV (Fig. 7). WSN consists of a set of N sensors, end devices (ED), located with the coordinates $(x_i, y_i), i = \overline{1, N}$, in some limited area with size r, which is often called by sensor field. In order to collect data from sensors and transmit them to the gateway it is used a set of K relays (routers) based on miniature UAVs located at a height h with the coordinates of the projection $(x_{0k}, y_{0k}), k = \overline{1, K}$ and radio coverage radius R. If sensor is located in the UAV radio coverage, data via repeaters network get to gateway and then via global networks to the data processing center (DPC). For data it can be used one of existing network protocols such as IEEE 802.15.4/ZigBee.

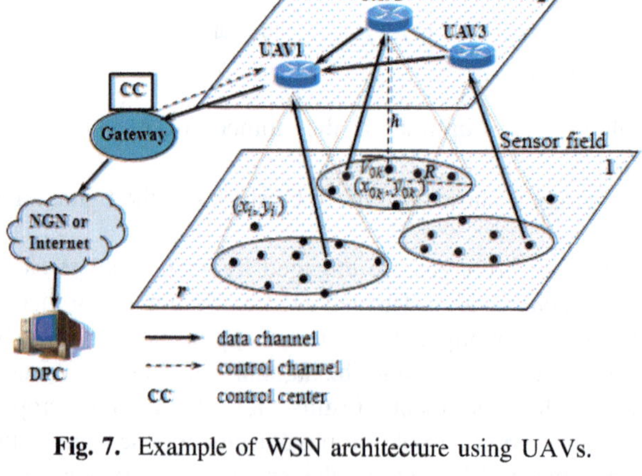

Fig. 7. Example of WSN architecture using UAVs.

Under the network topology we mean a set of geometric location of nodes and the probability of use the communications between them to deliver messages: $(\|R_i\|, \|p_{ij}\|), i,j = \overline{1,N}$, where $\|R_i\|$ is a set of WSN nodes; $\|p_{ij}\|$ is a set of probabilities of using the communications between nodes. Thus, the WSN can be represented as a weighted directed graph consisting of a set of vertices (nodes) and edges (channels) (Fig. 8).

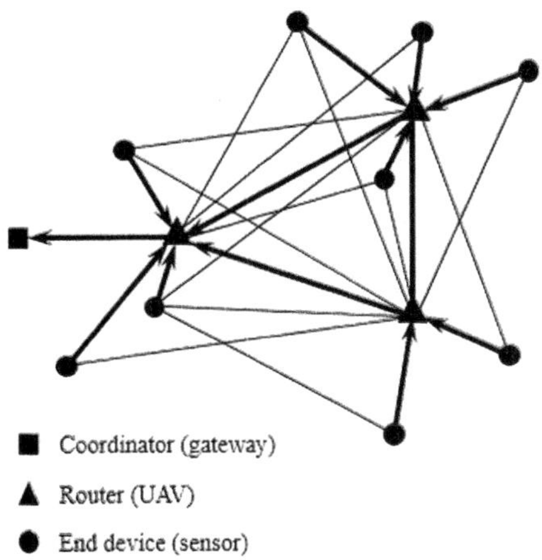

Fig. 8. Example of WSN network graph

Construction the routes from ED to the gateway is realized by using a dynamic programming algorithm - the method of Bellman-Ford. Criterion function of this algorithm (Bellman function) defines the conditional cost of data flow transporting between adjacent nodes of route:

$$W = C_m + C_r \underset{\|R_i\|}{\rightarrow} \min, \text{ where } C_m = \begin{cases} C_m^{nom} \\ 0 \end{cases},$$

$$C_r = \begin{cases} \frac{U_{RX}^* + U_{TX}^*}{U_{\max} - U_w^*} C_m^{nom}, (U_{RX}^* + U_{TX}^*) \leq (U_{\max} - U_w^*) \\ \infty, (U_{RX}^* + U_{TX}^*) > (U_{\max} - U_w^*) \end{cases},$$

where C_m is conditional cost of WSN node, which receives zero value with the repeated use of node in the topology (it allows to minimize the number of repeaters),

C_m^{nom} is conditional nominal cost of node,
C_r is conditional cost of relaying,
U_{RX}^*, U_{TX}^* is actual traffic through the node taking into account repeated relaying,
$U_{\max}$ is maximum traffic through the node,
U_w^* is equivalent density of data flow emitted by adjacent nodes taking into account competition for multiple access to radio channel.

Thus, we can formulate the following mathematical statement of the problem - to find a network topology (UAV repeaters location), that minimizes the cost of transporting the data flows from sensors to gateway, when the constraints on network resources, structural connectivity and network operation parameters are performed:

$$X_0 = \arg \min_{X_0 \in \Omega_{1,2}} C(X) = \arg \min_{X_0 \in \Omega_{1,2}} \sum_{i=1}^{N+K} \sum_{j=1}^{N+K} W_{ij}\ , i \neq j, \tag{12}$$

$$\Omega_1 : \{X \in r,\ p_{ij} \leq PER^0, P \leq P^0, N \leq 1000, K \leq 100\}$$

$$\Omega_2 : \{s(m_{ab}) \geq s^0,\ t_3(m_{ab}) \leq t_3^0, \quad a, b = \overline{1, N}\},$$

where X is vector of projections coordinates of UAV set placement; PER_0 is threshold of packet error rate in the channel; P, P^0 are network reliability and its threshold; $s(m_{ab}),\ t_3(m_{ab}), s^0, t_3^0$ are throughput and delay in the route m_{ab} between sensor a and b and the corresponding thresholds.

The problem of designing a rational WSN topology from a formal point of view is similar to the problem of forming the topological structure of any wireless network, which is one of principal in its design and lies in the choice the optimal scheme of connection between switching units, choice of channels bandwidth and optimal routes of data transmission.

There are currently approaches to solving such problems based on the use of algorithms for combinatorial analysis. These algorithms are based on the representation of the data network in a finite graph without loops and multiple edges, whose vertices correspond to network nodes and edges - links. Application of the theory of graph enumeration for solving the problem of topological optimization recently finds increasingly widespread use, that is connected with an increase in performance of computers, the development of new highly efficient algorithms for generating graphs with desired properties. The main advantage of this approach is the possibility of obtaining the exact solution for small networks therefore this approach is effective for small networks. In addition, the presence of the exact solution allows us to estimate the quality of existing and developed approaches of topology optimization.

The main disadvantage of algorithms for constructive enumeration of graphs is the inability of their application for building large-scale network topology, because the number of generated graphs grows exponentially with the growing of network nodes number. It determines the feasibility of developing an alternative approach to the design of a rational WSN topology.

To solve this problem it is developed a method, described in detail in [16], which represents a computational procedure that includes the following steps:

4. Evaluation of network connectivity. In the case of mobile sensors initially it is forecasted the duration of their stay in the radio visibility zone of UAV repeaters, and in the case of fixed – immediately it is assessed the degree of sensors coverage by UAVs repeaters. If conditions on the connectivity and network reliability is not performed, it performs the procedure of placing the UAV so that to cover all of nodes.

5. Evaluation of given functional parameters (throughput and delay of routes).
6. Improved search algorithm for a rational UAVs topology.

The essence of improvement lies in the fact that for reducing the exhaustive search of topologies, it is used a set of rules for UAVs placement that enables to find a rational solution and minimizes the time of its search. Set of rules for UAVs placement is described in detail in [4].

Thus, the application of the method will reduce the enumeration of possible graphs and get a rational solution (which differ from the optimum on 5–7%) in real time (units/tens of seconds).

Section 2 Conclusion

There was formulated a mathematical statement of the problem of finding a rational topology WSN using UAVs, which consists of problem of sensor localization in monitoring zone and problem of UAV relays localization in the space for the organization of connected network with desired characteristics. Rational WSN topology, built according to these principles, will reduce traffic and reduce energy consumption at the nodes, which in turn will increase network lifetime and reduce the total cost of its maintenance.

References

1. Conti, M., Giordano, S.: Mobile ad hoc networking: milestones, challenges, and new research directions. IEEE Commun. Mag. **52**(1), 85–96 (2014)
2. Sova, O., Romanyuk, V., Zhuk, P., Romanyuk, A.: Concept of hierarchical design of intelligent control systems of MANET class tactical radio networks. In: 22nd International Conference "UHF and Telecommunication Technologies", pp. 265–266. Crimico, Sevastopol (2012)
3. Romayuk, V.: Target functions of operational control of tactical radionetworks. Collect. Sci. Work MITI NTUU "KPI" **1**, 109–117 (2012)
4. Russel, S., Norvig, P.: Artificial Intelligence. A Modern Approach, 1408 p. Wiliams (2007)
5. Bugaichenko, D.: Design and implementation of methods considering formal and logical specifications of self-tuning multiagent systems with time constraints. Ph.D. thesis: 05.13.11, 259 p. SPb (2007)
6. Sova, O., Simonenko, O., Romanyuk, V., Umanec, Y.: Analysis of possibilities of intelligent agents usage for building the node control systems for MANET. Collect. Sci. Work MITI NTUU "KPI" **1**, 76–84 (2013)
7. Gavrilova, T., Horoshevski, V.: Knowledge Bases of Intelligent Systems, 384 p. Piter, Saint Petersburg (2000)
8. Romayuk, V., Sova, O., Zhuk, O.: Architecture of MANET control systems. In: Conference "Problems of Telecommunications", p. 77, K.:ITS NTUU "KPI" (2011)
9. Minochkin, A., Shacilo, P.: Objective, multiagent model of operational control of mobile component of a new generation military network. Collect. Sci. Work MITI NTUU "KPI" **3**, 107–118 (2008)
10. Minochkin, A., Romanyuk, V.: Methods of decision making in a mobile radio network control system. Collect. Sci. Work MITI NTUU "KPI" **1**, 66–71 (2006)

11. Sova, O., Romanyuk, V., Minochkin, D., Romanyuk, A.: Methods of processing the knowledge about a situation in MANET network for building intelligent node control systems. Collect. Sci. Work MTI STU **1**, 97 (2014)
12. Blumin, S., Shuikova, I.: Methods of Decision Making in an Uncertain Environment, p. 138. LEGI, Lipetsk (2001)
13. Lorincz, K., Malan, D., Fulford-Jones, T., et al.: Sensor networks for emergency response: challenges and opportunities. Pervasive Computing **10–11**, 16–23 (2004)
14. Ramesh, M., Kumar, S., Rangan P.: Wireless sensor network for landslide detection. In: International Conference on Wireless Networks (ICWN 2009), Las Vegas, pp. 1–7 (2009)
15. Sun, Z., Wang, P., Vuran, M., et al.: BorderSense: border patrol through advanced wireless sensor networks. Ad Hoc Netw. **9**, 468–477 (2011)
16. Lysenko, O., Valuiskyi, S.: Capacity increasing of sensor telecommunication networks. Telecommun. Sci. **3**(1), 5–11 (2012)

Project Management and Business Informatics

Public Environmental Protection Project Management Practices in Ukraine

O. Zhykhor[1], O. Iafinovych[2], N. Pohribna[2(✉)], and N. Miedviedkova[2]

[1] National Aerospace University «Kharkiv Aviation Institute», Kharkiv, Ukraine
abyss.olena@gmail.com

[2] Taras Shevchenko National University of Kyiv, Kyiv, Ukraine
{olenayafinovych, nataliapohribna, nsmedvedkova}@knu.ua

Abstract. The paper follows the implementation of "Good practices of public environmental expenditure management in transition economies" suggested by UN Economic Commission for Europe in Ukrainian context. The study illustrates that Ukraine follows the UNECE's good practices in some points, however its actual performance when it comes to environmental effectiveness, management efficiency and fiscal prudence is half way less than the best possible performance. The full success of copying UNECE's good practices has not been observed in the Ukrainian case. However, the plausible scope conditions for successful copying for developing roadmap for remedying the situation on environmental protection project management in Ukraine can be suggested.

This paper seeks to contribute to the call for more research on the internal and external drivers and barriers to modernization of public financial management in the Central and Eastern European countries. Previous research, discussing the context of Central Europe shows that the issue of the relation between economic growth and environmental protection at the state and municipal levels has become increasingly important in recent years, e.g. in case of Czech Republic. However, since its independence, Ukraine has been a major receiver of external aid not only financial one, but also in the form of knowledge, advices, recommendations, good practices, etc. This articles builds up on Røvik's translation perspective on knowledge transfer to illustrate how "Good practices of public environmental expenditure management in transition economies" suggested by UN Economic Commission for Europe have been reproduced by Ukrainian government and how this policy translation can be explained in Ukrainian context.

Keywords: Public expenditures · Environmental protection · Environmental protection project management · Efficiency of public environmental protection projects · Fiscal prudence

1 Introduction

Since its independence in 1991, Ukraine has taken important steps in establishing the system of environmental protection. In particular, the country has developed a comprehensive regulatory framework for environmental protection and has become a part of major international conventions. Also, the Ministry of Ecology and Natural

M. Nechyporuk et al. (Eds.): *Integrated Computer Technologies in Mechanical Engineering*, AISC 1113, pp. 405–416, 2020.
https://doi.org/10.1007/978-3-030-37618-5_34

Resources of Ukraine (Ministry of Environmental Protection) has been created in addition to other institutions, empowering with environmental authorities, tools for environmental management and funds for environmental protection and nature conservation. However, despite all these developments, today there are still unresolved issues in the field of environmental protection project management in Ukraine.

This paper represents the results of research within the international project Norwegian-Ukrainian Cooperation in Public Sector Economy Education: Accounting, Budgeting and Finance (NUPSEE) to address some current issues of the public sector transformation. The research results were discussed at the Workshop on "Modernization, Westernization, and Democratization of Public Financial Management" and International Conference "World Trends and Prospects for Development of the Financial System of Ukraine" in 2018.

Analysis of Recent Research and Publications. The theory of organizational translation of practices and ideas represents a promising advance in organization research [1]. Translation is 'the spread in time or place of anything – claims, artefacts, goods….' or is an iterative process by which ideas are materialized, turned into slogans, objects or actions in practice and then turned again into ideas as they are communicated [2, p. 267]. Czarniawska and colleagues introduced the 'sociology of translation' to organization theory, focusing on the circulation, or 'travel', of various organizational ideas between actors and places [3, 4]. For recent theorists of translation in public policy, the central research question is to understand the effect of language and meaning in politics [5].

"Good practices of public environmental expenditure management in transition economies" suggested by UN Economic Commission for Europe (UNECE, 2003) represent an interesting case to analyze how these policy recommendations have been implemented in Ukraine. Mukhtarov states [5, p.71] that "[…] little is known about what makes certain policies spread widely while others remain limited in mobility. It remains a mystery why some policy ideas produce an impact on the ground as they travel across countries while others are formally adopted but rarely produce policy change".

UNECE's good practices [6] is the result of Fifth ministerial conference conducted in Kiev 21–23 of May, 2003. So, it represents form of knowledge transfer to the transition economies' countries on how to build up good and sound practices of public environmental expenditure management. It seems that translation theory can be a useful framework to discuss knowledge transfer as it is often referred as knowledge translation [7]. This paper aims to contribute to the growing literature about knowledge translation as a "phenomenon in its own right, opening up the possibility that knowledge may change its content, form and appearance as it moves from one context to another" [7, p. 244].

Methodology. To fulfill the aim of describing and analyzing the environmental protection project management practices in the case of Ukraine, the article has used the legislatures and regulations of forms and levels of budget spending on environmental programs in Ukraine. Analysis of legal, professional and regulatory texts has been conducted. These documents provided an important source to analyze the state and local funds for environmental protection. Also, the analysis of environmental protection project management practices in Ukraine was done based upon mass media and information available on the websites of state authorities, with existing analytical reports on the state on financing the environmental protection in Ukraine.

2 Research Results

2.1 UNECE's Good Practices of Public Environmental Protection Project Management in Transition Economies

The world practice conceptually defines two basic sectors in the structure of environmental protection projects: public and private. The public is connected with the budget system and is especially sensitive to citizens given the large number of competing applicants among the various social programs on limited public resources. The private sector concerns the funds of economic entities, which affects their income and competitiveness [6].

The quality of management of the environmental protection projects, in the context of achieving the basic level, depends on various factors, in particular: on selected principles, procedures, institutional mechanisms for implementation of environmental programs. Thus, the implementation of environmental spending programs is carried out with a wide range of institutional structures that are selected by the environmental departments of the government at different levels. At the same time, both public and private agents can be involved for the implementation of environmental measures.

In addition, international practice [6, p.10–11] has conveyed that the necessary elements for creating a quality system of government spending on environmental protection include the following: fiscal discipline, effective distribution of public funds, cost efficiency.

It is recognized [6, p. 11] that the prerequisites for creating the above-mentioned elements are transparency as it provides inexpensive access to relevant information; accountability as it is the ability of public officers to be responsible for their actions and their implementation; and integrity of the budget as ideally means the pooling of all public sector income into the general fund, with the legislature and the executive body allocating these total resources to public expenditure programs in order to equalize the marginal social benefits for each program.

2.2 Practices of Environmental Protection Project Management in Ukraine

The disclosure of environmental protection project management practices in Ukraine should begin with consideration of the environmental management system, which, in turn, depends on the functioning of the overall system of public administration in the country. At the present stage, the overall system of public administration has undergone a change during reforms in the area of decentralization of power, which partially began in 2010–2012, and mainly continued after the Dignity Revolution of 2013–2014. The current environmental management system in Ukraine is characterized by certain features of the former system, and therefore faces a number of problems and is characterized by a low level of environmental costs. So, in Ukraine a large number of government structures that manage the environmental protection in their specific sectors at the central level is involved in the system of environmental protection.

Considering the system of environmental protection project management, it is necessary to discuss the legislation in the field of ecology and environment protection in Ukraine. After all, the modern society of developed countries of the world is characterized not only by the high level of material welfare and technical development, but also by the sound legislation in the field of environmental activity and the readiness of citizens to adhere to these laws. The EU-Ukraine Association Agreement ratified by Ukraine in 2014, defines the principles of cooperation in various domains, including the area of environmental protection and rational use of natural resources.

The issue is that the ecology tax is the source of filling both special and general funds of the appropriate budget. The ecology tax was the source of special funds of state and local budgets only in 2012 and 2013. Since 2014 it is a source but of the general fund of the budget, not of the special fund, under the relevant amendments to the Budget Code. This means that directing the revenues of ecology tax to the general fund of the State and local budgets reduces the level of financial support of protection activities since the money from general fund can be spent on other needs, quite different from the environment.

The process of reforming the economy in the direction of fiscal decentralization that began in Ukraine in 2014, caused a legislative initiative to change the structure of distribution of ecology tax in the intergovernmental vertical and horizontal plane. Regulatory provisions of 2012–2018 dramatically changed the proportions of ecology tax allocation between the levels of the budget system, as well as the funds of environmental protection at various levels - national, regional, and village, town, and city budgets, budgets of united communities (Table 1).

Table 1. The distribution of ecology tax between the budgets of different levels in Ukraine.

Years	The level of budget		
	State	Regional	Village, town, city budgets, budgets of united territorial communities
2012	30%**	20%**	50%**
2013	53%**	13.5%**	33.5%**
2014	65%*	10%**	25%**
2015	20%*	55%*	25%*
2016	20%*	55%**	25%**
2017	45%*	30%**	25%**
2018	45%*	30%**	25%**

Note: except for budgets of Kyiv and Sebastopol; except for the tax charged for the generation of radioactive wastes and/or the temporary storage of radioactive wastes by manufacturers beyond the established terms of the license

* general fund of the budget;

** special fund of the budget

Source: compiled by the authors according to [8]

Starting from 2014, at the state level, this tax has partially lost its importance as a source of financial resources for funding measures to protect and restore the environment, i.e. it is a source of filling the general fund of the State Budget. Then, for local (rural, town, city, united territorial communities), as well as regional and equivalent budgets, it is a source of filling a special fund (except 2015), which in effect obliges to spent the amount from paid ecological tax on financing the necessary measures for the restoration of the environment.

Thus, the concept of decentralization that is observed in the current approach to the distribution of the environmental tax between the levels of the budget system of Ukraine, on the one hand, allows the use of a wide range of mechanisms for the implementation of projects aimed at rehabilitation of the environment and cooperation of interested parties at the local level, and on the other hand, makes it impossible to implement effective environmental projects at the state level.

During the years of independence Ukraine changed the number and types of payments for the use of natural resources, as well as for pollution of the environment. Also, the range of users of natural resources has been extended, who are obliged to make environmental payments. However, it did not allow the state to provide over 2 percent of GDP on environment protection, as do foreign countries.

Recognizing the insufficiency of financial resources to finance the environmental area in Ukraine, it is necessary to determine the reserves for increasing the amount of the source base for financing the environmental protection measures. Rentals can be a

Table 2. The structure of the consolidated budget environmental payments in Ukraine in 2012–2018.

Year	Environmental payments (actual amounts)									
	Rent payments, charges for fuel and energy resources		Ecology tax		Fee for pollution of environment*		Fines for damage caused by breach of legislation on environmental protection as a result of economic and other activities		Other revenues to fund environmental protection	
	mln. UAH.	%	mln. UAH.	%	mln. UAH.	%	mln. UAH.	%	mln. UAH.	%
2012	35191.7	92.3	2816.0	7.4	42.7	0.1	61.5	0.2	3.9	0.0
2013	34563.3	89.7	3899.5	10.1	16.9	0.0	56.6	0.2	3.9	0.0
2014	39584.2	88.9	4830.9	10.9	6.2	0.0	91.8	0.2	5.2	0.0
2015	49203.6	94.7	2691.0	5.2	8.7	0.0	57.3	0.1	3.4	0.0
2016	46887.4	90.1	4987.4	9.6	12.0	0.0	154.0	0.3	5.8	0.0
2017	51120.2	91.6	4698.4	8.4	2.0	0.0	11.4	0.0	7.7	0.1
2018	50080.7	90.8	4921.5	8.9	1.0	0.0	123.3	0.2	9.2	0.0

* Fees for environmental pollution existed before 2010.
Source: calculated by the authors according to [9].

powerful source of filling different environmental protection funds in Ukraine. To do this, we will analyze their dynamics and the share in the structure of all eco-resource revenues of the Consolidated Budget of Ukraine over the past 7 years (Table 2).

Data shows that revenues from rent, fuel and energy taxes, as well as environmental taxes demonstrate a practically constant upward trend. At the same time, a dominant role in 2012–2018 is given to rent, together with fees for fuel and energy resources, which make up 89–95% of the volume of revenues from eco-resource payments. The second position is taken by the environmental tax, which amounts to 5.18% in 2015 to 10.85% in 2014. Fines for damages caused to the environment have the smallest share varied from 0.02% in 2017 to 0.22% in 2018 of all environmental payments during 2012–2018.

The growth of this kind of fiscal payments in 2012–2018 is due to a radical revision of rental rates and fees for the use of subsoil. In general, for the analyzed period, the main reason for the annual growth of eco-resource payments was the mechanical growth of environmental taxes (environmental pollution charges) and rent payments. If we compare the environmental expenditures with revenues from eco-resource payments, then the following picture is observed: in 2014, the state and regions funded environmental protection expenditures amounted to 12.8 UAH of eco-resource payments, in 2017 and 2018 – 9.4 UAH and 8.3 UAH, respectively (Fig. 1). According to calculations, the level of coverage of expenses of the state and regions by revenues from eco-resources payments for the analyzed period has a downward trend, partly attributable to the rule-making provisions of 2014–2018, which radically changed not only the proportions of the distribution of environmental taxes between the levels of the budget system, but also between the funds to which they were attributed - general or special.

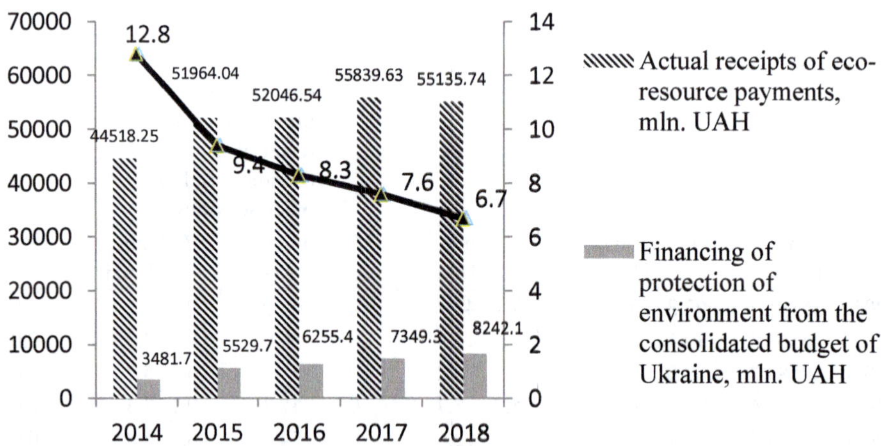

Fig. 1. The level of coverage of environmental protection expenditures of the state and regions by revenues from eco-resources payments in Ukraine in 2014–2018. Source: calculated by the authors on the basis of [9]

This indicates that the state, in case of appropriate changes to the budget legislation, can significantly increase the amount of environmental investments and current environmental costs, which will minimize the negative technogenic impact on the environment, more efficiently reproduce certain types of natural resources, and also approximate the indicators of financing of environmental measures in Ukraine to world standards. So far, Ukraine spent only 0.26, 0.25, and 0.23% of GDP on environmental projects in 2016, 2017 and 2018, respectively.

The effectiveness of environmental protection project management also depends on the volume of private sector investment in the environment at the expense of budget funds. World practice proves that the need for public financing of environmental protection in the business environment is determined on the basis of the polluter pays principle, which is based on the ecological policy of the OECD countries.

In this regard, it is worth paying attention to the structure of environmental costs of business entities in Ukraine by sources of funding in 2013–2017. According to Table 3, the main source of financing environmental expenditures by economic entities during the years 2013–2017 was the own funds of enterprises and organizations. Its share ranges from 85.4% in 2013 to 67.9% in 2016. It is worth noting a clear tendency to reduce the share of financing of environmental protection measures at the expense of own funds of enterprises, institutions and organizations in the total expenses for the environmental protection, while simultaneously increasing the absolute volumes of this source by 1.4 times in 2017 compared to 2013. Thus, the share of budget financing of environmental programs among business entities in 2017 reached 8.2% (compared to only 4.6% in 2013), which suggests a more detailed analysis of the need for public funding of individual private sector programs. In addition, in recent years there has been a significant increase in the share of environmental investments (from 10% in 2013 to 26.1% in 2016) financed by other sources, in particular, the funds of international financial and credit organizations and interstate associations.

Table 3. The structure of expenditures on environmental protection by enterprises, organizations and institutions of Ukraine by sources of funding in 2013–2017.

Year	Actually spent, total		incl.:							
			From the State and local budgets		At the expense of own funds of enterprises, organizations, institutions				From other sources	
					Total		incl.:			
							Capital Investments, mln. UAH	Current expenses, mln. UAH		
	mln. UAH	% of GDP	mln. UAH	%	mln. UAH	%			mln. UAH	%
2013	20377.8	1.4	942.2	5	17408.9	85	3593.5	13815.5	2026.7	10
2014	21925.6	1.4	1162.1	5	17433.7	80	3924.5	13509.2	3329.8	15
2015	24591.1	1.2	1401.7	6	19074.4	78	2692.3	16382.1	4115.6	17
2016	32488.7	0.9	1949.3	6	22069.6	68	3896.9	18172.7	8479.6	26
2017	31492	1.06	2582.3	8	24246.6	77	5132.1	19114.5	4660.8	15

Source: compiled by the authors on the basis of [10].

The previous empirical data makes it possible to conduct coherent assessment of the effectiveness of state projects on environmental protection in Ukraine. The model, based on UNECE [6], assesses the effectiveness of public expenditures on the environment in three main areas: environmental efficiency, fiscal prudence and management efficiency.

Each of the three dimensions for assessing the quality of public expenditure management for environmental protection consists of five principles. Each principle is evaluated by the following points: "zero", if no good practices are applied, "one", if some but not all good practices are applied, and "two" if all good practices are applied. Thus, each component of the performance evaluation can be estimated from 0 to 10 points. Transferring the results to the graphic plane will allow determining the actual performance of each component, as well as seeing the range of improvement and developing a roadmap for remedying the situation. In this way, an assessment of the effectiveness of an integrated system of environmental protection project management can be performed, as well as the effectiveness of a separate program, or a separate agency operating in this area (for example, the Environmental Protection Fund) can be assessed. After analyzing the productivity of the Ukrainian system of public environmental protection projects at the present stage for the above-mentioned model, we obtain the results presented in Table 4.

Table 4. Efficiency of Public Environmental Protection Projects in Ukraine (based on UNECE, 2003).

Environmental effectiveness		Fiscal prudence		Management efficiency	
Additionality and consistency with other environmental policy instruments	1	Fiscal integrity of revenue	2	Sound governance	1
Sound and well-defined programming framework	2	Negative efficiency impacts of earmarking are minimized	1	Professional executive management	0
Sound consideration of environmental effects	1	High standards of fiscal discipline and transparency	1	Sound project cycle management	1
Maximizing environmental effect from available funds	0	Accountability and transparency	0	Fair and unbiased relations with external stakeholders	1
Leveraging additional private and foreign finance for the environment	1	Collection of revenues and public procurement separated from expenditure management	1	Effective management of financial products and related risks	1
$\sum$	5	$\sum$	5	$\sum$	4

If the results obtained above are arranged using a graphic image, then we obtain the triangle of the efficiency of public environmental protection projects in Ukraine (Fig. 2).

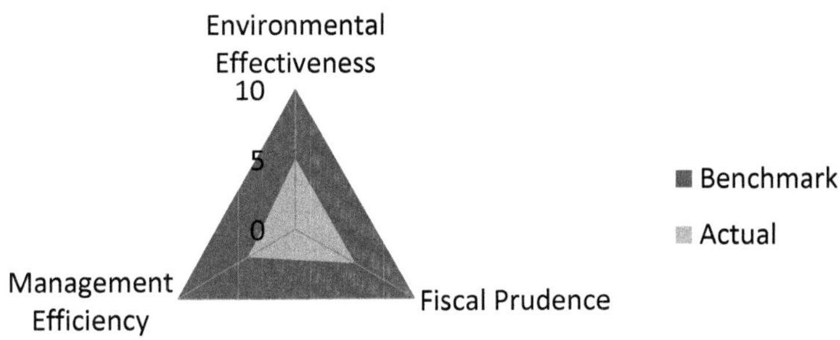

Fig. 2. Triangle of efficiency of public environmental protection projects in Ukraine.

Empirical findings show that not all of the Good Practices by UN Economic Commission for Europe [6] are used in Ukraine. The mechanism for financing environmental projects in Ukraine is carried out according to the traditional scheme - both through general budgets and through state environmental funds, which is controlled by the Ministry of Ecology and Natural Resources of Ukraine.

3 Concluding Remarks

Public spending on environmental protection is an important issue for the sustainable development of the society. The changing institutional and socio-economic landscape in the transition economies requires a profound reassessment of institutions, actors, and roles of those involved in the public environmental expenditures management practices. The actual performance in Ukraine when it comes to public environmental expenditures programs and management agencies is half way towards best possible performance. As it can be observed such elements as accountability and transparency, professional executive management and maximizing environmental effect from available funds represent “zero” meaning no good practices are applied. It confirms Røvik’s [1] theorization of scope conditions for copying when it comes that the contexts of source and recipient should resemble each other on critical variables, such as country, culture, governmental regulations, etc. (p. 301). UNECE’s good practices represent as the best practices available for the transition countries, however, historical, cultural and governmental developments make it challenging to provide safer strategy of replication.

Roadmap for remedying the situation on public environmental expenditure management should contain recommendations on well-functioning of state and local environmental funds in Ukraine. It should be necessary to amend the legislation in the part of the renewal of the receipt of environmental taxes in the revenues of the special fund of the state and local budgets and use them exclusively for environmental protection measures. In addition, the activities of key institutions engaged in environmental protection activities in Ukraine at the central level, has to have proper coordination between these institutions and to avoid certain overlapping of functions. There is a need to review functions and responsibilities in the field of environmental management between the Ministry of Environment and Natural Resources and other ministries, departments, regional and local authorities as their functions are not clearly defined.

Under the limited financial resources for Ukraine, it is important to develop various instruments for the implementation of the state financial policy in the field of environmental protection projects. To solve the problems of greening the economy, the developed countries have developed a certain toolbox, which provides for the application of methods and means of state management of environmentally oriented projects. These include: the adoption of laws aimed at protecting the natural environment; the introduction of a system of incentives for enterprises to comply with legislative standards (tax incentives and concessional targeted lending); the construction of an effective system of environmental taxes, fees and charges; creation of environmental funds; encouraging the implementation of environmental legislation through environmental auditing and environmental insurance; the introduction of environmental certification of products etc.

First of all, it is an urgent need to restore the functioning of the State Environmental Protection Fund of Ukraine. Moreover, it would be advisable to create such an extrabudgetary fund, like the Pension Fund of Ukraine, the source of which would be the ecological tax, part of rental payments and other fees, which would be used exclusively for environmental projects. Indeed, according to the current Ukrainian legislation, the State Environmental Protection Fund is not currently liquidated, however, it is not able to fully function due to the direct allocation of part of the environmental tax to the general fund of the state budget.

It is worth paying attention to the creation of other funds in the country in 2018, in particular the State Fund for Radioactive Waste Management and the State Fund for Development of Water Management, whose activities are also partially related to the implementation of projects to recreate the environment in these areas. At the same time, the main managers of the budget funds of these funds are various departments (the State Agency of Ukraine on Exclusion Zone Management - the first, the Ministry of Ecology and Natural Resources - the second), and the source of filling the first one is the environmental tax that is levied for the generation of radioactive waste (including already accumulated) and/or temporary storage of radioactive waste by their producers in excess of the period established by the special conditions of the license, the second - 10% of the rent for special use of water (except for rent for special use of water of water bodies of local importance). This situation allows us to conclude about the dispersion of resource payments between different ministries and departments, as well as a certain duplication of functions in the field of environmental protection, that also does not contribute to the effectiveness of environmental projects.

Secondly, the distribution of funds between projects aimed at protecting the environment occurs with a low level of transparency and publicity. The Ministry of

Ecology and Natural Resources is conducting a procedure for selecting events that will be financed under the budget program "Environmental Protection Activities", without attracting observers, without online broadcasting the meeting of the working group on project selection, without publishing the evaluation of all submitted programs.

Thirdly, an important step towards improving the efficiency of environmental project management is to determine the criteria by which projects are selected to finance them. Currently, by order of the Ministry of Ecology and Natural Resources of Ukraine No. 194 dated 12 June 2015 "On Approval of Planning and Financing of Environmental Measures", the following criteria have been defined: compliance with the general goal, objectives of the budget program and areas of activity that ensure the implementation of the program; readiness of environmental protection measures at the time of the request; environmental effect; economic efficiency (payback period); the availability of own funds as a source of financing; the availability of guaranteed financing from other sources; deadline for the implementation of environmental protection measures. However, are they all based on the philosophy of greening production? In particular, the criterion of economic efficiency (payback period) and the readiness of the environmental measure at the time of the request etc. raises questions. In addition, as the domestic practice of using the project selection system for financing shows, these criteria may vary depending on the region for which the selection and implementation of the environmental project will be carried out.

Fourth, medium-term budget planning in Ukraine was officially introduced in 2019, which is partially explaining the instability and inconsistency in budget financing of environmental projects in the past practice of allocating funds for such events. Indeed, many projects require long-term, or even constant funding, which directly determines the result of environmental measures. And due to the fact that fund managers are trying to satisfy the interests of all persons who have submitted documents for financing environmental protection measures, funds are being sprayed. Their amount is so scanty that it makes it impossible to carry out high-quality work to recreate the environment, and in fact, simply "patching holes" occurs.

Fifthly, although domestic legislation provides for liability for its violation in the field of financing environmental projects, gaps in the regulatory framework enable contractors to misuse funds from environmental funds, that is, even those small amounts of funds that should be directed to environmental programs are used for other purposes. The existence of such a problem in Ukraine is evidenced by the reports of the Accounting Chamber of Ukraine, which annually conducts audits that are directly related to the problems of protecting the environment, environmental safety and preventing emergency situations. Thus, according to the conclusion of the most independent regulatory body of Ukraine in 2018, the Ministry of Ecology and Natural Resources did not provide control over the implementation of environmental measures aimed at the protection and rational use of water resources. In addition, the audits showed that 43.5% of the allocated funds from the provided budget allocations for the implementation of measures for the protection and rational use of water resources were not used, and also noted the facts of inefficient use of funds.

Sixth, the solution to the problem of the imbalance of the revenue and expenditure parts of the state system of financing environmental protection measures is possible through encouraging enterprises and organizations to use their own funds to finance the protection of environment, in particular, to strengthen public-private partnerships in the field of environmental protection. Indeed, due to the imperfection of legislation and the mismatch of certain institutions with modern tasks, we have an insufficient level of interaction between government bodies and local self-government with civil society institutions. Through the general distrust of the parties, there are problems in the joint actions of the authorities, non-governmental public organizations and private business, which, in turn, makes it difficult to co-finance and implement generally useful initiatives in the field of environmental protection.

Seventhly, verification of the environmental effect of the implementation of projects aimed at reproduction of the environment, greening production is not carried out in Ukraine. Consideration of such a weighty criterion when making a decision on project financing takes place only at the initial stage of the environmental protection measure and does not find its continuation at the stage of project completion. Unfortunately, Ukraine does not currently monitor the state of the environment after the project is completed, it would significantly increase the efficiency of the use of funds for environmental protection measures and improve the management system for such projects.

References

1. Røvik, K.: Knowledge transfer as translation: review and elements of an instrumental theory. Int. J. Manag. Rev. **18**, 290–310 (2016)
2. Latour, B.: The powers of association. In: Law, J. (ed.) Power, Action and Belief: A New Sociology of Knowledge. Routledge, London (1986)
3. Czarniawska, B., Joerges, B.: Travels of ideas. In: Czarniawska, B., Sevón, G. (eds.) Translating Organizational change (1996)
4. Czarniawska, B., Sevón, G.: Introduction. In: Czarniawska, B. and Sevón, G. (eds.) Translating Organizational change (1996)
5. Mukhtarov, F.: Rethinking the travel of ideas: policy translation in the water sector. Pol. Polit. **42**(1), 71–88 (2014)
6. United Nations Economic Commission for Europe/UNECE: Good practices of public environmental expenditure management in transition economies. In: Fifth Ministerial Conference "Environment for Europe," Kiev, Ukraine (2003). https://www.unece.org/fileadmin/DAM/env/documents/2003/ece/cep/ece.cep.96.e.pdf
7. Wæraas, A., Nielsen, J.A.: Translation theory 'Translated': three perspectives on translation in organizational research. Int. J. Manag. Rev. **18**, 236–270 (2016)
8. Budget Code of Ukraine from 08.07.2010 No 2456-VI, Bulletin of the Supreme Council of Ukraine, no. 50–51, p. 572 (2010). http://zakon2.rada.gov.ua/laws/show/2456-17. [in Ukrainian]
9. Report on budget execution in 2012–2018, State Treasury Service of Ukraine. http://www.treasury.gov.ua/main/uk/doccatalog/list?currDir=146477. [in Ukrainian]
10. Environment of Ukraine: Statistical collection. In: Prokopenko, O. (ed.). State Statistics Service of Ukraine, Kyiv, Ukraine, p. 225 (2018). [in Ukrainian]
11. Soukopová, J., Bakoš, E.: Assessing the efficiency of municipal expenditures regarding environmental protection. Environmental Economics and Investment Assessment III, Cyprus, WIT Press, pp. 107–119 (2010)

Models of Achieving Communicative Competence in English as a Foreign Language

T. P. Starovoyt(✉), T. O. Hryhorenko, A. O. Makarenko, and N. L. Kalaytan

National Aerospace University «KhAI», Kharkiv, Ukraine
t.starovoyt@khai.edu

Abstract. The article presents models of achieving communicative competence in English as a foreign language. It also deals with the views on the essence of the notion communicative competence and its structure. In the context of studying English as a foreign language it has been proposed to consider communicative competence in two main aspects: as the result of studying English and as the ability to successfully implement communicative behavior in a certain situation. Communicative competence as the result of studying is achieved by practice and achievement motivation. Communicative competence as the ability to successfully implement communicative behavior can be explained by the Yerkes–Dodson law.

Keywords: Communicative competence; English as a foreign language · Language learning · Motivation · The Yerkes–Dodson law

1 Introduction

Formulation of the Problem. According to the Order # 13 About the adoption of the Procedure of assigning scientific titles to the scientific and scientific and pedagogical staff of 14.01.2016 of the Ministry of Education and Science of Ukraine [1], one of the requirements to scientific and scientific and pedagogical staff of Ukraine is having the certificate which confirms a rather high level of knowledge of a foreign language, i.e. a certificate that confirms the knowledge of one of the European Union languages in compliance with the Common European Framework of Reference for Languages (at the level not lower than B2) [1]. European Commission put Multilingual competence, which it defines as the ability to use various languages appropriately and effectively for communication in the list of Key competences, i.e. such competences, which are needed foremost for personal fulfilment and development, employability, social inclusion and active citizenship, for lifelong learning [2]. Wherein the level of language competence, for instance, the English language competence in Ukraine is estimated to be within the range from moderate to low. In particular, Ukraine is currently taking 43rd place in the EF English Proficiency Index rating, which corresponds to moderate English language competence [3]. Therefore, Ukrainian society has high necessity in improving English language skills, which automatically rises an issue of finding the recourses to increase English teaching efficiency, especially when we deal with

M. Nechyporuk et al. (Eds.): *Integrated Computer Technologies in Mechanical Engineering*, AISC 1113, pp. 417–426, 2020.
https://doi.org/10.1007/978-3-030-37618-5_35

teaching adults, who as a rule do not have enough time to devote to doing homework and consolidating the learnt material and are characterized by having experience of studying English language, by and large not very successful or even frustrating.

Analysis of Recent Research and Publications. In Ukrainian and Russian-speaking scientific environment there is a terminological lack of clarity in the notion "competence" alongside which the notion "professionally essential qualities" is used. Taking this into account, Russian psychologists Bazarov, Erofeyev and Shmelyov conducted a survey among 45 expert-psychologists as the resulted of which they concluded that a competence is such a combination of knowledge, abilities, skills, motivational factors, personal features and situational intentions which ensures effective solution by the executer of tasks of certain class at a certain workplace, in a certain production team [4]. In the context of learning a foreign language there are such notions as "communicative competence", "language competence", "foreign language communicative competence", etc. The term "communicative competence" itself was first introduced by Hymes in 1966 based on the notion "linguistic competence" defined by Chomsky [5]. Hymes wrote about linguistic competence and linguistic performance. Linguistic competence, according to his words, is understood as the which deals with the tacit knowledge of language structure, that is, knowledge that is commonly not conscious or available for spontaneous report, but necessarily implicit in what the (ideal) speaker-listener can say. In its turn linguistic performance is most commonly understood as the one which deals with the processes often called encoding and decoding [5].

Therefore, communicative competence is a term in linguistics which refers to a language user's grammatical knowledge of syntax, morphology, phonology and the like, as well as social knowledge about how and when to use utterances appropriately. The concept of communicative competence is one of the theories which is basis for the communicative approach to teaching a foreign language. Communicative language teaching (CLT) is a modern language teaching approach that highlights teaching languages holistically, based on meaningful communication and interaction. In CLT, languages are learned not only as skills but also as social behavior, by using tasks that learners would also do in the real world [6].

In the Common European Framework of reference for languages learning, teaching assessment there are distinguished two types of competences: General competences and Communicative Language Competences [7]. The competences are defined as «the sum of knowledge, skills and characteristics that allow a person to perform actions» [7]. General competences are those not specific to language, but which are called upon for actions of all kinds, including language activities. Communicative language competences are those which empower a person to act using specifically linguistic means [7, p. 9]. Zimniaia considers communicative competence as a specific goal-result of language teaching, as the ability of a subject to carry out speech activity, realizing speech behavior, consistent with various communicative tasks and situations [8].

Together with the notion "competence" one of the basic notions in the studied field is the notion "competency", which is often confused with "competence". The authors are inclined to think that competency means the ability to successfully carry out certain activity, and competence is a significant component of this ability in the form of knowledge, abilities, skills, etc. Along with that competence is sometimes defined as the ability to implement knowledge and abilities, successfully act based on practical experience when solving basic everyday tasks as well as professional ones.

According to the of Key competences for lifelong learning by the European Commission multilingual competence is based on the ability to understand, express and interpret concepts, thoughts, feelings, facts and opinions in both oral and written form (listening, speaking, reading and writing) in an appropriate range of societal and cultural contexts according to one's wants or needs. Languages competences integrate a historical dimension and intercultural competences. It is based on the ability to switch between different languages and media, as outlined in the Common European Framework of Reference. As appropriate, it can include maintaining and further developing mother tongue competences, as well as the acquisition of a country's official language(s) [2].

This competence requires knowledge of vocabulary and functional grammar of different languages and an understanding of the main types of verbal interaction and styles of languages. Knowledge of societal conventions, and the cultural aspect and variability of languages is important [2].

Vital skills for this competence consist of the ability to understand spoken messages, to start, maintain and conclude conversations and to read, understand and write texts, with different levels of proficiency in different languages, according to the individual's needs. Individuals should be able to use tools appropriately and learn languages formally, non-formally and informally throughout life [2].

A positive attitude involves the appreciation of cultural diversity, an interest and curiosity about different languages and intercultural communication. It also involves respect for each person's individual linguistic profile, including both respect for the mother tongue of persons belonging to minorities and/or with a migrant background and appreciation for a country's official language(s) as a common framework for communication [2].

There exists different understanding of communicative competence structure. For instance, Canale and Swain [9] determine four main kinds of competence:

Grammatical competence: vocabulary, pronunciation, spelling, semantics and sentence formation;

Sociolinguistic competence: correspondence of statements in form and connotation in a specific situation to the context-dependent background.

Discourse competence is the ability to form complete, coherent and logical statements in written and oral communication;

Strategic competence: compensation by certain means of insufficient knowledge of a language, spoken and social communication experience in a foreign language environment [10].

According to Jack C. Richards the competence comprises the following aspects of language knowledge [10]:

- knowledge about how to use language for a variety of different purposes and functions
- understanding of how to vary our use of language according to the setting and the participants (e.g., knowing when to use formal and informal speech or when to use language appropriately for written or spoken interaction)

- knowing how to produce and understand different types of texts (e.g., narratives, reports, interviews, conversations)
- knowing how to keep communication despite having limitations in one's language knowledge (e.g., through using different kinds of communication strategies) [10].

In pedagogical and linguistic literature the main attention is given to the structure of communicative competence, its formation with the help of various pedagogical impacts and less attention is paid to the individual and psychological factors which have an influence in the process of its development. The latter is mainly the focus of psychologists' works. For instance, in the research conducted by Nigel Mantou Lou and Kimberly A. Noels 150 language course students were randomly divided into two groups in accordance with the primed to them understanding of how language intelligence is developed. One group of students was encouraged to think that language intelligence is fixed – entity language theory, and the other was motivated to think that language intelligence can be improved – incremental language theory. In the result of the study it was determined that in the incremental condition, learners more strongly supported learning goals regardless of their perceived language competence, and in turn reported more mastery-oriented responses in failure situations and stronger intention to continue learning the target language. Those who perceived themselves having strong language skills advocated performance-approach goals and in turn reported more helpless-oriented responses and fear of failure [11].

The analysis of the literature demonstrates that although there are different opinions regarding definition of communicative competence and its structure, there is no doubt that it is essential for a modern person. At the same time there is little analysis in scientific literature of communicative competence is achieved by adult learners in the process of studying a foreign language and which factors are vital for its achievement. Therewith understanding of these factors could help the teachers make the process of education more efficient.

The Aim of the Study. The aim of the study is to develop models of achieving communicative competence in English as a foreign language.

2 Models of Achieving Communicative Competence When Studying English Language

With regard to communicative competence in the context of studying English language, it is reasonable to consider it in two main aspects, as they are connected to the impact of various psychological patterns.

The first aspect is the communicative competence as the result of learning process. The competence, which a person achieves in the process of learning English, and which is the result of this process, can be presented with the following formula:

$$C = f\,(Ma, P).$$

where

C – competence,
Ma – achievements motivation,
P – practice.

There have been conducted several researches of motivation special features among students with non-linguistic majors of National Aerospace University "KhAI", who took English course. Teaching was conducted within the communicative approach to teaching a foreign language. Within one of the researches [12] 76 students answered the questions of the questionnaire "Scale of academic motivation" by Gordeeva, Sychov, Osin [14], which enabled the authors to assess the intensity level of the three internal motives types – cognitive, achievement, and self-development, and three types of external motives – self-respect, duty and practical, as well as a motivation. This questionnaire is based on the understanding of internal and external motivational activity, suggested in self-determination theory by Deci and Ryan and the needs model of motivation, developed by Gordeeva applied to educational activity of students [13].

The questionnaire instructions were slightly modified to meet the research tasks. Students were asked to answer the questions in the questionnaire regarding the English language learning process, and not the process of studying in University in general. In the result there have been determined seven motives for studying English. The first group of motives consists of internal motives, to which belong cognitive, achievement and self-development motives. The cognitive motive meant the aspiration to learn something new, understand English. It is also connected with experiencing interest and enjoyment in the process of learning. The achievement motive showed the desire to achieve ultimately the highest results in learning English, enjoy the process of overcoming challenges. The self-development motive is the ambition to develop one's own abilities, open the potential in English, achieve the sense of mastering the language and competence.

The second group of motives constitutes the external personality motives, the first of which is the motive of self-respect. It was based on the basic need of a human for respect and self-respect and assessed the ambition to achieve self-respect by achieving high results in learning English. It also reflected the urge to study to boost self-esteem and self-worth by achieving high results in education. The second is the duty and it means the urge to study English conditioned by the sense of shame and the sense of duty both to oneself and to the other meaningful people. The third is a practical motive assessed the ambition to study English conditioned by the need to meet the requirements set by the society, i.e. to study English to receive rewards for achieving high results, to avoid possible problems or punishments or to earn the praise of meaningful people.

The scale of a motivation demonstrated the absence of interest and of the sense of meaningfulness in studying a foreign language.

The results of the study showed that among the seven motives the highest level was taken by the practical motive of studying English, i.e. the urge to study a language to achieve one's own goals which are least dependent on the process of studying a language. Thus, Ukrainian students understand that knowledge of English is important for their future. This was all demonstrated by the low level of a motivation.

For a more detailed study among the surveyed students two opposing groups were distinguished based on their average semester grade. It reflected their level of knowing English. One group consisted of 24 students with high average grade, the other was comprised of 10 students with low average grade. It has proven to be that among the seven motives statistically credible differences were found only in a higher level of achievement motive ($p = 0.003$). Therefore, the students with a higher level of communicative competence were characterized by a higher level of achievement motive. That means that it can be suggested that in the process of learning a language they took pleasure from ascendance over themselves in achievements and in solving complex problems for themselves.

In addition to that, the students answered the questionnaire of the basic psychological needs in university, developed by Gordeeva, Sychov and Osin [14], based on the principles of the self-determination theory by Deci and Ryan. The achievement motive demonstrated correlation with the basic psychological needs, such as the need for competence and relatedness need. Based on the obtained results the authors can suggest that the higher level of satisfying the need for competence the students have – i.e. the sense of ability to cope with difference educational situations and achieve success in learning English language – the higher is their achievement motive and the average semester grade. Satisfying the relatedness need in terms of English language means that the students feel the teacher treats their interests with respect, considers their desires and feelings and they have good rapport with the teacher.

Berges-Puyó in his study of Motivational factors in learning determined that the that among the top 10 reasons for learning 6 were based on intrinsic factors and 4 on extrinsic ones. This data was justified by 173 high school L2 students learning process [15]. Also, these results showed that the students mostly preferred materials based on the audio or video. What concerns the teachers' motivational practices, the results showed that learners see teacher's personality traits as the most important motivational factors [15].

The authors have found statistically credible differences in the percent of missed classes among the students with the high and low average semester grades in English ($p = 0.001$). Another study demonstrated interconnection between the successful mastering of a foreign language and the practice, which can be understood in broad sense as different means of obtaining experience in using a foreign language, and in a narrow sense as the attendance of English classes in university [16]. Within the sampled 39 students which had a course of English language it was determined that the number of missed classes negatively correlates with the average grade in English for semester ($p = 0.05$).

Therefore, practice is significantly interconnected with the average grade, i.e. with the assessment of the students' communicative competence in English.

Nevertheless, based only on the presence and amount of practice we cannot forecast with the high level of probability that the students will have high communicative competence, as the practice itself or attending classes without motivation will not lead, or will, but will take much more time, to the achievement of high level of communicative competence. Internal motivation alone, and namely the achievement motive, does not determine the success in mastering English. However, it should be noted that high urge for achievement and satisfaction of the need for competence and relatedness need can contribute to the fact the student spends a lot more time on studying English and misses classes more rarely.

Savignon has described comparative study of three groups of beginner college French learners at the University of Illinois, in which she found that to develop communicative competence it was essential to devote time to practice unprepared communication in class, regardless of the mistakes students make [17].

The second aspect of studying communicative competence concerns the understanding of competence as the ability of a person to successfully implement speech behavior in a specific communicative act, i.e. in a situation of social interaction in a foreign language. In this case there is a different psychological pattern, and motivation, for instance, can on the contrary become a factor which is negatively correlated with successfulness. Here the authors are referring to the known in psychology the Yerkes-m law, described in 1908 by Yerkes and Dodson [18]. The essence of this law lies in the fact that the task performance increases with the rise in the level of physiological or mental arousal but only to a certain extent. With the growth of stimuli intensity the successfulness in studying first increases, then reaches the optimal level, and then decreases. If the study problem is challenging, this level is somewhat lower, and if the task is easy, it is higher. When the arousal level becomes too high for a person, performance decreases (Fig. 1).

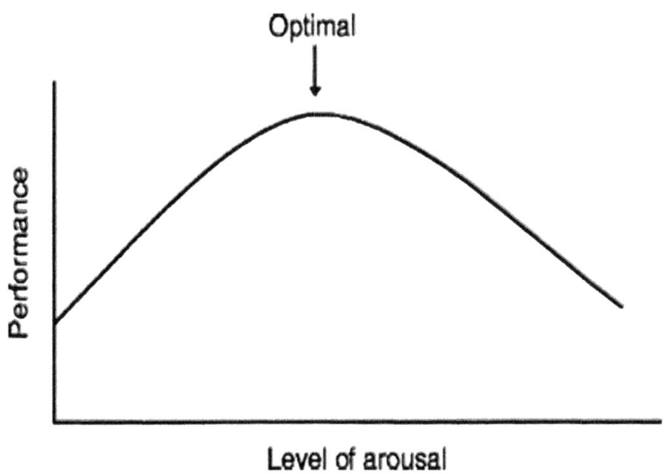

Fig. 1. The Yerkes–Dodson law [18]

Initially, stimulation was seen as an independent variable, which had an influence on performance. Later motivation or the activation level was considered to be an independent variable. When the intensity of stimuli increases, performance first rises, then reaches the optimal level and later falls. If there is a difficult study task, this level is somewhat decreased, and if the task is easy, it is increased. Fraisse and Piaget in "Experimental psychology" stated that "intensifying the arousal, or in a more broad sense motivation, leads to the decreased level of performance" [19].

The appearance of Yerkes-Dodson effects in psychology is explained in relation to the notion of the psychic systems, responsible for education. It is supposed that if the stimulation is low, the systems do not work at their full capacity, thus the learning process is not completely successful. If the stimulation is on the contrary too high, these very systems are overloaded and thus fail. Raudis and Yustitskis see their source in the special features of the chosen (intentionally or unintentionally) when solving a certain study task or the learning algorithm [20].

Therefore, competence, if we understand it as an ability to successfully implement speech behavior in a specific communicative act can comply with the Yerkes-Dodson law. At that it is important to consider that the difficulty of the task is a subjective notion, thus it is more appropriate to talk about perceived difficulty of the task. When the task difficulty crosses a certain threshold if there is a strong desire to be competent, a person will on the contrary face failure. Regarding motivation an important parameter is also the level of significance of results for the student. The higher is this significance, the higher is the motivation, and thus the level of nervousness. Relating to the study of English language it concerns the situations when the knowledge is assessed, and especially exams and any form of stressful interaction for a student. Typically the most stressful for students is the speech practice, especially with native speakers. Considering the competence in such a way, it can be presented in such a way:

$$\mathrm{C} = \mathrm{f}\,(KA,\ \mathrm{M}, D).$$

where

KA – knowledge, abilities,
M – motivation,
D – perceived difficulty of the task.

Knowledge of English and abilities are essential for achievement of "situational competence" in English. However, the presence of these variables only does not explain why some students are more successful in realizing their speech behavior in a specific situation that the others. Along with that we cannot expect high competence in English if a person has low level of knowledge and abilities.

The second variable is motivation. In this case the notion motivation most likely is understood the value which is given by a person to the result or the intention to achieve the expected positive result or the intention to avoid the expected negative result. For instance, in the situation with the English language exam the student may want to be successful in passing the exam or may mostly fear failing this exam. When we deal with a real and not hypothetical person, it is quite likely that both of these trends are present but in the individual correlation. Psychological factor which leads to that fact that motivation "crosses" its optimal point may be inadequate self-esteem. It happens because based on the self-esteem a person anticipates his result, i.e. forms his expectations. For example, a person with the low self-esteem may want to avoid the negative result, which he has anticipated for himself thus making his level of activation not optimal. Another factor is personal anxiety. For a person with the higher level of anxiety the motivational optimum is lower than for the person with the lower level.

And the third variable is the perceived difficulty of the task. When a person faces the need to solve certain task, he assesses two parameters – the present recourses to solve this task and difficulty of the task itself. If the difficulty of the task significantly exceeds the present recourses, he can even not start doing it, as the positive result is unlikely. If the task is too easy and the positive result can be obtained with the high level of probability, he will lose interest in solving it.

3 Conclusions

The purpose of studying English language in the modern society is not only mastering linguistic norms and rules but also the development of ability to successfully adapt through understanding of another culture, one of the main features of which is language. From the point of view of teaching English as a foreign language, communicative competence in this work is seen, on the one hand, as the result of teaching, and on the other, as the student's ability to successfully implement communicative behavior in the situation of communication in English. In the first case, the achievement of communicative competence is connected to the level of achievement motive that a student has as well as the quantity of practical hours in English. In the second case, it is connected to having special knowledge and abilities alongside with motivation, which in this given context is understood as the intention to achieve the desired result or to avoid the undesired, taking into consideration the perceived difficulty of the task.

References

1. Order # 13 About the adoption of the Procedure of assigning scientific titles to the scientific and scientific and pedagogical staff of 14 January 2016 of the Ministry of Education and Science of Ukraine. https://zakon.rada.gov.ua/laws/show/z0183-16. (Ukrainian)
2. Proposal for a Council Recommendation on Key Competences for LifeLong Learning. https://eur-lex.europa.eu/legal-content/EN/TXT/?uri=CELEX%3A52018SC0014
3. EF English Proficiency Index. https://www.ef.com/wwen/epi/regions/europe/ukraine/
4. Bazarov, T.Yu., Erofeyev, A.K., Shmelyov, A.G.: Collective definition of the notion "competence": an attempt to acquire semantic regularities from fuzzy expert knowledge. Moscow University Vestnik (series 14: Psychology), no. 1, pp. 87–102 (2017). http://msupsyj.ru/pdf/vestnik_2014_1/vestnik_2014-1_87-102.Pdf. (Russian)
5. Hymes, D.H.: On communicative competence. In: Pride, J.B., Holmes, J. (eds.) Sociolinguistics: Selected Readings, pp. 269–293. Penguin, Harmondsworth (1972)
6. Lana Loumbourdi Communicative Language Teaching (CLT) (2018). https://doi.org/10.1002/9781118784235.eelt0167
7. The Common European Framework of Reference for Languages Learning, Teaching Assessment. Cambridge University Press (2001). https://rm.coe.int/16802fc1bf
8. Zimniaia, I.A.: Key competencies - a new paradigm of education outcome. Eksperiment i innovaczii v shkole **2**, 7–14 (2009). (Russian)
9. Canale, M., Swain, M.: Theoretical bases of communicative approaches to second language teaching and testing. Appl. Linguist. (1980). https://www.researchgate.net/profile/Merrill_Swain/publication/31260438_Theoretical_Bases_of_Communicative_Approaches_to_

Second_Language_Teaching_and_Testing/links/0c960516b1dadad753000000/Theoretical-Bases-of-Communicative-Approaches-to-Second-Language-Teaching-and-Testing.pdf
10. Richards, J.C.: Communicative Language Teaching Today. Cambridge University Press (2006). http://www.professorjackrichards.com/wp-content/uploads/Richards-Communicative-Language.pdf
11. Lou, N.M., Noels, K.A.: Changing language mindsets: implications for goal orientations and responses to failure in and outside the second language classroom. Contemp. Educ. Psychol. **46**, 22–33 (2016). https://www.sciencedirect.com/science/article/abs/pii/S0361476X16000217
12. Starovoyt, T.P., Hryhorenko, T.O.: The specific features of motivation for English language studying among the students of «unlinguistic» specialities. Young Sci. **1**(28), Part 2, 150–153 (2016). http://molodyvcheny.in.ua/files/journal/2016/1/80.pdf. (Ukrainian)
13. Gordeeva, T.O., Sychev, O.A., Osin, E.N.: "Academic Motivation Scales" questionnaire. Psikhologicheskii Zhurnal **35**(4), 96–107 (2014). https://publications.hse.ru/mirror/pubs/share/folder/y93jdtmioo/direct/122549995. (Russian)
14. Gordeeva, T.O.: Motivation of educational activity of schoolchildren and students: structure, mechanisms, conditions of development. MSU, Moscow, 444 p. (2013). http://www.psy.msu.ru/science/autoref/doc/gordeeva/gordeeva_diss.pdf. (Russian)
15. Berges-Puyó, J.: Motivational factors in learning an L2: a study on intrinsic/extrinsic motivation, classroom materials and teacher's behaviors (2018). https://www.researchgate.net/publication/325204332_Motivational_factors_in_learning_an_L2_a_study_on_intrinsicextrinsic_motivation_classroom_materials_and_teacher's_behaviors
16. Starovoyt, T.P., Hryhorenko, T.O.: Motivation factors which encourage success in studying English language by the students of "unlinguistic" specialities. Young Sci. **8**(48), 175–179 (2017). http://molodyvcheny.in.ua/files/journal/2017/8/40.pdf. (Ukrainian)
17. Savignon, S.J.: Communicative Competence. The TESOL Encyclopedia of English Language Teaching (2017). https://doi.org/10.1002/9781118784235.eelt0047
18. Cohen, R.A.: Yerkes–Dodson law. In: Kreutzer, J.S., DeLuca, J., Caplan, B. (eds.) Encyclopedia of Clinical Neuropsychology. Springer, New York (2011). https://link.springer.com/referenceworkentry/10.1007%2F978-0-387-79948-3_1340
19. Fraisse, P., Piaget, J.: Experimental psychology, vol. 5, pp. 119–125 (1975). http://flogiston.ru/library/piazhe1. (Russian)
20. Raudis, Sh., Yustitskis, V.: The Yerkes-Dodson law: the link between stimulation and learning success. Voprosy Psikhologii, **3**, 119–126 (2008). http://www.maf.vu.lt/katedros/cs/Asmen/SRaudys_21.pdf. (Russian)

Project Risk Management Methodology

Anastasia Romanskaya(✉) and Anatoliy Berdnikov

CSD, V. N. Karazin Kharkiv National University, Kharkiv, Ukraine
anastasimore@gmail.com

Abstract. The proposed methodology can be used to more accurately analyze and predict project results. The methodology is based on the interaction of the method of evaluating the project's timing and budget - PERT and the method of constructing a multivariate forecast of the dynamics of the external environment - the decision tree. The combination of these methods excludes their shortcomings separately. Thus, the proposed methodology visualizes the development of the project in a graphic way and includes a probabilistic assessment of the project scenarios. The relevance of the work is due to the fact that many projects are currently being implemented in various spheres of human life, but not all projects achieve their goals, including IT projects. According to The Standish Group, around the world last year $ 750 billion was spent on software development and implementation projects, and only 36% of projects were completed successfully, while 48% were only partially successful, i.e. went beyond budget or on time, and the remaining 16% failed. As a result, $ 120 billion was spent on failed projects of $ 750 billion.

Keywords: Risk · Risk management · Project management · Decision tree method · PERT method

Decision-making processes in project management occur, as a rule, in conditions of uncertainty, which depends on the following factors:

- incomplete knowledge of all the parameters of the management process or circumstances in the situation of decision making;
- the presence of the probabilistic nature of environmental behavior and, accordingly, the need to take into account random factors;
- the presence of subjective factors of opposition, when decision-making takes place in a situation of interaction of partners with opposite or conflicting interests.

Risk is understood as the potential, numerically measurable possibility of adverse situations and related consequences in the form of losses, damage, losses [1].

Thus, under these broad definitions of risks and quite general descriptions of risk management, it is clear that risk avoidance/risk mitigation programs must be multi-dimensional. These programs often include good management practice, leadership and human resource issues, as well as scheduling, contingency planning and buffer management (buffer sizing and placement).

The focus of this paper is this latter set of programs in that we analyze the way that a project team can utilize quantitative planning tools to contain project risk and to hedge against its impact on success.

M. Nechyporuk et al. (Eds.): *Integrated Computer Technologies in Mechanical Engineering*,
AISC 1113, pp. 427–436, 2020.
https://doi.org/10.1007/978-3-030-37618-5_36

Seven sources of schedule risk:

1. Lack of a realistic schedule developed to a level of detail that accurately reflects how the work will be done, with fully developed work scopes and sequential logic.
2. Inherent uncertainty of the work arising from advanced technology, design and manufacturing challenges, and external factors including labor relations, etc.
3. Complexity of projects, which requires coordination of many contractors, suppliers, government entities, etc.
4. Estimates prepared in early stages of a project with inadequate definition of the work to be performed, and inaccuracies or optimistic bias in estimating activity durations.
5. Over-use of directed (constraint) dates, perhaps in response to competitive pressures to develop aggressive, unrealistic schedules.
6. Project management strategies favoring late start scheduling or fast track implementation.
7. Lack of adequate float or management reserve.

The key to risk management is the quantification of risk, and the use of software tools to reduce the impact of risk on project schedules. Indeed, with the increasing power of the computer, with better and easier-to-use software, with more and better data available, and with increasing pressures to manage projects effectively. Using decision technologies to manage risk in projects is now an important part of project risk management.

There are two methods for determining the likelihood of unwanted events: objective and subjective.

An objective method for determining the probability of unwanted events is based on calculating the frequency with which a particular result was obtained under similar conditions [2].

The subjective likelihood of adverse events is an assumption regarding a specific outcome. This method is based on a heuristic judgment and the personal experience of the decision maker. In other words, based on personal experience and intuition, the decision maker needs to make a quantitative assumption about the likelihood of an undesirable event.

Thus, risk measurement is the determination of the probability of a risk event.

Risk management in projects includes identification, quantification and risk management. All projects have some degree of risk. Projects that use new technologies face the prospect that these technologies do not live up to expectations, very complex projects solve the problem of accurate estimation of time and cost, and even the smallest and simplest projects have some element of risk.

In today's world economy, the competitive position of the economy is determined by its ability to create and apply high technology, produce competitive high-tech products, including information and communication technologies and software. In many cases, the cost and duration of software development projects far exceed the estimated time, and the quality characteristics do not meet the required ones, which is detrimental to customers, users and developers.

The risk management of the project for the implementation of information technologies (IT projects) consists in identifying all possible risks in advance and conducting a set of preventive measures to avoid serious problems during the implementation of the project.

The problem of researching risks in software development processes is complicated by the increasing diversity and complexity of software products being developed. Modern software development projects are characterized by the inability to clearly describe the project product in the initial stages of its implementation [3].

For large software manufacturers developing hundreds of software products, the inability to properly manage risks can lead to tangible economic losses, which could be significantly reduced by timely analysis, forecasting and risk reduction, which makes this direction one of the most relevant [1]. The use of risk management in practice will increase the validity of management decisions and the economic efficiency of the activities of software manufacturing enterprises.

The creation of economic and mathematical methods of risk management in the field of software engineering is today an important and insufficiently studied task, which indicates the relevance of the study on this topic [4].

Risk management has been practiced since the mid-1980s. Effective risk management can lead to a number of benefits to the project manager, such as identifying a favorable alternative course of action, increasing confidence in the project goal, improving the chances of success, fewer surprises, more accurate estimates (due to reduced uncertainty), etc.

The basis of the theory of risk management was laid by scientists who tried to develop methods of solving certain types of problems associated with emerging risks in the industrial and commercial world. The founder of risk management is Wayne Snyder, who in 1956 first described the profession of "risk manager". The first risk management textbook was published in 1963 by Robert Irwin Mayor and Bob Atkinson Hedges, entitled "Risk Management and Business." The main goal of risk management was to maximize the efficiency of the enterprise.

All risks cannot be eliminated, so you need to identify and manage them to prevent project failure. Risk management is the only way to get project approval because it presents risks as clearly identified and therefore controlled.

The goal of creating a methodology is to increase the confidence of quantifying project risk assessment and speed up decision-making by creating a risk management methodology. To achieve this goal, the following goals need to be met:

Form a concept for the risks of an innovative project.

To analyze existing methods of project risk assessment.

Explore traditional approaches to making innovative decisions in the face of uncertainty and the global financial crisis.

To form and classify significant risk factors for the project at all stages of its life cycle.

Combine the advantages of two different project risk management methods - the decision tree and the PERT method - in one methodology in order to improve the quality of the project risk management.

A quantitative risk analysis is carried out in relation to those risks that, in the process of a qualitative risk analysis, were qualified as potentially or significantly affecting the competitive properties of the project [1]. Quantitative assessment can be applied both separately from qualitative and in conjunction with it. If time and budget allow, and if both types of valuation are needed, sharing is the best choice. The purpose of the analysis is to determine the probabilities of achieving specific goals of the project, identify risks that require special attention, determine realistic and achievable goals for the cost, schedule or content of the project taking into account the risks of the project, determine the best solution for project management when some conditions remain undefined. This analysis presents a quantitative approach to decision making under uncertainty in project management.

At the stage of risk analysis, it is necessary to assess them on the basis of qualitative and quantitative analysis.

As part of the analysis, it is necessary to find the most effective ways to apply the methods for: avoiding, reducing, making decisions about risks, or transferring the situation to another person, and complete these steps with the development of an optimal action plan for the best combination of effectiveness.

In the process of a qualitative risk analysis, the sources of risks are determined, as well as the reasons for their occurrence:

Identification of possible risk areas;

Identification of risks associated with the work of the company;

Modeling of probabilistic positive and negative results of manifestation of identified risks.

With a qualitative analysis, the possible types of risks of the investment project are identified and identified, and the reasons and factors affecting the level of each type of risk are also identified and described.

A quantitative risk analysis of an investment project involves a numerical determination of the magnitude of individual risks and the risk of the project as a whole. Quantitative analysis is based on probability theory, mathematical statistics, and theory of operations research [1, 2].

To carry out a quantitative analysis of project risks, two conditions are necessary: the presence of a basic calculation of the project and a full analysis.

The task of quantitative risk analysis is to numerically measure the impact of changes in risk factors of the project on the behavior of the criteria for project effectiveness. When applying quantitative methods, the person responsible for risk assessment has the opportunity to present the risks in the form of a formal system.

The methods used to assess the magnitude of the risk are usually quantitative. However, a complete quantitative analysis is not always possible due to a lack of information about the system or activity being analyzed. Under such circumstances, comparative quantitative or qualitative risk ranking may be effective by specialists well-informed in the field and systems. In those cases, when a high-quality ranking is carried out, it is necessary to have a clear explanation of all the terms used and the justification of all classifications of probabilities and damages should be recorded. In the case when a complete quantitative assessment of the risk value is carried out, it is necessary to take into account that the calculated risk values are approximate estimates and care should be taken that their accuracy corresponds to the accuracy of the initial data used and analytical methods.

Elements of the risk assessment process are common to all types of risk. First of all, the possible causes of a negative event are analyzed in order to determine the frequency of occurrence of such events, their duration, as well as nature. In the process of analysis, it may be necessary to determine the assessment of the probability of danger causing negative consequences, and conduct analyzes of the sequence of causing events.

There are a number of different methods to quantify risks. The vast majority of them are devoted to planning and scheduling, taking into account the risk of delays in the implementation of various stages of the project, as well as the risk of exceeding the project budget (in particular, assessing the necessary reserves). This is a network planning method (CPM, PERT), a decision tree analysis, a Monte Carlo simulation method. Let us analyze the decision tree method. This is a graphical tool for analyzing design situations that are under the influence of risk factors. The decision tree displays sequential decisions in the form of tree branches, located from left to right. Branches originate from the starting point of decision making and "grow" until the final results are obtained. A path along the branches of a tree consists of a sequence of individual decisions and random events. In order to evaluate solutions, it is necessary to calculate the expected value of each path by "folding" the tree in the opposite direction - from the end points to the source. This method allows you to visualize even quite complex structures of risks and decisions. The disadvantage of this method is the difficulty of accurately determining the probabilities and losses in the event of a risk event [5, 6].

One of the greatest advantages of using decision trees is the intrinsic transparent attributes. Unlike other management tools, decision trees are explicit, evaluate accurately all potential options, and links each option to its termination, permitting simple evaluation among the numerous possible decisions. The employment of discrete links to represent customer classified assessment, offers additional precision and lucidity to the process of decision making.

Program (Project) Evaluation and Review Technique (abbreviated PERT) - a technique for evaluating and analyzing programs (projects), which is used in project management. PERT is a way to analyze the tasks needed to complete a project. In particular, analysis of the time it takes to complete each individual task, as well as determining the minimum time needed to complete the entire project.

PERT was designed primarily to simplify paper planning and scheduling large and complex projects. PERT is designed for very large-scale, one-time, complex, non-routine projects. The technique implied the presence of uncertainty, making it possible to develop a project work schedule without accurate knowledge of the details and the necessary time for all its components. The PERT network diagram is most effective for modeling large projects in which the risk of a change in duration is high enough. Experts have the opportunity to give three estimates of the duration of work, which allows to take into account the risks affecting their implementation to varying degrees [5].

When criteria for the project's effectiveness for the so-called "base case" have been obtained, its assessment cannot yet be considered completed due to the uncertainty of future events and the inaccuracy of the information collected. The cost of the project, which is planned for the near future, and even more so the prices for the products or services of the project accepted for calculations, may not be confirmed over time. Therefore, during the analysis of quantitative indicators of the project, it is necessary to determine not only their specific values for the calculation model, but also the possible area of their changes. Consequently, the tasks of project preparation also include analysis of the remaining 50%, which objectively cannot be foreseen, or risk analysis [5, 7].

Despite the complexity of the project categories and project risk, today at most enterprises there is no structured and centralized risk management system. Risk management at the enterprise, including project risk management, is integrated into the enterprise activity management system and project management system by providing risk management status to daily crisis prevention actions. However, insufficient attention is paid to the procedural aspects of risk management, as well as to issues of risk assessment.

In order to improve and increase the effectiveness of the risk management system at the enterprise, it is necessary to create a unified and centralized risk management system using all possible methods of analysis and assessment of risk situations, as well as all risk management tools with subsequent monitoring and control [7].

Decision making under uncertainty is based on the fact that the probabilities of various scenarios are unknown, that is, due to the lack of necessary information, such a probability cannot be established. Decision making under risk is based on the fact that in each situation, the probability of occurrence of individual events that affect the final result with varying degrees of accuracy can be established. This allows you to take into account each of the effectiveness values and choose the situation with the lowest risk level for implementation.

Suppose that there is a project for which it is necessary to calculate the possible risks when obtaining a loan for the development of the project and not getting a loan. We will calculate the main indicators of economic efficiency of the project according to the 1.

$$NPV = \sum\nolimits_{t}^{T} \frac{\Pi_t - O_t}{(1+r)^t}. \tag{1}$$

NPV (Net present value) is calculated as the difference between the discounted cash inflows (Пt) and outflows (Ot) for current (operating) and investment activities for the entire period of the project. A positive value of NPV (NPV > 0) indicates that the project is breaking even [6, 8].

NPV of the first and second options for the development of the project is 24524.56 UAH and 30,086.33 UAH, respectively. When analyzing risks using the decision tree method, it was concluded that in terms of commercial efficiency, the project has the greatest risk when applying for a loan (Fig. 1). The effectiveness of participation in a project in which equipment is purchased on credit, on the contrary, is less risky. Thus, we can conclude that the loan is a good solution in the framework of the project. Having examined a number of methods associated with the project risk, it can be emphasized that the company should strive to prevent risks or minimize their negative impact, it is also necessary to remember the rational use of project resources, as well as monitor activities for successful implementation.

Next, we will consider the method of calculating the PERT schedule to determine the expected duration of the project. The results of the calculation of the expected time, variance and standard deviation are presented in Table 1.

The expected time is calculated according to the 2.

$$Te = \frac{O + 4M + P}{6}, \tag{2}$$

where Te - Estimated Time

O - Optimistic assessment
M - Most likely score
P - Pessimistic assessment

The calculation of the dispersion of the critical path is presented in 3:

$$\sqrt{\sigma^2} = \sum \left(\frac{P - O}{6} \right)^2 = 18,22. \tag{3}$$

The standard deviation is calculated according to the 4:

$$\sqrt{\sigma^2} = \sqrt{\sum \left(\frac{P - O}{6} \right)^2} = \sqrt{18,22} = 4,27. \tag{4}$$

Within one standard deviation from *Te* with a probability of 68.26%, the duration of the project can vary from 278 to 295 days. Within two standard deviations from *Te* with a probability of 94.44%, the duration of the project can vary from 267 to 304 days.

Table 1. Project expected duration

№	Predecessor	Duration estimation works			Expected time	Dispersion	Standard deviation
		O	M	P			
1	**2**	**3**	**4**	**5**	**6**	**7**	**8**
1	–	4	5	7	5	0,25	0,5
2	1	3	7	10	7	1,36	1,17
3	1	14	24	28	23	5,44	2,33
4	2	7	10	14	10	1,36	1,17
5	3	15	20	23	20	1,78	1,33
6	5	6	7	10	7	0,44	0,67
7	6	3	6	10	6	1,36	1,17
8	6	7	8	12	9	0,69	0,83
9	7, 8	3	4	7	4	0,44	0,67
10	7, 8	5	6	7	6	0,11	0,33
11	4	5	5	7	5	0,11	0,33
12	9, 10, 11	7	10	12	10	0,69	0,83
13	12	8	10	13	10	0,69	0,83
14	13	10	13	14	13	0,44	0,67
15	14	5	7	10	7	0,69	0,83

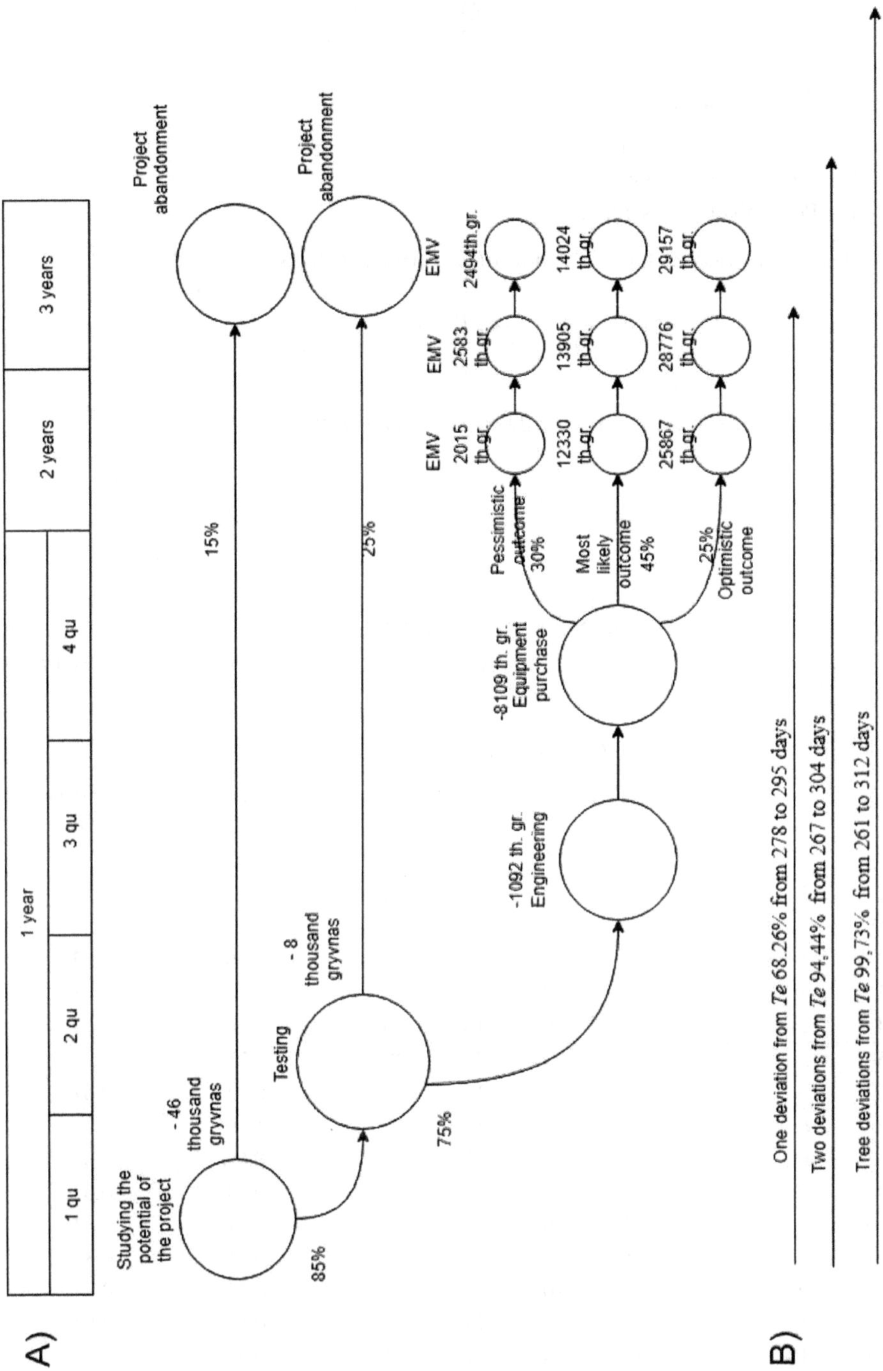

Fig. 1. Combined methods PERT and decision tree

Within three standard deviations from *Te* with a probability of 99.73%, the duration of the project can vary from 261 to 312 days.

We visualize the data obtained by combining them with the decision tree in Fig. 1.

The developed methodology shows that the PERT method in combination with the decision tree method is a convenient way to work with risk events that are considered to be independent of each other (which may not correspond to reality), not only in small but also in large projects, formally or informally.

The methodology is integrated into the generally accepted risk management procedure and is applied at the stage of risk assessment and analysis. This technique is effective due to the fact that it connects the most dangerous and possible risks with the ability to manage them.

The main procedure of the risk management stage is the choice of a risk management method and its further application. The company considers all groups of risk management methods and applies them both in the planning phase and in the project implementation phase. These methods include methods of risk aversion, localization, diversification and risk compensation. Directly during the application of risk management methods, monitoring of this process takes place, which may reveal the need for various adjustments.

In this paper we have advocated the use of readily available decision technologies to manage schedule risk in projects. Although our focus has been somewhat narrow, i.e., confined to schedule, but the use of modern decision technology tools can be useful to a project team as it plans and executes a project.

Thus, the use of the considered model of project risk management at the enterprise will allow to structure the risk management system itself, to take into account the procedural aspect of risk management by developing an algorithm of actions at each stage of the project. The introduction of a centralized risk management system can also solve the problem of unforeseen risks for the enterprise and its projects, simplify risk decisions and reduce risk management costs compared to the existing level.

References

1. Kachalov, R.M.: Management of economic risk: theoretical foundations and applications: monograph. Nestor-Istoriya, St. Petersburg, 248 p. (2012)
2. Risk analysis of an investment project. Risk Management, no. 3. pp. 18–20 (2002)
3. Chernov, V.B.: Risk analysis of an integrated investment project. Risk Management, no. 3. pp. 56–64 (2003)
4. Smolyak, S.A.: Assessment of the effectiveness of investment projects in conditions of risk and uncertainty. Theory of the Expected Effect, Nauka, 182 p. (2002)
5. Lepeshkina, M.N.: Methodological aspects of risk assessment. Management in Russia and Abroad, no. 6, pp. 88–98 (2001)
6. Sorokina, O.V.: Simulation modeling as a method of optimizing investment risks. In: Traditions, Innovations and Investments of a Modern Market Economy: Materials International Conference on Kazan, Part 2, pp. 351–355. Academy of Management "TISBI", Kazan (2004)

7. Sekerin, A.B., Selyutin, V.D., Stroyev, S.P.: A fuzzy-plural model for managing the risk of economic insolvency of a manufacturing enterprise. Risk Management, no. 2, pp. 28–35 (2008)
8. Studenikina, S.A., Zaslavskaya, S.E.: The use of mathematical models in risk management. Modern problems of textile and light industry: abstracts of the Interuniversity. Scientific Conference Part 2, RosZITLP, 160 p. (2004)

Classification Features of International Projects

Bondarieva Tetiana and Sariieva Anastasiia(✉)

Department of Management, Faculty of Software Engineering and Business,
National Aerospace University Kharkiv Aviation Institute,
Kharkiv 61070, Ukraine
bond_tat@ukr.net, sariieva.anastasiia@gmail.com

Abstract. At the current level of development of various forms of foreign economic activity, an organization of any level of management may be faced with the need for cooperation with foreign entities. In most cases, such cooperation occurs in the form of project activities, has its own characteristics and significant differences, which, in turn, depend on the scope of activities and goals. Therefore, for the successful implementation of a specific project, it is necessary to identify the distinctive features of this type and form of international project, which will allow choosing the right project management approach. In order to identify classifiers that will optimize the methods and tools of specific project management and improve the efficiency of their implementation, the theoretical issues of approaches to the definition of "international project" have been studied in the article. The project classifications had been also analysed, and the main classification features and categories that are most commonly used in the scientific literature and practical activities had been considered. In the course of the study it was found that the presented approaches to the classification of projects do not take into account the specifics of the implementation of international projects. As a result, the authors proposed classification features specific to international projects.

Keywords: International project · Classification · International cooperation

1 Articulation of Issue

In the modern world, any business entity in one way or another is faced with the need to engage in various forms of foreign economic activity. Most often, such relationships are built in the form of various projects. Therefore, project management in the implementation of international operations is becoming increasingly important. And since the variety of projects is extremely large, it becomes necessary to determine their distinguishing classification features, which will allow to choose the most effective tools and the optimal approach to international project management.

For the present, there are a fairly large number of project definitions presented both in the standards and project management body of knowledge, and by individual authors.

M. Nechyporuk et al. (Eds.): *Integrated Computer Technologies in Mechanical Engineering*,
AISC 1113, pp. 437–443, 2020.
https://doi.org/10.1007/978-3-030-37618-5_37

2 Analysis of Recent Research and Publications

PMBoK's Project Management Knowledge Body defines a "project" a temporary endeavor undertaken to create a unique product or service [1].

Phil Begley in [2] defines a project as a sequence of interrelated events that occur over a limited period of time and aim to achieve a unique but at the same time definite result.

In [4], a project is a time-limited, purposeful change of a separate system with goals, which from the very beginning are clearly defined, and the achievement of these goals determines the completion of the project, with established requirements for the timing, results, risk, framework for spending funds and resources and the organizational structure.

V.N. Funtov defines the project as a purposeful, time-limited activity carried out to meet specific needs with external and internal constraints and limited resources.

As regards the definition of an international project, it varies significantly. A survey conducted among 85 companies by Cockpit Consulting within the Pool2Business project in 2009 had found that 57.6% of participants (companies actively involved in carrying out international activities) identify projects as international projects if the project members are coordinated in different countries to achieve a common goal.

Almost a quarter of them consider a cooperation with foreign project partners as the crucial aspect of international projects.

The lowest representation was demonstrated by the last two definitions of international projects - only 5.9% of the companies surveyed consider it a multicultural association of team members within the project team, and 4.7% regard a project principal as the main characteristics of the international project.

This wide spreading of the answers shows a very broad range of definitions of international project management. International does not always mean international.

3 Separation of Previously Unresolved Parts of a Common Issue

An international project is one of the most complex and costly types of project activities. This type of project, as a rule, has the highest cost. International projects include all projects in which organizations from different countries take part. In addition, these projects are distinguished by an important role in the economy and politics of the countries for which they are developed [13].

Typically, such projects are based on complementary relationships and opportunities of partners. It is not uncommon that the joint ventures are created to solve problems within the framework of similar projects, with cooperation of two or more participants to achieve certain commercial goals under the appropriate common control. In this case, each partner makes contributions and in a certain way participates in profits.

The purpose of the article was to analyze and highlight the features of international projects, with a view to developing a classification system allowing to give broader characteristics to such projects, taking into account the specifics of international activities.

4 Statement of Basic Materials

To date, most experts have no disagreement on classification criteria that consider types, kinds, classes of projects [3, 5–7]. However, the commonly used classifiers do not consider international projects as a separate type of organization activity, which also requires the allocation of specific categories. The absence of such typification contributes to the occurrence of widespread international project management failures and the use of inefficient approaches. Since it seems that international projects appear to be indistinguishable from standard projects. Indeed, international projects are not too different from standard projects when it comes to the nature of organization, industry, place in the value chain and duration. However, there are significant differences.

The main such differences are:

- purposes;
- scope;
- participants (stakeholders);
- risk intensity.

Purposes of International Projects. For the successful implementation of any project, it is necessary to correctly formulate its purpose. According to the problems to be solved, the following main purposes of international projects can be distinguished.

Search for new territories for geographical presence or new foreign stakeholders. Non-profit organizations have an interest in gaining new supporters on an international level, thereby expanding their fundraising base. With a large number of international members acquired by new local branches, non-profit organizations also tend to have greater influence over different governments or supranational decision makers that play an important role in their respective areas. An example is Greenpeace, which was founded in Canada in 1971. In 1979, it became a centralized international organization. Today Greenpeace is present in more than forty countries.

Increase of global market share, market power, global political powers, or the effectiveness of global activity. Government-run international projects often aim to increase the political power of one government on a global level. The motives vary widely: examples are wars led by a superpower against small countries, for instance, to ensure the supply of natural resources. International projects of governmental agency may also aim to help other countries recover from natural disasters like the 2004 tsunami in Southeast Asia. There are also plenty of projects initiated by governmental agencies of industrialized countries to improve living conditions in the poorer regions of the world. An example is a project financed by the German Ministry for Economic Collaboration to help Egyptian peasants to manage irrigation more efficiently (G:'I'Z, 2006).

Realization of efficiency gains. To reduce manufacturing costs, US car manufacturers have relocated their factories from the USA to Mexico, where the output of assembled cars for the American market has increased sharply in 2006, in contrast to the reduced output in the domestic market.

According to a report from the research firm ISuppli from 2006, 82.6% of laptops sold by multinationals such as Hewlett Packard or Dell Corporation are manufactured by Taiwanese companies, which in turn host their production sites mainly in China.

Access to scarce and/or unique resources. A growing number of organizations are trying to develop new products and services using foreign labor. In so-called transnational projects, involving members from several corporate units located in different countries, including headquarters and subsidiaries, companies such as the European Aeronautics Corporation EADS develop new products, such as satellite equipment.

Many non-profit organizations with limited material resources seek volunteers from all over the world to recruit talented and dedicated employees. An example is a small orphanage in Brazil, near Rio de Janeiro, where a multinational team of volunteers works with local staff to improve children's education there.

Risk reduction. An example of risk reduction through risk sharing is the development by Boeing of its new aircraft the 787, the so-called Dreamliner. Several "risk-sharing partners" scattered across the globe supply 75% of all parts and components of the aircraft.

Scope of International Projects. Another significant difference between standard and international projects is scope. By definition, the scope of an international project goes beyond the usual conditions of the internal market of an organization. For example, several objects will be involved there, which, as a rule, are located in different countries. In many international projects, several different structural units of an organization may be involved.

International Project Stakeholders. Unlike standard projects, international projects usually involve foreign stakeholders, especially customers, who are often non-domestic. As the number of collaborative international projects increases, there tend to be more stakeholders outside an organization. IT consulting company located in USA had to work with 14 subcontractors from Austria, Germany, Lebanon, Dubai, Greece and Cyprus to implement an IT-infrastructure project for a bank in the Gulf region.

Risk Intensity of International Projects. International projects tend to be more affected by risks and uncertainties than standard projects. One of the reasons is the dynamism of the international environment, which is difficult to analyze. Changes often occur suddenly and unpredictably. Another reason is the complexity of the project organization structure with numerous interfaces and a large number of stakeholders [12].

There are also factors that determine some specific features in the sphere of international project management (see Table 1).

Having regard to the above it is necessary to highlight the main classification categories, which will greatly simplify the management and implementation of international projects.

Table 1. Differences between international and standard projects

Criteria	Standard (domestic) projects	International projects
Customer	One or more customers with specified and clear purposes	Presence of several customers from different countries and branches with their own interests
Systems and Technologies	Unified	Various systems and technologies adapted to local conditions
Stakeholders	Representatives of one culture	Representatives of different cultures
Corporate culture	Single or related	Various corporate cultures with their own characteristics and company history
Organization	Focused on project implementation	Takes into account the main activities of the company, depending on resource requirements
Self-interest	As a rule, understandable to each participant	Each participant has his own interest, not always understood by others

One of the solutions to this problem is to use the key areas of modern international cooperation as a classification category: health sector; education sector; environmental protection and safety; economic and social area; peacekeeping and anti-terrorism activity; scientific and technological innovation (see Table 2).

Table 2. Classification criteria for international projects

Classification features	Type of project	Project parameters
By field (area of international cooperation)	Health care	Various projects
	Ecological	Various projects
	Peacemaking and anti-terrorism	Various projects
	Educational	Strategic
	Socio-economic	Short-term social projects
	Science and Technology	Various commercial projects
By degree of difficulty (by class)	Monoprojects	Simple projects
	Multiprojects	Complex (organizational, technical, resource) projects
	Megaprojects	Cost of more than 1 billion US dollars and the duration of 5–7 or more years
By scope (size)	Small	Cost is up to 10 million US dollars and labor effort up to 40–50 thousand m/h
	Medium	Cost of 10–50 million US dollars
	Significant	Cost of 50–100 million US dollars
	Over significant	Cost of more than 100 million US dollars and labor effort of 20 million m/h

(*continued*)

Table 2. (*continued*)

Classification features	Type of project	Project parameters
By duration (terms of implementation)	Short term	Duration up to 3 years
	Medium term	Duration 3–5 years
	Long term	Duration over 5 years
By nature of change	Operational	Small, simple projects
	Strategic	Medium, more complex and costly projects
By main purpose of implementation	Commercial	Various projects
	Non-profit	Various projects
By power relations	Mutually exclusive (alternative)	Various projects
	Substitute	Various projects
	Complementary	Various projects
	Independent	Various projects
	Synergistic	Various projects

The use of such a classification will make it possible to adapt the already existing conceptual framework of international cooperation for project management, simplify the understanding of the specifics of projects by participants and understanding among the participants themselves.

5 Conclusions and Proposals

Thus, the identification of international projects as a special kind will simplify the identification and adaptation of certain parameters of projects by participants in international relations.

References

1. A Guide to the project management body of knowledge (PMBOK). Project management Institute, USA (2004)
2. Bagueley, P.: Project Management. FAIR-PRESS, Moscow, Russia (2002)
3. Derenska Ya, N.: Project management in schemes: a training manual. – Moscow. NUoF Publishing House: Golden Pages, Kharkiv, Ukraine (2007)
4. Ivasenko, A.G., Nikonova, Y.I., Karkavin, M.V.: Project management. Phoenix, Rostov-on-Don, Russia (2009)
5. Mirzoyan, N.V.: Project cost management. Moscow University of Industry and Finance "Synergy", Russia (2007)
6. Popov Yu, I., Yakovenko, A.V.: Project management. INFRA-M, Moscow, Russia (2005)
7. Popov, V.M., Kurakov, L.P., Lyapunov, S.I., Mingazov, H.Kh.: Business plan. Foreign and domestic experience. New development practice and documentation, vol. 2, 2nd edn. Finance and Statistics, Moscow, Russia (2005)

8. Funtov, V.N.: Fundamentals of project management in the company. 2nd edn., Peter, St. Petersburg, Russia (2008)
9. Martin, P., Thein, K.: Project Management. Peter, St. Petersburg, Russia (2006)
10. Clark, A. Campbell.: Project Management on One Page. Williams Publishing House, Moscow, Russia(2009)
11. Batmanova, V.V.: Improving the management of large-scale investment projects (megaproject). Extended abstract of dissertation, Volgograd, Russia (2017)
12. Köster, K.: International Project Management. TJ International Ltd., Padstow (Great Britain, Cornwall) (2010)
13. Mazur, I.I., Shapiro, V.D., Olderogge, N.G.: Project management. 2nd edn. Omega-L, Moscow, Russia (2004)

Robotics and UAV

Assessing Unmanned Traffic Bandwidth

Olha Pohudina(✉), Dmitriy Kritskiy, Serhii Koba, and Andrii Pohudin

National Aerospace University, Kharkiv Aviation Institute, Kharkiv, Ukraine
ok.pogudina@gmail.com

Abstract. The subject of the study in the article is the processes of assessing the airspace in controlling small unmanned aerial vehicles (sUAV) traffic management. The goal is to improve the quality of air traffic control, taking into account the avoidance of conflicts involving three or more sUAV. Problems: to develop a mathematical model of the probabilistic traffic map, as well as to formalize the construction of a random geometric graph model for the estimation of alleged sUAVs conflicts and collisions; To implement algorithms given models construction for airspace automation. The models used: Poisson process whose intensity model is used for building a probabilistic traffic map, random geometric graph model is used for calculate the number of possible conflicts involving the sUAV. The following results are obtained. A formalized model of the sUAV location map has been created taking into account: the given region with the specified population density and the expected number of operations during the specified time interval. This model was used in the construction of a random geometric graph, in which, taking into account the minimum distance possible for the approximation of two sUAVs, an estimation of the probability of conflicts and collisions was conducted. The model is the basis for obtaining an algorithm for estimating the factors limiting the capacity of the airspace, as a result of the occurrence of difficult solvable conflicts. The scientific novelty of the obtained results is as follows: The random geometric graph model is improved by formalizing the position of the vertices. The vertices, taking into account the law of the Poisson process, are placed in the cells of a given region. This allows us to obtain an objective picture of the location of the sUAV in the city's airspace. Two-dimensional models of probabilistic traffic maps (Dutch model Metropolis, model Cal) have been further developed, due to the formalization of the initial sUAV placement, taking into account the law of the Poisson process. This will help to determine the technical requirements for ensuring uninterrupted operation of small unmanned aerial vehicles in the urban airspace.

Keywords: Small unmanned aerial vehicle · Airspace · Unmanned traffic control · Probabilistic traffic map · Random geometric graph

1 Introduction

The increasing number of small unmanned aerial vehicles (sUAV) proves the importance of unmanned traffic management (UTM) task. One of the main tasks of UTM is to assess a traffic bandwidth. It allows to answer the question how many sUAVs can be

M. Nechyporuk et al. (Eds.): *Integrated Computer Technologies in Mechanical Engineering*, AISC 1113, pp. 447–458, 2020.
https://doi.org/10.1007/978-3-030-37618-5_38

safely placed in a given airspace with the possibility of successful control. It is important to take into account the next factors:

- airspace restriction, in case of hard solvable conflicts (if their probability is high, then the definition of conflict management measures is required);
- excessive acoustic noise from sUAV;
- operator communication interference (taking into account the cyber security, because encryption protocols require more bandwidth).

In this paper, we consider the first item, which most affects the number of sUAVs.

Airspace UTM bandwidth assessment is based on aviation traffic control models and methods [1–3], which are mainly planning flights from airport to airport. The differences between the UTM model and the aircraft flight control are: a large number of aircrafts and persons operating them, a variety of flight missions and the ability to take off and land on unprepared sites; the UTM model is limited by sUAV flight allowed zones and height above ground level (150 m) [4].

The stages of sUAV introducing into a single airspace are considered. According to NASA recommendations, there are four main stages, corresponding to the levels of technological ability:

(1) sUAV flight control technologies for agricultural monitoring, fire fighting and infrastructure monitoring;
(2) technologies for dynamic airspace access and contingency management;
(3) technologies for determining the safe distance between private sUAVs in moderately populated areas;
(4) sUAV technologies in higher density urban areas for tasks such as collecting news and delivering packages.

The results of sUAV stages integration into a single airspace are presented on Fig. 1.

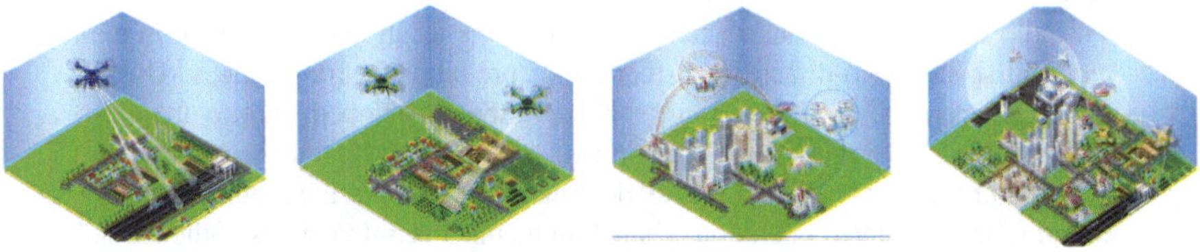

a) stage 1 b) stage 2 c) stage 3 d) stage 4

Fig. 1. The results of sUAV stages integration into a single airspace

Let's examine the goals of the stages.

Stage 1. Testing of single sUAV flights. Study of the demand for flights in sparsely populated areas.

Stage 2. Initial integration of sUAV into non-segregated airspace, implementation of the basic UTM model.

Stage 3. Integration of sUAV into the manned aircraft airspace, implementation of regional UTM.

Stage 4. Transparency in the management of sUAV and situational awareness of all air traffic participants, the creation of a centralized UTM.

Various information subsystems are used to determine the zones allowed for flight, as well as to estimate current traffic [5]. For example, on the website of the Federal Aviation Administration of the United States maps are available indicating areas allowed for flight, areas with flight height restrictions, and it is also possible to register and obtain an electronic flight license.

In order to create such a subsystem in Ukraine, it is necessary to estimate probabilistic traffic, to collect information on areas prohibited for flight.

The concept of the developed sUAV probabilistic traffic map model (LiU - Likely UTM) is based on the following works:

- dutch model «Metropolis» [6] – probabilistic model, where the aircrafts are distributed evenly in a given airspace. In general, the flight direction of the sUAV is evenly distributed in a given airspace circle (at the same time, it is considered that sUAVs fly at the same height, i.e. two-dimensional model). This model allows you to get the likelihood of conflict, but the uniform distribution of sUAV over a given territory will not correspond to reality;
- Cal model [7] – improves the previous model, in terms of the flights destinations. They are selected based on population density.

The aim of this work is to improve the quality of UTM due to the rational choice of bandwidth in a given airspace, limited by such number of sUAVs, at which their safety decreases due to approaching an unsafe distance at the same time by three and more sUAVs.

To achieve the goal, simulation methods were used with the help of the Poisson process model to set the motion intensity. also a model for constructing and estimating a random geometric graph was used [7].

2 Building UAV Location Map

To build a location map of the studied airspace, the following input data are required [9, 10]:

- region of interest Reg;
- population density D(gd), given for each point gd ϵ Reg;
- observation time T (T = 12 h);
- expected number N of sUAV operations during time T (parameter N changes during numerical experiments).

Considering the extremely small heights of the airspace, suppose that the demand for airspace is generated by the model Cal: the start time of the flight from any point $a_d \in Reg$ is formed by Poisson process law, according to which the intensity is proportional to the population density at a given point:

$$\lambda_s(a_d) = \frac{N}{T}\frac{D(a_d)}{\int\limits_{Re\,g} D da_d}.$$

b_d is randomly selected as the end point of the flight. Based on a given population density, the probability that a flight ends at a point $b_d \in Re\,g$ is equal to

$$p(b_d) = \frac{D(b_d)}{\int\limits_{Re\,g} D db_d}.$$

The developed model calculates pointwise distribution of traffic. The intensity of the sUAV flying from ad to bd emergence at the point $g_d \in a_d b_d$ according to the law of the Poisson process equals to

$$\lambda_{a_d b_d} = \lambda_s(a_d)p(b_d) = \frac{N}{T}\frac{D(a_d)D(b_d)}{(\int\limits_{Re\,g} D da_d)^2}.$$

Integration over all pairs of departure-destination, allows you to determine the intensity of sUAV occurrence at any point g_d

$$\lambda(g_d) = \int\limits_{a_d b_d \ni g_d} \lambda_{a_d b_d} da_d db_d = \frac{1}{(\int\limits_{Re\,g} D da_d)^2}\frac{N}{T}\int\limits_{a_d b_d \ni g_d} D(a_d)D(b_d) da_d db_d,$$

where $a_d b_d \ni g_d = \{(a_d, b_d) \in Re\,g^2 : g_d \in a_d b_d\}$ - the set of end points of all segments containing gd.

To create a simulation model we should discretely present the model specified above: on Reg we fix the grid L and the population density $D_{dp}(g)$ is set for each grid cell $g \in L$. The start time of flights from any grid cell $a \in L$ forms a Poisson process, the intensity of which

$$\lambda_s(a) = \frac{N}{T}\frac{D_{dp}(a)}{\sum_{x \in L} D_{dp}(x)}$$

is proportional to the population density in the cell, and the target cell of flight b is randomly selected based on the density of the same probability

$$p(b) = \frac{D_{dp}(b)}{\sum_{x \in L} D_{dp}(x)}.$$

The straight-line path of the sUAV between cells a and b is represented by a sequence of grid cells through which the sUAV passes (Fig. 2). Suppose that a sUAV spends the same time t = l/V in each cell of the path, where l is the length of the cell side, V is the speed of the sUAV. We assume that l = 150 m and V = 25 m/s, therefore, t = 6 s.

Similar to the continuous case for any cell $g \in ab$ of a sUAV, flying from a to b fall into cell g in accordance with a Poisson process with intensity:

$$\lambda_{ab} = \lambda_s(a)p(b) = \frac{N}{T}\frac{D(a)D(b)}{\sum_{x \in L} D(x)}.$$

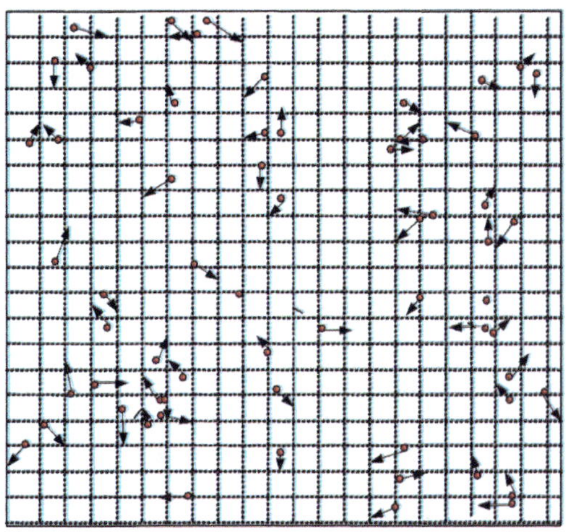

Fig. 2. sUAV location map

Summing up all the pairs of departure-destination, we obtain that, in general, at each moment in time, each sUAV is in an arbitrary cell g of segment a-b with intensity

$$\lambda(g) = \sum\nolimits_{ab \ni g} \lambda_{ab} = \frac{1}{\left(\sum_{x \in L} D_{dp}(x)\right)^2}\frac{N}{T}\sum\nolimits_{ab \ni g} D_{dp}(a)D_{dp}(b),$$

where $ab \ni g = \{(a, b) \in L^2 : g \in ab\}$ is the set of end points of all segments containing g.

Let's call the function graph $m(g) = \sum_{ab \ni g} D(a)D(b)$ a sUAV's map, which shows the probability of sUAV appearing in different cells.

In general, the number of sUAVs n(g) that can be viewed simultaneously in the g cell during the time t is calculated as:

$$\overline{\lambda(g)} = \lambda(g)t = \frac{t}{T\left(\sum_{x \in L} D(X)\right)^2} Nm(g).$$

In the experiments, the values of t and T do not change. In order to simplify the formulas, we normalize the density in such a way that

$$\frac{t}{T(\sum_{x \in L} D(X))^2} = 1,$$

therefore:

$$\overline{\lambda(g)} = Nm(g). \quad (1)$$

After obtaining the number of sUAVs, which can be seen simultaneously in the g cell, their coordinates are generated according to the uniform distribution law. In this case we can go from the two-dimensional matrix representation to the three-dimensional, taking into account the flight height restrictions.

3 SUAV Conflict Assessment Using Random Geometric Graph Model

The model of constructing a random geometric graph (RGG) will be used to simulate the number of sUAV conflicts and collisions (within a given amount of traffic). RGG is a graph G(S, R), with a set of vertices S, which is obtained by placing randomly n vertices, with two vertices connected by an edge, if the Euclidean distance between them is not greater than R (Fig. 3). The number $k > 0$, where $p_k(S, R)$ denotes the probability that G (S, R) has a coherence of at least k, is given:

- $p_1(S, R) = 1$, since any vertex is a connected component of size 1,
- $p_2(S, R)$ - is the probability that G(S, R) has an edge,
- $p_n(S, R)$ - is the probability that the graph is connected,
- $p_k(S, R) = 0$ when $k > N$ (because G(S, R) has only N vertices).

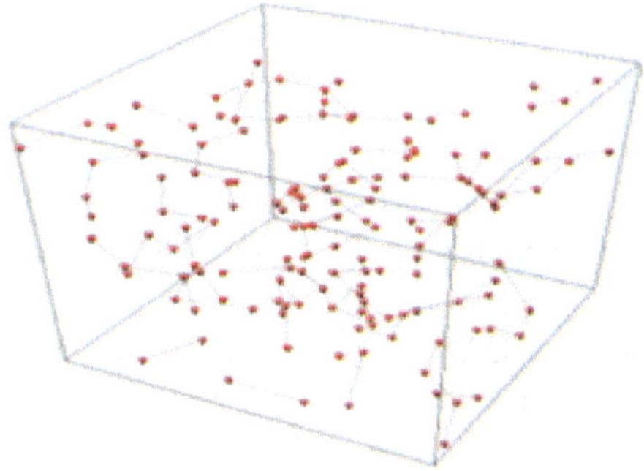

Fig. 3. Random geometric graph

Suppose R is a conflict event (two sUAV conflict with each other at a distance R or collided when R is small enough). Then for N randomly distributed sUAVs, the existence in G (S, R) of a subnet with a connectedness size k means a conflict of k sUAVs. For small subnet sizes (k = 2), the conflict can be resolved using simple rules (for example, maneuver to the right). For k > 2 a conflict may indicate a security violation event with $p_k(S, R)$ probability.

The R_{real} parameter represents the technical capabilities of the sUAV - communication quality, navigation accuracy, autopilot quality, etc. Therefore, for the practical implementation of the RGG model in the UTM system, it is necessary to find a distance R that will provide a low probability $p_k(S, R)$. At the same time, technical characteristics of sUAV should be indicated allowing to provide this distance.

Let's calculate the number of conflicts in the RGG (related components) for two sUAVs (k = 2). The total expected number of such conflicts is the sum of conflicts over all pairs of pixels:

$$C = \sum_{g,g\prime \in L^2} C(g, g') + \sum_{g \in L} C(g), \tag{2}$$

where $C(g, g')$ - expected number of edges between nodes (sUAVs) in cells g and g′, and $C(g)$ - expected number of edges between sUAVs in cell g.

If we consider the expected number of flight hours $H = N\tau$, then we can see that this indicator will grow linearly, while the number of collisions C_d will grow quadratically. For example, for N = 2457 the expected number of conflicts will be C_d 0,001. In this way, it is possible to determine an acceptable value for the movement intensity for which UTM measures are required.

4 Collect Information on Areas Prohibited for Flight

The next step to the developed model is the conflict probability assessment of sUAVs, taking into account the flight allowed zones and geometric variability of the existing static obstacles such as buildings and location. To develop the flight allowed zones map we built the ontological model of the State Aviation Administration of Ukraine directives named "Current legal order of Ukraine's airspace usage".

To build the ontology was used the Protégé platform which is free open-source ontology editor and framework for building intelligent systems.

In the document [8] structural elements are placed in hierarchical order, up-to-date information which will be necessary is placed at the lowest level of the hierarchy.

Suppose that there are:

- the given finite set of categories is $C = c1,\ c2\ c3,\ c4|C|$, where c1 stands for prohibited flight zones, c2 stands for the zones restricted for flights, c3 means dangerous zones and c4 stands for temporarily reserved zones.
- the given attribute space $P = P1 \times P2 \times \ldots \times PN$, where Pi is attribute set of values of i-th attribute;
- the given finite set of order elements $St = d1,\ d2,\ldots,\ d|D|$.
- the given attribute function $f\colon D \to P,\ (di) = (p1,\ p2,\ldots,\ pN)$ is the attribute description of structural element di;

- an undetermined function Φ: $D \times C \rightarrow \{0, 1\}$, which for each pair (*di*, *cj*) determines if the given element di, which has the attribute description *f*(*di*), refers to the category *cj*;

Further we present the attributes to determine the categories:

[Airspace][categories list][is classified according to][*] → c1

For *di* which was determined for a certain category, we need to define the list of attributes, which will constitute subclasses or category notions:

Further we describe attributes to determine notions and subclasses of certain categories:

[*] [flights are carried out] [without | not] [list of attributes] → 1
[*][flights are carried out] [list of attributes] → 1
[Restricted flight zones] [list of attributes] → 2
[Dangerous zones] [list of attributes] → 3

where d1 stands for the prohibited zones attributes, d2 is for restricted zones, d3 contains attributes of dangerous zones.

Further we give attributes for determining categories subclasses:

[*] [central streets] [list of attributes] → 1
[*] [operation zones] [list of attributes] → 2
[*] [national roads] [list of attributes] → 3
[*] [determined objects] [list of attributes] → 4

These notions were divided into categories subclasses, because every notion focuses on one separate object, but they are connected by one common attribute.

Taking into account the aforementioned attributes, we have identified the main categories and categories subclasses, based on which the ontological model will be developed:

- prohibited zones: storage facilities of fuel, oil, gas and other hazardous substances and liquids, etc.; custodial correctional establishments; airdrome taxiways; places of accidents and disasters; electric transmission lines, detention centers, railways, product pipelines, other important state and potentially hazardous objects; power plants, runways, helicopter airdrome taxiways; state borders, industrial zones; sea ports, railways stations; central streets (in the cities, towns and villages); zones of operations (special, police, counterterrorist); national roads (international, national, regional, local); objects which are determined by state bodies (Ministry of Defense of Ukraine, Ministry of Interior of Ukraine, State Border Guard Service of Ukraine, Security Service of Ukraine, National Police of Ukraine, National Guard of Ukraine, State Guard of Ukraine, other military and law enforcement establishments, formed in compliance with the laws of Ukraine with regard to which the guard is carried out).
- restricted flight zones: training areas (bomb dropping, shooting, rocket launch, use of military explosives, destruction of ammunition by detonation, airdrop missions);

areas with shootings conducted to impact hydrometeorological processes in the atmosphere.

- dangerous zones: open sea area.

All the categories and attributes are hierarchically connected by the IS-A ratio as every attribute of the higher level constitutes consolidation of the lower level.

Based on the given categories and notions we have built an ontological model which is given in Fig. 4.

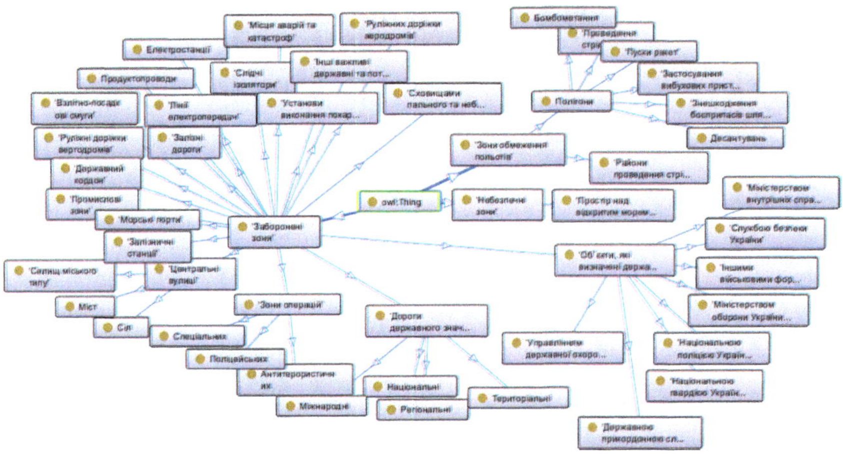

Fig. 4. Ontological model of flight zones categories

Using the built ontological model, we have developed software which shows prohibited, restricted and dangerous zones on the map Fig. 5. The main software operating scenario in generalized view consists of the following steps:

Step 1. Retrieve data from the ontological model and save them in the software database;

Step 2. Process data using requests to maps. Essential is the question if the existing electronic map can provide the data on flight zones and how up to date will this information be. At the development stage of the software testing was carried out only on available maps.

- save data;
- output the data using external service.

External tools in this case are Google API and database server. Figure 6 shows the software algorithm.

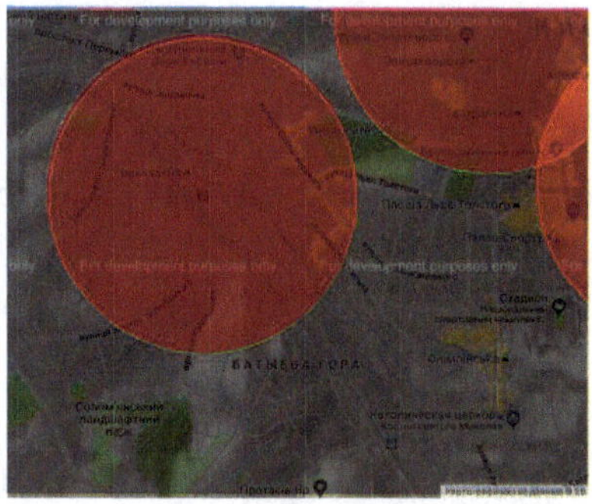

Fig. 5. Flight zones categories map

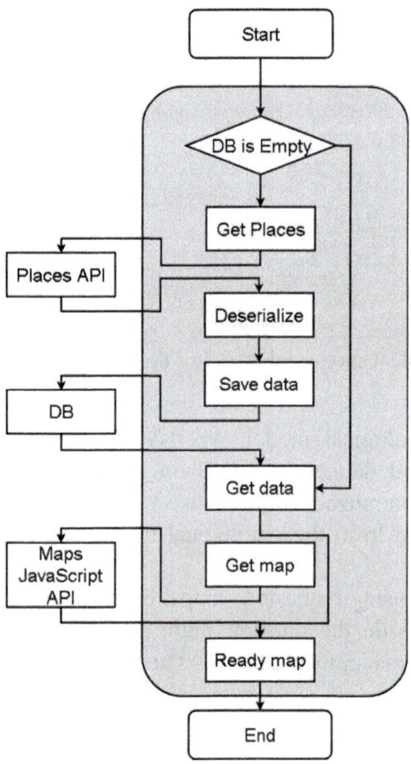

Fig. 6. Software algorithm

The obtained map will enable to build a model of random geometric graph to model the number of conflicts and collisions of sUAVs within the set traffic volume considering the zones where flights are allowed. Furthermore, using the model built in [11], we can evaluate the probability of sUAVs conflicts in heavily built-up area (Fig. 7).

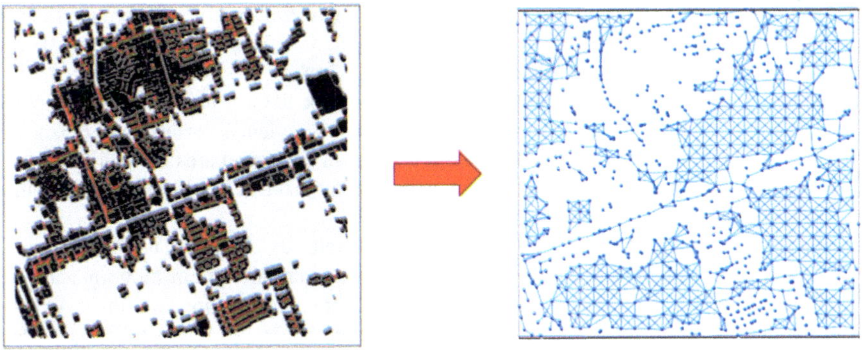

Fig. 7. Structured urban airspace design

Acknowledgment. To date, the use of sUAVs in the airspace of Ukraine is actively discussed by the aviation community. The concept of the airspace usage is being discussed, a temporary usage order (TUO) of the sUAV is defined [8].

The most controversial issues are the following:

- restrictions related to the flight height of the sUAV without registration in the state aviation service. The sUAV TUO document specifies a flight height limit of 50 m, although in most countries there is a limit of 150 m;

- the sUAV TUO document indicates a number of restrictions in the flight zones, which concern not only the borders of the state and airports, but also other infrastructure facilities (roads, railways, etc.),

- there is a suggestion to rank the sUAV usage by risk zones (for example, when used in sparsely populated areas and in regional centers).

The suggested model can be used to solve the third issue - the allocation of risk zones, taking into account the intensity of traffic in various localities. In this case, it will be necessary to modify the sUAV location map with regard to the approved flight zones.

This article has developed a method for assessing the capacity of the airspace when controlling the traffic of unmanned aerial vehicles. Examples of the sUAV conflicts expected number and intensity of their movement calculations are provided, taking into account a given population density [12].

To build the model the next tools and models were used: an intensity model for poisson process applications, model for constructing a random geometric graph, as well as MATLAB toolkits for displaying research results and calculations.

As a result, the sUAV collisions probability values graph was obtained, taking into account the expected number of sUAVs, as well as the distance allowed for approaching other aircraft.

References

1. Skal'ko, Ja.I.: Sovershenstvovanie sistemy upravlenija vozdushnym dvizheniem [Improving the air unmanned traffic management]. Radiojelektronika i informatika – Kharkiv **4**(21), 139–142 (2008)

2. Kageyama, K.: ATC procedures modeling for capacity estimation of Japanese airspace. In: AIAA Modeling and Simulation Technologies Conference, AIAA SciTech Forum, 08 May 2017, pp. 1–10 (2017). https://doi.org/10.2514/6.2017-0805
3. Danilov, Ju.A., Obidin, D.N., Timochko, A.A., Berdnik, P.G.: Razrabotka modeli potoka vozdushnyh obektov v rajone ajeroporta dlja sistemy upravlenija vozdushnym dvizheniem [Development of a model for the flow of air facilities in the airport area for an air unmanned traffic management]. Sistemi upravlinnja, navigacii̇̈ ta zv'jazku **2**(38), 14–20 (2016). Poltava, PNTU Publication
4. Johnson, M., Jung, J., Rios, J., Mercer, J., Homola, J., Prevot, Th., Mulfinger, D., Kopardekar, P.: Flight test evaluation of an unmanned aircraft system traffic management (UTM) concept for multiple beyond-visual-line-of-sight operations. In: Twelfth USA/Europe Air Traffic Management Research and Development Seminar, pp. 1–10 (2017). https://ntrs.nasa.gov/archive/nasa/casi.ntrs.nasa.gov/20170011344.pdf. Accessed 31 May 2019
5. Visualize it: See FAA UAS Data on a Map. http://faa.maps.arcgis.com. Accessed 31 May 2018
6. Sunil, E., Hoekstra, J., Ellerbroek, J., Bussink, F., Nieuwenhuisen, D., Vidosavljevic, A., Kern, S.: Metropolis: relating airspace structure and capacity for extreme traffic densities. In: 11th USA/EUROPE Air Traffic Management R&D Seminar, Lisbon, Portugal, June 23–26 2015, pp. 1–10. https://repository.tudelft.nl/islandora/object/uuid%3A1019a338-5f69-409a-a06c-c8346cd343e. Accessed 31 May 2018
7. Sedov, L., Polishchuk, V., Bulusu, V.: Sampling-based capacity estimation for unmanned traffic management. In: 36th Digital Avionics Systems Conference (DASC), pp. 1–10 (2017). http://weber.itn.liu.se/~valpo40/pages/liu.pdf. Accessed 31 May 2018
8. Ty`mchasovy`j poryadok vy`kory`stannya povitryanogo prostoru Ukrayiny` [Temporary use of airspace of Ukraine]. Postanova Derzhavnoyi aviacijnoyi sluzhby` Ukrayiny, 31 May 2018. https://zakon.rada.gov.ua/rada/show/n0001763–18. Accessed 31 May 2019
9. Kritsky, D.N., Ovsiannik, V.M., Pogudina, O.K., Shevel, V.V., Druzhinin, E.A.: Model for intercepting targets by the unmanned aerial vehicle. In: Palagin, A., Anisimov, A., Morozov, A., Shkarlet, S. (eds.) Mathematical Modeling and Simulation of Systems. MODS 2019. Advances in Intelligent Systems and Computing, vol. 1019. Springer, Cham (2020)
10. Kritsky, D.N., Druzhinin, E.A., Pogudina, O.K., Kritskaya, O.S.: Decision making by the analysis of project risks based on the FMEA method. In: Proceedings of IEEE 13th International Scientific and Technical Conference on Computer Sciences and Information Technologies, CSIT 2018 (2018)
11. Cho, J., Yoon, Y.: How to assess the capacity of urban airspace: a topological approach using keep-in and keep-out geofence. Transp. Res. Part C: Emerg. Technol. **92**, 137–149 (2018)
12. Kritsky, D.N., Druzhinin, E.A., Pogudina, O.K., Kritskaya, O.S.: A method for assessing the impact of technical risks on the aerospace product development projects. Advances in Intelligent Systems and Computing (2019)

Smart Energy and Grids

Increasing the Operation Efficiency of Railway Air Conditioning System on the Base of Its Simulation Along the Route Line

Mykola Radchenko(✉), Roman Radchenko, Veniamin Tkachenko, Serhiy Kantor, and Evgeniy Smolyanoy

Admiral Makarov National University of Shipbuilding, Heroes of Ukraine Avenue 9, Mykolayiv, Ukraine
nirad50@gmail.com

Abstract. The operation of air coolers of railway air conditioning (AC) systems is characterized by considerable variations in current heat loads according to actual climatic conditions on the route lines. This causes increased changes of refrigerant flows. Over filling the air cooler coils by liquid refrigerant recirculation enables excluding a decrease in heat flux within variations in current heat loads and provides increasing the heat efficiency of air coolers compared with conventional air coolers with complete refrigerant evaporation and superheated vapor at the exit. Thus a larger deviation of current heat load on railway route lines are permited without falling air cooler heat efficiency. The method to determine the rational design heat load on air coolers of railway AC systems, providing closed to maximum refrigeration output generation over considered time period, was developed.

Keywords: Railway air conditioner · Changeable heat load · Liquid refrigerant recirculation

1 Introduction

The performance of railway AC systems is characterized by considerable variations in current heat loads on their air coolers according to actual climatic conditions on the route line. So, the problem is to determine the rational design heat load on air coolers of railway conditioners, providing closed to maximum refrigeration output generation over considered time period.

The system of over filling the air cooler coils by liquid refrigerant recirculation enables a large deviation of current heat loads from their rational design value without considerable falling air cooler heat efficiency. The system of refrigerant circulation in air coolers by injector that enables excluding the final dry-out stage of refrigerant evaporation with extremely low intensity of heat transfer and as result provides increasing the heat efficiency of air coolers (overall heat flux) by 20–30% compared with conventional air coolers with complete refrigerant evaporation and superheated vapor at the exit might be proposed [1]. The injector uses a potential energy of high pressure liquid refrigerant,

M. Nechyporuk et al. (Eds.): *Integrated Computer Technologies in Mechanical Engineering*, AISC 1113, pp. 461–467, 2020.
https://doi.org/10.1007/978-3-030-37618-5_39

leaving a condenser, which is conventionally lost while it throttling to evaporation pressure in expansion valve.

2 Literature Review

A lot of researches deal with improving the performance of AC systems by intensification of heat transfer processes in heat exchangers [1–3], application of differ refrigerant circulation schemes [4–7], waste heat recovery technics [8–11], modern methods of modelling, experimental, monitoring and statistical methods [12–14].

As modern trend in AC systems the application of Variable Refrigerant Flow (VRF) system is considered to modulate heat load by varying refrigerant feed to air coolers [15–17]. The VRF system maintains the zone comfort by supplying adequate amount of refrigerant to air coils to meet cooling duties. The performance evaluations showed that the VRF system reduced energy consumption by 40% to 60% compared to that of central AC systems [18]. But the problem of inefficient operation of air coolers in VRF system caused by dry-out of inner walls at the final stage of inside tube refrigerant evaporation followed by dropping the intensity of heat transfer remains unsolved.

As alternative approach of the heat load modulation in AC systems the concept of incomplete refrigerant evaporation [1, 19] with overfilling air coils that leads to excluding a dry-out of inner surface of air coils is developed through liquid refrigerant recirculation by injector (jet pump).

Considerable changes in the current heat loads q_0 on the air cooler need choosing its rational design value, providing maximum refrigeration output generation over considered time period [20–22].

The basic approaches to determine a design heat load on air coolers of AC system with taking into account the currentchangeable climatic conditions were developed in [24–26] and quite acceptable small deviations of current heat loads from adesign heat load value was shown to prove the results [25] as well as expedience of over filling the air coolers by liquidrefrigerant recirculation [1] that enables large current cooling load fluctuations on railway route lines without considerablefalling air cooler heat efficiency as in present investigation.

The aim of the study is to develop the method to determine the rational design heat load on the air coolers of railway AC systems, providing closed to maximum refrigeration capacity generation under changeable actual heat loads during railway routs.

3 Research Methodology

The operation of railway AC systems is characterized by considerable changes in the current heat loads Q_0 on the route lines and in corresponding specific heat loads i.e. specific cooling capacity – related to the unit of air mass flow: $q_0 = Q_0/G_a$, were G_a – ambient air mass flow in air cooler, kg/s. The specific cooling capacity is calculated as $q_0 = \xi \cdot c_a \cdot (t_{amb} - t_{a2})$, kJ/kg, were ξ – coefficient of water vapor condensation heat, determined as ratio of the overall heat, removed from the air being cooled, including the latent heat of water vapor condensed from the wet ambient air, to the sensible heat removed; t_{amb} – ambient air temperature, t_{a2} – air temperature at the air cooler outlet, c_a – specific heat of ambient humid air.

The current heat loads are calculated according to varying actual ambient air parameters on the route lines with using the Meteomanz program [23] or others.

So as the efficiency of AC systems and their refrigeration machine performance depends on their cooling loading (current cooling capacities) q_0 and a duration τ of their operation, the summarised refrigeration capacity $\sum(q_0 \cdot \tau)$ generated during railway routes over the most hot month, might be considered as a primary criterion for the choice of a rational design cooling load of AC system. For this the current refrigeration capacities, generated by the refrigeration machine in response to the cooling duties for cooling ambient air to the target leaving air temperature, have been summarized over the summer month to determine the rational design cooling load of AC system.

4 Results of Investigation

The current values of temperature t_{amb} and relative humidity φ_{amb} of ambient air and temperature decrease Δt_{a} within cooling ambient air from current ambient temperatures t_{amb} to the temperature $t_{\mathrm{a2}} = 15$ °C and corresponding current specific refrigeration capacity (specific heat load on the air cooler) q_0, kW/(kg/s), or kJ/kg (at air mass flow $G_{\mathrm{a}} = 1$ kg/s), during direct route Kherson-Lviv (Kh-Lv) and return route Lviv-Kherson (Lv-Kh) per day for 1.08–3.08. 2018 are presented in Figs. 1 and 2.

As Fig. 1 shows the behavior of the curves corresponding to current values of specific refrigeration capacity q_0 and temperature decrease Δt_{a} within cooling ambient air to the temperature $t_{\mathrm{a2}} = 15$ °C does not coincide because of variation in relative humidity φ_{amb} of ambient air and corresponding latent heat.

The results of summarizing the specific refrigeration capacity values $\sum(q_0 \cdot \tau)_{\mathrm{r1}}$ (at air mass flow $G_{\mathrm{a}} = 1$ kg/s) for cooling ambient air to the temperature $t_{\mathrm{a2}} = 15$ °C during direct Kherson-Lviv (Kh-Lv) and return Lviv-Kherson (Lv-Kh) routes and their summarized value $\sum(q_0 \cdot \tau)$ for 1.08–3.08. 2018 through summarizing their values $\sum(q_0 \cdot \tau)_{\mathrm{r1}}$ for each route are presented in Fig. 2.

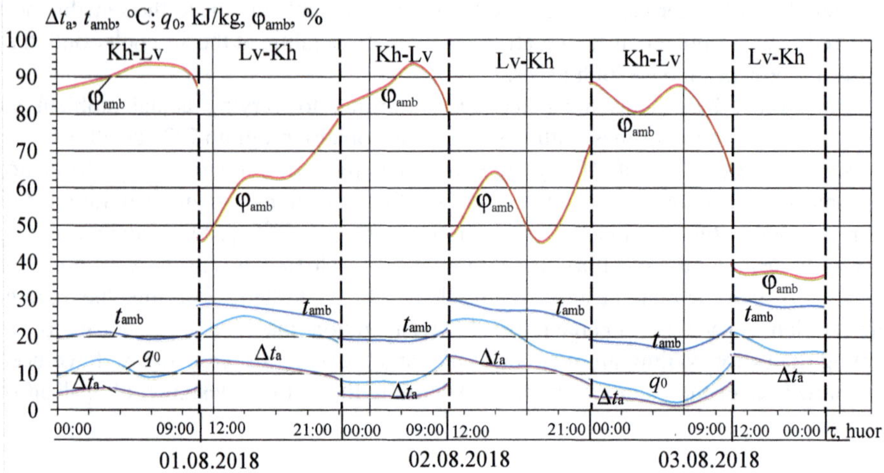

Fig. 1. Current values of temperature t_{amb} and relative humidity φ_{amb} of ambient air, temperature decrease Δt_a due to cooling ambient air to t_{a2} = 15 °C and corresponding current specific refrigeration capacity q_0 during direct routes Kherson-Lviv (Kh-Lv) and return routes Lviv-Kherson (Lv-Kh) for 1.08–3.08. 2018

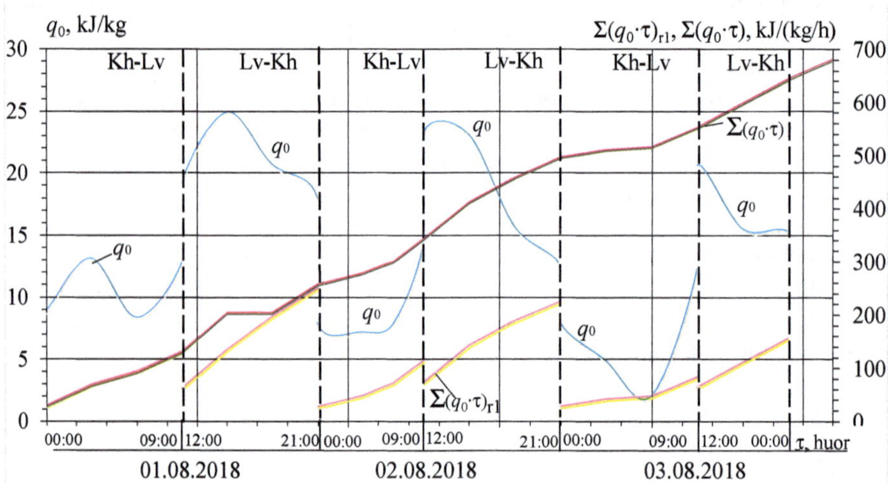

Fig. 2. Current values of specific refrigeration capacity q_0 and summarized values of specific refrigeration capacity $\sum(q_0 \cdot \tau)_{r1}$ for cooling ambient air to the temperature t_{a2} = 15 °C within each route (direct Kherson-Lviv (Kh-Lv) and return Lviv-Kherson (Lv-Kh) routes) and their summarized value $\sum(q_0 \cdot \tau)$ for 1.08–3.08. 2018

As Fig. 2 shows, the summarized values of specific refrigeration capacity $\sum(q_0 \cdot \tau)_{r1}$ for air conditioning in direct (Kh-Lv) and return (Lv-Kh) routes are nearly the same that is confirmed by monotonous rate of their increments $\sum(q_0 \cdot \tau)$ for 1.08–3.08. 2018.

Considerable changes in the current heat loads q_0 on the air cooler need choosing its rational design value, providing maximum refrigeration capacity generation over considered time period. The monthly refrigeration output in relative values $\sum(q_0 \cdot \tau)$ (at air mass flow G_a = 1 kg/s) against design specific refrigeration capacity $q_0 = Q_0/G_a$ of refrigeration machine for cooling ambient air to the temperature t_{a2} = 15 °C and climatic conditions on the route lines Kherson-Lviv and Lviv-Kherson for August, 2018 year, are presented in Fig. 3.

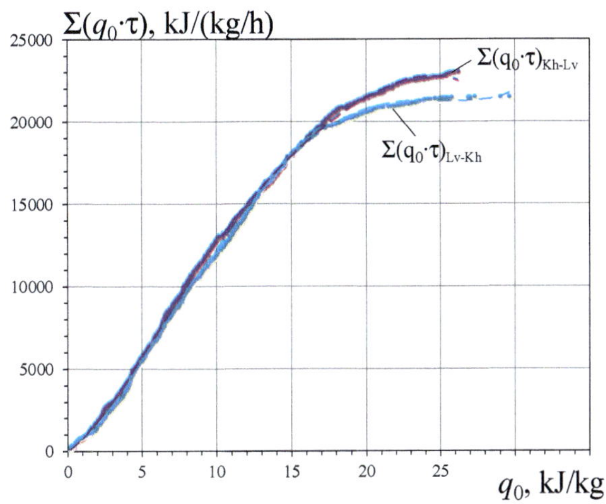

Fig. 3. The monthly refrigeration output in relative values $\sum(q_0 \cdot \tau)$ for ambient air cooling to the temperature t_{a2} = 15 °C against designed specific refrigeration capacity $q_0 = Q_0/G_a$: $\sum(q_0 \cdot \tau)_{Kh\text{-}Lv}$ – summarized for all direct railway routes Kherson-Lviv; $\sum(q_0 \cdot \tau)_{Lv\text{-}Kh}$ – summarized for all return railway routes Lviv-Kherson, August 2018

As Fig. 3 shows, the monthly (August) specific refrigeration output $\sum(q_0 \cdot \tau)$ for cooling ambient air to the temperature t_{a2} = 15 °C at specific refrigeration capacity q_0 = 30 kJ/kg, or kW/(kg/s), is evaluated as $\sum(q_0 \cdot \tau) \approx$ 23 MJ/(kg/h) for all direct railway routes Kherson-Lviv as well as $\sum(q_0 \cdot \tau) \approx$ 22 MJ/(kg/h) for all return railway routes Lviv-Kherson in August and achieved with monotonous rate of their monthly increments $\sum(q_0 \cdot \tau)$ with increasing the specific refrigeration capacity q_0 up to 30 kJ/kg.

Because of negligible rate of the monthly increments $\sum(q_0 \cdot \tau)$ the further increase in specific refrigeration capacity q_0 from 30 to 35 kJ/kg does not result in appreciable increment in the monthly refrigeration output $\sum(q_0 \cdot \tau)$ for July, but causes oversizing refrigeration machine, that leads to increasing its cost. Thus, the specific refrigeration capacity q_0 = 30 kJ/kg, or kW/(kg/s), is accepted as rational one to calculate a total designed refrigeration capacity Q_0 of refrigeration machine according to the total air mass flow G_a, kg/s: $Q_0 = G_a \cdot q_0$, kW.

5 Conclusions

The method to determine the rational design heat load on air coolers of railway AC systems, matching current changeable climatic conditions and providing closed to maximum refrigeration output generation over any considered time period of performance, was developed.

References

1. Radchenko, M., Trushliakov, E., Radchenko, A.: Enhancing heat efficiency of air coolers of air conditioning systems by injector refrigerant circulation. Contemp. Issues Heat Mass **2** (360), 619–639 (2019). Monography № 360 of the Faculty of Mechanical Engineering of Koszalin University of Technology
2. Bohdal, T., Sikora, M., Widomska, K., Radchenko, A.M.: Investigation of flow structures during HFE-7100 refrigerant condensation. Arch. Thermodyn.: Pol. Acad. Sci. **36**(4), 25–34 (2015)
3. Khovalyg, D.M., Baranenko, A.V.: Dynamics of two-phase flow with boiling refrigerant R134a in minichannels (in Russian). J. Tech. Phys. **85**(3), 34–41 (2015)
4. Butrymowicz, D., Gagan, J., Śmierciew, K., Łukaszuk, M., Dudar, A., Pawluczuk, A., Łapiński, A., Kuryłowicz, A.: Investigations of prototype ejection refrigeration system driven by low grade heat. In: HTRSE-2018, E3S Web of Conferences, vol. 70, p. 03002, 7 p. HTRSE-2018 (2018)
5. Smierciew, K., Gagan, J., Butrymowicz, D., Karwacki, J.: Experimental investigations of solar driven ejector air-conditioning system. Energy Build. **80**, 260–267 (2014)
6. Elbel, S., Lawrence, N.: Review of recent developments in advanced ejector technology. Int. J. Refrig. **62**, 1–18 (2016)
7. Radchenko, R., Radchenko, A., Serbin, S., Kantor, S., Portnoi, B.: Gas turbine unite inlet air cooling by using an excessive refrigeration capacity of absorption-ejector chiller in booster air cooler. In: HTRSE-2018, E3S Web of Conferences, vol. 70, p. 03012, 6 p. (2018)
8. Radchenko, A., Radchenko, M., Konovalov, A., Zubarev, A.: Increasing electrical power output and fuel efficiency of gas engines in integrated energy system by absorption chiller scavenge air cooling on the base of monitoring data treatment. In: HTRSE-2018, E3S Web of Conferences, vol. 70, p. 03011, 6 p. (2018) (2018)
9. Radchenko, M., Radchenko, R., Ostapenko, O., Zubarev, A., Hrych, A.: Enhancing the utilization of gas engine module exhaust heat by two-stage chillers for combined electricity, heat and refrigeration. In: 5th International Conference on Systems and Informatics, ICSAI 2018, Jiangsu, Nanjing, China, pp. 240–244 (2019)
10. Konovalov, D., Kobalava, H.: Efficiency analysis of gas turbine plant cycles with water injection by the aerothermopressor. In: Ivanov, V., et al. (eds.) Advances in Design, Simulation and Manufacturing II. DSMIE 2019. Lecture Notes in Mechanical Engineering, pp. 581–591. Springer, Cham (2020)
11. Goetzler, W.: Variable refrigerant flow systems. ASHRAE J. **49**(4), 24–31 (2007)
12. Im, P., Malhotra, M., Munk, J.D., Lee, J.: Cooling season full and part load performance evaluation of variable refrigerant flow (VRF) system using an occupancy simulated research building. In: Proceedings of the 16th International Refrigeration and Air Conditioning Conference at Purdue, West Lafayette, USA, 11–14 July 2016
13. Khatri, R., Joshi, A.: Energy performance comparison of inverter based variable refrigerant flow unitary AC with constant volume unitary AC. Energy Procedia **109**, 18–26 (2017)

14. Lee, J.H., Yoon, H.J., Im, P., Song, Y.-H.: Verification of energy reduction effect through control optimization of supply air temperature in VRF-OAP system. Energies **11**(1), 49 (2018)
15. Park, D.Y., Yun, G., Kim, K.S.: Experimental evaluation and simulation of a variable refrigerant-flow (VRF) air-conditioning system with outdoor air processing unit. Energy Build. **146**, 122–140 (2017)
16. Zhang, L., Wang, Y., Meng, X.: Qualitative analysis of the cooling load in the typical room under continuous and intermittent runnings of air-conditioning. Procedia Eng. **205**, 405–409 (2017)
17. Zhu, Y., Jin, X., Du, Z., Fang, X., Fan, B.: Control and energy simulation of variable refrigerant flow air conditioning system combined with outdoor air processing unit. Appl. Therm. Eng. **64**, 385–395 (2014)
18. Liu, C., Zhao, T., Zhang, J.: Operational electricity consumption analyze of VRF air conditioning system and centralized air conditioning system based on building energy monitoring and management system. Procedia Eng. **121**, 1856–1863 (2015)
19. Radchenko, N.: A concept of the design and operation of heat exchangers with change of phase. Arch. Thermodyn.: Pol. Acad. Sci. **25**(4), 3–19 (2004)
20. Radchenko, A., Bohdal, L., Zongming, Y., Portnoi, B., Tkachenko, V.: Rational designing of gas turbine inlet air cooling system. In: Tonkonogyi, V., et al. (eds.) Grabchenko's International Conference on Advanced Manufacturing Processes. InterPartner-2019. Lecture Notes in Mechanical Engineering, 10 p. Springer, Cham (2020)
21. Forduy, S., Radchenko, A., Kuczynski, W., Zubarev, A., Konovalov, D.: Enhancing the fuel efficiency of gas engines in integrated energy system by chilling cyclic air. In: Tonkonogyi, V., et al. (eds.) Grabchenko's International Conference on Advanced Manufacturing Processes. InterPartner-2019. Lecture Notes in Mechanical Engineering, 10 p. Springer, Cham (2020)
22. Konovalov, D., Trushliakov, E., Radchenko, M., Kobalava, G., Maksymov, V.: Research of the aerothermopresor cooling system of charge air of a marine internal combustion engine under variable climatic conditions of operation. In: Tonkonogyi, V., et al. (eds.) Grabchenko's International Conference on Advanced Manufacturing Processes. InterPartner-2019. Lecture Notes in Mechanical Engineering, 10 p. Springer, Cham (2020)
23. Meteomanz Homepage. http://www.meteomanz.com/. Accessed 21 May 2019
24. Radchenko, A., Radchenko, M., Trushliakov, E., Kantor, S., Tkachenko, V.: Statistical method to define rational heat loads on railway air conditioning system for changeable climatic conditions. In: 5th International Conference on Systems and Informatics: ICSAI 2018, Jiangsu, Nanjing, China, pp. 1308–1312 (2018)
25. Trushliakov, E., Radchenko, M., Radchenko, A., Kantor, S., Zongming, Y.: Statistical approach to improve the efficiency of air conditioning system performance in changeable climatic conditions. In: 5th International Conference on Systems and Informatics: ICSAI 2018, Jiangsu, Nanjing, China, pp. 1303–1307 (2018)
26. Trushliakov, E., Radchenko, M., Bohdal, T., Radchenko, R., Kantor, S.: An innovative air conditioning system for changeable heat loads. In: Tonkonogyi, V., et al. (eds.) Grabchenko's International Conference on Advanced Manufacturing Processes. InterPartner-2019. Lecture Notes in Mechanical Engineering, 10 p. Springer, Cham (2020)

Software Engineering and IT-infrastructure

Improvement of the Reliability of Speech Input Systems by Taking into Account the Emotional State of the Operator

Y. I. Gulyi, O. D. Nauchitel, O. M. Tynkov(✉), and Y. M. Yakusheva

National Aerospace University «KhAI», Kharkiv, Ukraine
tam54@ukr.net, o.tinkov@khai.edu

Abstract. To increase the reliability of speech input systems, the emotional state of the operator shall be taken into account. The article presents a method and device to compensate for deviations of the displacement of the components of the speech spectrum. The essence of such compensation is that in the process of voice input of information, the operator has the ability to visually control their own condition. Therewith, it becomes possible to correct the spectrum by the operators in real time by self-monitoring their condition, and subsequent hardware elimination of the spectrum deviation from the "reference standard". This "reference standard" represents a pre-spoken series (alphabet) of words used by operator. Visualization can be organized in the form of LED matrix, where each of the three gradations of the functional state – "large deviation", "average deviation", and "reference standard" – will correspond to a certain color of the matrix.

Keywords: Operator's state · Compensation · Reflection · Voice input · Visualization

1 Introduction

Formulation of the Problem. The paper deals with the issue of making speech recognition independent of the speaker (speakers). The complexity of this recognition is associated not only with differences in the voices of the speakers, but with the emotional state of the speaker.

The problem of speech recognition independent of the speaker can be partially solved by creating an established series of words taken as standards. In case of several speakers, a phrase bank can be used that reflects the characteristics of the frequency components of the speech signals of each speaker individually. However, this solution is insufficient and it is necessary to take into account the emotional state of the operator.

The operator's activities occur in a specific environment, such as physical and social. In addition, an internal psychological environment is taken into account. This complex "environment" can be represented as a multifactor model. Factors of the external physical environment are as follows: atmospheric pressure, environmental temperature data, air composition, wind speed, moon phases, and magnetic storms.

M. Nechyporuk et al. (Eds.): *Integrated Computer Technologies in Mechanical Engineering*, AISC 1113, pp. 471–481, 2020.
https://doi.org/10.1007/978-3-030-37618-5_40

Factors of the external social environment are as follows: cohesion and compatibility of the team, conflict, psychological health of the team, sociability, corporate culture. In addition, the functional state of the operator is affected by the following factors: the intensity and duration of the activity, the situations that may arise (accidents, malfunctions of the equipment). The following factors can be named as internal: motivation to work, character, temperament, mental processes (attention, memory, thinking, will). Under the influence of these factors, the functional state of the operator changes. The speech of the operator varies in many ways, which results in the reduction of the reliability of speech input systems.

Technical Aspect. Technical systems for speech recognition are mainly used to provide interactive mode in systems with artificial intelligence, for instance, formation of text files; information and reference systems; voice dialing of numbers, codes in access systems; dispatch type control systems; voice messaging systems; training systems and others.

Recognition accuracy is a key indicator of the reliability of these systems and it decreases as the number of words used in the operator's activity increases. In addition, a decrease in recognition accuracy is associated with a change in the spectrum of the speech signal of the current operator, which inevitably occurs due to a change in their psychological state. Thus, in his research K.R. Scherer [1] noted the fact that operators under stress showed an extension of the range of variation of the fundamental frequency, both downward and upward.

If any of such conditions as euphoria, confusion, anxiety, fear, fatigue, emotional tension, boredom, apathy, monotony occurs (in the process of activity), this indicator changes significantly. These changes can be aimed at both increasing and decreasing the frequency and they are perceived by people (human auditory analyzer) even in a normal, everyday environment. A person easily determines the emotional state of the interlocutor without any special technical means, but simply "by ear". As it follows from the above, the conclusion may be drawn that the use of hardware for analyzing the spectrum of a speech signal will allow such changes to be recorded with great accuracy.

In quantitative terms, recognition accuracy decreases down to six percent with the number of words in the word bank equal to ten [2]. The recognition accuracy can amount to ninety percent with a vocabulary length of sixty words, provided that there is no adjustment of the vocabulary to a specific operator. Should this be adjustment is implemented, the number of words recognized by the system may increase to five hundred [3]. Currently, a great arsenal of methods and technical means for recognizing human speech has been accumulated. Further the most promising in our opinion approaches are highlighted.

Thus, a method has been proposed for suppressing interference arising in the channel of a speech signal, based on the suppression of interference in the modulation area. This method can significantly reduce the interference compared with the RASTA (RelAtiveSpecTrA) algorithm widely used for these purposes.

Most studies on speaker recognition use the calculation of the cepstrum coefficient, which is determined by the envelope of the spectrum. This calculation assumes a fast Fourier transformation and is implemented by applying bandpass filters. This approach provides simplicity of calculations and takes into account the individual characteristics of the speaker's voice. In addition to the above, Gaussian mixtures models (GMM) and support-vector machines (SVM) are gaining popularity. Recognition methods using artificial neural networks and hidden Markov chains are also used.

The development of automatic speech recognition technology has a long history. Studies in this direction has been conducted over the past 75 years. The purpose of the first studies was to create a system that is close in characteristics to the speech capabilities of a person. It was not possible to solve the problem as a whole. It turned out that the human speech signal is very variable. It became clear that the continuous speech recognition of a random speaker is an extremely difficult task.

However, after the development of recognition systems for business tasks solving that provided a sufficiently high recognition reliability, the interest in such systems has increased dramatically. In such systems, recognition algorithms were applied using mathematical methods, such as hidden Markov chains. The speech recordings arrays (data banks) of various people were created. At the same time, developers have the opportunity to test newly created systems on a large data array. The recognition tasks were also facilitated by the emergence of a high-quality elemental base: microphones of special telephone headsets, large RAM and productive processors for personal computers.

The function of speech in human communications has several meanings. By means of speech, people transmit information containing a certain meaning. This is the semantic function of speech. But there is an emotional function of speech, such as intonation, loudness of statements, gestures (prosody of speech). Prosody contains data on the emotional state of the speaker. The speaker emotionally refers to both his own statements and the statements of other people. In emotional speech, the speaker uses jokes, irony, thereby expanding the semantic field of the speech message, which leads to a difficulty in understanding the main meaning of the statement. All this complicates the search for speech processing algorithms and, thus, the development of reliable means of speech recognition. However, the purpose of recognizing speech commands in automated control systems is to recognize the semantic component of speech. Other components (pitch, speech pace) are presented as interference in the way of the recognition problem solving.

One of the speech recognition systems developers' tasks is the conversion of analog sound into digital code for further computer processing. An acoustic sound wave is captured by a microphone, which converts the mechanical vibrations of the membrane into complex electrical signals (analog signal).

For subsequent processing by a computer, it is necessary to convert the analog signal to a discrete (digital) one. Such conversion is performed by an analog-to-digital converter (ADC), which samples the speech signal.

The process of a speech signal sampling is carried out by dividing the analog signal into temporary sections, each of which corresponds to a certain value (phase, amplitude) of the analog signal. The discretization reduces the processing time and the amount of processed data, which contributes to the simplification of technical processing tools and their speed. There are some difficulties in choosing a sampling rate. The sampling frequency should be such as to exclude possible omissions of significant changes in the analog signal. In accordance with the Kotelnikov theorem, the sampling frequency should be at least twice as high as the maximum frequency of the converted signal. If this condition is not fulfilled, the necessary information for recognition in the analog signal may be lost.

The frequency range of human speech is in the range of 100–6000 Hz, much narrower than that which it can perceive. For practical purposes, a much narrower range is sufficient. For example, in communication systems (telephony), a speech signal is transmitted in the range often 300–3000 Hz. In this case, the sampling frequency is 8 kHz. An increase in the sampling rate leads to an increase in the amount of noise (interference) during recognition. It is necessary to take into account the fact that the maximum power of a speech signal falls on frequencies below 1 kHz, and about 80% of the total power falls on frequencies below 500 Hz. There are few sounds with high frequencies in speech, but almost all the power of consonant sounds falls at frequencies above 1 kHz, so the loss of high frequencies can lead to a decrease in speech intelligibility [4].

One of the main tasks solved by the developers of human speech signal recognition systems is to determine the characteristics of bandpass filters. Bandpass filters are necessary to separate a speech stream having different frequency and amplitude characteristics into channels for further processing.

If passive RC or LC filters were used previously, now digital filters are used.

Psychological Aspect. The human speech signal has components related not only to the individual characteristics of the speaker, but to their emotional and physiological state. The signs that convey information about the psychological state of the speaker include changes that occur in [4–6]:

- dynamic range of speech;
- energy components of the speech spectrum;
- formant frequencies;
- fundamental frequency;
- temporal characteristics of the speech signal.

Energy components have low information content and are used only in combination with other characteristics of the speech signal.

Unlike energetic, the spectral characteristics of a speech signal are more informative, their changes are less dependent on the state of the operator; however, the practical implementation of this speech analysis is complicated.

Regarding the temporal characteristics of the speech signal, they are said to be used to diagnose human fatigue. In this case, a decrease in the activity of the functional physiological systems of the body may be observed [5].

The published papers [6, 7, 9] contain the information about speech indicators of the psychological state of a person. It is well known that a person can distinguish between the gradations of the state of another speaker, focusing only on their speech. This confirms the assumption of the existence of regular and stable speech correlates of psychological state of a person [10].

In situations when a person shows anxiety, fear or depression, a negative state inevitably arises, i.e. emotional stress. In this state a simplification, impoverishment of speech, and decrease in its semantic load may be observed.

Moreover, it contains semantically unrelated words and phrases, which make it difficult for a person to choose words. The total number of parasite words also increases, and long pauses between words and phrases may be observed [5]. In the process of activity, the operator may experience a state of emotional arousal associated with bright events. These events can be the successful solution of a professional problem, promotion, completing testing of new technical products [10]. In this case an "excess of words", an increase in the expressiveness of speech with a predominance of words spoken in superlatives may be observed. In the presence of high information content of the linguistic characteristics of speech, it shall be remembered that the human auditory apparatus is able to distinguish detailed gradations of the speaker's emotional state against the background of a lack of analysis of the semantic and lexical components of speech. This allows formalization of the process associated with the acoustic shifts of emotionally colored speech and abandonment of the more complex processing of lexical and grammatical characteristics of speech [10].

Phonetic signs of speech under the influence of the emotions of the speaker are perceived by the listener as altered in volume, tone, intonation, and tempo. These shifts can be detected by analyzing the spectrum, intonations, and temporal characteristics of speech.

Studies of the spectrum of the speech signal [7–9] showed similarities in changes in the spectrum of speech during emotional stress in humans, which found expression in the shift of formants, as well as energy shifts that make up both the low-frequency and high-frequency ranges of the speech signal. The emotions of operators performing professional activities are associated with joy, anger, fear, in contrast to operators who are inactive. In this case, the working operators have an expansion of the speech spectrum; they have an increase in the average frequency of the spectral maximum, an increase in the high-frequency components of the spectrum.

The depressive nature of emotions ("anguish" or "sadness") is associated with a reduction in the width of the spectrum, as well as with its shift towards low frequencies. In both the first and second cases, the formant scheme of the speech signal has variations associated with the corresponding shifts in the spectrum [4].

As the amount of emotional tension increases, it can be observed the increase of the dispersion of the amplitude values of the speech signal, the frequency shift of the fundamental frequency and the first formants of vowels, and the increase of the power of the high-frequency area of the spectrum (2000–4000 Hz) relative to the low-frequency (100–500 Hz).

Interesting results were obtained in [5]. The author concludes that the perception of an emotionally colored speech signal does not deteriorate in case of narrowing the speech spectrum in the range 200–I500 Hz. In this case, the spectrum of the speech signal and the amplitude (volume of spoken words) has no significant shifts, and therefore it does not depend on the characteristics of the speaker's voice. However, the question of the practical application of this method (this pattern of speech) arises. Indeed, narrowing the spectrum of human speech to this extent will make it impossible for other people to recognize it. However, from a theoretical point of view, this feature of speech can lead to the creation of promising cybernetic recognition systems.

Additional information about the functional state of a person is conveyed by the intonation of speech. It is perceived by a person as a change in the pitch and volume of voice. These characteristics are taken into account when creating speech input systems, in particular, the fundamental frequency, since they turned out to be sensitive to the manifestation of sthenic and asthenic emotional states of a person [4].

Currently, the parameters of changing the fundamental frequency, as an indicator of the variability of the speech signal, are fairly well studied. For operators in a state of inaction (operational comfort), the pitch frequency has a small dispersion and "slow" changes in time (transition from rise to fall and vice versa). Sthenic emotions are accompanied by an increase in the average fundamental frequency. The development of asthenic emotional states leads to decrease in the fundamental frequency and, accordingly, to decrease in dispersion thereof [1, 2]. A detailed discussion of changes in the fundamental frequency and other characteristics of the speech spectrum of a person in a state of fatigue are presented in paper [1].

From the above analysis, it should be concluded that the reliability of these methods is associated with the psychological state of the speaker. It is also important to emphasize that their functional state, which changes during the course of professional activity of the operators, has a significant impact on the frequency characteristics of the speech signal. Given that the technical means of inputting information work using programs have rigorous algorithms for analyzing the speech signal, it inevitably leads to recognition errors.

Objective. Closest to the proposed method is a voice recognition method that determines the position of the center of the frequency spectrum of a person's voice signal [4]. The technical implementation of this method is a system for periodically determining the average frequency and amplitude of a speech signal. Additionally, the system enables determination of the position of the center of the spectrum of the speech signal. Therewith, data on the localization of the center of the spectrum of the speech signal is entered into the RAM system. If a signal belonging to another person or a signal belonging to the same person, but with a changed emotional state, appears the data on the localization of the center of the spectrum of the speech signal is sent to a computing device that determines the difference in localizations. The difference signal is obtained by comparing the frequency centers. The signal received at the system input is compared with the signal stored in the system memory. If the difference between these signals (spectrum centers) is less than the established value, it is concluded that both centers of the spectrum belong to the same person.

Additionally, a speaker's voice recognition method was developed [5], based on a comparison of the sequence of speech signal standards over a certain period of time. These standards can be measurements of a number of characteristics of the signal spectrum for each phoneme. As a result of measurements, phonemic speech assessments can be obtained. Evaluation of speech is performed by highlighting the phonemic segments of speech. According to the characteristics of the phonemic segments of speech, various speakers can be determined and identified. This method is also comparable to the proposed method.

2 Presentation of Principal Materials of the Research

The article presents a method and device to compensate for deviations of the displacement of the components of the speech spectrum. The essence of such compensation is that in the process of voice input of information, the operator has the ability to visually control their own condition. Therewith, it becomes possible to correct the spectrum by the operators in real time by self-monitoring their condition, and subsequent hardware elimination of the spectrum deviation from the "reference standard". This "reference standard" represents a pre-spoken series (alphabet) of words used by operator. Visualization can be organized in the form of LED matrix, where each of the three gradations of the functional state – "large deviation", "average deviation", and "reference standard" – will correspond to a certain color of the matrix.

The implementation of the proposed method consists in the fact that the period of the fragment of the speech signal is recorded. The registration period is 2 s (by time). During this period, 100 signal segments are measured. Next, the average frequency of the first formant and the variance of the fundamental frequency of the speech signal are calculated. Then, the difference between the pitch frequency stored in the device memory and the current value thereof is calculated. In this case, data on the frequency stored in the memory are evaluated by the operator in a state of functional rest (whose value is taken as "background"). Under the current value it is meant the value of the frequency of the fundamental frequency obtained in the process of the operator's activity. Subsequently, the magnitude of the difference of the fundamental frequency is calculated by the criterion of Student's t-distribution.

If the difference turns out to be significant, then taking into account the reliability of the statistical criterion, it is concluded that the operator is in a state different from the "background" one.

The prototype of the developed method is the method described in [4]. The author of this method proposes to send to the human auditory analyzer (via telephone) a synchronous recording of their own voice, which exceeds the natural strength of the voice by 40 dB with a playback delay of about 0.15–0.3 s. This results in a state of mental tension. This "feedback acoustic connection" allows making change to the functional state of a person. However, the content of this condition is negative. The state of mental tension negatively affects the efficiency of the operator. They will inevitably make mistakes and increase the time to solve professional problems. In other words, the reliability of activities will decrease. In contrast to the developed one, in the proposed method this relationship has a visual character, a different form of implementation. The proposed method allows the operator to switch from a negative current state to a positive one independently (using the proposed device).

The proposed device with the ability to visually control own state contains the following: a microphone, a low-frequency amplifier, a compressor, six band-pass filters, a control unit, six analog-to-digital converters (ADC), a calculator unit with random access memory (RAM), a comparison unit, a differential frequency amplifier, thresholds unit and display unit (visualization). The visualization unit serves to graphically display the difference value (level) of the functional state of "rest" and "current" state of the acting operator. The formula of the device can be represented in the following way. The device of speech input of information transforms the "current" functional state of the operator into a predetermined state of functional rest. This transformation (control) occurs due to the internal psychological resource of a person, their ability to self-regulation. In this case, the self-regulation is performed by the speaker themselves, who focuses on color changes in the display unit (a system for displaying information about a person's state). Hence, the speaker's goal is not only the speech information necessary to solve a professional problem, but also independent restructuring of the voice. The indicator of the rate of this adjustment in this case is the color of LED indicators of the display unit.

The device operates as follows. The operator's speech signal through a microphone enters the input of the compressor, the output whereof is connected to the input of the low-frequency amplifier, the output whereof is connected to the input of the bandpass filter unit, consisting of six bandpass filters. The output of the filter unit is connected to the input of the ADC unit, the outputs whereof are connected to the input of the control unit, the outputs whereof are connected to the input of the memory unit and the input of the differential frequency calculator, the outputs whereof are connected to the inputs of the threshold unit, the outputs whereof are connected to the indication (visualization) unit.

The block diagram of the proposed device is shown in Fig. 1.

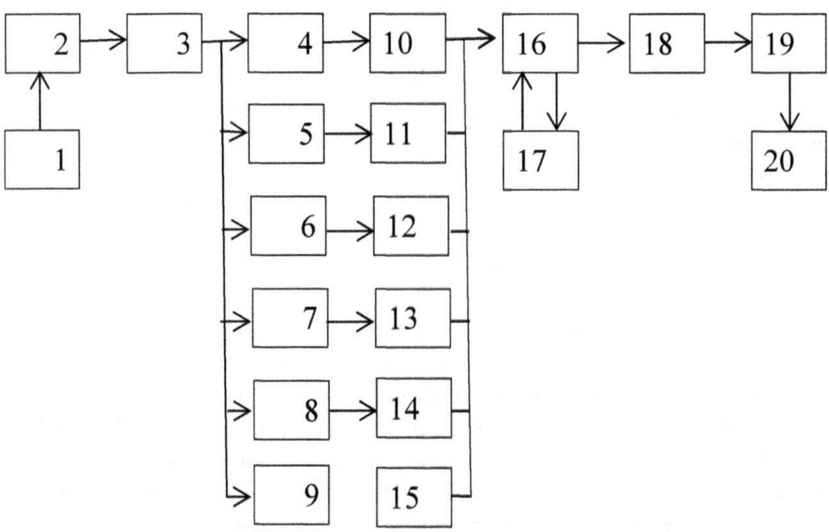

Fig. 1. Block diagram of the device

1. Microphone.
2. Compressor.
3. Low frequency amplifier.
4–9. Band-pass filters.
9. Control unit.

10–15. Analog-to-digital converters (ADC).
16. Control unit.
17. Memory unit (RAM).
18. Calculator of the difference frequency.
19. Thresholds unit.
20. Display unit (visualization).

The device operates as follows.

Block 1 converts the acoustic signals of the operator's speech into electrical impulses.

The compressor (block 2) has a passband from 250 Hz to 4500 Hz. Its application is due to the need to stabilize the amplitude changes in the signal arriving at its input. This solution is a compromise, since the level of interference increases due to the large amplification of the compressor. However, this fact is not significant to the objective function of using this device.

The low-frequency amplifier (block 3) amplifies the signal to a value sufficient for the normal operation of the ADC (blocks 10–15).

Blocks 4–9 are passive bandpass filters of the RC type, with central transmission frequencies of 300 Hz, 700 Hz, 1200 Hz, 1800 Hz, 2300 Hz, 3000 Hz, 3800 Hz, respectively.

The control unit 9 provides the following functions:

- synchronization of the ADC with the presence (absence) of a speech signal at the input of the device (from the microphone);
- splitting the speech signal into segments of a fixed length of 2 s (implementation period). The implementation period of a speech fragment, in turn, is divided into segments of a fixed length (20 ms) so that the segment duration exceeds the maximum possible period of the fundamental frequency.
- generation of a stop signal after counting 100 segments;
- recording information (in digital code) of the speech signal in RAM (block 17);
- permission of calculations in block 18 (subtraction operation).

Blocks 10–15 – ADC convert the speech signal into a digital code, respectively, each in a frequency band defined by bandpass filters 4–9.

Block 18 – the computer performs the subtraction, shift of the signal spectrum. The computer "compares" the information stored in RAM (data on the spectrum of the operator's speech signal, which was in the "operative rest" state, i.e. prior to performing activities) and information on the "current" state of the acting operator.

The display unit 20 is used to display information about the offset of the frequency spectrum of the speech signal. The magnitude of this shift is associated with certain colors of the glow of the screen observed by a person. The difference in screen color is related to the set thresholds defined in the threshold unit 19.

Block 19 has threshold values for the characteristics of the spectrum of the speech signal for each of the six filters and the ADC.

It should be noted that the proposed method and device for its implementation are an autonomous complement to speech input systems. The device can be used in the process of activity of operators, and in laboratory conditions, to study the psychological functional state of operators. In addition, it can be used in access control systems. However, certain restrictions on the practical application exist in case of the direct implementation of professional actions by the operator. The shorter the time limits for decision-making by operators, the more difficult it will be to go through the process of "self-tuning" of the operator to a "predetermined" (reference, rest state) functional state. The operator may lack the time. However, an attempt to enter a speech command by an operator in a negative state (fatigue, emotional tension) will inevitably lead to an increase in input errors. Thus, on the one hand it will lead to a decrease in performance, and on the other hand it will result in an increase in the number of errors. This can be considered as mutually exclusive factors. The reliability of recognition (input) will be the same. A practical conclusion from the above is that the additional time spent on introducing a voice command will not significantly reduce the reliability of recognition systems. In case of application of the proposed device in access control systems, when the input time of the voice command is not strictly limited, the reliability of recognition systems will be high. It should be noted that the proposed method and device for the implementation thereof are an autonomous complement to speech input systems.

3 Conclusions

Using the proposed method, it is possible to control the psychological state of the operator by his speech signal.

The developed device provides continuous, remote monitoring of the psychological state of operators and is an autonomous complement to speech input systems.

Visualization of the functional state of the operators can be organized in the form of LED matrix, where each of the three gradations of the functional state - "large deviation", "average deviation", "reference standard", will correspond to a certain color of the LED matrix.

The proposed method of increasing the reliability of information input (recognition) systems is based on psychological self-reflection, a person's ability to self-regulation.

References

1. Scherer, K.R.: Emotion effects on voice and speech: paradigms and approaches to evaluation. In: ISCA Workshop on Speech and Emotion, Belfast (2000). http://www.qub.ac.uk/en/isca/proceedings/pdfs/scherer.pdf
2. Smirnov, B.A., Tinkov, A.M.: Methods of Engineering Psychology, 528 p. Humanitarian Center, Kharkov (2008)
3. Lebedeva, N.N., Karimova, E.D.: Acoustic characteristics of a speech signal as an indicator of a person's functional state. Adv. Physiol. Sci. **45**(1), 57–95 (2014)
4. Kucheryavyy, A.A.: Airborne Information Systems: Lecture course, edited by Mishina, V. A., Klyueva, G.I., 2nd enlarged edn., 504 p. UlSTU, Ulyanovsk (2004)
5. Galunov, V.I., Pikturna, V.V., Yanushavichyus, V.Y.: Acoustic correlates of emotional speech. In: Mater. of Reports and Messages of the 5th All-Union Meeting-Symposium of the Cycle "Acoustics of Speech and Hearing": Emotions and Automation. Speech Recognition, Odessa, pp. 16–25 (1989)

6. Gomina, T.G.: The influence of various emotional states on the change in the spectrum of English vowels. In: Mater. and Message All-Union Symposium "Speech, Emotion and Personality", pp. 94–96. Nauka, London (1978)
7. Nosenko, E.L.: Changes in the characteristics of speech with emotional tension. In: Smirnov, A.A., Konopkin, O.A. (ed.) Psychology Issues: Published Since 1955, no. 6, pp. 76–85 (1978)
8. Leonova, A.B.: Psychodiagnostics of functional states of a person, 200 p. Publishing House of Mosk. University, Moscow (1984)
9. Frolov, M.V., Milovanova, G.B.: Features of monitoring the human-operator state in terms of the basic tone and spectrum of his speech. Hum. Physiol. **35**(2), 136–138 (2009)
10. Sidorov, K.V., Filatova, N.N.: Analysis of signs of emotionally-colored speech. Bull. TvSTU (20), 26–32 (2012)

A Two-Step Approach to Providing a Desired Quality of Lossy Compressed Images

Sergey Krivenko, Dmytro Demchenko, Igor Dyogtev, and Vladimir Lukin(✉)

Department of Information-Communication Technologies, National Aerospace University, 17 Chkalova St, Kharkiv 61070, Ukraine
krivenkos@ieee.org, lukin@ai.kharkov.com

Abstract. A problem of lossy image compression with providing a desired quality according to a given quality metric is considered. Several approaches to its solving are discussed. A two-step approach based on using the averaged rate-distortion curve is proposed and tested for a coder AGU based on discrete cosine transform. It is shown that it is often possible to use two iteration steps instead of sufficierntly larger number of iterations that makes compression with providing a desired quality faster. Meanwhile, there are ranges of visual quality where further modifications are needed to produce a desired accuracy of quality providing.

Keywords: Full reference metrics · Visual quality · Color image database · Mean opinion score · Threshold values

1 Introduction

There is a huge amount of images and video produced each day that should be stored, processed and/or transferred via communication channels [1–3]. Due to this, image and video compression has become a standard operation in image processing chain [3–5]. There are lossless and lossy compression techniques where the latter ones become more and more useful since compression ratio (CR) provided by lossless compression does not meet requirement frequently [5, 6]. Meanwhile, there is a lot of unsolved questions for lossy image compression so far including what coder to use (choose), how to provide a desired CR (for some coders [5, 7]), how to provide fast compression and decompression [8], etc. But one of the most important task is to provide a desired quality of compressed images since, otherwise, lossy compression might lose sense [4–8].

Saying image (or video) quality, one immediately comes to a question how to characterize it. Standard metrics such as mean square error (MSE) and peak signal-to-noise ratio (PSNR) have been criticized by many authors [4, 8–12] since they are not adequate enough with respect to human perception. Because of this, MSE and PSNR are still used in performance analysis and design alongside with numerous so-called HVS-metrics [12–15] that are capable of taking into account many important features of human vision system (HVS). Note that lossy image compression is an area where full-reference metrics (i.e., the metrics calculated for a pair of distorted (compressed)

M. Nechyporuk et al. (Eds.): *Integrated Computer Technologies in Mechanical Engineering*, AISC 1113, pp. 482–491, 2020.
https://doi.org/10.1007/978-3-030-37618-5_41

and a sample (original) images) are mainly used to control introduced losses and to characterize quality of the compressed image.

According to image quality that has to be provided by lossy compression, several requirements are possible. In particular, it is possible that visually lossless compression is desired [16, 17]. Then, one has to apply an adequate visual quality metric for which a threshold of invisibility of distortions should be known in advance. A task to be solved at compression stage is to produce a value of the used metric not worse than this threshold [18]. Another case is that introduced distortions can be visible but they should not be annoying. Then, a metric sgould be chosen and a threshold that corresponds to "not annoying" distortions should be a priori known. At compression stage, the metric value not worse than this threshold should is to be provided.

As one can see, several tasks have to be solved. First, a good metric should be chosen. Second, the aforementioned thresholds have to be known. Third, a procedure able to provide these thresholds has to be available and it should be fast and accurate enough.

Fortunately, the former two tasks have been partly solved recently [18]. In particular, for the visual quality metrics PSNR-HVS-M [14], FSIM [13] and some others that are rather adequate for lossy image compression, the threshold values have been established using the database TID2013 (Tampere image database, available at http://ponomarenko.info/tid2013.htm). Concerning the third task, two main approaches exist [7, 17, 19–21]. According to one of them [7, 17], an iterative procedure is applied. It is based on assumption that rate-distortion curve (RDC), i.e., dependence of a considered quality metric on a parameter that controls compression (PCC) for a used coder is monotonous. Exploiting this property, multiple compression/decompression is done where each time the used full reference metric is calculated and compared to a desired value (threshold), then PCC is changed to approach the desired value and compression/decompression is carried out again [7]. The main drawback of this approach is that the number of iterations unknown a priori and it can be rather large (up to 10...12 depending upon parameters of the iterative procedure). Due to certain means, the number of steps can be decreased by two-three times [7] but it remains a priori unknown anyway.

The approach [19–21] is based on metric prediction before compression. An advantage of this approach is that no iterations are needed and prediction is considerably faster than compression itself (this is because, a limited statistics in DCT domain is analyzed). However, compression with a desired quality based on prediction can be not enough accurate. Thus, analysis and design of other possible solutions is crusial.

The goal of this paper is to design and analyze a hybrid solution which we call the two-step approach. It includes preliminary compression/decompression with calculation of a used full reference quality metric (PSNR-HVS-M [14] as one possible case). Then, using the earlier obtained averaged RDC, it is determined how PCC has to be changed. After this, the final compression is accomplished. The DCT based coder AGU [22] is analyzed as a case study. Peculiarities of the proposed procedure, its advantages and drawbacks are considered and discussed.

2 AGU Coder and Its Main Characteristics

The AGU coder [22] has been proposed as an alternative to JPEG and JPEG2000. It performs data processing (two-dimensional DCT) in 32 × 32 pixel blocks, uses improved coding of quantized DCT coefficients and employs image deblocking after decompression. Due to these modifications, it considerably outperforms JPEG and slightly outperforms JPEG2000. To get initial imagination about AGU performance, Table 1 presents PSNR values for AGU and JPEG2000 for five standard test images and four values of CR.

Table 1. PSNR values (dB) for different test images and CR

Test image	CR = 8		CR = 16		CR = 32		CR = 64	
	JPEG2000	AGU	JPEG2000	AGU	JPEG2000	AGU	JPEG2000	AGU
Lenna	40.33	40.52	37.27	37.46	34.15	34.51	31.02	31.50
Barbara	38.07	39.26	32.87	34.65	28.89	30.77	25.87	27.55
Baboon	29.11	29.70	25.57	26.12	23.18	23.69	21.68	22.01
Goldhill	36.54	37.03	33.24	33.65	30.53	31.09	28.49	28.97
Peppers	38.17	38.33	35.80	35.55	33.54	33.32	30.79	30.90

As one can see, AGU provides slightly better PSNR values compared to JPEG2000 for all test images for all considered CRs. In parallel, data in Table 1 clearly show one problem in lossy compression of still images – for a given CR, quality of compressed images sufficiently depends upon image properties (content, complexity). For example, for CR = 8, PSNR varies from 38…41 dB (invisible distortions) for images of quite simple structure (Lenna, Peppers) till 29…30 dB (visible distortions) for the test image Baboon that is highly textural. Similarly, for CR = 32, PSNR vary from ≈34.5 dB (almost invisible distortions) for Lenna till ≈23 dB (annoying distortions) for Baboon. In fact, limits of these variations can be even wider. But it is clear that CR should be set individually (in adaptive manner) for each image to be compressed with a desired quality.

In addition, it can be supposed from analysis of data in Table 1 that RDCs (dependences of PSNR on CR for the considered case) are monotonous. Let us demonstrate that this is true (at least, in many practical cases). Figure 1 (taken from [7]) presents dependences of the visual quality metric PSNR-HVS-M on CR for the test image Pole that has a quite simple structure. Dependences are represented for nine different coders including well known JPEG (the version with non-uniform quantization of DCT-coeffciients) and JPEG2000 (the version without visual quality optimization), SPIHT which is freely available modification of JPEG2000, DCT-based coders AGU, AGU-M, ADCT, ADCT-M, and modern image compression techniques WebP and H265 (its version for still images). It is worth recalling here that PSNR-HVS-M is the visual quality metric for grayscale images that takes into account two important peculiarities of HVS – less sensitivity to distortions in high spatial frequencies and masking effect of texture. Similarly to PSNR, PSNR-HVS-M is expressed in dB where the threshold about 41 dB corresponds to invisibility of introduced distortions. Also note that the versions of DCT-based coders AGU-M are specially intended on providing high visual quality of compressed images.

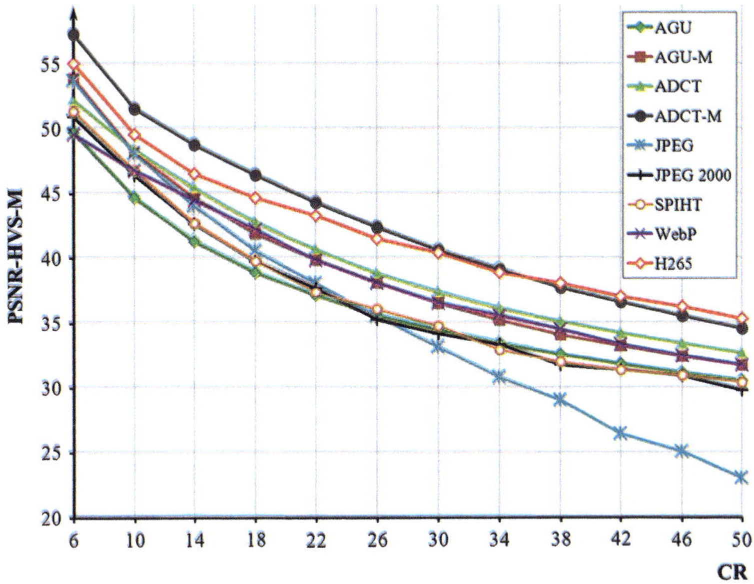

Fig. 1. Dependences of PSNR-HVS-M on CR for the test image Pole compressed by 9 different compression techniques

Analysis of dependences in Fig. 1 results in several conclusions. First, it is seen that all dependences are monotonically decreasing and the reasons for this are clear – a larger CR is associated with more distortions introduced. Second, coder modifications intended on providing qood visual quality (ADCT-M, H265) really perform in the best way. JPEG performs well for small values of CR but its performance decreases in the fastest way if CR increases. Third, all dependences are "smooth" functions and this property has been already exploited at final stages of iterative procedure [7] where linear interpolation of the latest results is used to calculate the final PCC.

The dependences in Fig. 1 are functions of CR (this way of representation is explained by necessity to compare data for different coders for which PCC are different; in particular, for JPEG2000, bits per pixel (bpp) is employed as PCC). For DCT-based coders, compression parameters are varied by quantization step (QS) or scaling factor (SF) [7, 16, 22]. So, it is interesting to see and analyze RDCs for these arguments (QS or SF). An example of such dependences is given in Fig. 2 for nine standard grayscale test images of different complexity. The test images Lena and Zelda are the "simplest" whilst the test images Baboon and Mountain are the most complex. The test image Patterns is artificial and has specific structure (see http://ponomarenko.info/agu.htm).

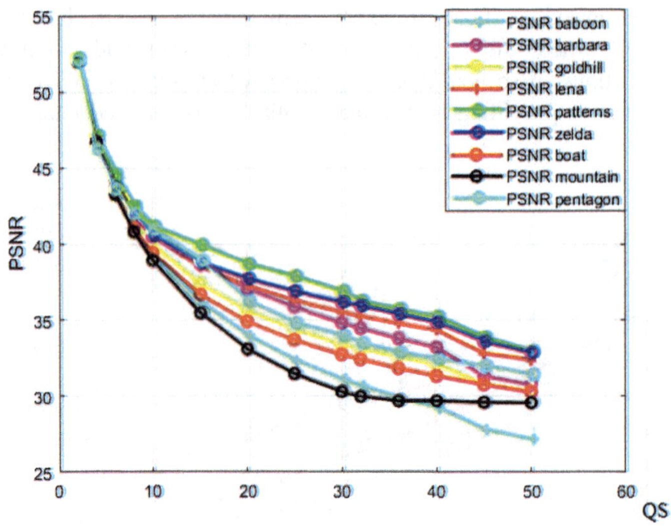

Fig. 2. Dependences of PSNR on QS for nine test images compressed by AGU

Analysis of the obtained dependences shows the following. First, the curves are placed in a very compact manner for small QS (let us say, QS < 10). However, they diverge for larger QS and, as it can be expected, curve behavior depends upon complexity of a compressed image. This leads to the fact that sufficiently different values of QS are needed to provide a given PSNR for different images. Foir example, to provide PSNR $\approx$35 dB (threshold of distortion invisibility [18], one has to set QS about 14 for the test image Mountain and about 40 for the test images Patterns and Zelda. So, we again come to the conclusion that PCC (QS in the considered case) should be set individually for each image that has to be compressed.

Second. Almost all curves go "almost in parallel" for QS in the limits from 15 to 50, i.e., in the area where compression is usually applied. This property will be exploited by us in the proposed two-step procedure.

3 Known and Proposed Compression Procedures

An improved iterative procedure proposed in [7] performs in the following manner. As initial step, it uses PCC (QS_1 in the considered case) that corresponds to averaged rate-distortion curve (it is shown in Fig. 3 by blue color). Then, compression and decompression are done, a full-reference metric M_l is determined and compared to a desired value M_{des}, then QS is changed to QS_2 to approach to providing M_{des} and the same operations are repeated until M_{des} occurs to be between M_l and M_{l+1} where l is iteration index. As one can guess, "convergence" depends upon $\Delta QS = |QS_{l+1} - QS_l|$. An empirical proposition in [7] was to set $\Delta QS = 2$. Then, if $PSNR_{des} = 35$ dB, QS_1 has to be about 25 (see the averaged cirve in Fig. 3). For most test images (see data in Fig. 2), a few iterations are needed but for some of them up to 8 iterations are required. This example proves the main drawbacks of iterative procedure [7] mentioned in Introduction. These deawbacks occur to be especially serious if compression and/or decompression take a lot of time. In particular, this takes place for the coders ADCT and ADCT-M [7] that employ optimized partition scheme at compression stage. One advantage of this procedure that has to be mentioned here is that it provides M_{des} accurately. In particular, $PSNR_{des}$ is provided with errors less than 0.2 dB which is very good for practical applications.

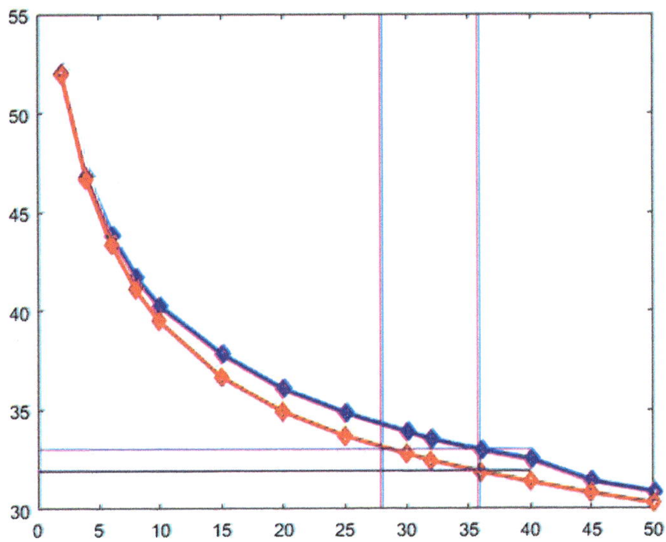

Fig. 3. Averaged rate-distortion curve (PSNR vs QS) (blue color) and RDC for the test image Boat (indicated by red color)

In turn, procedures of QS setting based on prediction [19–21] employ known (approximated) dependences of a metric on QS (or SF) and some other parameter characterizing image complexity in indirect way. They do not perform compression and decompression. Instead, at initial stage, they collect DCT coefficient statistics form a limited number of 8x8 pixel blocks. Based on the obtained data and approximation dependences obtained in advance in offline mode, the methods find QS to provide M_{des} and use this QS for compression.

The drawback of this method is that errors of providing Mdes can be too (inappropriately) large. The paper [20] reports on errors of providing PSNRdes for JPEG up to 1 dB, similar or even larger errors are possible for other DCT-based coders for the metrics PSNR and PSNR-HVS-M, especially for compression with visible distortions.

These were the reasons for design of compromise variant presented below. Look at Fig. 3. It shows the dependence for a particular test image (Boat in our case). Certainly, it differs from the averaged dependence but these dependences are very similar and for small ranges of QS variation they go “almost in parallel”.

These observations allow proposing the following procedure. The first step is to compress and decompress an image using initial PCC (QS_1 for AGU and ADCT). This initial PCC_{init} can be determined from the averaged dependence of a used metric on a used PCC (as PSNR on QS presented in Fig. 3) for a given coder for M_{des}. Such averaged dependences (see examples in [7]) can be obtained in advance in off-line mode for a large set of test image. In addition, derivative of the metric dM/dPCC = M′ for the initial value of PCC should be determined. Then, calculate M_{init} produced by compression with PCC_{init}.

Certainly, for a particular image to be compressed with M_{des} the values of M_{des} and M_{init} differ. Then, assuming linear dependence of M on PCC for a considered part of dependence, it is possible to get an estimate of PCC_{des} as

$$PCC_{des} = PCC_{init} + (M_{des} - M_{init})/M'.$$

In other words, at the second step, PCC_{des} is estimated immediately using linear approximation of the averaged RDC. After this, compression is carried out using PCC_{des}.

An example is shown in Fig. 3. Suppose that we need to provide $PSNR_{des}$ = 33 dB. Then, according to the averaged curve (blue), QS_{init} has to be set equal to 35,74. But for the test image Boat this leads to $PSNR_{init}$ about 31.92 dB, i.e. less than desired. Then, QS is corrected and set equal to 27.95. After compressing this image with this QS, PSNR occurs to be equal to 33.17, i.e. considerably closer to 33 dB than initially.

4 Quantitative Analysis

The example presented above deals with a particular test image and particular $PSNR_{des}$. So, we need to carry out statistical experiment to analyze what is positive (or, maybe, negative) effect of the second step. Such experiments have been carried out and below we present some results.

To characterize accuracy, let us determine variance values of the initial and finally provided metrics ($PSNR_{init}$ and $PSNR_{prov}$ in our case) after the first and second steps, respectively. Table 2 also presents the values of ΔQS and recommended QS_{rec} used at the second step. As one can see, $PSNR_{init}$ after the first step vary in rather wide limits around the desired $PSNR_{des}$ = 34 dB. For simple structure images they are larger than $PSNR_{des}$ whilst for complex structure images the situation is the opposite. Due to this, ΔQS can be both positive and negative. As the result, QS_{rec} varies in wide limits from 14.73 for the test image Baboon to more than 40 for the test images Zelda and Patterns. Due to step 2, $PSNR_{prov}$ become closer to $PSNR_{des}$ than $PSNR_{init}$ for eight out of nine test images. As the result, variance after two step compression is sufficiently smaller than after one step. However, there is one problem. Even after two steps, variance is quite large and errors in some cases exceed 1 dB which is undesired.

Table 2. Statistics and parameters of providing $PSNR_{des}$ = 34 dB

Test image	QS_{init}	$PSNR_{init}$	ΔQS	QS_{rec}	$PSNR_{prov}$
Baboon	29.45	31.18	−14.72	14.73	36.23
Barbara	29.45	34.96	5.07	34.53	34.08
Goldhill	29.45	33.57	−2.23	27.22	34.43
Lenna	29.45	35.66	8.69	38.15	34.59
Patterns	29.45	36.95	15.50	44.96	33.86
Mountain	29.45	30.43	−18.67	10.78	38.35
Zelda	29.45	36.27	11.91	41.36	34.53
Boat	29.45	32.89	−5.77	23.68	34.03
Pentagon	29.45	34.03	0.20	29.66	34.01
Variance	4.46		2.79		

This happens for complex structure images for which $PSNR_{prov}$ occurs to be sufficiently larger than $PSNR_{des}$. The reasons for this are the following. First, real values of derivatives for the corresponding RDCs can sufficiently differ from M' for the averaged curve. Second, linear approximation is too rough and does not work properly (ΔQS for complex structure images are about twice smaller than QS_{init}).

Let us now analyze data for $PSNR_{des}$ = 37 dB. They are given in Table 3. It is seen that due to step 2 the results have improved for all nine test images - $PSNR_{prov}$ are closer to $PSNR_{des}$ than $PSNR_{init}$. Variance has sufficiently decreased as well. Only for two complex structure images, the errors of providing the desired PSNR exceed 1 dB. However, the positive moment is that $PSNR_{prov}$ exceeds $PSNR_{des}$ for practically all test images. This means that our task to provide image quality not worse that desired is reached.

Table 3. Statistics and parameters of providing $PSNR_{des}$ = 37 dB

Test image	QS_{init}	$PSNR_{init}$	ΔQS	QS_{rec}	$PSNR_{prov}$
Baboon	17.30	35.07	−5.46	11.84	38.06
Barbara	17.30	37.92	2.63	19.94	37.1
Goldhill	17.30	36.59	−1.14	16.16	36.99
Lenna	17.30	37.92	2.62	19.92	37.28
Patterns	17.30	39.33	6.64	23.95	38.025
Mountain	17.30	34.35	−7.53	9.77	39.11
Zelda	17.30	38.25	3.56	20.87	37.55
Boat	17.30	35.87	−3.20	14.10	37.19
Pentagon	17.30	37.66	1.88	19.19	36.66
Variance	2.36		0.96		

Finally, Table 4 presents the obtained data for $PSNR_{des}$ = 40 dB. Due to the second step the results have become better for all nine images and the errors of providing the required PSNR are acceptable for all test images except Baboon and Mountain for which the provided quality is essentially higher than desired. So, our task is reached for the case of providing $PSNR_{des}$ = 40 dB.

Summarizing the presented results, it is possible to state the following. The proposed two-step procedure works well enough if it is desired to provide PSNR larger than 37 dB, i.e. if introduced distortions have to be invisible. Note that similar results and practically the same conclusions have been obtained for the metric PSNR for the coder ADCT [7]. Meanwhile, the errors of providing desired PSNR occur to be too large for $PSNR_{des}$ smaller than 35…37 dB. The reasons for this have been explained above.

Table 4. Statistics and parameters of providing $PSNR_{des}$ = 40 dB

Test image	QS_{init}	$PSNR_{init}$	ΔQS	QS_{rec}	$PSNR_{prov}$
Baboon	10.44	38.93	−2.21	8.23	40.82
Barbara	10.44	40.79	1.64	12.09	40.02
Goldhill	10.44	39.82	−0.37	10.07	40.02
Lenna	10.44	40.39	0.81	11.26	40.05
Patterns	10.44	41.06	2.22	12.67	40.51
Mountain	10.44	38.58	−2.95	7.49	41.40
Zelda	10.44	40.38	0.80	11.25	40.09
Boat	10.44	40.52	−2.87	7.67	40.43
Pentagon	10.44	40.79	1.65	12.09	40.41
Variance	0.68		0.38		

5 Conclusions

We have proposed the two-step procedure of lossy image compression for providing a desired quality of lossy image compression according to PSNR (preliminary experiments show that the approach also works for some other metrics). Its performance has been analyzed for the set of test grayscale images and it has been shown that the approach performs well enough for compressing images with providing invisible distortions. However, if one needs to provide not-annoying distortions, the approach needs to be modified (further improved). The advantage of this approach is that it is faster than iterative procedures [7] with fixed step of PCC changing. We also expect that the approach is quite universal and it can be used for different compression techniques and different metrics of image quality including visual quality metrics.

References

1. Symes, P.D.: Digital Video Compression. McGraw Hill Professional, New York (2004)
2. Carli, M.: Perceptual aspects in data hiding. Thesis for the degree of Doctor of Technology, Tampere University of Technology (2008)
3. Chen, Z., Guillemot, C.: Perceptually-friendly H.264/AVC video coding based on foveated just-noticeable distortion model. IEEE Trans. Circ. Syst. Video Technol. **20**, 806–819 (2010)

4. Wu, H.R., Lin, W., Karam, L.: An overview of perceptual processing for digital pictures. In: Proceedings of International Conference on Multimedia and Expo Workshops, Melbourne, pp. 113–120 (2012)
5. Taubman, D., Marcellin, M.: Standards and Practice JPEG2000: Image Compression Fundamentals. Kluwer, Boston (2002)
6. Ponomarenko, N., Krivenko, S., Lukin, V., Egiazarian, K.: Lossy compression of noisy images based on visual quality: a comprehensive study. EURASIP J. Adv. Signal Process. 13 p. (2010). https://doi.org/10.1155/2010/976436
7. Zemliachenko, A., Ponomarenko, N., Lukin, V., Egiazarian, K., Astola, J.: Still image/video frame lossy compression providing a desired visual quality. Multidimension. Syst. Signal Process. 22 p. (2015)
8. Mittal, A., Moorthy, A. K., Bovik, A.C.: Visually lossless H.264 compression of natural videos. Comput. J. (2012). https://doi.org/10.1093/comjnl/bxs105
9. Larson, E.C., Chandler, D.M.: Most apparent distortion: full-reference image quality assessment and the role of strategy. J. Electron. Imaging **19**(1), 011006 (2010)
10. Moorthy, A.K., Bovik, A.C.: Visual quality assessment algorithms: what does the future hold? Multimed. Tools Appl. **51**(2), 675–696 (2011)
11. Jin, L., Egiazarian, K., Jay Kuo, C-C.: Perceptual image quality assessment using block-based milti-metric fusion (BMMF). In: Proceedings of ICASSP, Kyoto, Japan, pp. 1145–1148 (2012)
12. Jin, L., Egiazarian, K., Jay Kuo, C.-C.: Performance comparison of decision fusion strategies in BMMF-based image quality assessment. In: Proceedings of APSIPA, Hollywood, pp. 1–4 (2012)
13. Zhang, L., Mou, X., Zhang, D.: FSIM: a feature similarity index for image quality assessment. IEEE Trans. Image Process. **20**(5), 2378–2386 (2011)
14. Ponomarenko, N., Silvestri, F., Egiazarian, K., Carli, M., Astola, J., Lukin, V.: On between-coefficient contrast masking of DCT basis functions. In: Proceedings of the Third International Workshop on Video Processing and Quality Metrics, USA, 4 p. (2007)
15. Zhang, L., Li, H.: SR-SIM: a fast and high performance IQA index based on spectral residual. In: 19th IEEE International Conference on Image Processing (ICIP), Orlando, USA, pp. 1473–1476 (2012)
16. Lukin, V., Zriakhov, M., Krivenko, S., Ponomarenko, N., Miao, Z.: Lossy compression of images without visible distortions and its applications. In: Proceedings of ICSP 2010, Beijing, China, pp. 694–697 (2010)
17. Ponomarenko, N., Zemlyachenko, A., Lukin, V., Egiazarian, K., Astola, J.: Performance analysis of visually lossless image compression. In: Proceedings of VPQM, Scottsdale, USA, 6 p. (2012)
18. Lukin, V., Ponomarenko, N., Egiazarian, K., Astola, J.: Analysis of HVS-metrics' properties using color image database TID2013. In: Proceedings of ACIVS, Italy, pp. 613–624 (2015)
19. Krivenko, S.S., Krylova, O., Bataeva, E., Lukin, V.V.: Smart lossy compression of images based on distortion prediction. Telecommun. Radio Eng. **77**(17), 1535–1554 (2018)
20. Minguillon, J., Pujol, J.: JPEG standard uniform quantization error modeling with applications to sequential and progressive operation modes. Electron. Imaging **10**(2), 475–485 (2001)
21. Krivenko, S., Zriakhov, M., Kussul, N., Lukin, V.: Prediction of visual quality for lossy compressed images. In: Proceedings of CADSM, Svalyava, Ukraine, 5 p. (2019)
22. Ponomarenko, N.N., Lukin, V.V., Egiazarian, K., Astola, J.: DCT based high quality image compression. In: Proceedings of 14th Scandinavian Conference on Image Analysis, Joensuu, Finland, pp. 1177–1185 (2005)

Discrete Atomic Compression of Digital Images: A Way to Reduce Memory Expenses

Vladimir Lukin, Iryna Brysina, and Victor Makarichev(✉)

National Aerospace University "Kharkiv Aviation Institute",
Chkalov Street 17, Kharkiv 61070, Ukraine
lukin@ai.kharkov.com, iryna.brysina@gmail.com,
victor.makarichev@gmail.com

Abstract. Digital image compression technology based on the application of atomic functions is presented. The fundamentals of this algorithm and the results of its comparison to the algorithm JPEG are given. It is shown that Discrete Atomic Compression saves more memory than JPEG with the same quality of results. It follows that application of Discrete Atomic Compression makes it possible in the future to reduce the cost of storing, processing and transferring a huge number of digital images via communication networks.

Keywords: Lossy image compression · JPEG · Discrete atomic transform · Atomic function

1 Introduction

In recent years, the rapid development of information technology and the widespread use of various gadgets have led to an explosive increase in data, especially graphics. According to some estimates, more than 1.2 trillion digital photos were taken in 2017 [1]. It is natural to assume that this number will increase in the future. Certainly, the costs of data storage, processing and transmission via networks will increase as well. Therefore, the question of possibilities to reduce the cost of resources can be considered as very important.

One way to solve the problem of data storage is to use compression algorithms. There is a wide variety of data compression methods [2–4]. A distinctive feature of digital image compression is the possibility to use lossy compression algorithms. Algorithm JPEG, which is de facto a standard for compression of full color digital images, is one of these technologies. The effectiveness of JPEG method is largely due to the application of such orthogonal transform as discrete cosine transform (DCT). Since the system of trigonometric functions has good approximation properties, JPEG compression allows to get a combination of high compression ratio with rather high quality of results.

Despite the high efficiency and widespread use of this technique, attempts to develop a more optimal image compression technology do not stop. There is a large number of attempts to apply digital transforms that are based on non-trigonometric systems of functions. One of such systems is atomic functions. In [5–8], approximation

M. Nechyporuk et al. (Eds.): *Integrated Computer Technologies in Mechanical Engineering*, AISC 1113, pp. 492–502, 2020.
https://doi.org/10.1007/978-3-030-37618-5_42

properties of atomic functions $up(x)$, $mup_s(x)$ and their generalizations were investigated. The main result of these works is the following: the atomic functions are as effective constructive tools as trigonometric functions. Hence, their application to data processing, in particular, to image compression is quite reasonable.

Discrete atomic compression (DAC) is an algorithm for compression of full color digital images using atomic functions. In this paper, we present results of the research of its efficiency and show the prospects of its further application.

This paper is organized as follows. First, we give a description of DAC and key features of this algorithm. Then we compare DAC and JPEG. In the final part, we discuss the results and describe the directions for further development and application of DAC.

2 Discrete Atomic Compression

2.1 Fundamentals

Atomic functions are solutions with a compact support of linear functional differential equation with constant coefficients and linear transformations of argument, i.e.

$$y^{(n)} + a_1 y^{(n-1)} + \ldots + a_n y = c_1 y(ax + b_1) + \ldots + c_m y(ax + b_m). \quad (1)$$

Necessary and sufficient conditions of existence of compactly supported solutions of this equation were obtained by Rvachev in [5]. The so-called refinement equations, which are widely used for construction of wavelets, are a special case of the Eq. (1). The function $up(x)$ is one of the most studied atomic functions. It is a solution of the equation

$$y'(x) = 2y(2x + 1) - 2y(2x - 1).$$

There are generalizations of *up*-function. For instance, $up_s(x)$ is a solution with a compact support of the equation

$$y' = 2\sum_{k=1}^{s} (y(2sx + 2s - 2k + 1) - y(2sx - 2k + 1)), s \geq 2.$$

In [9], another generalizations of $up(x)$ and $up_s(x)$ were constructed.

Atomic functions have a number of useful properties. The fundamentals of application of atomic functions and their generalizations to data processing were obtained and summarized in [10]. In this paper, discrete atomic transform (DAT), which is analogous to DCT, was introduced. The DAT-procedure is a transform that associates to the data $\{d_i\}_{i=0}^{N}$ a set of DAT-coefficients $\{\omega_i\}_{i=0}^{N}$, which can be used to reconstruct the original data. By processing the DAT-coefficient, the source data can be processed. It was shown in [10] that such an approach has the following advantages:

1. Complexity of DAT- and the inverse DAT-procedures is linear in the size of data, i.e. it equals $O(N)$. This is crucial for big data processing.

2. DAT uses smooth functions that are not analytic. This feature is useful when processing digital images with contrasting color changes, large number of small objects etc. Aerial photos are examples of such images.
3. Original data can be well described using a small number of the corresponding DAT-coefficients. It is especially important when compressing data. DCT-coefficients also have this feature.
4. It is not necessary to use block splitting procedure. By contrast, application of DCT in JPEG compression requires this step (actually, this causes unwanted artifacts at high ratio compression).
5. In compression of digital images, gradual drawing effect can be achieved if we use atomic wavelets [11] or generalized atomic wavelets [12] in DAT.

Discrete atomic transform is the key step of discrete atomic compression.

2.2 Algorithm

Consider the following scheme of data processing (Fig. 1):

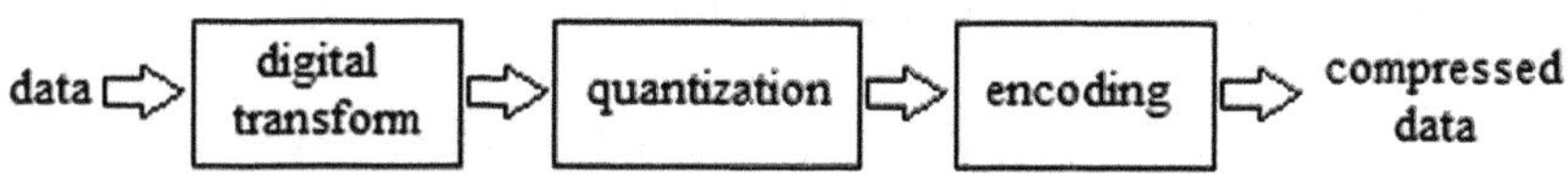

Fig. 1. Lossy compression of data.

This classic approach is used in different compression algorithms, for example, in JPEG [13] and JPEG2000 [14]. It should be mentioned that it is not possible to reconstruct the original data using only their compressed version. Mainly losses occur at quantization stage. For this reason, algorithms based on this scheme are called lossy compression algorithms. Despite the loss of quality, such algorithms are widely used due to the possibility of obtaining a high degree of compression in combination with a required quality.

DAC is an algorithm based on the presented lossy compression scheme. Its main idea consists in application of DAT. Discrete atomic compression of 24-bit full color digital images is as follows (Fig. 2):

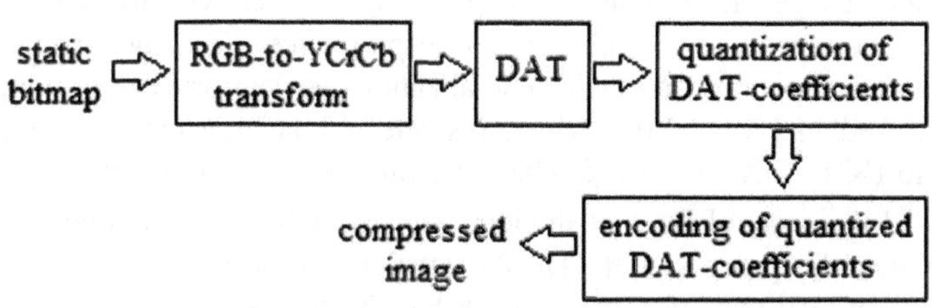

Fig. 2. Discrete atomic compression of full color digital image.

The input of this algorithm is static bitmap. In general, we assume that this image is a matrix of RGB-components.

In the first stage, an image is converted into YCrCb color space. We get three matrices: Y, Cr and Cb. It is clear that this stage can be changed depending on the form of the image representation. For example, this step is not needed when compressing grayscale images.

Next, DAT is applied to each matrix Y, Cr and Cb. After that we get three matrices of the corresponding DAT-coefficients: W_Y, W_{Cr} and W_{Cb}.

The next step is quantization of DAT-coefficients.

At the last stage, the quantized DAT-coefficients are encoded using lossless compression algorithms.

The output of the algorithm is a compressed image recorded in the corresponding dac-file.

In order to reconstruct the image, we use the following scheme (Fig. 3):

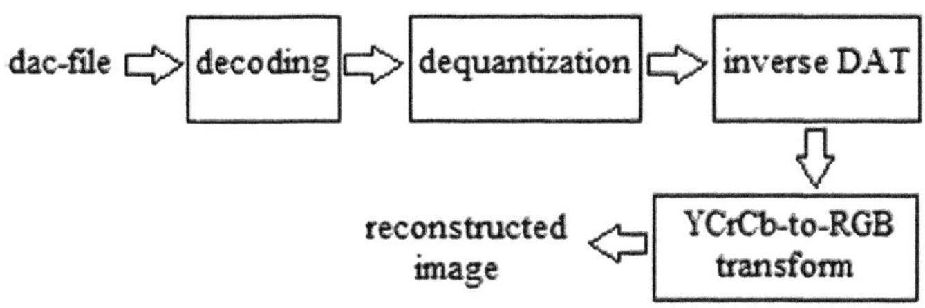

Fig. 3. Decompression of DAC-image.

2.3 Key Features

Flexibility of Settings. DAC is a lossy compression algorithm. We note that major quality losses are obtained at the quantization stage. However, it is this conversion that makes possible to save memory. By varying the coefficients of quantization, one can get results of different quality and compression ratio. The following question naturally arises: what quantization coefficients should be used? Clearly, that it depends on the requirements to the result of compression. There is another question: how can optimal quantization coefficients be found? This is a non-trivial problem. There are several solutions. We can search for optimal coefficients for each image that should be compressed. This approach has a high computational complexity, which may be unacceptable when processing a large number of images. For this reason, a unified approach based on the standard version of the applied algorithm is often used. In addition, various methods of image preprocessing are used to achieve a more optimal compression result (for example, procedure of color palette change).

By fixing the quantization coefficients, we can get different modes of the algorithm DAC (see Fig. 4). We consider a mode called "Optimal" in this paper. Using this mode of DAC, it is possible to get a combination of a high compression with visually indistinguishable changes. In the next section, we compare DAC "Optimal" with the corresponding mode of the algorithm JPEG.

Fig. 4. Test image “Lena”: a - original image, 768 kB; b - high quality result, 164 kB; c - low quality result, 3 kB.

Progressive Reconstruction. If atomic wavelets or generalized atomic wavelets are used in DAT, then matrices W_Y, W_{Cr} and W_{Cb} of DAT-coefficients have block structure. One of these blocks contains information that allows to restore a “rough” approximation of the original data, and the elements of the other blocks can be used to obtain a more accurate approximation. Therefore, the structure of the resulting dac-file is as follows: the header contains a small copy of the image, and the following information that can be used to obtain a larger copy of the image (see Fig. 5).

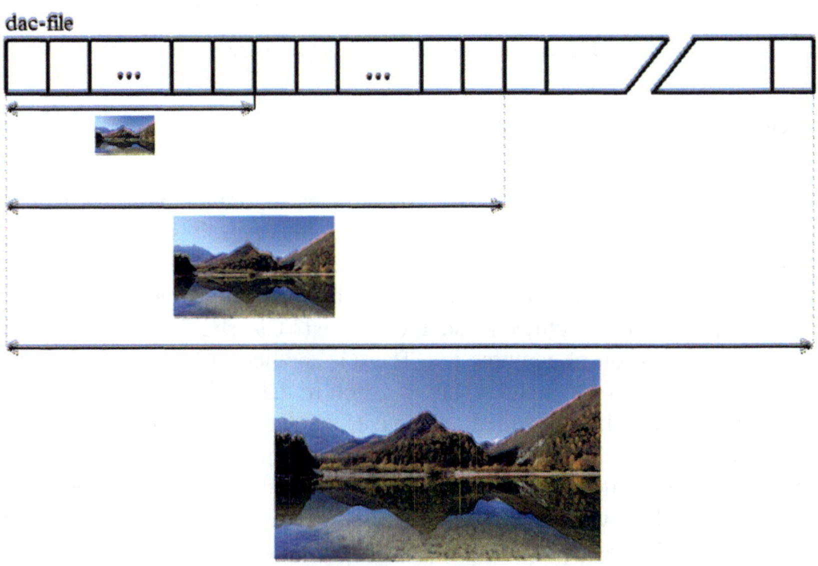

Fig. 5. Progressive reconstruction of the image.

Such a possibility is extremely important, for example, in the process of previewing, when in order to get an image icon it is sufficient to download a small part of the dac-file.

Also, it is useful when sharing digital images between users (for instance, it is not necessary to send the entire file over the network to view a small copy of some digital photo). Obviously, this feature saves resources.

Quality Loss Control. There are many metrics for assessing loss of quality. In the current research, we use uniform metric (U), root mean square (RMS) and peak signal-to-noise ratio (PSNR). Note that uniform metric is the most sensitive to any changes. Therefore, the assessment of this metric is extremely important in case of high requirements to the quality of the result.

Discrete atomic transform is a kind of discrete wavelet transforms. Hence, matrices W_Y, W_{Cr} and W_{Cb} of DAT-coefficients have block structure. Each block B_{ij} of these matrices is quantized, using the coefficient q_{ij}. It follows from the properties of atomic functions that quality losses measured by U-metric do not exceed the sum of the values of the quantization coefficients q_{ij}. This yields that uniform loss of quality can be controlled.

3 Comparison of DAC with JPEG

The aim of this Section is to show that application of DAC to digital image compression is reasonable and perspective. For this purpose, we consider several sets of different test images and compare the results of their compression using DAC and JPEG.

3.1 Description of the Comparison Procedure

Consider the following sets of test images:

- classic test images ("Baboon", "Lenna", "Peppers" etc.; these images were downloaded from the USC-SIPI image database [15]);
- aerial test images [15];
- screenshots (images containing pictures and text);
- small copies of Canon test photos (initially, each of Canon test photos was 5760×3840 JPEG-image; we made 1024×683 copies of them);
- small copies of Nikon test photos (initially, each of Nikon test photos was 6000×4000 JPEG-image; we made 1024×683 copies of them);
- test images TID2013 [16];
- classic grayscale images and test texture images [15];
- test texture images [15].

Comparison of algorithms DAC and JPEG was made on this test data. In our experiments, we used DAC "Optimal" in the computer program "Discrete Atomic Compression: User Kit" [17]. JPEG compression was performed using MS Picture Manager.

In our experiments, we use bmp-version of each test image.

For each test image we applied the following actions:

1. DAC and JPEG compression.
2. Computation of compression ratio (CR), which is defined as the ratio of the size of the source image file to the size of the compressed image file. Notice that a higher value of CR means more memory savings. We also compute percentage of saved memory (PSM) for each file.
3. Decompression.
4. Assessment of quality loss.

To assess the loss of quality, the following metrics were used:

- uniform metric (U)

$$U = \max_{i=1,\ldots,N} |x_i - y_i|;$$

- root mean square metric (RMS)

$$RMS = \sqrt{\frac{1}{N}\sum\nolimits_{i=1}^{N} (x_i - y_i)^2};$$

- peak signal-to-noise ratio (PSNR)

$$PSNR = 20\log_{10}\left(\frac{MAX}{RMS}\right).$$

Here, N is the size of the data; $\{x_i\}_{i=1}^{N}$ and $\{y_i\}_{i=1}^{N}$ are original and recovered data, respectively; *MAX* is the maximum value of the elements.

These metrics make it possible to estimate in some way the deviation of the obtained data from the original. It should be mentioned that an adequate numerical way to assess the loss of quality has not yet been found. Each of the existing methods has its advantages and disadvantages, but none of them is reliable. In this research, we use several quite different metrics to estimate both local (U-metrics) and global (RMS- and PSNR-metrics) changes. Besides, we checked results visually, but it is too subjective and unreliable.

3.2 DAC vs JPEG

In this Section, we compare DAC with JPEG. We use the following sets of test images: (1) full color 24-bit digital images: classic test images ("Baboon", "Lenna", "Peppers" etc.; these images were downloaded from the USC-SIPI image database [14]), aerial test images [14], screenshots (images containing pictures and text), small copies of Canon and Nikon test photos, TID2013 [15]; (2) grayscale test images: classic grayscale images and test texture images [14]. The total number of test images is 208.

Tables 1 and 2 show average, minimal and maximal values of *U*-metric, *RMS*-metric, *PSNR*-metric as well as average, minimal and maximal values of compression ratio (*CR*), which is defined as the ratio of the size of the source image file to the size of the compressed image file, and percentage of saved memory (PSM).

Table 1. DAC vs JPEG: full color test images

Indicator		Algorithm	
		DAC	JPEG
PSNR-metric	Average	39.094	39.2696
	Minimal	36.3409	26.4239
	Maximal	43.0052	49.6249
RMS-metric	Average	2.8988	3.2091
	Minimal	1.8042	0.8419
	Maximal	3.8859	12.1715
U-metric	Average	19.9375	66.3661
	Minimal	13	11
	Maximal	26	174
CR	Average	5.966	3.6799
	Minimal	2.2203	2.0715
	Maximal	21.8085	7.3301
PSM	Average	79.9035	71.0369
	Minimal	54.9609	51.7255
	Maximal	95.4146	86.3577

Table 2. DAC vs JPEG: grayscale test images

Indicator		Algorithm	
		DAC	JPEG
PSNR-metric	Average	40.7583	41.0558
	Minimal	40.465	40.6487
	Maximal	44.1316	48.7353
RMS-metric	Average	2.34095	2.2738
	Minimal	1.5848	0.9327
	Maximal	2.4171	2.3664
U-metric	Average	11.6354	11.4583
	Minimal	9	7
	Maximal	14	13
CR	Average	2.1199	1.8804
	Minimal	1.219	1.1743
	Maximal	5.3333	4.4912
PSM	Average	48.7884	43.4529
	Minimal	17.9687	14.8437
	Maximal	81.25	77.7343

As it can be seen, average values of PSNR for DAC and JPEG are almost equal. At the same time, DAC provides higher compression ratio. Hence, DAC saves more memory than JPEG with the same quality of the result.

Besides, *U*-metric for DAC is significantly lower than *U*-metric for JPEG (this is important if further recognition technologies are applied to compressed image).

We have also compared the compression results for each specific image. In more detail, for any test image the following differences were computed: *U*(JPEG) – *U* (DAC), *RMS*(JPEG) – *RMS*(DAC), *PSNR*(JPEG) – *PSNR*(DAC), *CR*(JPEG) – *CR* (DAC) and *PSM*(JPEG) – *PSM*(DAC). Average, minimal and maximal values of these differences are presented in Tables 3 and 4.

Table 3. Comparison of differences: classic test images

Indicator	Value		
	Average	Minimal	Maximal
PSNR-metric	0.1756	−13.9705	7.4638
RMS-metric	0.3103	−1.5048	9.7347
U-metric	46.4286	−3	157
CR	−2.2861	−16.0982	0.2637
PSM	−8.8666	−15.5979	1.7331

Table 4. Comparison of differences: grayscale test images

Indicator	Value		
	Average	Minimal	Maximal
PSNR-metric	0.2975	0.0777	4.9046
RMS-metric	−0.0671	−0.7078	−0.0212
U-metric	−0.1771	−3	2
CR	−0.2395	−0.8421	0.7385
PSM	−5.3355	−12.8906	9.375

Table 5. Data size, kB

Test data	Info	Files		
		BMP (source)	JPEG	DAC
Full color test Images	Total size	218013	63490	46502
	Size of zip-archive	157665	62947	42944
Grayscale test images	Total size	43776	23854	21048
	Size of zip-archive	18598	23704	19825

Table 5 presents comparison of size of source data with the size of compressed data. Also, each set of files was compressed using the algorithm ZIP.

It follows that ZIP-compression of JPEG files provides a slight reduction in memory, but it requires additional resources. Also, we see that this lossless compression algorithm provides at least 5% reduction in memory when processing DAC files.

In Tables 1 and 2, general results are presented. More detailed analysis is available on the link to Google drive:

https://drive.google.com/open?id=1Y6UGi2iBxeETDQHffHQqN_d2O45acIvM.

4 Conclusions

It follows from the results presented in this paper that further research of discrete atomic compression is promising. It is clear that DAC should be compared with JPEG and other algorithms with different quality settings. Also, other metrics of compressed image quality should be used. However, the hypothesis, which was stated in [18], of the possibility of effective application of special classes of atomic functions to image compression is confirmed.

There are many directions for further development of the technology presented in this paper. Here are some of them:

- Optimization of DAC. First, as it can be seen from the results presented in the previous Section, the use of additional lossless compression of DAC-files is effective. Therefore, encoding procedure of quantized DAT-coefficients can be improved. Second, in the current research, the fixed set of quantization coefficients was used for each test image. It is clear that by varying these parameters depending on the content of compressed image, we can get better results.
- Application of DAC to processing of some specific types of images (for example, medical images).
- Application of DAC to compression of video. It seems to be promising, since DAC of digital images provides the combination of high compression ratio with high quality of result.
- Pattern recognition. DAC provides low local loss of quality. Whence, the following question is of significant interest: is it possible to increase efficiency of recognition using DAC-files?

These will be the object of other research.

References

1. People will take 1.2 trillion digital photos this year – thanks to smartphones. https://www.businessinsider.com. Accessed 20 Nov 2018
2. Salomon, D., Motta, G., Bryant, D.: Handbook of Data Compression, 5th edn. Springer, Heidelberg (2010)
3. Pearlman, W.A., Said, A.: Digital Signal Compression: Principals and Practice. Cambridge University Press, Cambridge (2011)
4. Sayood, K.: Introduction to data compression, 5th edn. Morgan Kaufman, Burlington (2017)
5. Rvachev, V.L., Rvachev, V.A.: Nonclassical methods of approximation theory in boundary value problems. "Naukova dumka" Publ., Kyiv (1979). (in Russian)
6. Rvachev, V.A.: Compactly supported solutions of functional-differential equations and their applications. Russ. Math. Surv. **45**(1), 87–120 (1990). https://doi.org/10.1070/RM1990v045n01ABEH002324

7. Makarichev, V.A.: Approximation of periodic functions by $mup_s(x)$. Math. Notes **93**(6), 858–880 (2013). https://doi.org/10.1134/S0001434613050258
8. Brysina, I.V., Makarichev, V.A.: Approximation properties of generalized Fup-functions. Visnyk of V.N. Karazin Kharkiv National University. Ser. Mathematics, Applied Mathematics and Mechanics, vol. 84, pp. 61–92 (2016). http://vestnik-math.univer.kharkov.ua/Vestnik-KhNU-84-2016-makarich.pdf
9. Brysina, I.V., Makarichev, V.A.: On the asymptotics of the generalized Fup-functions. Adv. Pure Appl. Math. **5**(3), 131–138 (2014). https://doi.org/10.1515/apam-2014-0009
10. Brysina, I.V., Makarichev, V.O.: Atomic functions and their generalizations in data processing: function theory approach. Radioelectron. Comput. Syst. **3**(87), 4–10 (2018). https://doi.org/10.32620/reks.2018.3.01
11. Brysina, I.V., Makarichev, V.O.: Atomic wavelets. Radioelectron. Comput. Syst. **1**(53), 37–45 (2012). http://irbis-nbuv.gov.ua/cgi-bin/irbis_nbuv/cgiirbis_64.exe?C21COM=2&I21DBN=UJRN&P21DBN=UJRN&IMAGE_FILE_DOWNLOAD=1&Image_file_name=PDF/recs_2012_1_7.pdf
12. Brysina, I.V., Makarichev, V.O.: Generalized atomic wavelets. Radioelectron. Comput. Syst. **1**(85), 23–31 (2018). https://doi.org/10.32620/reks.2018.1.03
13. Pennebaker, W.B.: JPEG: Still Image Data Compression Standard. Springer, Heidelberg (1993)
14. Taubman, D., Marcelin, M.: JPEG2000: Image Compression Fundamentals. Standards and Practice. Springer, Heidelberg (2002)
15. The USC-SIPI image database. http://sipi.usc.edu/database/. Accessed 10 Oct 2018
16. Ponomarenko, N., Jin, L., Ieremeiev, O., Lukin, V., Egiazarian, K., Astola, J., Vozel, V., Chehdi, K., Carli, M., Battisti, F., Jay Kuo, C.-C.: Image database TID2013: peculiarities, results and perspectives. Sig. Process. Image Commun. **30**, 57–77 (2015)
17. Makarichev, V.O.: Discrete atomic compression: user kit. The Certificate on official registration of the computer program copyright, no. 83047 (2018)
18. Makarichev, V.O.: Application of atomic functions to lossy image compression. Theoretical and applied aspects of cybernetics. In: Proceedings of the 5th International Scientific Conference of Students and Young Scientists, pp. 166–175. "Bukrek" Publ., Kyiv (2015)

Development of Game Modules with Support for Synchronous Multiplayer Based on Unreal Engine 4 Using Artificial Intelligence Approach

Bohdan Levchenko, Andrii Chukhray, and Dmytro Chumachenko(✉)

National Aerospace University "Kharkiv Aviation Institute", Chkalow Street, 17, Kharkiv, Ukraine
bodya.levchenko20@gmail.com, achukhray@gmail.com, dichumachenko@gmail.com

Abstract. The subject of the research is the processes of constructing the optimal structure of game modules and the methodology for synchronizing them on the Internet. The aim of the work is to create prototypes of game modules with synchronous multiplayer. The physics of a bullet and a car was developed from scratch. The basic methods and tools of building gaming systems are considered. It is proposed to use property replication and remote procedural calls to create multiplayer. The configuration of Internet packages and calculations on CPU/GPU are optimized.

Keywords: Game module · RPC · Replication · Voronoi cell · Linetrace · Internet roles · Blueprints · Lag · CPU · GPU

1 Project Relevance (Introduction)

The field of Artificial Intelligence has seen a rapid boom over recent years [1–3], as the set of techniques and technologies behind AI promises to improve almost every aspect of our lives [4–7]. Hence, AI is used for enhancing manufacturing efficiency [8–11], creating self-driving systems for cars, predicting stock prices, scheduling repairs for industrial goods, and for other useful things that could directly make our lives better [12–16]. However, there is another application of AI algorithms that does not seem to bear any economic benefits but is still researched widely—AI that plays video games.

Virtual reality is used to teach professions where the operation of real devices and mechanisms involves increased risk, or high costs (aircraft pilot, train driver, public transport driver, soldier, dispatcher, etc.) [17].

Since there are no similar domestic developments in the modern market, and the audience of consumers is only growing every year, the topic of the project can be considered relevant.

M. Nechyporuk et al. (Eds.): *Integrated Computer Technologies in Mechanical Engineering*, AISC 1113, pp. 503–513, 2020.
https://doi.org/10.1007/978-3-030-37618-5_43

2 Analysis of the Subject Area

As a result of the analysis of the subject area it was decided to choose the following tools to solve the tasks.

Line tracing - line point-to-point linear check for any obstacles. This is one of the most important and powerful tools you will need when developing modules. Properly used, it does not cause heavy CPU loads (so it is best not to use it in every frame) [18].

The big advantage of using Voronoi cells, among other methods, is that the algorithm compilation takes place during the level load, not when the gameplay begins. This increases the performance of your computer [19].

Using interfaces reduces the amount of written code, and the enumerated data type, as discussed below, will help you build optimal structural models. Interpolations will be used for smoothing, and the finite state machine model will be used for the logic of transitioning between animations. On the trace in general, built every in-game mechanic.

Then we will set up network update frequency.

Actors will observe a maximum update frequency set in their NetUpdateFrequency variable. By reducing this variable on less-important or less-frequently-changing Actors, network updates can be made more efficient, potentially leading to smoother play experiences in limited-bandwidth scenarios. Common update frequency values are 10 (updating every 0.1 s) for important, unpredictable Actors like player-controlled characters in a shooter, 5 (updating every 0.2 s) for slower-moving characters like AI-controlled monsters in cooperative games, or 2 (updating every 0.5 s) for background Actors that are not very important to gameplay, but are still synced over the network and/or are controlled by server-side logic and thus need replication. For maximum network performance for our car we become value 1 (updating every 1 s).

3 Choosing the AI Behavior Model

Most artificial intelligence developers know that for simple systems, an abstract model called the finite state machine is ideally suited (Fig. 1). By simply defining the state and conditions for changing that state, artificial intelligence can be put into an endless cycle of decision-making. That is, the machine will continue to perform the current state until the conditions for switching to another.

The ability to move from one state to another by defining the right conditions allows you to set up a finite state machine for artificial intelligence behavior without unnecessary difficulty. However, this is a disadvantage of this approach. In complex games, a finite state machine can hold hundreds of states, and at such scales, debugging becomes extremely difficult.

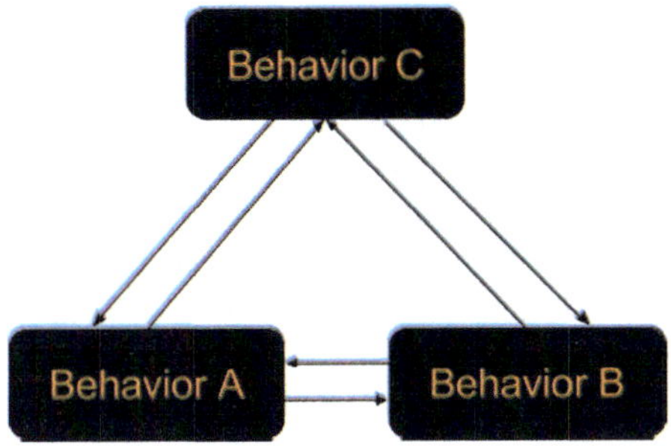

Fig. 1. Model of a finite state machine

As the number of states of a finite state machine increases, its complexity increases sharply. The number of transitions k in SM with the number of states n is determined by the formula:

$$k = n \cdot (n - 1). \tag{1}$$

If we take $n = 4$, then we get 12 possible transitions. If we add another state, the conversions will be 20, another will be 30.

In part, this problem is solved by the so-called hierarchical finite state machines (Fig. 2). The idea of a hierarchy has led to an increase in the number of states available, but even so, it does not allow coping with too large and complex finite state machines.

At present, this technology has not been tested in time and is no longer used in robotics, especially in the development of computer games.

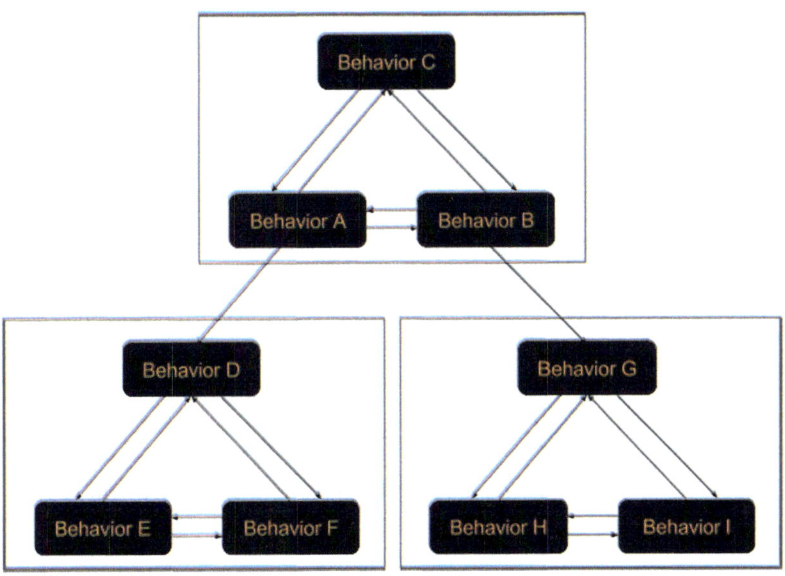

Fig. 2. Model of hierarchical finite state machine

As a result, the idea of hierarchical organization led to the idea of organizing tasks into tree structures - they are the trees of behavior (Fig. 3). They work in such a way that the behaviors are either performed one at a time or in sequence. Current behavior is viewed at a certain frequency, such as once a second. In this case, each tree is evaluated each time, and if a behavior other than the current one is selected, the executed state changes, otherwise the previous behavior is executed. At this point, a number of disadvantages of a finite state machine are overcome, such as hang-ups in one state or erroneous change of states. Unreal Engine includes a convenient implementation of behavioral trees, which are commonly referred to when talking about this technology.

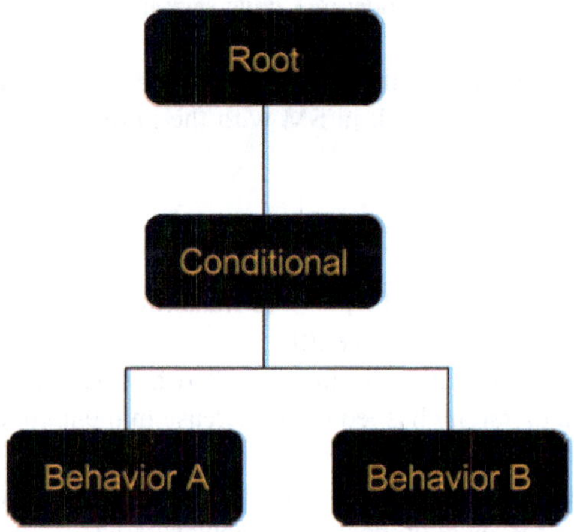

Fig. 3. Behavior tree

If the SM spells out for each state its decision-making logic, then in the behavior tree it is derived beyond their limits. This allows you to add and remove nodes even while the program is running: just write new code to call the node or delete the old one.

In addition, a tree with a large number of states can be split into small subtrees - this further simplifies the orientation in the code, which improves the search for flaws and errors.

In this project, two prototype robots were created:

- bot-IAdam (decision-making system uses a finite state machine model);
- bot-Mutant (the decision-making system uses a behavior tree model).

4 The Main Actor's Physics

Physics engine calculations are not enough to satisfy our goals. We will have to write bullet ballistics using the physical capabilities of the engine, as well as machine physics from scratch.

4.1 Bullet Ballistics

Bullet ballistics involves several tasks that need to be addressed in the course of this work. First, you need to understand whether the projectile will punch through, or whether the bullet will remain in the material. Secondly, you need to calculate the angle at which the ball will reflect from the object and how its velocity will change. Trying to solve these issues only creates new ones. Support functions will also need to be implemented to control the correctness of the actions taken.

A pure function called ComputeExitLocation created in the project performs tracing (multispherical association for objects). Initial distance - 500 cm before collision, in the direction of impact velocity. The final part of the retreat is at the point of hit. The function goes through each collision in reverse order and returns the first hit whose "shock actor" coincides with the original "hit actor". Then move the exit point a short distance from the point of impact to prevent a collision. Finally, the check is made that the exit does not interfere with another object and, if so, the next exit is by recursion. The function returns location vectors, boolean, and HitResult (Fig. 4) structure. Objects greater than 500 cm in depth will not return the exit point. We will use the HitResult structure later.

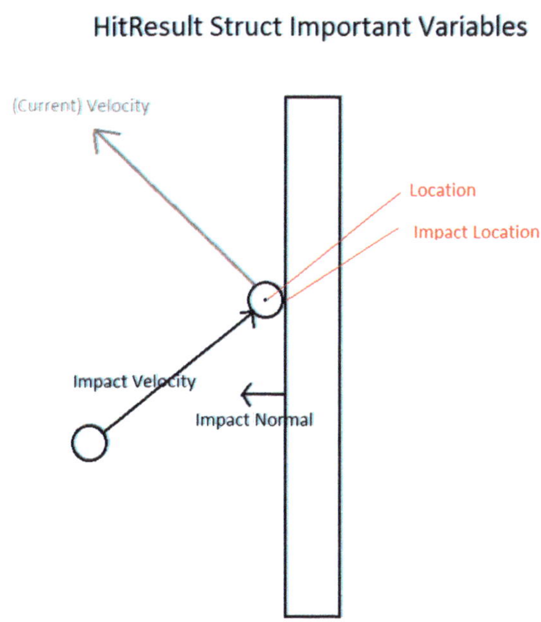

Fig. 4. HitResult structure rendering

Back in the event graph, add a placeholder area for a 'DecideWhetherToPenetrate' function. At the moment we will simply check that the projectile's velocity is over a certain amount.

If the projectile has enough velocity, and an exit location was found, spawn a new FirstPersonProjectile.

We cannot teleport the current projectile as Unreal's ProjectileMovementComponent precludes this. A cleaner but more time-consuming alternative would be to create a clone of Unreal's ProjectileMovementComponent and change its behavior so teleporting the object at this point is possible.

After spawning a new projectile, set it's velocity to half the impact velocity. We will create a more complex function to replace this later.

Finally call the destroy function to get rid of the original projectile. Also destroy the original projectile if no exit location was found.

You should now be able to test that the projectile penetrates objects.

If desired, open YourProjectProjectile.cpp and comment out the "Destroy();" line so that projectiles can bounce off physics objects.

Next we create a GetNewSpeed function (pure) with inputs for the hit, exit location, and impact velocity. It should return a float. Energy lost is calculated by multiplying the distance penetrated and material dependant constant:

$$E_1 = S \cdot c. \tag{2}$$

The energy of the ball can be calculated by the formula:

$$E_2 = \frac{mv^2}{2}, \tag{3}$$

We get a new value of energy:

$$E = E_2 - E_1. \tag{4}$$

The new speed is calculated by redistributing this:

$$V = \sqrt{\frac{E}{\frac{1}{2} \cdot m}}. \tag{5}$$

Making sure that the new velocity is greater than zero (if so, the new velocity of the projectile is set, and if not, the original projectile is destroyed), it is possible to verify that the projectile is losing velocity proportional to the penetration distance.

4.2 Car Physics

One of the key points in simplifying vehicle physics is to handle the longtitudinal and lateral forces separately. Longtitudinal forces operate in the direction of the car body (or in the exact opposite direction). These are wheel force, braking force, rolling resistance and drag (air resistance). Together these forces control the acceleration or deceleration of the car and therefore the speed of the car. Lateral forces allow the car to turn. These forces are caused by sideways friction on the wheels.

First let's consider a car driving in a straight line. Which forces are at play here? First of all there's what the tractive force, i.e. the force delivered by the engine via the rear wheels. The engine turns the wheels forward (actually it applies a torque on the wheel), the wheels push backwards on the road surface and, in reaction, the road surface pushes back in a forward direction. For now, we'll just say that the tractive force is equivalent in magnitude to the variable Engine force, which is controlled directly by the user:

$$F_{traction} = u \cdot EngineForce, \tag{6}$$

where u is a unit vector in the direction of the car's heading.

If this were the only force, the car would accelerate to infinite speeds. Clearly, this is not the case in real life. Enter the resistance forces. The first and usually the most important one is air resistance, a.k.a. aerodynamic drag. This force is so important because it is proportional to the square of the velocity. When we're driving fast (and which game doesn't involve high speeds?) this becomes the most important resistance force:

$$F_{drag} = -C_{drag} \cdot V \cdot |V|, \tag{7}$$

where C_{drag} is a constant and V is the velocity vector and the notation $|V|$ refers to the magnitude of vector V.

Then there is the rolling resistance. This is caused by friction between the rubber and road surface as the wheels roll along and friction in the axles, etc. We'll approximate this with a force that's proportional to the velocity using another constant.

$$F_{rr} = -C_{rr} \cdot V, \tag{8}$$

where C_{rr} is a constant and V is the velocity vector.

At low speeds the rolling resistance is the main resistance force, at high speeds the drag takes over in magnitude. At approx. 100 km/h (60 mph, 30 m/s) they are equal. This means C_{rr} must be approximately 30 times the value of $\mathrm{C_{drag}}$.

The total longtitudinal force is the vector sum of these three forces:

$$F_{long} = F_{traction} + F_{drag} + F_{rr}. \tag{9}$$

Note that if you're driving in a straight line the drag and rolling resistance forces will be in the opposite direction from the traction force. So in terms of magnitude, you're subtracting the resistance force from the traction force. When the car is cruising at a constant speed the forces are in equilibrium and $\mathbf{F}_{\mathrm{long}}$ is zero.

The acceleration (a) of the car (in meters per second squared) is determined by the net force on the car (in Newton) and the car's mass M (in kilogram) via Newton's second law:

$$a = \frac{F}{m}. \tag{10}$$

The car's velocity (in meters per second) is determined by integrating the acceleration over time. This sounds more complicated than it is, usually the following equation does the trick. This is known as the Euler method for numerical integration:

$$V = V + a \cdot dt, \tag{11}$$

where *dt* is the time increment in seconds between subsequent calls of the physics engine.

The car's position is in turn determined by integrating the velocity over time:

$$p = p + V \cdot dt. \tag{12}$$

With these three forces we can simulate car acceleration fairly accurately. Together they also determine the top speed of the car for a given engine power. There is no need to put a maximum speed anywhere in the code, it's just something that follows from the equations. This is because the equations form a kind of negative feedback loop. If the traction force exceeds all other forces, the car accelerates. This means the velocity increases which causes the resistance forces to increase. The net force decreases and therefore the acceleration decreases. At some point the resistance forces and the engine force cancel each other out and the car has reached its top speed for that engine power.

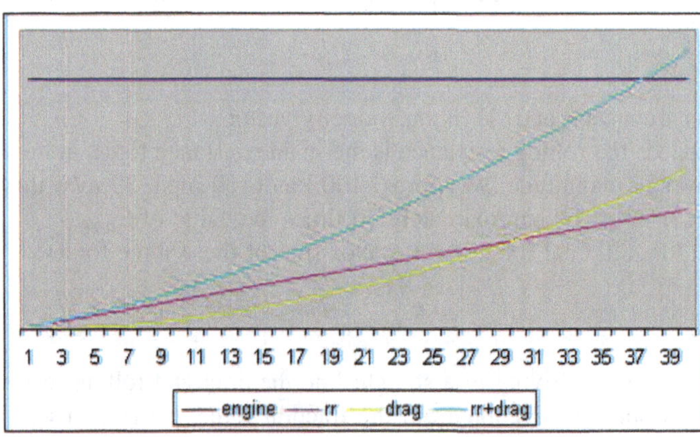

Fig. 5. Dependency diagram

In this diagram (Fig. 5) the X-axis denotes car velocity in meters per second and force values are set out along the Y-axis. The traction force (dark blue) is set at an arbitrary value, it does not depend on the car velocity. The rolling resistance (purple line) is a linear function of velocity and the drag (yellow curve) is a quadratic function of velocity. At low speed the rolling resistance exceeds the drag. At 30 m/s these two functions cross. At higher speeds the drag is the larger resistance force. The sum of the two resistance forces is shown as a light blue curve. At 37 m/s this curve crosses the horizontal traction force line. This is the top speed for this particular value of the engine power (37 m/s = 133 km/h = 83 mph).

5 Network Synchronization and Cheat Protection

Unreal Engine 4 uses a standard Server-Client architecture. This means, that the Server is authoritative and all data must be send from Client to Server first. Then the Server validates the data and reacts depending on your code.

When we move our car, as a Client, in a Multiplayer Match, we don't actually move our car by yourself, but tell the Server that you want to move it. The Server then updates the location of the car for everyone else, including you.

To prevent a feeling of "lag" for the local Client, Coders often, in addition, let this Player directly control their Character locally (car for us), although the Server still might override the Character's Location when the Client starts cheating! This means, that the Client will (almost) never 'talk' to other Clients directly.

When sending a Chat-Message to another Client, you are actually sending it to the Server, which then passes it to the Client you want to reach. This could also be a team, guild, group, etc.

Never trust the Client! Trusting the Client means, you don't test the Clients actions before executing them. This allows cheating! A simple example would be Shooting: Make sure to test, on the Server, if the Client actually has Ammo and is allowed to shoot again instead of directly processing the shot!

Programly, marking the function as Server, Client, or NetMulticast (RPC's calls) makes the function replicated. E.g., when a Server function is called on the client the call is replicated to the server and the server executes the body of code (in an _Implementation function. E.g, if we marked SetCarMoveForward as Server, we need to do our work in SetCarMoveForward_Implementation). So just marking that function as Server is probably not what we want.

BlueprintAuthorityOnly is closer to what we want. Marking a function BlueprintAuthorityOnly means that in blueprints, only the server may execute the function (if a client tries to call it, nothing will happen and execution will continue).

Having the Role check in the function body may still be a good idea, since SetCarMoveForward could be called directly from C++ as well.

Property replication does only ever happen server - client. Clients cannot set properties and have them propagate to the server. The only way clients communicate with the server is by calling Server replicated functions. A hacker could in theory send bogus calls to (only) Server functions (they cannot trick the server into executing non Server functions remotely). Any parameters to Server replicated functions could be compromised as well.

6 Conclusion

For the implementation of the tasks posed, we carried out an analysis of the subject area, including some optimal methods of implementation. We used the basic methods of building gaming systems. Optimized calculations on the graphic and central processors. We added the physical capabilities of the engine to implement the physics of the car and bullets, chose the optimal model of artificial intelligence of the characters of the opponents. Synchronized actions on the Internet. As a result, we got two game modules with realistic shooting and controlled transport.

References

1. Meniailov, I., et al.: Using the K-means method for diagnosing cancer stage using the Pandas library. In: CEUR Workshop Proceedings, vol. 2386, pp. 107–116 (2019)
2. Chumachenko, D., et al.: Development of an intelligent agent-based model of the epidemic process of syphilis. In: 2019 IEEE 14th International Scientific and Technical Conference on Computer Sciences and Information Technologies (CSIT), Lviv, vol. 1, pp. 42–45 (2019)
3. Chumachenko, D., et al.: On agent-based approach to influenza and acute respiratory virus infection simulation. In: Proceedings of 14th International Conference on Advanced Trends in Radioelectronics, Telecommunications and Computer Engineering, TCSET 2018, pp. 192–195 (2018)
4. Polyvianna, Y., et al.: Computer aided system of time series analysis methods for forecasting the epidemics outbreaks. In: 2019 15th International Conference on the Experience of Designing and Application of CAD Systems (CADSM), pp. 7.1–7.4 (2019)
5. Chumachenko, D., Chumachenko, T.: Intelligent agent-based simulation of HIV epidemic process. In: Advances in Intelligent Systems and Computing, vol. 1020, pp. 175–188 (2019)
6. Chumachenko, D., Chumachenko, K., Yakovlev, S.: Intelligent simulation of network worm propagation using the code red as an example. Telecommun. Radio Eng. **78**(5), 443–464 (2019)
7. Chumachenko, D., Yakovlev, S.: On intelligent agent-based simulation of network worms propagation. In: 2019 15th International Conference on the Experience of Designing and Application of CAD Systems (CADSM), pp. 3.11–3.13 (2019)
8. Dotsenko, N., Chumachenko, D., Chumachenko, I.: Modeling of the processes of stakeholder involvement in command management in a multi-project environment. In: Proceedings of 2018 IEEE 13th International Scientific and Technical Conference on Computer Sciences and Information Technologies, CSIT 2018, pp. 29–32 (2018)
9. Dotsenko, N., Chumachenko, D., Chumachenko, I.: Project-oriented management of adaptive teams' formation resources in multi-project environment. In: CEUR Workshop Proceedings, vol. 2353, pp. 911–923 (2019)
10. Dotsenko, N., Chumachenko, D., Chumachenko, I.: Management of critical competencies in a multi-project environment. In: CEUR Workshop Proceedings, vol. 2387, pp. 495–500 (2019)
11. Dotsenko, N., Chumachenko, D., Chumachenko, I.: Modeling of the process of critical competencies management in the multi-project environment. In: Proceedings of 2019 IEEE 13th International Scientific and Technical Conference on Computer Sciences and Information Technologies, CSIT 2019, vol. 3, pp. 89–93 (2019)

12. Bazilevych, K., et al.: Stochastic modelling of cash flow for personal insurance fund using the cloud data storage. Int. J. Comput. **17**(3), 153–162 (2018)
13. Chumachenko, D., et al.: Intelligent expert system of knowledge examination of medical staff regarding infections associated with the provision of medical care. In: CEUR Workshop Proceedings, vol. 2386, pp. 321–330 (2019)
14. Chumachenko, D.: On intelligent multiagent approach to Viral Hepatitis B epidemic processes simulation. In: Proceedings of the 2018 IEEE 2nd International Conference on Data Stream Mining and Processing, DSMP 2018, pp. 415–419 (2018)
15. Mazorchuck, M., Dobriak, V., Chumachenko, D.: Web-application development for tasks of prediction in medical domain. In: 2018 IEEE 13th International Scientific and Technical Conference on Computer Sciences and Information Technologies (CSIT), Lviv, pp. 5–8 (2018)
16. Chumachenko, D., Yakovlev, S.: Development of deterministic models of malicious software distribution in heterogeneous networks. In: 2019 3rd International Conference on Advanced Information and Communications Technologies (AICT), Lviv, Ukraine, pp. 439–442 (2019)
17. Rossokhin, A., Izmagurova, V.: Virtual happiness or virtual addiction. Textbook **1**, 516–517 (2004)
18. Unreal Engine [Electronic resource]. Wikipedia - the free encyclopedia. https://ru.wikipedia.org/wiki/Unreal_Engine. Accessed 29 Sept 2017
19. Modeling [Electronic resource]. Wikipedia - the free encyclopedia. https://ru.wikipedia.org/wiki/Modeling. Accessed 09 Dec 2017
20. Samarsky, A., Mikhailov, A.: Mathematical modeling: ideas, methods, examples. Textbook **2**, 320–322 (1997)

Application of Parallel Computing in Robust Optimization Design

Ievgen Meniailov(✉), Serhii Krivtsov, Mykhaylo Ugryumov, Kseniia Bazilevich, and Irina Trofymova

National Aerospace University, Kharkiv, Ukraine
evgeniimenyailov@gmail.com, krivtsovpro@gmail.com, ugryumov.mykhaylo52@gmail.com, ksenia.bazilevich@gmail.com, i.trofymova@khai.edu

Abstract. In the process of research, the authors described a computational method for synthesizing solutions to problems of multi-criteria optimization (based on the use of parallel computing) and decision-making with a priori data uncertainty, based on a memetic algorithm that implements the joint use of an evolutionary method with parameters changing from generation to generation: coding, fitness and relaxation functions, Decremental Neighborhood Method, and the random path generation method.

Keywords: Stochastic programming · Computational mathematics · Numerical analysis and programming (parallel computing) · Memetic algorithm

1 Introduction

When mass production of objects of new technology is required to take into account a consistent set of design parameters of system elements, as well as issues of strength, which together is the basis for the formation of technological processes of manufacturing. Therefore, one of the current problems in creating new technology objects is the problem of reducing the cost of fine-tuning and the operation of systems and processes. The solution of this technical problem is possible by introducing into practice the methods of robust optimal design and intelligent diagnostics of systems and processes [1–6].

There are only a few software systems that provide robust optimal design capabilities in the world. Let's look closer on them.

Dakota optimization toolkit [7] provides a flexible and extensible interface between models and iterative analysis. Methods Dakota contains optimization algorithms using gradient and non-gradient methods, least squares methods; and sensitivity analysis/dispersion analysis with experimental design and parameter research methods. These capabilities can be used on their own or as components in advanced strategies, such as surrogate optimization, mixed integer non-linear programming, or optimization under uncertainty. Using object-oriented design to implement the abstractions of key components needed for iterative system analysis. Dakota provides a flexible and extensible problem solving environment for designing and performing analysis of computational models on high-performance computers. Dakota provides the user with an overview of the features and procedures for using the software, as well as various research examples.

M. Nechyporuk et al. (Eds.): *Integrated Computer Technologies in Mechanical Engineering*, AISC 1113, pp. 514–522, 2020.
https://doi.org/10.1007/978-3-030-37618-5_44

Ipopt (Interior Point OPTimizer) [8] is a software package for large-scale non-linear optimization. Designed to search for (local) solutions to problems of mathematical optimization of the form. Ipopt is written in C++ and released as open source code under the Eclipse Public License (EPL). The package is designed to solve optimization problems arising in a number of important engineering, financial, scientific and medical applications, ranging from the optimal control of industrial processes and the design of digital circuits, also to optimize in the design of complex technical systems. From the identification of parameters in systems biology to cancer treatment planning, etc.

Sherpa [9] is a free open source hyperparameter optimization library for machine learning models. It is intended for problems with computationally expensive estimates of iterative functions, such as setting up hyper parameters of deep neural networks. With Sherpa, you can quickly optimize hyperparameters using various powerful and interchangeable algorithms. In addition, the infrastructure makes it easy to implement custom algorithms.

The disadvantages of the considered libraries are the following: the lack of self-adaptation in the process of work, their low efficiency at the final stage of optimization. The large influence on the efficiency of the algorithms implementing these methods shows the choice of free parameters used when setting up the algorithms.

Thus, there is a need to improve the existing mathematical models and the algorithms implementing them in multi-criteria optimization and decision-making with a priori data uncertainty. The purpose of this study is to develop a model and method for synthesizing solutions of multicriteria stochastic optimization problems with mixed conditions (MV problems) using parallel computations.

Examples of the implementation of the proposed method are presented for solving test problems in deterministic and stochastic formulations.

2 Setting a Research Problem

Let X^0 be the vector of random variables of dimension M (model parameters, control variables, state variables), F^0 be the vector of random variables of dimension I (measurement data, objective functions). F^0 values can be found using the original mathematical model (IMM) of the object of study, presented in the form $F^0 = F(X^0)$, where F is a vector function.

We define the projections of X^0 and F^0 as random variables with the normal distribution law, specifying their mathematical expectations, standard deviations and correlation matrices. The given input data allow us to proceed to the representation of X^0 and F^0 as systems of several random variables with a multidimensional normal distribution law.

In accordance with the concept of power averages A.N. Kolmogorov, we will use the criteria for testing the hypothesis about the equality of the distribution centers for representative samples from two multidimensional general sets of t - Student statistics, and the hypothesis about the equality of covariance matrices—a multidimensional V.I. Romanovsky Ro:

$$t = \sqrt{\frac{n_\alpha}{2} MD^2}, \tag{1}$$

where n_α is a dimension of samples from general populations;

MD - Mahalanobis distance;

$$Ro = \frac{|\chi^2 - k|}{\sqrt{2k}}, k = n_\alpha - 3, \tag{2}$$

where $\chi^2 = \frac{n_\alpha}{N}(\sigma^0)^T R\sigma^0$ is a multidimensional analogue of Pearson's consent criterion;

N – dimension of X^0 (or F^0);
$\sigma^0 = \left\{\frac{\sigma_n}{\sigma_n^*}\right\}, n = 1..N$;
σ_n, σ_n^* are standard deviations of variables $x_n \in X^0$ (*index - desired values);
R is a correlation matrix.

We define the logarithmic likelihood function. The final form of scalar convolution of objective functions for decision-making problems [10] using (1–2) has the form:

$$L(\hat{X}/t_F, Ro_F) = \frac{1}{2}\left(t_F^2 + Ro_F + t_X^2 + Ro_X\right) + C_L.$$

Hereinafter, as a scalar convolution of objective functions in MV problems, assuming that $R_X = R_F = E$, where R_X and R_F are correlation matrices, convolution was used:

$$\begin{aligned} E = {} & \frac{1}{2I}\sum_{i=1}^{I}\left\{ f_{fit}\left[\left(\mu_i(f_i^*)\frac{\Delta_{f_i}}{f_i^*}\right)^2\left(1+\sigma_{f_i}^0\right)^{-2}\right] + \beta_f \cdot f_{fit}\left(\frac{\left|\chi_{f_i}^2 - k\right|}{\sqrt{2k}}\right)\right\} \\ & + \gamma\frac{1}{2M}\sum\nolimits_{m=1}^{M}\left\{ f_{fit}\left[\left(\mu_m(x_m^*)\frac{\Delta_{xm}}{x_m^*}\right)^2\left(1+\sigma_{xm}^0\right)^{-2}\right] + \beta_x \cdot f_{fit}\left(\frac{\left|\chi_{xm}^2 - k\right|}{\sqrt{2k}}\right)\right\} \end{aligned} \tag{3}$$

where

$$\Delta_{f_i} = M_\alpha[f_i] - f_i^*, \chi_{f_i}^2 = n_\alpha \frac{M_\alpha\left[(f_i - M_\alpha[f_i])^2\right]}{\left(\sigma_{f_i}^*\right)^2}; \sigma_{f_i}^\circ = \left\{\frac{\sigma_{f_i}}{\sigma_{f_i}^*}\right\};$$

$$\frac{\left|\chi_{f_i}^2 - k\right|}{\sqrt{2k}} = \frac{n_\alpha}{\sqrt{2(n_\alpha - 3)}}\left|(\sigma_{f_i}^\circ)^2 - 1 + \frac{3}{n_\alpha}\right|;$$

$$\Delta_{xm} = M_\alpha[x_m] - x_m^*, \chi_{xm}^2 = n_\alpha \frac{M_\alpha\left[(x_m - M_\alpha[x_m])^2\right]}{\left(\sigma_m^*\right)^2}; \sigma_{xm}^\circ = \left\{\frac{\sigma_{xm}}{\sigma_{xm}^*}\right\};$$

$$\frac{\left|\chi_{xm}^2 - k\right|}{\sqrt{2k}} = \frac{n_\alpha}{\sqrt{2(n_\alpha - 3)}}\left|(\sigma_{xm}^\circ)^2 - 1 + \frac{3}{n_\alpha}\right|,$$

x_m^*, σ_m^* are values of mathematical expectation and standard deviation of a variable x_m for the prototype;
σ_{xm} is a standard deviation of variable $x_m \in X^0$.
f_i^*, $\sigma_{f_i}^*$ are values of the expectation and standard deviation of the criteria for the choice of solutions f_i for the prototype;
σ_{f_i} is a standard deviation of selection criteria for solutions $f_i \in F^0$.
f_{fit} is a fitness function (FF),
$f_{fit}(d) = 1 - exp(-C \cdot d)$, $C > 0$ (is selected from the condition that the initial value of $E_{av}^{(1)}$ was: $E_{av}^{(1)} < 1$), d - FF argument ($d > 0$);
$\mu_i(f_i^*)$, $\mu_m(x_m^*)$ are membership functions;
γ - regularization parameter ($\gamma = 0$ for identification, $\gamma = 1$ for optimization);
β_f, β_x - robustness parameters.

Thus, the task of evaluating $\hat{X} = \left(M[X^0], \sigma_X^0\right)$ can be reduced to Multicriteria problem of stochastic optimization with mixed conditions (in our case the MV problem), the quasi-solution of which, according to the maximum likelihood principle (M-estimate), is:

$$\hat{X} = arginf_{\hat{X} \in D_X} E\left(\hat{X}/t_F, Ro_F\right), \tag{4}$$

where D_X is a set of correctness, which is determined in the general case by the system of preferences of the decision maker. In this case, it was assumed to be a convex set.

3 Method for Solving the Problem

The “Master - Slave” (sometimes referred to as “client - server”) model is used in this work, since the expenses for calculating fitness function values are evenly distributed over all processors for which the same fitness function is used. Therefore, for n individuals and P (identical) processors we each processor include n/P individuals. The values of the fitness function are calculated by the corresponding (working) processors and sent to one processor (host), which collects all information, processes and transmits it again to the processor. The “Master” processor has information about the values of the fitness function for all individuals and can generate the next generation on this basis. Master-processor performs the central part (core) of the algorithm, while the “rogue work” - the calculation of the values of the fitness function for all individuals is implemented on the processor-workers. For balance, the set of treated individuals of the population is divided into approximately identical subsets.

At the end of each step, synchronization points are placed. When the master processor reaches these points, it goes into standby mode until all the processors fail to complete their tasks, which guarantees the global correctness of the algorithm. In this case, the work between the host processor and the workers is distributed as follows.

Master-processor:

1. performs all input-output operations with the user and the file system, reads the task and records the results;
2. distributes tasks to each working processor;
3. controls the process of finding a solution and, as appropriate, sends relevant activation reports to the working processors; after completion of the task by the processor, the processor-host takes the results and accordingly changes the global data structures (a general list of tasks, the value of the fitness function for individuals, etc.)

Each “slave” accepts the task from the “master” and determines the value of the fitness function for individuals, the resulting result sends to the owner and expects the next task. Since the size of the population is much larger than the number of processors, a good balance is achieved in downloading processors.

When using the model “slave-master”, the final results (finding a solution to this problem) are similar to those obtained on a single processor computer system using a similar algorithm. In this case, the quality of the solution is not lost and, in most cases, it improves somewhat, and the time of its search is significantly reduced. In general, this model allows you to quickly (with minimal modifications) perform a parallelization of GA and, above all, accelerates the process of finding a solution. This model is not difficult to implement on a local network using socket technology.

Algorithm:

1. Randomly create a population
2. Perform in parallel: Assess the fitness function for each person
3. Selection of the best persons
4. Execution of genetic crossover operators and mutations
5. Formation of the intermediate population (Parallel to all individuals)

6. Perform in parallel: Assess the fitness function for each person
7. Entering the best new people
8. Removing the worst old people
9. Formation of a new population
10. Criterion of stoppage of the algorithm is executed? If, no, then we return to item number 3
11. Finding the best person in the final population

In this paper we use a real crossover operator simulating a binary one [11].

As the real mutation operator, the operator of the nonuniform mutation of Mikhalevich [11] belonging to the class of non-stationary mutators is used. Which at the initial epochs of GA appears to be close to the operator of a random mutation, and at the final epochs, produces mutations ensuring the closeness of the values of the quantities $x_m^{(t)}$ and $x_m^{(t+1)}$.

After the crossover operations and mutations are performed, the most suitable person is selected (in the case of finding solutions to multi-criteria problems of parametric optimization (multicriteria decision-making) - a person with the most appropriate scalar convolution of decision selection criteria), which is also placed in a set of individuals for the next epoch of the algorithm.

Additionally, when creating a new population, elite selection was also used.

One of the means of increasing the convergence rate of GA is, as is known, clustering. In order to increase the rate of convergence and the accuracy of finding an extremum, the Decremental Neighborhood Method, implementing the ideas of clusterization, was developed.

4 Experiments and Results of the Modeling

The web application developed should be developed on the principle of multi-level client-server software architecture. "Client-server" is an architecture in which network load is distributed between clients -user services and servers-providers of services. The components interact with each other through a computer network and can be both software and different physical devices. The scheme of client-server architecture is shown in the Fig. 1.

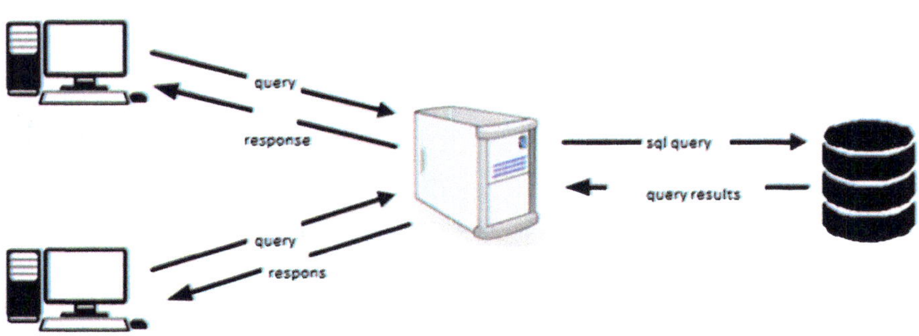

Fig. 1. Client-server software architecture scheme

Advantages of client-server architecture:

- simplifies the maintenance of the computer system by allocating the functions of the computer system between several independent computers on the network.

Disadvantages of client-server Architecture:

- the server failure can make the whole system unusable.

To develop the client architecture, a typical model of the MVC architecture is used. To implement the chosen architectural model, AngularJS framework is used. The architecture of the client part of the software is shown in Fig. 2.

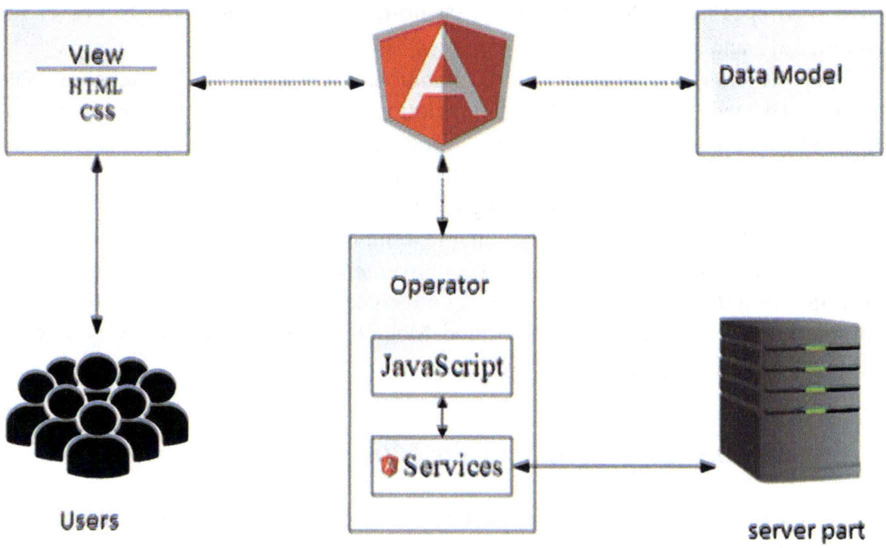

Fig. 2. Client-side software architecture scheme

To develop the architecture of the server part, a typical multi-level architecture model [12]. Umbraco framework is used to implement the chosen architecture model. The architecture of the server part of the software is shown in Fig. 3.

Fig. 3. Server-side software architecture scheme

Rustrigin's function is chosen for testing. The function is unconfined, which is used to test the effectiveness of optimization algorithms, a typical example of a nonlinear multimodal function. Finding a minimum of this function is a rather difficult task because of the large search area and a large number of local minima.

$$F(x) = 10 * N + \sum_{i=1}^{N} x_i^2 - 10\cos(2\pi x_i),$$
$$where\ \mathrm{A} = 10\ and\ \mathrm{x}_i \in [-5.12 \quad 5.12].\ Global\ minimum\ at\ the\ point\ x = 10,$$
$$where\, f(x) = 0$$

The results of the algorithm are shown in the Table 1.

Table 1. Comparison algorithm execution on different computer architectures

Number of individuals, N	Intel Core I7 HQ 2.8 Ghz, working time, sec	Intel Pentium 4 2.16 Ghz, working time, sec
10	1,1	3
30	3,3	5,2
50	5,1	7,8
200	21,3	25,9
300	33,1	39,2
500	54,9	65,8

5 Conclusion

The paper analyzes software solutions in the field of robust optimal design. A computational method for synthesizing solutions to multi-criteria optimization problems (based on using parallel computations) and making decisions with a priori data uncertainty is described. It is based on a memetic algorithm that implements the joint use of an evolutionary method with parameters varying from epoch to epoch: relaxation, as well as the method of narrowing neighborhoods and the randomized method of laying paths. The analysis of the algorithm on the test functions.

References

1. Tronchuk, A., Ugryumova, K.: Mathematical models and evolutionary method for solving stochastic optimization problems. J. Kharkiv Nat. Univ. Ser. Math. Model. Inf. Technol. Autom. Control Syst. **19**(1015), 292–305 (2012)
2. Ugryumov, M., Strelets, V., Tronchuk, A., Ugryumova, K., et al.: System improvement of elements of complex technical systems based on the concept of inverse problems. Nat. Aerosp. Univ. Kharkiv (2013). (Системное совершенствование элементов сложных технических систем на основе концепции обратных задач [Текст]: монография/ В.Е. Стрелец, А.А.Трончук, Е.М.Угрюмова и др.; под общ. ред. М. Л. Угрюмова. – Х.: Нац. аэрокосм. ун-т им. Н. Е. Жуковского « Харьк. авиац. ин-т » , 2013. – 148с. ISBN 978-966-662-312-9

3. Mazorchuk, M., Parfeniuk, Y., Dobriak, V., Meniailov, I., Chumachenko, D.: Stochastic modelling of cash flow for personal insurance fund using the cloud data storage. Int. J. Comput. **3**(17), 153–162 (2018)
4. Ugryumov, M., Tronchuk, A., Afanasjevska, V., Myenyaylov, A.: Gas turbine engine elements systematic improvement on the base of inverse problem concept by stochastic optimization methods. In: Abstracts Book and CD–ROM Proceedings of the 20-th ISABE Conference, Gothenburg, pp. 673–681 (2011)
5. Ugryumov, M., Afanasjevska, V., Tronchuk, A., Myenyaylov, A.: Stochastic optimization models and method in the turbomachines system improvement problem. In: Program and CD–ROM Proceedings of ASME-JSME-KSME Joint Fluids Engineering Conference (AJK2011-FED), Hamamatsu, pp. 755–761 (2011)
6. Ugryumova, K., Chernysh, S., Meniailov, I., Ugryumov, M.: A method synthesis of selection function scalar convolutions for the multi-objective decision-making problems with a priori uncertain data. J. Kharkiv Nat. Univ. Ser.: Math. Model. Inf. Technol. Autom. Control Syst. **27**, 172–180 (2015)
7. Dakota. https://dakota.sandia.gov/
8. Ipopt (Interior Point OPTimizer). https://projects.coin-or.org/Ipopt
9. SHERPA: A Python Hyperparameter Optimization Library. https://parameter-sherpa.readthedocs.io/en/latest/#
10. Meniailov, I., Khustochka, O., Ugryumova, K., Chernysh, S., Yepifanov, S., Ugryumov, M.: Mathematical models and methods of effective estimation in multi-objective optimization problems under uncertainties. In: Advances in Structural and Multidisciplinary Optimization: Proceedings of the 12th World Congress of Structural and Multidisciplinary Optimization (WCSMO12) By Axel Schumacher, pp. 411–427. Springer, Braunschweig (2018)
11. Karpenko, A.: Modern algorithms of search optimization. Algorithms inspired by nature. The Bauman University Publishing House, Moscow (2014). (А. П. Карпенко. Современные алгоритмы поисковой оптимизации. Алгоритмы, вдохновленные природой. Москва: Издательство МГТУ им. Н. Э. Баумана, 2014)
12. Ford, B., Ruebbelke, L.: AngularJS in Action. University of Michigan, Manning (2014)

Possibilities of Position Determination

Olha Pohudina(✉), Dmitriy Kritskiy, A. V. Karatanov, and A. N. Bykov

National Aerospace University «KhAI», Kharkiv, Ukraine
o.pogudina@gmail.com, d.krickiy@khai.edu

Abstract. The article presents an implementation of methods for obtaining the form and appearance of real objects. The analysis of existing approaches, as well as recognition algorithms. The best methods were selected for determining an object in three-dimensional space. Reviewed by method of object recognition to determine the distance from the camera to the marker and methods for determining the position of the object in space. Algorithm for positioning an object in space is developed with the using of the OpenCV library functions. As a result, calculations of relative coordinates of the found object are carried out. Information selection rules based on information are suggested.

Keywords: OpenCV library · Point cloud · Coordinate systems · Haar filter · Object recognition · Position determination

1 Possibilities of Position Determination Methods in Three-Dimensional Space

Machine learning is often used in 3D reconstruction tasks, i.e. in the processes of obtaining the shape and appearance of real objects. The process can be performed by passive or active methods [1, 2]. The active method involves interaction with the object being restored, mechanically or radiometrically. Passive 3D-reconstruction methods do not affect the object being repaired, but only use a sensor to measure the emanation reflected or emitted by the surface of the object to obtain its three-dimensional shape [3, 4]. Usually, camera photoreceptors sensitive to the visible range are used as a sensor. The input to the processing algorithm is either a set of several images (one, two or more) or a video stream. In this case, it is the reconstruction based on the images. In this subsection, three tasks of 3D reconstruction will be considered: visualizing a cloud of points using two cameras, using stereo vision methods for determining the position of an object in space, and the method of object recognition to determine the distance from the camera to the marker [5, 6].

1.1 Development of a Point Cloud Visualization Subsystem in Three-Dimensional Space

The task of recreating a cloud of points is related to the position of the cameras in space and is always associated with several coordinate systems. Consider a few examples. Above a cloud of points, which moves in space, two cameras observe. The position of

M. Nechyporuk et al. (Eds.): *Integrated Computer Technologies in Mechanical Engineering*, AISC 1113, pp. 523–537, 2020.
https://doi.org/10.1007/978-3-030-37618-5_45

the cameras is determined by reference to the global coordinate system, and the optical axes are located at a right angle to each other (Fig. 1).

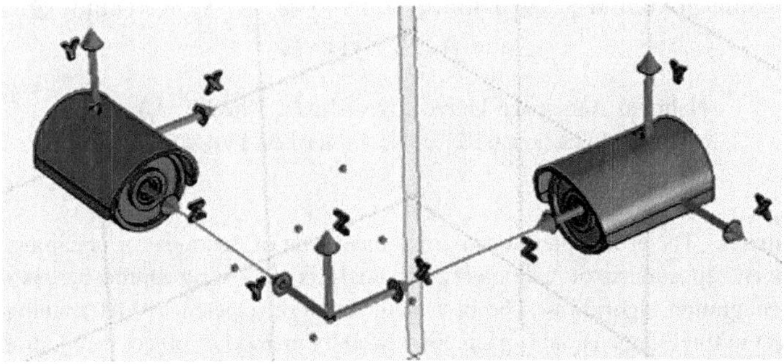

Fig. 1. Statement of the point cloud visualization task

Such values, as the focal length of both cameras, the size of the matrix, the resolution of the obtained images, are given. Test examples were considered, where in the process of movement all points were always in the camera's field of view and their coordinates in three-dimensional space did not coincide as they moved. The points were completely black in a completely white space. The size of the dots is 1 pixel.

The following algorithm was developed (Fig. 2).

Step 1. Loading of two images is performed, with verification of compliance with the input parameters
Step 2. Finding the contours of the points.
Step 3. The following steps of the algorithm are reduced to finding the coordinates of points for the further construction of the scene. Consider the coordinate system. The first system is the device's coordinate system (or screen coordinate system). The origin of coordinates is located in the upper left corner of the screen. The basic unit of measurement is the pixel. Points on the screen are characterized by pairs of coordinates y and x. The x coordinates increase to the right; the y coordinates increase from top to bottom. The second coordinate system is the world or mathematical (camera coordinate system). It is a Cartesian system (X, Y), defined by the programmer, and is independent of the specific graphic device:

$$\mathrm{X} = X_{MAX}\left(\frac{\mathrm{x} - xmin}{xmax - xmin}\right);$$

$$\mathrm{Y} = Y_{MAX}\left[1 - \left(\frac{\mathrm{y} - ymin}{ymax - ymin}\right)\right],$$

where xmin, ymin, xmax, ymax define a rectangular area in the coordinate system of the device, and Xmax, Ymax – in mathematical two-dimensional space.

To find the coordinates of a scene point in three-dimensional space, it is necessary to construct two straight lines passing through the points projected on the matrixes of the cameras and the optical centers of the cameras [7]. The intersection point of the two rays is the desired quantity. The equations of straight lines in space were considered (Fig. 3):

$$\begin{bmatrix} \frac{x - x_{f1}}{x_1 - x_{f1}} = \frac{y - y_{f1}}{y_1 - y_{f1}} = \frac{z - z_{f1}}{z_1 - z_{f1}} \\ \frac{x - x_{f2}}{x_2 - x_{f2}} = \frac{y - y_{f2}}{y_2 - y_{f2}} = \frac{z - z_{f2}}{z_2 - z_{f2}} \end{bmatrix}$$

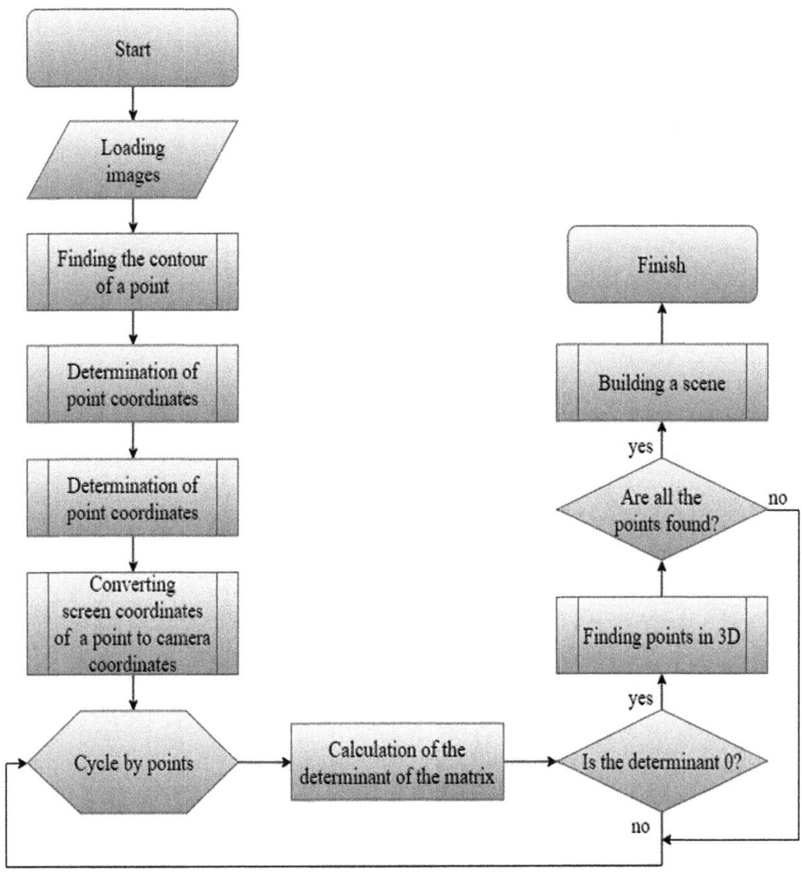

Fig. 2. Algorithm of searching for points on an image

Thus, the condition for the intersection of straight lines was considered:

$$\begin{vmatrix} x_{f2}-x_{f1} & y_{f2}-y_{f1} & z_{f2}-z_{f1} \\ x_1-x_{f1} & y_1-y_{f1} & z_1-z_{f1} \\ x_2-x_{f2} & y_2-y_{f2} & z_2-z_{f2} \end{vmatrix} = 0$$

After which the search was performed, the coordinates x0, y0, z0

$$\begin{cases} x_0=(x_1-x_{f1})t+x_1 & \vee \quad x_0=(x_2-x_{f2})\tau+x_2 \\ y_0=(y_1-y_{f1})t+y_1 & \vee \quad y_0=(y_2-y_{f2})\tau+y_2 \\ z_0=(z_1-z_{f1})t+z_1 & \vee \quad z_0=(z_2-z_{f2})\tau+z_2 \end{cases}$$

It is possible that due to camera defects and measurement errors, these rays will not intersect. Therefore, it is necessary to check nearby points.

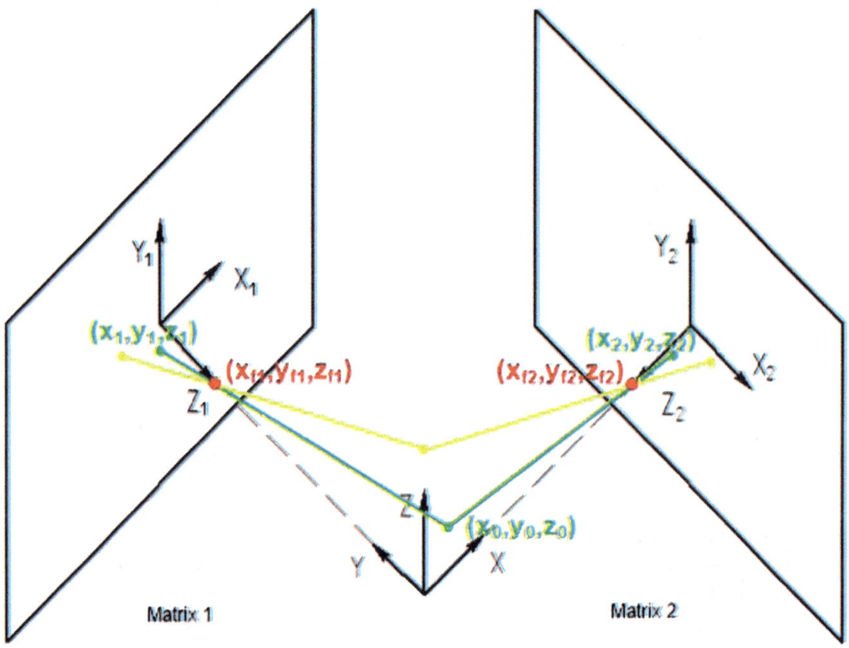

Fig. 3. Ratio of coordinate systems

1.2 Use of Stereovision Methods for Determining the Position of the Object in Space

Consider the stereovision method, which is used to recognize the three-dimensional coordinates of an object, to select its contours and to calculate the distance to the object. Stereo vision is a spatial (three-dimensional) vision, which causes the emergence of a three-dimensional visual image of the observed object with processing the image of the object from different sides in the objective space [8].

To obtain data of the depth of an object on the scene using images obtained when processing a video stream of two cameras, it is necessary to implement an algorithm that includes the following steps:

Step 1. Calibration of cameras or correction of effects of their distortion.

For calibration in OpenCV was used the method of ChessBoards (chessboard). The principle of the method is to use as an object calibration sheet with black and white squares. The advantage of the method is: lack of priority to the horizontal or vertical axis. The OpenCV function:

```
int cvFindChessboardCorners  (const void * image,
        CvSize pattern_size,
        CvPoint2D32f * corners)
```

is used, where image is the image containing the chessboard; `pattern_size` shows how many corners are there in each row and column of the board; corners value is a pointer to an array in which the location of the corners can be written. If the function has successfully found all the corners, then the return value will be the number of corners of the chessboard, in the other case it will be 0. The result of this step is the absence of any defects in the images obtained from the cameras. Next, two images are saved in separate variables of the `IplImage` structure.

Step 2. Rectification of images.

The resulting image variables `ImageA`, `ImageB` may contain a common point on the scene, located at the intersection of the rays `RayA`, `RayB`. Ideally, the rays should intersect in the coordinates of the 3D point, however, due to the error in determining the coordinates of the cameras, the coordinates of the points in the `ImageA`, `ImageB` images, they may be located at the midpoint of the segment of the ray connection (midpoint on the segment).

The goal of the second step is to minimize this segment (shortest segment connecting the rays). Also at this step, the images are aligned so that all the epipolar lines are parallel to the sides of the image (usually horizontal).

Step 3. Getting the depth map and displacement map.

A depth map is an image in which, for each pixel, its distance from the camera is stored instead of a color. Depth map is built on a stereo pair of images. For each point on one image, a pair point is searched for on another image. It is better to look for the pair point on the epipolar line, then the pair point corresponding to it can be found in the same line on the image from the second camera (Fig. 4).

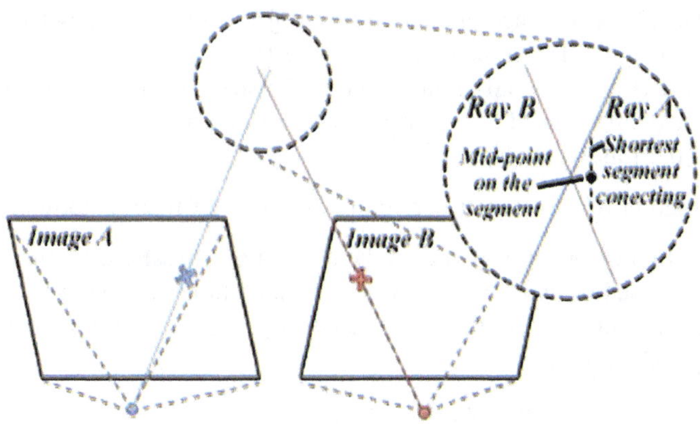

Fig. 4. Rectification of images [14]

For each pixel of the left picture with coordinates (x0, y0), a pixel is searched in the right picture. It is assumed that the pixel in the right picture should have the coordinates (x0 - d, y0), where d is an inconsistency/disparity.

Disparity sets the number of points segmented by planes. The planes are located at the different distances from the Cameras. The search for the corresponding pixel is performed by calculating the maximum of the response function, which can be, for example, the correlation of neighborhoods of pixels. The result is a disparity map.

Step 4. By a pair of corresponding points found in Step 3, you can determine the coordinates of the object in three-dimensional space.

For this we use the function

```
cvReprojectImageTo3D ( const CvArr * disparity,
        CvArr * _3dImage,
        const, CvMat * Q),
```

where

- disparity is the map of differences;
- _3dImage is an array, that contains a three-dimensional image;
- Q is a rebuilding matrix, which is obtained after matrix calibration.

Using the three-dimensional coordinates of the image, the depth is calculated as the distance to the camera plane.

In the course of research into the use of stereo vision methods with the OpenCV library, an algorithm was implemented to obtain data of the depth of an object on the scene using images received from a video stream of two cameras.

As a result, the information subsystem forms the final image on which the object is outlined by the contour. The distance from the camera to the object is displayed near the contour in cm. According to the test results, it was found out that the subsystem is highly sensitive to the displayed objects. They should be in contrast to the background and located near the camera at a distance of no more than 1 m.

1.3 Method of Object Recognition to Determine the Distance from the Camera to the Marker

In the next task, it is necessary to determine the distance from the camera to the marker that needs to be recognized. Then this information should be saved in some storage, which will later be used for other purposes. Unlike previous tasks, a single camera will be used to determine the distance.

To solve the problem, the pattern matching method (Fingerprinting) was taken as the basis – the location of the object is determined with comparing the currently measured power signatures of signals from available access points with signatures that are stored in a pre-established database. This method refers to the direction of visual odometry.

In robotics and computer vision, visual odometry (VO) is a process of determining position and orientation by analyzing sequential images obtained with the camera.

The method of comparison with the sample in the case of determining the distance to the object and its height is based on obtaining an image of the object, then rotating the optical axis of the camera lens horizontally by angle ϕ (with the condition that the object remains in the field of view of the lens after the specified rotation) and obtaining the second image (Fig. 5). Next, the offset di of the i-th object is determined by the difference in the location of the i-th object in the first and second image, as well as the distance Ri to the i-th object in accordance with the expression:

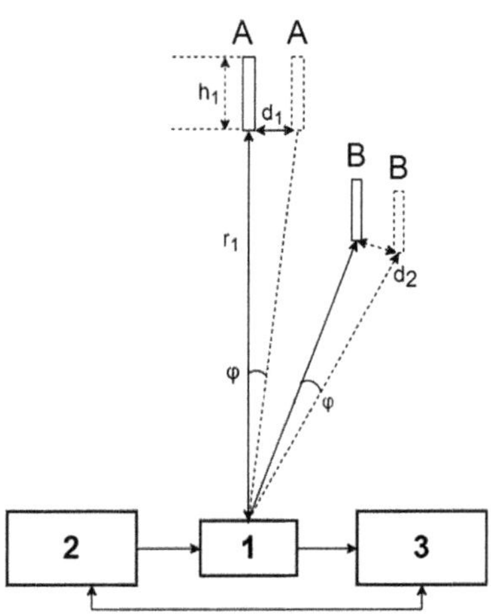

Fig. 5. The method of determining the distance to objects and their heights

$$R_i = d_i / tg\varphi,$$

and the height Hi of the i-th object in accordance with the expression:

$$H_i = R_i \times tg\left(\varphi \frac{h_i}{d_i}\right),$$

where hi is the height of the i-th object in digital photos.

To solve the problem, a red ball will be used as a marker, the parameters of which are entered into the program memory. Then, in automatic mode, the distance from the camera to the marker is determined (using the Fingerprinting method) [9]. This work uses the HP TrueVision HD webcam.

The task is to determine whether a sample is present on the stage, and in its localization. In this case, the sample on the scene can:

1. Have a different scale.
2. Be rotated in the image plane.
3. Be in an arbitrary place of the scene.
4. Be noisy, not fully visible, partially covered by other objects.
5. Have a different brightness and contrast from the sample.
6. Not be at all.

The main steps in the search for objects on the scene:

1. The formation of the section descriptions;
2. Search for all features, using the algorithm to search and describe the singular points of the image SURF (Speeded Up Robust Features);
3. Search for matching pairs of features in two images (if the pair is small, the object is not found);
4. Search for the location of the object: checking the transformation between the original image and the one in which the search is taking place, according to the obtained value of the transformation parameter, the coordinates of the rectangle characterizing the object are located;
5. Highlighting features in circles.

Let us consider in detail the model of searching for features in the image used in step 2. To obtain the features of an image, statistical samples based on local features or methods of extracting special points of the image are most often used. At the same time, methods for removing special image points are implemented in most libraries, such as OPENCV and OPENNI. Identification of objects – algorithms SURF [10] and SIFT. A comparison of the results of using these two algorithms shows the advantage of SURF over SIFT and its higher efficiency.

The SURF algorithm works on the basis of the Hesse matrix, but uses very simple approximations. This algorithm relies on an integral image to reduce computational costs and reduce time. Therefore, it is called the detector "Fast-Hessian". The descriptor describes the distribution of the "Haar-wavelet" within the percentage points. The input of the integral image I (x, y) with the location X = (x, y) is defined as:

$$I(x,y) = \sum_{i=0}^{i \le x} \sum_{j=0}^{j \le y} I(i,j),$$

where I (x, y) is the sum of all points in a rectangular area of the original image I, formed with X = (x, y). Calculate I (x, y) with four applications to calculate the sum of the intensities of any vertical rectangular area.

Detector "Fast-Hessian". The algorithm is based on the Hesse matrix and has high performance and good accuracy. Then the hessian is:

$$H(X,\sigma) = \begin{bmatrix} Lxx(X,\sigma) & Lxy(X,\sigma) \\ Lxy(X,\sigma) & Lyy(X,\sigma) \end{bmatrix},$$

where, Lxx (X, σ) - Gaussian convolution:

$$Lxx(X,\sigma) = \frac{\partial^2}{\partial x^2} g(\sigma).$$

Gaussians are optimal for spatial analysis [11]. However, in practice, Gaussian convolution must be discrete and even with Gaussian filters, smoothing occurs in such way that the resulting images are subdiscrete.

Finding the orientation of a singular point. The SURF algorithm provides the determination of the prevailing orientation of the differences of the singular points for calculating the invariance of their descriptors. SURF initially calculates gradients in adjacent pixels of a particular point using the Haar filter (Fig. 6). The filter size is 4s, where s is the scale of the singular point.

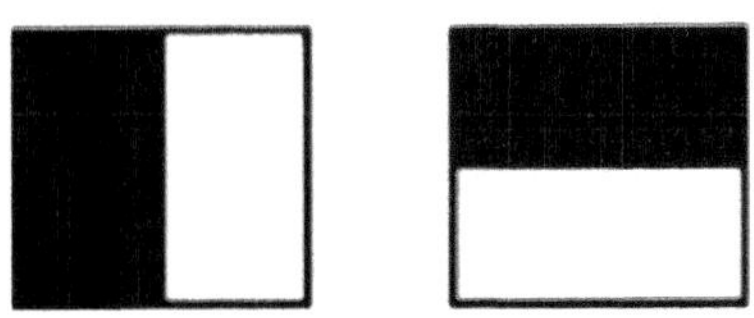

Fig. 6. Example of a Haar filter; black area value: −1; white area: +1

The point value of the brightness difference along the X and Y axis is obtained using the Haar filter. The filter shape is rectangular, the values of the filter parameters are calculated using an integral matrix and require only 6 operations to calculate one filter of arbitrary size [11].

Haar filters use dX and dY values (Fig. 7), which are multiplied by the weight for each point. Weight is a Gaussian value with a center at a singular point and a sigma equal to 2s (where s is the scale of the singular point).

Next, the corner window with a size of π/3 [12], which rotates around the center of coordinates, is used. The algorithm chooses the position of the windows in which the maximum length of common points for which noise is required, is obtained. The algorithm does not require the invariance of rotation descriptors, in which the orientation of the singular points is not calculated but identifies points that rotate by no more than ±15° [12].

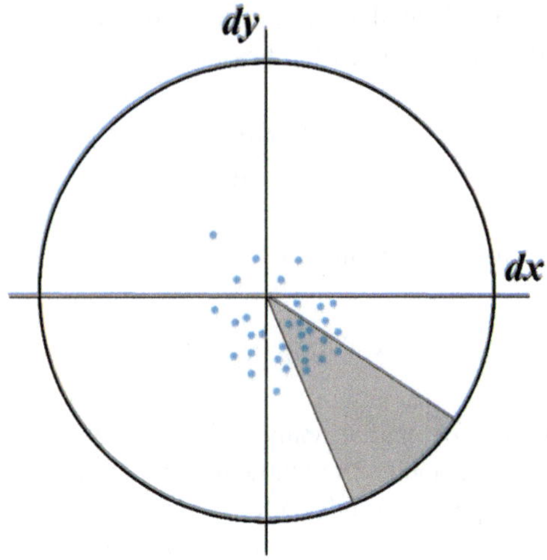

Fig. 7. Gradients of daughters in the dXdY plane

The design of descriptor. A descriptor is an array of 64 (or 128) numbers identifying a particular point. The rectangular area is used to compute a descriptor around a feature point of 20s. The size of the first area is 40 × 40 pixels, it is oriented along the direction calculated for a particular point. Descriptor is a specification of the gradient of 16 quadrants around a singular point (Fig. 8).

The square is divided into 16 subregions, formed by a regular 5 × 5 grid used to find the gradient on grid points using a Haar filter.

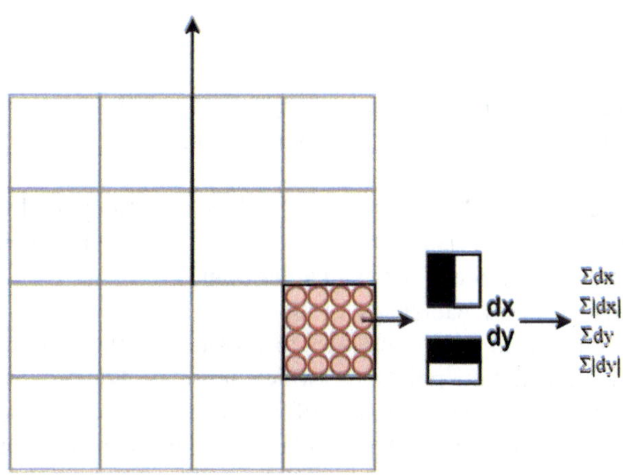

Fig. 8. Illustration of obtaining a gradient of a singular point [12]

It should be noted that when calculating the Haar filter, the image does not rotate, the filter processes the usual coordinates of the image. However, this gives the coordinates of the gradient (dX, dY), rotated by the corresponding angle as the orientation of the square. Thus, 400 Haar filters are used to calculate descriptors of a singular point (in each of the 16 subregions, 25 Haar filters are calculated).

After that, the algorithm calculates four values (Fig. 19), which are the descriptor components:

$$\sum dx \sum |dx|, \sum dy \sum |dy|,$$

where, dx and dy are summed over each region and

$$\sum |dx| \sum dy$$

are absolute values.

These values show the behavior of different parts of the image and will be saved in the vector

$$v = \left(\sum dx \sum |dx|, \sum dy \sum |dy|\right).$$

The SURF descriptor is obtained by concatenating 16 vectors calculated for each subregion when normalizing the corresponding 64-dimensional vector (Fig. 9). Therefore, its Euclidean norm is unitary. Thus, the descriptor is invariant to a change in affinity contrast [13].

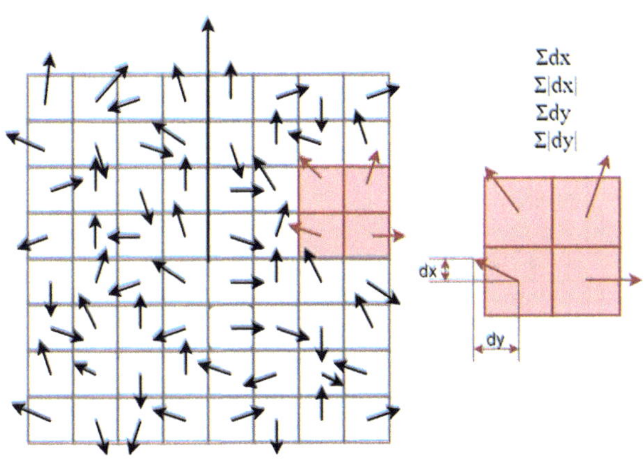

Fig. 9. Illustration of building a descriptor of a singular point [12]

A program was developed in which the following OpenCV classes were used (Fig. 10):

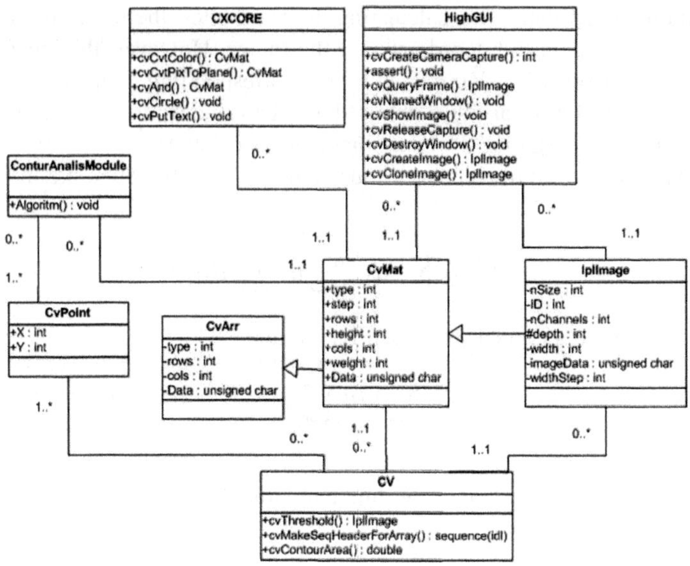

Fig. 10. Used OpenCV library classes

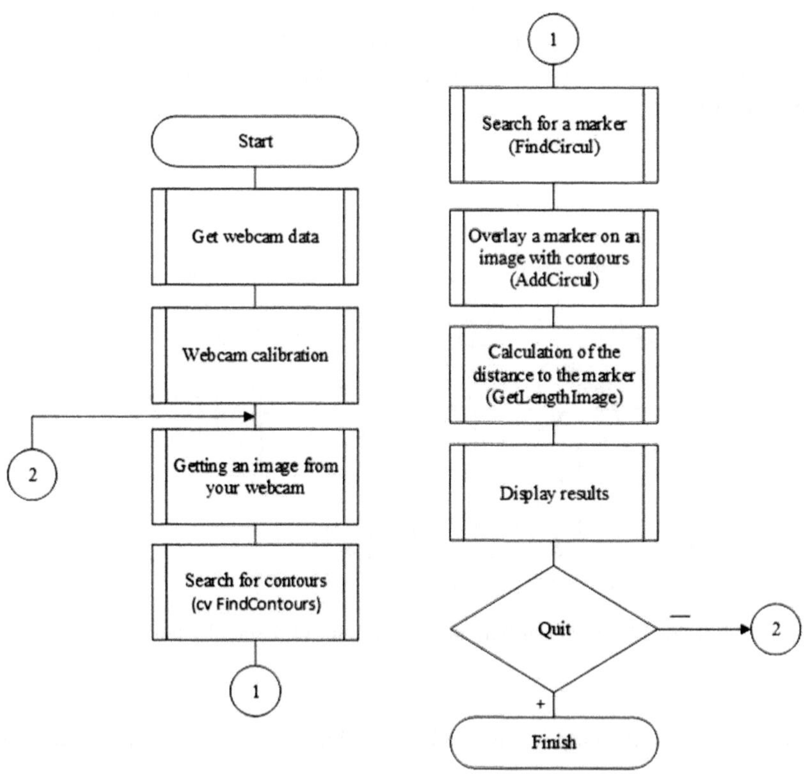

Fig. 11. Algorithm of the program

In Fig. 11 the algorithm of the software product is given. The algorithm presents the sequence of execution of the developed methods aimed at obtaining the distance to the obstacles.

The results of experimental testing.

Testing was carried out with the help of a laptop and a marker – a red object with a rounded shape – a red ball with a diameter of 7 cm (Fig. 12). First, the program is launched and adjusted to the selected marker, and the distance from the ball to the camera is measured in manual mode using a ruler. Then it is compared with the software calculation of the distance to the object (Fig. 12).

As it can be seen from Fig. 12, program is working properly and the error is approximately 1–3 cm.

Fig. 12. Setting up the program on the selected marker

After testing the software, it was determined that the program works with an error of approximately 0.01–0.03 m. In order to improve the accuracy, the appropriate lighting and the best equipment are needed.

The positioning of the object in space was realized using the methods of the OpenCV library. To do this, the marker should be fixed in the program memory (a round red object), after that the program automatically displays the distance from the camera to this marker. In order to fix the distance to an object, it is necessary that the image of the red ball is present on the scene.

2 Conclusions

Described how to develop cloud point visualization in three-dimensional space and determine the distance from the camera to the marker. The Stereo-Visions Method is considered which is used for: the Recognition of object 3D coordinates, highlight its contours and calculate the distance to the object. Algorithm and informational subsystem is developed with the using of the OpenCV library functions. The methods for constructing the depth map and camera calibration have been analyzed. The results that were obtained, characterized by a high sensitivity of the algorithm to recognize objects. Objects should contrast with the background, located near the camera at a distance within 1 m results were published.

The methods of calculating the relative coordinates of the found object are described. The rules of the automated process of selecting the area of interest based on the information received during the processing of the previous frame are proposed.

References

1. Breazeal, C., Brooks, A., Gray, J., Hoffman, G., Kidd, C., Lee, H., Lieberman, J., Lockerd, A., Mulanda, D.: Humanoid robots as cooperative partners for people. Int. J. Hum. Robot. **1** (2004), 1–34 (2004)
2. Bauer, A., Wollherr, D., Buss, M.: Human-robot collaboration: a survey. Int. J. Hum. Rob. **5**, 47–66 (2008)
3. Mitra, S., Acharya, T.: Gesture recognition: a survey. IEEE Trans. Syst. Man Cybern. Part C Appl. Rev. **37**, 311–324 (2007)
4. Parasuraman, R., Sheridan, T.B., Wickens, C.D.: A model for types and levels of human interaction with automation. IEEE Trans. Syst. Man Cybern. Part A Syst. Hum. **30**, 286–297 (2000)
5. Liu, H., Wang, L.: Gesture recognition for human-robot collaboration: a review. Int. J. Ind. Ergon. **68**, 355–367 (2018)
6. Kritsky, D.N., Ovsiannik, V.M., Pogudina, O.K., Shevel, V.V., Druzhinin, E.A.: Model for intercepting targets by the unmanned aerial vehicle. In: Palagin A., Anisimov A., Morozov A., Shkarlet S. (eds.) Mathematical Modeling and Simulation of Systems. MODS 2019. Advances in Intelligent Systems and Computing, vol 1019. Springer, Cham (2020)
7. Kritskiy, D., Alexander, K., Koba, S., Druzhinin, E.: Increasing the reliability of drones due to the use of quaternions in motion. In: Proceedings of 2018 IEEE 9th International Conference on Dependable Systems, Services and Technologies, DESSERT (2018)
8. Kritsky, D.N., Druzhinin, E.A., Pogudina, O.K. Kritskaya, O.S.: A method for assessing the impact of technical risks on the aerospace product development projects. In: Advances in Intelligent Systems and Computing, vol. 871, pp. 504–521. Springer, Heidelberg (2019)
9. Kaehler, A., Bradski, G.: Learning OpenCV 3: Computer Vision in C++ with the OpenCV Library. O'Reilly Media, Inc. (2016)
10. Quack, T., Mönich, U., Thiele, L., Manjunath, B.S.: Cortina: a system for large-scale, content-based web image retrieval. In: Proceedings of the 12th Annual ACM International Conference on Multimedia, pp. 508–511. ACM (2004)
11. Fleyeh, H.: Color detection and segmentation for road and traffic signs. In: IEEE Conference on Cybernetics and Intelligent Systems, 2004, vol. 2, pp. 809–814. IEEE (2004)

12. Tamura, H., Mori, S., Yamawaki, T.: Textural features corresponding to visual perception. IEEE Trans. Syst. Man Cybern. **8**(6), 460–473 (1978)
13. Kefer, M., Kubinger, W.: Evaluation of kinect depth sensor for use in mobile robotics. In: Annals of DAAAM & Proceedings, pp. 147–149 (2011)
14. Baggio, D.L.: Mastering OpenCV with Practical Computer Vision Projects. Packt Publishing Ltd. (2012)

Analysis of Modern Continuous Integration/Deployment Workflows Based on Virtualization Tools and Containerization Techniques

Yurii Vlasov[1](✉), Nadiia Khrystenko[2], and Dmytro Uzun[1]

[1] National Aerospace University "KhAI", Chkalova str. 17, Kharkiv 61070, Ukraine
y.vlasov@student.csn.khai.edu
[2] National Medical University, Nauki ave. 4, Kharkiv 61000, Ukraine

Abstract. Modern virtualization techniques and server architecting solutions are studied. In order to reduce development time and assets cost, there is a great need in full utilization of all calculation resources for both dedicated and virtual servers. Containerization allows shorten environment integration and software deployment time basing on virtualization and tighter integration on high abstract level. The detailed description and analysis of common containerization techniques is given. Comparative analysis of modern solutions shows that the containerization orchestration systems are the best solution which advantages allow optimizing resource utilization and addressing arisen challenges. The features of the joint use of container orchestration and management system together with different cloud providers are given. Comparison of modern CI/CD approaches is given. Environment quality and its configuration for active application deployments make major impact on needed calculation resources quantity. Common environment architecture approaches in context of containerization usage are described. CI/CD solutions allow significantly reduce deployment time and increase reliability.

Keywords: Prototyping tools · Virtualization techniques · Orchestration system · Docker · Podman · Server-side architecture · Virtual/dedicated resource utilization · Development economy · Continuous integration and delivery · GitOps CI/CD methodology · Jenkins · TeamCity · TravisCI · CircleCI · GitLab · On-premise clouds · Mixed clouds

1 Introduction

1.1 Motivation

Actively developing technologies of cloud computing provide easy access to distributed remote resources in a short period of time. The higher level of resource unification allows to reduce count of deployment dependencies and simply run code without setting up runtime environment while plain server-based system architecture needs less expenses for complete manual support of cloud infrastructure and tune of

M. Nechyporuk et al. (Eds.): *Integrated Computer Technologies in Mechanical Engineering*, AISC 1113, pp. 538–549, 2020.
https://doi.org/10.1007/978-3-030-37618-5_46

resources in order to get much more efficient runtime environment. Thus, serverless approach cannot always provide enough input/output operations per time range in order to handle high traffic load to databases, while dedicated servers need much more manual configuration, support, monitoring and audit checks.

In case of short development cycle and actively changing requirements, Agile has been created. Most popular framework from Agile methodology is Scrum. Scrum is an agile process that allows to focus on delivering the business value in the shortest time. It rapidly and repeatedly inspects actual working software, emphasizes accountability, teamwork, and iterative progress toward a well-defined goal. The Scrum Framework usually deals with the fact that the requirements are likely to change or most of the time not known at the start of the project [1].

System engineering emerged as a method to speed the process of building, testing and releasing software products providing additional layer between software engineers and system administrators. Reliable and flexible server infrastructure which can handle high load is the main task for system engineers who manage deployment of business products to the production environment and so involved in development of software integration operations architecture. Main target of modern business solutions is not to make stable product but create minimum reliable solution which can be sold and make a profit in shortest amount of time for involving investors and lead to sales explosive growth. One of the main responsibilities of the system engineer is creation of several stable environments for both development, testing, production traffic with basic continuous integration (CI) which would allow to simplify basic deployment tasks but at the same time there is no objective in complete continuous integration and deployment integration (CI/CD) for all layers of system architecture. Complete CI\CD configuration for all software product sub solutions flows is rarely required and paid with owner of project. Modern CI/CD pipelines are commonly use containerization because it allows to manage server infrastructure on higher abstraction level with guarantee of applications immutability in transitions between environments. Current trends of developing web applications, IoT systems say to use virtualization and separate containers for each layer of the system architecture. It is usually made for isolating layers between themselves and providing more secure and flexible environment for development. But the main advantage of such approach is the possibility to easily scale up services and even emulate (prototyping) some parts of the whole system with help of special containers [2].

For stable delivery of software products to target environment the orchestration system is required. Specially in case of need of making automated monitoring and management operations. Container orchestration and management system consists of composition of tools for stitching of software and hardware components together for providing of automated workflows which decrease system reaction time on predefined malfunction cases. The orchestration systems allow to control the lifecycle of a set of containers and provide important scaling tools which work in Infrastructure as Code (IaC) methodology [2].

Hence, using of virtualization techniques based on containerization of software lead to reducing infrastructure costs, increase resources utilization per server and minimize impact (downtime) of layers malfunction on main functionality thanks to ability of setting up automated workflows and container hypervision. Developers together with system engineers analyze development tools and compose one in order to simplify further software containerization. Virtualization allow configuration of test environment for execution of unit tests after build of the container. There is no need to run tests before each deployment of new container basing on the built image, but only once due to guaranteed *immutability*. Security point is that recreation of container completely reset its filesystem layers to state of source image removing all injections and new vulnerabilities. However, deployment of high load services and controlling of the cluster depends on stable and flexible orchestration system which allows building server infrastructure both on-premise, cloud or mixed types [2].

1.2 State-of-the-Art

Software can be containerized using special build tools which depends on chosen *container runtime engine environment*. Commonly, containerization engine provided embedded build mechanisms. Built software called an *image* while run instance basing on previously built image is named a *container*.

Mechanisms of containerization based on *control groups* and *namespaces* which minimally acceptable implementation was released in Linux kernel version 3.8 at February of 2013 [3]. That allows to isolate processes from host operation system (OS) in the way when process owns only dedicated resources and cannot impact on processes from other namespaces. All container processes can be easily spotted and monitored from host OS. Namespaces also allowed making processes hierarchy with detailed management of network, ipc and other resource on each level.

There is de facto standard solution for making containers on both Linux, MacOS and Windows based OS – *Docker container engine*. Docker is a platform for creating virtual containers which can even provide addressing every application across the hybrid cloud. Docker makes powerful tools for application creation and orchestration, accessible to everyone [4]. Docker provides user with rich functionality of making and mounting data volumes, creation of overlay networks, composing and publishing software images which can be pulled from public or private registries and deployed in several steps. Building Docker container is possible with *Dockerfiles* which describes future container *layers* basing on your product. Docker architecture is described on Fig. 1.

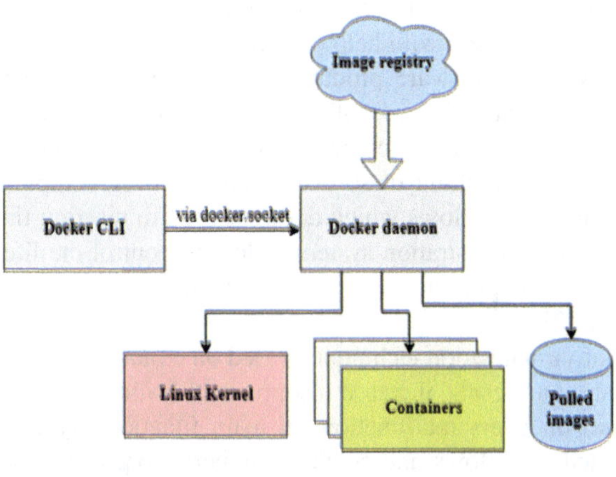

Fig. 1. High level abstraction diagram describing Docker daemon architecture.

Docker running daemon process on host OS which use to do all work. It has several cons:

- Single daemon process is single point of failure.
- Daemon process owns all the child processes of spawned containers. In case of daemon failure container subprocesses tracking will be complicated and merely impossible.
- All Docker daemon commands can be only executed by user with root privileges [5].

Considering Docker experience Red Hat together with open source community is developing *libpod* and *podman*. Libpod provides a library for applications looking to use the Container Pod concept, popularized by Kubernetes. Libpod also contains the Pod Manager tool which manages pods, containers, container images, and container volumes [6]. This project has come instead of Docker daemon to recent stable releases of CentOS and RedHat Enterprise Linux distributions, providing *rootless* functionality for spawning Docker-compatible containers with CLI tool which has no single point of failure as Docker has (see Fig. 2).

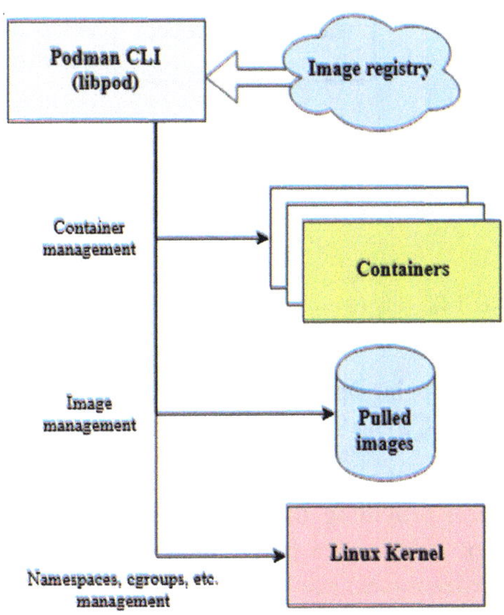

Fig. 2. High level abstraction diagram describing podman architecture.

At the present moment, libpod is on stage of active development. Community emphasizes backward compatibility with Docker CLI and docker-compose IaC approach, while compatibility is not full enough and many features were not yet implemented in libpod framework or work in different way as it was in case of Docker ecosystem. Hence, libpod would supplant Docker based containerization approach in next few years, but now podman-based deployments should be considered as less stable.

1.3 Continuous Integration, Delivery and Deployment

In order to reduce the time of working CI/CD pipelines, engineer have to find most proper solution for delivering product to target infrastructure. Moreover, in case of deploying to the on-premise or mixed cloud infrastructures there are more manual setting points to manage. In order to develop durable distributed system, it is needed to analyze main software product layers and aims which it trying to achieve.

Results of the analysis allow formalizing CI\CD system process structure by means of high-level IDEF0 diagram of product development which is designed and presented on Fig. 3.

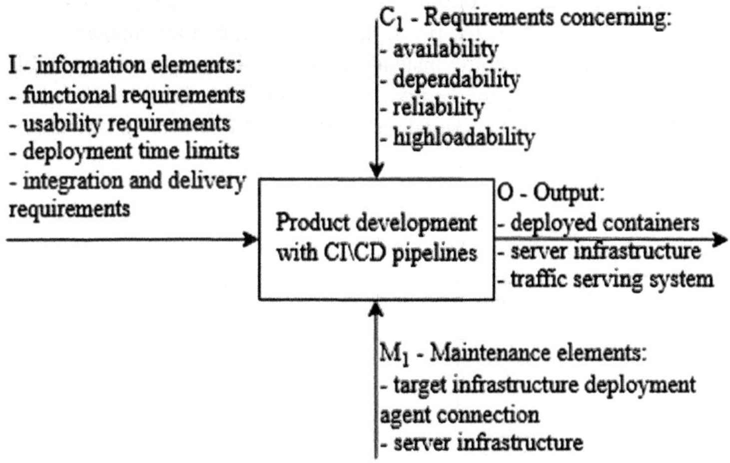

Fig. 3. High-level IDEF0 diagram of development process with CI/CD workflows.

There are few most popular approaches in building of server infrastructure for product placement. Further CI/CD setup is highly dependable on chosen cloud type:

- *On-premise.* Advantages: physical access 24/7/365 to all hardware, software and data storage resources; providing of confidentiality of the sensitive data. Disadvantages: necessity of high-skilled professionals for managing of hardware, software and data storage resources; need to ensure stable Internet connection; high cost of hardware, software and data storage resources as well as premises rent and electricity consumption.
- *Cloud deployment* relieves of the need to monitor and allows to spent less in case of need of fewer resources. So that one can work with the cloud on high abstraction level and see only the OS to deploy. However, in this case the customer depends on unique features and services provided by cloud provider.
- *Mixed* way allows to combine advantages and pay less but still lives opportunity to move some services to personal work stations. Reveals high severity tasks of settings stable and network connection between cloud and on-premise parts of mixed infrastructure [2].

Server-side architecture should meet cost-efficiency criteria for particular project on the one hand, and provide secure, flexible and scalable solution with opportunities of easy CI/CD workflows integration on the other hand. The most cost-effective approach for deploying will be on-premise architecture while most work-effective will be cloud solution. Result architecture highly depends on dedicated amount of time for system engineer, on expected traffic and geo-position of most of target auditory.

Thus, next steps of the research are dedicated to comparison of CI/CD techniques considering chosen server architecture to be either on-premise or mixed cloud because of absent vendor lock-in of technologies stack and lower price because of lower abstraction level.

1.4 Structure of the Paper

The structure of the paper is as follows. Section 2.1 presents description of main CI/CD tasks and an overview of server-side requirements in case of setting up of CI/CD pipelines using containerization tools and virtualization techniques. Section 2.2 presents the analysis and overview of the GitOps methodology in management, versioning and delivery of software products to the target server infrastructure. Section 2.3 describes common solutions using configured on-premise CI/CD servers like Jenkins [7] or TeamCity [8]. Section 2.4 is comparison between GitOps methodology and dedicated CI/CD server approach. Section 3 concludes, describes future research plans, describe acknowledgments and appreciations.

2 General Approaches and Methodologies in Setting up and Configuration of CI/CD Pipelines

2.1 Cloud Infrastructure Saving in Comparison to On-Premise Solutions

Migration to cloud from dedicated server infrastructure can significantly decrease total costs. By using the cloud solution, the airport estimates that it is saving 60%–70% on costs compared to their previous hosting solution. In case of usage on-demand instances on AWS without paying for upfront reserving of instances saving can rise up to 80%. Price general server setup for medium scale web application with unpredictable load can rise up $5152 per month while the same setup on AWS will cost $1051 [9].

High level of resource unification makes management easier and CI/CD pipelines based on cloud provider's mechanisms are cheaper than manually supported. For CodeDeploy on EC2/Lambda: There is no additional charge for code deployments to Amazon EC2 or AWS Lambda through AWS CodeDeploy [10]. Cloud providers develop tools with native support of deployment to their resources which makes pipelines configuration much easier and simplify support of CI/CD workflows.

2.2 Requirements to Server-Side and CI/CD System

Server infrastructure must provide enough resources for hosting all primary and secondary services, store databases with main data, handle gathering resource consumption and productivity metrics. In case of CI/CD workflows server infrastructure must satisfy basic requirements, which are described on Fig. 4 and in text below:

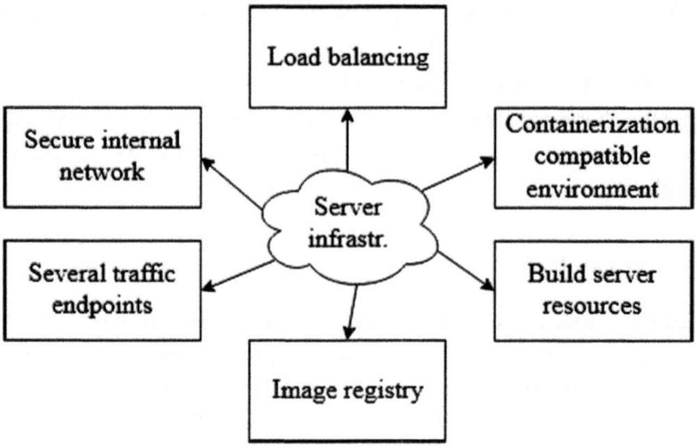

Fig. 4. Server infrastructure requirements with CI/CD workflows.

- *Several traffic serving endpoints.* Despite all efforts, the fact remains that even the most durable of NAS devices have average work period from 3 to 5 years [11]. In case of single endpoint for launching software delivery pipelines, system can lose not only user data but also make impossible fast restoration and redeployment of fallen services to other nodes (servers).
- *Secure internal networking.* In case of on-premise or mixed cloud solution setup, there is a need in secure networking between all servers in order to prevent Man-in-the-Middle attacks and interception of data no matter either it depends to user traffic serving or CI/CD pipelines execution flows.
- *Load balancing.* In order to spread working load, there is a need to exactly determine target auditory in view of geo replication strategy and spread traffic endpoints all over the world. Such approach allows distribute work load evenly and more accurately predict services consumption in case of planning of deployment targets. Also, very important to envisage possible failure of one or several nodes and ensure that health checks mechanisms will prevent traffic routing to offline servers – in order not to lose user requests.
- *Containerization compatible environment.* Nowadays, Docker containerization engine is compatible with both Linux, Windows and MacOS operation systems, but the most stable solution is using of CentOS or RedHat Enterprise Linux distributions on either virtual machines or dedicated servers. Docker was natively developed using namespaces and control groups of Linux kernel and in case of Windows there is undershelf virtual machine used for running containers which is less stable solution of plain server Linux distribution setup usage.
- *Build server resources.* There are many free solutions, runners, which can handle build process and result with built image to user. But in case of need in securing

source code flows or speeding up of building processes – best solution will be hosting own build server which would be configured for all CI/CD features according to current project needs.

- *Image registry.* Mirroring of all dependent images from all third-party registries will guarantee reduce probability of complete pipelines inactivity in case of dependencies failure. On the other hand, hosting personal registry will need additional configuration and management of used disk drive space, cleaning of old images and replication/backups of registry's data.

CI/CD workflows must provide stable, secure and fast way of building and deployment project's software in automated or semi-automatic mode (depending on security setup of pipelines) requirements described on Fig. 5 and text below:

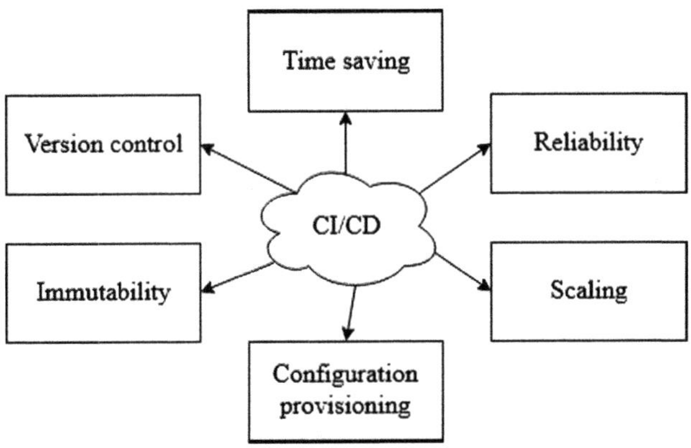

Fig. 5. CI/CD workflows requirements.

- *Time saving.* All build process must be executed in automated mode triggered either by some event or manual developer request. Build stage is most resource consumptive one, hence there must be enough resources for running builds.
- *Version control.* The Semantic Versioning specification is authored by Tom Preston-Werner, inventor of Gravatar and cofounder of GitHub [12]. Versioning of images must allow anytime downgrade or upgrade to desired version which was build earlier. Also, existent CI/CD pipelines should obviously describe what version is executed on each environment starting from development one and ending with all production environments.
- *Immutability.* Docker containers are in fact immutable. This means that a running container never changes because in case of need to be updated, the best practice is to create a new container with the updated version of the application and delete the old one. Once build application from source code commit would not match build from the same commit but which was made after some time. Some floating dependencies may change result versions of dependencies which will be installed on stage of build. Hence application rebuilds will lead to probable failures in case of missed floating dependencies.

- *Configuration provisioning.* Often, configuration is gathered by application from container environment variables, while there are many other ways to insert configuration values. In any case, CI/CD configuration should allow easy and massive change of result deployment configuration with splitting among environment because deployment of the same immutable image to different environments may require different configuration of running container to match database connection strings, data storage paths, etc. CI/CD could not provide configuration mechanisms only in case of usage of separated configuration server as HashiCorp Vault [13].
- *Scaling.* CI/CD pipelines must allow using of both simultaneous and sequential deployment strategies in case of requirements from system administrators or developers. Simultaneous deployment save time between commit of source code to source code version control system (SVC) and deployment of build image to the target environment, while sequential approach allows redirect real traffic to new version of software step-b-step and rollback in case of failure without losing user data.
- *Reliability.* Deployment pipelines must be informative and stable. Once triggered pipeline must not break current setup of application and deploy the new version with minimum or absent down time. Hence, CI/CD system as monitoring, backups and audit systems must be secured and most stable among all project's subsystems.

2.3 Git Operations as Most Native CI/CD Approach

Development nowadays is impossible without SVC. Git is most popular SVC nowadays [14]. SVC can be used for storing both source code of the project and sources of CI/CD workflows in case of using of Infrastructure as a Code (IaC) approach. Main points for using Git repository as base for all infrastructure operations are follows:

- Simultaneous versioning of project code and CI/CD code.
- Git as the single source of truth of a system.
- Git as the single place for operations (CRUD) for all environments.
- All changes made to Git repository are observable and verifiable.

All changes which was made to Git repository are signed by developers, hence build process can be triggered basing on events from SVC. Most common approaches are:

- Trigger build and/or deployment on commit to predefined branch.
- Trigger build and/or deployment on tag assignment with version number.
- Trigger build and/or deployment on merge request approval from feature to main branch.

Common development cycle is described on Fig. 6.

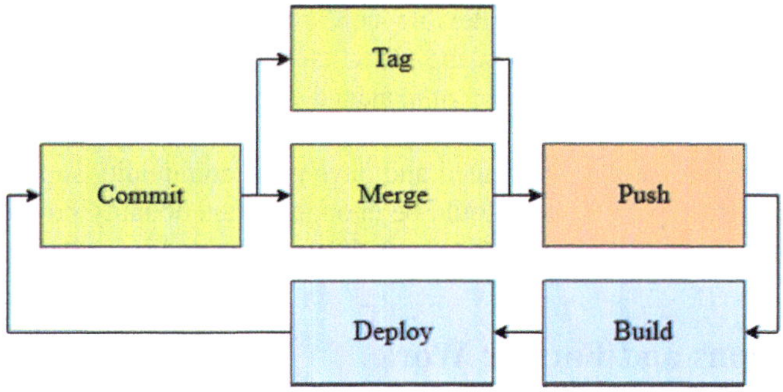

Fig. 6. General Git development cycle.

2.4 CI/CD Cloud and Dedicated Solutions

There are two ways of building CI/CD pipelines: using of cloud native solutions like GitLab-CI [15], CircleCI [16], TravisCI [17] etc. or setting up dedicated CI/CD server for project needs. Dedicated CI/CD server approach means installing system like Jenkins [7] or TeamCity [8] on separated server. There are currently more than 12.500 projects supported by Jenkins [18]. Comparison of both solutions are described in Table 1.

Table 1. Comparison of CI/CD cloud and dedicated server solutions.

Criteria	Cloud pipelines	Dedicated CI/CD server
Need build server resources on target infrastructure	Most of cloud CI/CD providers support building of images on their side. Build server on the side of the client is optional	Needs dedicated VM or server for server itself and additional VMs in case of need in connecting of build agents
Support IaC approach	Natively supported describing of pipelines in YAML/JSON formats by most of providers	Pure support of IaC approach. Used in Jenkins Gradle pipelines. Not available in TeamCity
Security, updates, maintenance	No need to support build server. In case of usage of build server on client side – need to update runner agent. Security of sourse code and builds artifacts are guaranteed only by CI/CD cloud provider	Build server updates must be controlled manually or scheduled. Security of build environment is guaranteed by user itself and fully managed
Documentation and community support	Fully documented pipelines and interfaces with many examples	Many plugins, good documentation, a lot of solved issues because of long existence time
Flexibility	No extensions and plugins, cannot be modified by user. Limited to provided functionality	Completely configurable solution with many community plugins and extensions

Dedicated setup allows more flexible configuration but need maintenance and additional calculation resources. On the other hand, cloud solutions support IaC but builds and image storing are made on unmanaged environments which security cannot be proofed.

Both solutions are well documented and have great community support. In case of setting up standard pipelines there could be enough of functionality provided by cloud solutions, while advanced tuning more natively implemented in dedicated servers.

3 Conclusions and Further Work

Using of containerized software together with flexible and stable CI/CD workflows helps increase resource utilization up to 50–85% and decrease deployment time in 2–5 times so developers can deploy builds more often [9].

On one hand, using of containerization make services easier to scale and distribute using CI/CD pipelines. On other hand, using of CI/CD pipelines stands for regular monitoring and support of workflows with additional time spending on bootstrap packing of services into images. Nevertheless, using of CI/CD in development and production process allows to automate successfully major part of building deployment process decreasing deployment time and speeding up deployment cycle.

Further steps should be connected with assessment of efficiency of CI/CD for fast growing projects and solving problems of deployment to multi OS environments.

Acknowledgments. The authors very appreciated to scientific society of consortium and in particularly the staff of Department of Computer Systems, Networks and Cybersecurity of National Aerospace University n. a. N. E. Zhukovsky «KhAI» for invaluable inspiration, hardworking and creative analysis during preparation of this paper.

References

1. What is Agile? | Atlassian. https://www.atlassian.com/agile/. Accessed 19 Oct 2019
2. Vlasov, Y., Illiashenko, O., Uzun, D., Haimanov, O.: Prototyping tools for IoT systems based on virtualization techniques. In: The 9th IEEE International Conference on Dependable Systems, Services and Technologies, DESSERT (2018)
3. Namespaces in operation, part 5: User namespaces [LWN.net]. https://lwn.net/Articles/532593/. Accessed 19 Oct 2019
4. Why Docker? | Docker. https://www.docker.com/what-docker/. Accessed 19 Oct 2019
5. Replacing Docker with Podman - Power of Podman - Cloudnweb. https://cloudnweb.dev/2019/06/replacing-docker-with-podman-power-of-podman/. Accessed 19 Oct 2019
6. GitHub - containers/libpod: libpod is a library used to create container pods. Home of Podman. https://github.com/containers/libpod/. Accessed 19 Oct 2019
7. Jenkins. https://jenkins.io/. Accessed 19 Oct 2019
8. TeamCity: the Hassle-Free CI and CD Server by JetBrains. https://www.jetbrains.com/teamcity/. Accessed 19 Oct 2019
9. Varia, J.: Amazon web services – the total cost of (Non) ownership of web applications in the cloud (2012)

10. Code Deploy Pricing | Amazon Web Services. https://aws.amazon.com/codedeploy/pricing/. Accessed 19 Oct 2019
11. What's the Life Expectancy of a NAS Device? – StorageCraft. https://blog.storagecraft.com/life-expectancy-nas-device/. Accessed 19 Oct 2019
12. Semantic Versioning 2.0.0 | Semantic Versioning. https://semver.org/. Accessed 19 Oct 2019
13. Vault by HashiCorp. https://www.vaultproject.io/. Accessed 19 Oct 2019
14. Compare Repositories - Open Hub. https://www.openhub.net/repositories/compare/. Accessed 19 Oct 2019
15. GitLab CI/CD | GitLab. https://docs.gitlab.com/ee/ci/. Accessed 19 Oct 2019
16. Continuous Integration and Delivery – CircleCI. https://circleci.com/. Accessed 19 Oct 2019
17. Travis CI - Test and Deploy Your Code with Confidence. https://travis-ci.org/. Accessed 19 Oct 2019
18. What are the best Continuous Integration Tools? https://stackshare.io/continuous-integration/. Accessed 19 Oct 2019

Author Index

M. Nechyporuk et al. (Eds.): *Integrated Computer Technologies in Mechanical Engineering*, AISC 1113, pp. 551–553, 2020.
https://doi.org/10.1007/978-3-030-37618-5

MIX
Papier aus verantwortungsvollen Quellen
Paper from responsible sources
FSC® C105338

Printed by Books on Demand, Germany